BRIAN CARROLL

AF690041

BASIC REAL AND ABSTRACT ANALYSIS

BASIC REAL AND ABSTRACT ANALYSIS

John F. Randolph

DEPARTMENT OF MATHEMATICS

UNIVERSITY OF ROCHESTER

2-7

ACADEMIC PRESS New York and London

COPYRIGHT © 1968, BY ACADEMIC PRESS INC.
ALL RIGHTS RESERVED.
NO PART OF THIS BOOK MAY BE REPRODUCED IN ANY FORM,
BY PHOTOSTAT, MICROFILM, OR ANY OTHER MEANS, WITHOUT
WRITTEN PERMISSION FROM THE PUBLISHERS.

ACADEMIC PRESS INC.
111 Fifth Avenue, New York, New York 10003

United Kingdom Edition published by
ACADEMIC PRESS INC. (LONDON) LTD.
Berkeley Square House, London W.1

LIBRARY OF CONGRESS CATALOG CARD NUMBER: 68-18691

PRINTED IN THE UNITED STATES OF AMERICA

PREFACE

The keen precision toward which mathematics constantly strives is greatly enhanced by abstraction. Without sharp awareness of the sources of abstraction, however, there is danger of fostering an impression that abstraction is complication instead of simplification. Hence the minds visualized as maturing on this presentation are led from the concrete to the abstract rather than the other way around.

Chapter 1 (Orientation) emerged last and was the hardest to write, since few, if any, instructors agree on an appropriate starting point for a course which could follow the two-year calculus sequence. Contact with past mathematical experience is made, the stage is set, abstractions to come are foretold by a taste of Boolean algebra, and careful reasoning by Dedekind cuts. Chapter 2 (Sets and Spaces) begins with the familiar, eases into generalizations, and includes just enough about transfinite cardinals to prove (in Chapter 5) that some Lebesgue measurable sets are not Borel sets. The classical material of Chapter 3 (Sequences and Series) shows how readily previous work may be tightened and extended. Limits inferior and superior are defined with care and their merit demonstrated by repeated use, not the least of which is l'Hospital's rule in Chapter 7. The first three chapters provide sufficient preparation for Chapter 4 (Measure and Integration).

One guiding light for this book was a strong conviction that time is overdue for Lebesgue theory to come into its own at an early stage without initial high abstraction. By starting with sets on a line, generalizing length, then sloughing off all except pertinent properties, Carathéodory's definition of measurability is revealed as a natural and intuitive requirement. Meaningful and significant theorems on integration are then proved with minimal effort and never a whisper of mystical visions. Those who continue to more advanced work will be equipped to fathom the thought processes that led to the Riesz or Daniell development of integration, either of which is excellent for a second exposure.

Classical Stieltjes integrals are kept for their role in the procession of ideas from Riemann to Radon-Nikodym, even though they are subsumed (in Sec. 8–6) by Lebesgue integrals upon replacing the awkward integrator functions with measures. Riemann integration takes its place as an antique in the true sense of the word: not a piece of junk, but a prized relic from a former era honored for inspiring replacements of greater comfort and utility, and still in working order for many uses.

Problems are scattered throughout the text and each chapter ends with a set of problems. The text problems encourage immediate checks on comprehension. Some of these problems are part of the development and are stated in theorem form with neither command nor entreaty to work them; a subsequent discussion may justify a step by referring to such a problem as readily as to any other theorem. The only reference to an end-of-chapter problem is in some problem at the end of a later chapter.

An "if" in a definition should be interpreted as if and only if, but "if" in a theorem means if. If a theorem is of the "if $\cdots$, then $\cdots$" form, then the proof does not "Let the hypotheses be satisfied," but starts right in as if they were. No theorem is stated as "Suppose $\cdots$. Then $\cdots$." In fact any "suppose" or "assume" is a signal that a contradiction will be reached.

The student's mathematical preparation is neither ignored nor relied upon. A proof of the Weierstrass approximation theorem, for example, is based on simple facts about derivatives of polynomials although derivatives are defined later. Trigonometric functions are used in examples and problems prior to their definition and theory in Chapter 7. The disparaging attitude that the student brings little ability to prove anything from any previous course is offset by trusting him to see that integration before differentiation (or even continuity) is realistic at this level and that methods yielding numerical answers need not immediately follow the definition.

Built-in flexibility for a two-semester course may be judged by the variety of topics tucked in where their omission will not dilute the course, but they are available for enrichment as time and inclination permit; for example, infinite products, Jensen's inequalities, and the Stone-Weierstrass theorem. A suggested hard-core one-semester course includes sections 1.1 to 1.10, 2.1 to 2.8, 3.1 to 3.9, 3.12, 4.1 to 4.5, 5.1, 5.2, 6.1 to 6.8, 7.1 to 7.10, and 8.1 to 8.4.

Rochester, New York J. F. R.
January, 1968

CONTENTS

Chapter 3. SEQUENCES AND SERIES

Chapter 4. MEASURE AND INTEGRATION

Chapter 5. MEASURE THEORY

Chapter 6. CONTINUITY

Chapter 7. DERIVATIVES

Chapter 8. STIELTJES INTEGRALS

BASIC REAL AND ABSTRACT ANALYSIS

Chapter 1

ORIENTATION

1–1 REAL NUMBERS

Classical mathematical analysis is based directly on the real number system and abstract analysis consists largely of structures inspired by either specializations or generalizations of real numbers. The real number system will therefore be subjected to close scrutiny in Section 1–6. We do not, however, ignore experience with real numbers and shall consider that antecedents for the early work are:

1. Ordinary arithmetic.
2. The use of $<$ and $>$ so that if a and b are real numbers, then one and only one of

   $$a < b, \quad a = b, \quad a > b$$

   holds. Also, for c any real number, then

   $$a + c < b + c \qquad \text{if and only if } a < b,$$

   and whenever $c > 0$, then

   $$ac < bc \qquad \text{if and only if } a < b,$$

   but whenever $c < 0$, then

   $$ac < bc \qquad \text{if and only if } a > b.$$

3. The integers consist of all natural (whole) numbers, their negatives, and zero.
4. If x is a given real number, then there is a unique integer denoted by $[x]$ such that

   $$[x] \leqslant x < [x] + 1$$

   or equivalently $x - 1 < [x] \leqslant x$.

5. Any number equal to the quotient of integers (denominator not zero) is a rational number. All other real numbers are irrational.

The following two theorems illustrate how these facts and their immediate consequences are considered as starting points of proofs.

Theorem 1. (*Archimedean Principle*). *If a and b are positive numbers, then there is a positive integer q such that*

$$qa > b.$$

Proof. Since $b/a > 0$, then $1 + b/a > 1$ and the integer q defined by

$$q = \left[1 + \frac{b}{a}\right]$$

is such that $q \geqslant 1$. Hence, from 4,

$$q > \left(1 + \frac{b}{a}\right) - 1 = \frac{b}{a}$$

and multiplication by the positive number a yields the desired result. ∎

Corollary. *If $a > 0$, then there is a positive integer q such that*

$$\frac{1}{q} < a.$$

Theorem 2. *Between any two real numbers there is a rational number.*

Proof. Let $a < b$ be real numbers. Since $b - a$ and 1 are positive, let q be a positive integer such that

$$q(b - a) > 1 \qquad \text{and hence } qb > qa + 1.$$

With $p = [qa + 1]$, then $(qa + 1) - 1 < p \leqslant qa + 1$ so that

$$qa < p < qb.$$

Upon dividing by the positive integer q, the result

$$a < \frac{p}{q} < b$$

shows the rational number p/q between a and b. ∎

A rigorous development of trigonometric and logarithmic functions is given in Sections 7–3 and 7–9. In the meantime we may use elementary properties of such functions for illustrative purposes and in problems, but shall not base any of the intermediate theory on these functions.

The following problem takes for granted that among a finite set of numbers there is a smallest number, although finite and infinite sets are not discussed until

Section 2–2. Thus, if q is a positive integer, if $0 < a_1 < a_2 < \cdots < a_{q+1} < 1$ and we let

$$s = \min_{1 \leqslant i \leqslant q} (a_{i+1} - a_i), \qquad \text{then } 0 < s < \frac{1}{q},$$

since

$$1 > a_{q+1} - a_1 = \sum_{i=1}^{q} (a_{i+1} - a_i) \geqslant qs.$$

Also, the method of contrapositive proof may be used in the solution of the following problem prior to the discussion of this method in Section 1–4.

PROBLEM 1. (a). If ξ is an irrational number and $0 < c < 1$, then there are integers m and n such that

$$0 < m + n\xi < c.$$

Note: ξ and n may be positive or negative and m may be zero, positive, or negative.

[*Hint:* For $k = 1, 2, 3, \cdots$ set $x_k = k\xi - [k\xi]$. Hence $0 \leqslant x_k < 1$, but from the irrationality of ξ, $0 < x_k < 1$, and also if $l \neq k$, then $x_l \neq x_k$. Take $q = [1/c] + 1$. Among the numbers $x_1, x_2, \cdots, x_{q+1}$ there are at least two which differ by less than $1/q < c$. A difference of such a pair may be written in the desired form.]

(b). If ξ is irrational and $a < b$, then there are integers m and n such that $a < m + n\xi < b$.

[*Hint:* Choose m_1 and n_1 such that $0 < m_1 + n_1\xi < b - a$ and let $p = [1 + a/(m_1 + n_1\xi)]$. Then $a < pm_1 + pn_1\xi < b$.]

(c). If $0 < a < b$, then there are integers m and n such that

$$a < 2^m 3^n < b.$$

[*Hint:* Show that $\xi = (\ln 3)/(\ln 2)$ is irrational by assuming it is rational and getting a contradiction. Choose m and n such that

$$\frac{\ln a}{\ln 2} < m + n\frac{\ln 3}{\ln 2} < \frac{\ln b}{\ln 2}.\Bigg]$$

1–2 SETS

The elements of set theory are considered as in the background. Since notation and terminology are not completely standard, however, the following expressions will establish unanimity.

$$a \in A$$

means that a is an element of the set A; $a \notin A$ means a is not an element of A. The set whose only elements are a, b, c, d is denoted by

$$\{a, b, c, d\} \qquad \text{or} \qquad \{b, c, a, d\},$$

or any order of writing these elements between brackets. Thus $\{a\}$ denotes the set whose single element is a.

$$\{x \mid \text{statement about } x\}$$

is the set of all elements for which the statement is true. Thus

$$\left\{x \,\middle|\, x = \frac{m}{n} \text{ with } n = 2m \neq 0\right\}$$

and $\{\frac{1}{2}\}$ are the same.

The **union** of sets A and B is†

$$A \cup B = \{x \mid x \in A \text{ or } x \in B\}$$

and the **intersection** is

$$A \cap B = \{x \mid x \in A \text{ and } x \in B\}.$$

The union of A and B will always be designated by using $\cup$, but $\cap$ may be omitted and the intersection denoted by AB.

Both $A \subset B$ and $B \supset A$ mean that whenever $a \in A$ then also $a \in B$; both are read "A is a subset of B," or "B is a superset of A," or "B covers A." $A = B$ if and only if both $A \subset B$ and $B \subset A$.

The empty set is denoted by $\varnothing$. For any set A,

$$\varnothing \subset A, \qquad A \cup \varnothing = A, \qquad \text{and} \qquad A \cap \varnothing = A\varnothing = \varnothing.$$

No symbol will be reserved for the universal set. If in some discourse the universal set is denoted by I, then

$$A \subset I, \qquad A \cup I = I, \qquad AI = A,$$

and the **complement** of A is denoted by A^c where

$$A^c = \{x \mid x \in I, \text{ but } x \notin A\}.$$

Even though the universal set is not denoted, the set BA^c means the set of all points of B not in A and the alternate notation $B - A$ may be used:

$$B - A = B \cap A^c = \{x \mid x \in B \text{ but } x \notin A\} = BA^c.$$

The following laws hold:

Commutative. $A \cup B = B \cup A, \qquad A \cap B = B \cap A.$

Associative. $(A \cup B) \cup C = A \cup (B \cup C), \qquad (AB)C = A(BC).$

Distributive.

$$\begin{aligned} A \cap (B \cup C) &= (A \cap B) \cup (A \cap C) \\ &= AB \cup AC; \end{aligned}$$

$$\begin{aligned} A \cup (B \cap C) &= (A \cup B) \cap (A \cup C) \\ &= (A \cup B)(A \cup C). \end{aligned}$$

de Morgan's laws.

$$\begin{aligned} (A \cup B)^c &= A^c \cap B^c; \\ (A \cap B)^c &= A^c \cup B^c. \end{aligned}$$

† The "or" is the "inclusive or," meaning "and/or."

A crutch for visualizing set operations is the Venn diagram scheme. The points of the plane are identified with the universal set and the interiors of simple closed curves with sets A, B, C, D, etc.

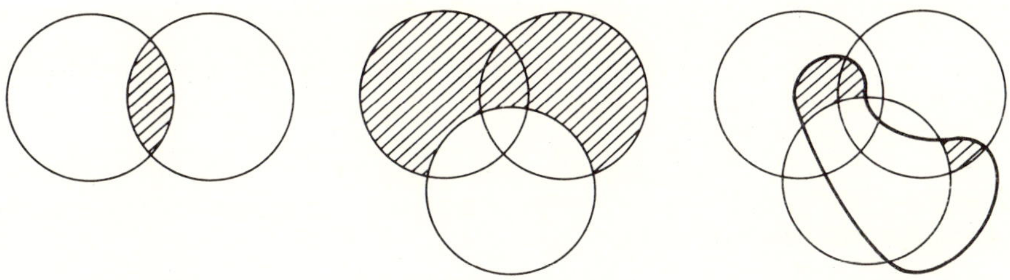

$A \cap B$ $\quad$ $(A \cup B) \cap C^c$ $\quad$ $(A \cup B) \cap C^c \cap D$

FIG. 1

Definition 1. *Let $\mathscr{B}$ be a set, of at least two elements, having two binary operations $\vee$ and $\wedge$ such that whenever $a \in \mathscr{B}$ and $b \in \mathscr{B}$, then also $a \vee b \in \mathscr{B}$ and $a \wedge b \in \mathscr{B}$. Then $\mathscr{B}$, together with these two operations, is said to be a* **Boolean algebra** *if the following conditions hold for all $a \in \mathscr{B}$, $b \in \mathscr{B}$,* and *$c \in \mathscr{B}$:*

(i) $\quad a \vee b = b \vee a \qquad a \wedge b = b \wedge a.$

(ii) $\quad (a \vee b) \vee c = a \vee (b \vee c) \qquad (a \wedge b) \wedge c = a \wedge (b \wedge c).$

(iii)
$$a \wedge (b \vee c) = (a \wedge b) \vee (a \wedge c) \qquad \text{and}$$
$$a \vee (b \wedge c) = (a \vee b) \wedge (a \vee c).$$

(iv) *There are elements of $\mathscr{B}$ denoted by* 0 *and* 1 *such that*

$$a \vee 0 = a \qquad \textit{and} \qquad a \wedge 1 = a \qquad \textit{for all } a \in \mathscr{B}.$$

(v) *Corresponding to each element a there is at least one element a' such that*

$$a \vee a' = 1 \qquad \textit{and} \qquad a \wedge a' = 0.$$

[*Note:* In some situations we read $\vee$ as "join," $\wedge$ as "meet," and the prime as "complement." If, however, the setting is in logic, $\vee$ is read as "or," $\wedge$ as "and," and a' as "not a."]

Select any nonempty set I. Let $\mathscr{B}$ be the collection of all subsets of I. Hence $A \in \mathscr{B}$ means that $A \subset I$. Then $\mathscr{B}$ has at least two elements, I and $\varnothing$. By interpreting $\vee$ to mean $\cup$, $\wedge$ to mean $\cap$, 1 to mean I, 0 to mean $\varnothing$, and A' to mean A^c, it is seen that $\mathscr{B}$ is a Boolean algebra.

PROBLEM 1. Let $\mathscr{B}$ have the two integers 0 and 1 as its only elements. For $a \in \mathscr{B}$ and $b \in \mathscr{B}$ interpret

$$a \vee b \text{ to mean } a + b + ab \pmod 2,$$
$$a \wedge b \text{ to mean } ab.$$

Show that $\mathscr{B}$ is a Boolean algebra. [Note that if $a \vee b$ means $a + b \pmod 2$, then the second part of (iii) does not hold when $a = b = 1$ and $c = 0$.]

Theorem 1. *Let* $(\mathscr{B}, \vee, \wedge)$ *be a Boolean algebra. Then:*

1. 0 *and* 1 *are unique.*
2. *Each* $a \in \mathscr{B}$ *has a unique complement.*
3. $1' = 0$ *and* $0' = 1$.

Proof. We are given that $a \vee 0 = a$ for all $a \in \mathscr{B}$. Let $\bar{0}$ be any element of $\mathscr{B}$ having the property that

$$a \vee \bar{0} = a \qquad \text{for all } a \in \mathscr{B}.$$

In particular we then have both

$$\bar{0} \vee 0 = \bar{0} \qquad \text{and} \qquad 0 \vee \bar{0} = 0.$$

Hence, from the commutativity expressed by the first equation in (i),

$$\bar{0} = \bar{0} \vee 0 = 0 \vee \bar{0} = 0.$$

It is in this sense that the element 0 is said to be unique.

Let $\bar{1} \in \mathscr{B}$ be such that $a \wedge \bar{1} = a$ for all $a \in \mathscr{B}$. Putting $a = 1$ here and $a = \bar{1}$ in (iv) we have

$$1 = 1 \wedge \bar{1} = \bar{1} \wedge 1 = \bar{1}.$$

Given any $a \in \mathscr{B}$, let $a^* \in \mathscr{B}$ be such that

$$a \vee a^* = 1 \qquad \text{and} \qquad a \wedge a^* = 0.$$

[There is at least one element having this property, namely, a'.] Then

$$\begin{aligned} a^* &= a^* \vee 0 \quad \text{by (iv)} \\ &= a^* \vee (a \wedge a') \quad \text{by (v)} \\ &= (a^* \vee a) \wedge (a^* \vee a') \quad \text{by second equation (iii)} \\ &= (a \vee a^*) \wedge (a^* \vee a') \quad \text{by (i)} \\ &= 1 \wedge (a^* \vee a') = a^* \vee a' \quad \text{by (iv).} \end{aligned}$$

On the other hand, $a' = a' \vee 0 = a' \vee (a \wedge a^*) =$ etc., as above, leads to $a' = a' \vee a^*$. Thus

$$a^* = a^* \vee a' = a' \vee a^* = a'.$$

Hence each $a \in \mathscr{B}$ has a' as its unique complement.

As to 3, we know that $1'$, the complement of 1, is the unique element of $\mathscr{B}$ such that $1 \vee 1' = 1$. But the first equation of (iv) with $a = 1$ is

$$1 \vee 0 = 1,$$

which says that 0 is a complement of 1. Thus $0 = 1'$.

In a similar way, $1 = 0'$. ▮

PROBLEM 2. If $(\mathscr{B}, \vee, \wedge)$ is a Boolean algebra, then for each $a \in \mathscr{B}$,

4. $(a')' = a$;
5. $a \vee a = a$ and $a \wedge a = a$;
6. $a \vee 1 = 1$ and $a \wedge 0 = 0$;
   and for all $b \in \mathscr{B}$;
7. $a \vee (a \wedge b) = a$ and $a \wedge (a \vee b) = a$;
8. $(a \vee b)' = a' \wedge b'$ and $(a \wedge b)' = a' \vee b'$.

[*Hint for* 5: $a = a \vee 0 = a \vee (a \wedge a') =$ etc.]
[*Hint for* 8: Check $(a \vee b) \vee (a' \wedge b') = 1$ and $(a \vee b) \wedge (a' \wedge b') = 0$.]

1–3 MATHEMATICAL INDUCTION

In some developments of the number system, mathematical induction is an axiom, in others it is a theorem, but here it is accepted as:

The Induction Principle. *If S is a set of positive integers having the two properties*

(i) $1 \in S$;
(ii) *Whenever $n \in S$ also $n + 1 \in S$;*
then S is the set of all positive integers.

The following is an example of "definition by induction."

Definition. *With x a real number, define*

$$x^1 = x, \tag{1}$$

and with n a positive integer for which x^n is defined, set

$$x^{n+1} = x^n x. \tag{2}$$

Of course, by definition,

$$x^0 = 1 \qquad \text{(even if } x = 0) \tag{3}$$

$$x^{-n} = \frac{1}{x^n}, \qquad x \neq 0, \quad n \text{ a positive integer.} \tag{4}$$

The next theorem illustrates "proof by induction."

Theorem 1. *If $y > -1$, then for each positive integer n,*

$$(1 + y)^n \geqslant 1 + ny.$$

Proof. Let

$$S = \{n \mid n \text{ is a positive integer and } (1 + y)^n \geqslant 1 + ny\}.$$

First, $(1 + y)^1 = 1 + 1y$ so that $1 \in S$, and (i) is satisfied.

Next take any $n \in S$. Hence for this n

$$(1 + y)^n \geqslant 1 + ny.$$

Since $1 + y > 0$, then

$$\begin{aligned}(1 + y)^{n+1} &= (1 + y)^n(1 + y) \quad \text{by (2)}\\ &\geqslant (1 + ny)(1 + y) = 1 + y + ny + ny^2\\ &\geqslant 1 + (n + 1)y\end{aligned}$$

because $ny^2 \geqslant 0$. Hence $(n + 1) \in S$ and (ii) is satisfied. Thus, by the induction principle, S consists of all positive integers. ∎

Corollary 1. *If $x > 1$ and n is a positive integer, then*

(5) $$x^n > n(x - 1).$$

Proof. $x^n = (1 + (x - 1))^n \geqslant 1 + n(x - 1) > n(x - 1)$. ∎

Corollary 2. *Given $x > 1$ and G a positive number, there is a positive integer N such that if n is an integer satisfying*

$$n \geqslant N, \qquad \textit{then } x^n > G.$$

Proof. Since $x - 1$ and G are positive, let N be a positive integer such that (by the Archimedean principle)

$$N(x - 1) > G.$$

With n a positive integer satisfying $n \geqslant N$, then by (5) and $x - 1 > 0$,

$$x^n > n(x - 1) \geqslant N(x - 1) > G. \quad ∎$$

Theorem 2. *For x real and m and n positive integers,*

(6) $$x^{m+n} = x^m x^n \qquad \text{and}$$

(7) $$(x^m)^n = x^{mn} = (x^n)^m.$$

Proof. Let m be a positive integer and set

$$S_m = \{n \mid n \text{ is a positive integer and } x^{m+n} = x^m x^n\}.$$

First, $x^{m+1} = x^m x$ by (2) so that $1 \in S_m$. Next, let $n \in S_m$ so that

(8) $$x^{m+n} = x^m x^n.$$

Then

$$\begin{aligned}x^{m+(n+1)} = x^{(m+n)+1} &= x^{m+n}x \quad && \text{by (2)}\\ &= (x^m x^n)x && \text{by (8)}\\ &= x^m(x^n x) && \text{by associativity}\\ &= x^m x^{n+1} && \text{by (2).}\end{aligned}$$

Hence $(n + 1) \in S_m$ so that S_m consists of all positive integers. Since m was any positive integer, (6) holds for all positive integers m and n.

A similar proof establishes (7). ∎

If x and y are real and m and n are integers it should be seen (by using (6) and (7) and the definitions (3) and (4)) that

$$x^m x^n = x^{m+n} \quad \text{unless } x = 0 \text{ and } m \text{ or } n \text{ is negative.}$$

$$\frac{x^m}{x^n} = x^{m-n} \quad \text{unless } x = 0.$$

$$(x^m)^n = x^{mn} \quad \text{unless } x = 0 \text{ and } m < 0 \text{ or } n < 0.$$

$$(xy)^n = x^n y^n \quad \text{unless } x \text{ or } y \text{ is zero and } n < 0.$$

$$\left(\frac{x}{y}\right)^n = \frac{x^n}{y^n} \quad \text{unless } y = 0 \text{ or unless } x = 0 \text{ and } n < 0.$$

PROBLEM 1

(a). If $x > 1$ and p is a positive integer, then $x^p > 1$.
(b). If $x > 1$ and $m > n$ are positive integers, then $x^m > x^n$.
(c). If $0 < x < 1$ and $m > n$ are positive integers, then $x^m < x^n$.
(d). If $x > y > 0$ and n is an integer, then

$$\begin{aligned} x^n &> y^n \quad \text{for } n > 0, \\ x^n &< y^n \quad \text{for } n < 0. \end{aligned}$$

We then have

Corollary 3. *Given $0 < x < 1$ and a number $\varepsilon > 0$, there is a positive integer N such that if n is an integer satisfying*

$$n \geqslant N, \quad \text{then } x^n < \varepsilon.$$

Proof. Since $1/x > 1$ and $1/\varepsilon > 0$, let N be a positive integer such that (by Corollary 2) if n is an integer satisfying

$$n \geqslant N, \quad \text{then } \frac{1}{x^n} > \frac{1}{\varepsilon}; \qquad \text{that is, } x^n < \varepsilon. \blacksquare$$

PROBLEM 2. Any nonempty set of positive integers has a smallest element.

1–4 LOGIC

Suppose there is a rational number whose square is 2. Write this rational number as p/q, with p and q positive integers having no common factor other than 1 and such that

$$\left(\frac{p}{q}\right)^2 = 2.$$

Hence $p^2 = 2q^2$, showing that p^2 is an even integer. Thus p itself is an even integer (for if p were odd, then p^2 would be odd). Write $p = 2p_1$, where p_1 is an integer.

Then

$$(2p_1)^2 = 2q^2 \quad \text{gives} \quad 2p_1{}^2 = q^2.$$

Thus q^2 is even so that q is even. Hence p and q have 2 as a common factor, which contradicts the choice of p and q with no common factor.

We have thus proved

Theorem 1. *There is no rational number whose square is* 2.

The above is an illustration of "proof by contradiction." A supposition is made and from then on only logic and/or mathematics is used. In case a contradiction to something known or accepted is obtained, then the statement made in the supposition is rejected and only this is rejected. For example, the above proof does not establish that some irrational number has its square equal to 2. Also if a supopsition is made and from then on only facts, known or accepted, are obtained, it cannot be concluded that the supposition is correct. Hence we shall never state a theorem in the form "Suppose that $\cdots$" or "Assuming that $\cdots$." Any "suppose" or "assume" is a signal that a contradiction will be reached.

Notice that the above contrapositive proof contains a subsidiary contrapositive proof. This is all right, since the parenthetic statement could have been given previously and then cited when needed; it relied in no way upon the original supposition.

Even though training in formal logic is not a prerequisite of this course, the capacity to think straight and make valid deductions is assumed. Some of the notation and terminology indigenous to logic has filtered into mathematics, sometimes lavishly. In this book the backward E, $\exists$, for "there exists," and the double arrow $\Rightarrow$, for "material implication," will be used. For example: With a and b real numbers

$$a < b \Rightarrow \exists \text{ rational } r \text{ such that } a < r < b$$

and

$$a < b + \varepsilon \text{ for each } \varepsilon > 0 \Rightarrow a \leqslant b. \tag{1}$$

In (1), a and b are to be thought of as fixed definite† real numbers, although their numerical values are not known, such that

$$a < b + 1, \qquad a < b + \tfrac{1}{2}, \qquad a < b + \tfrac{1}{10}, \qquad a < b + \tfrac{1}{100}, \qquad a < b + 100$$

and no matter how small (or large) a positive number is considered, a is actually less than b plus that number. All of this forces the conclusion that a is no larger than b. In ordinary mathematical language:

Theorem 2. *If a and b are numbers such that for each positive number ε*

$$a < b + \varepsilon, \quad \textit{then } a \leqslant b.$$

† These adjectives are, of course, redundant.

Proof. Suppose $a > b$. Then $a - b > 0$ so that (by using $\varepsilon = a - b$),

$$a < b + (a - b); \qquad \text{that is, } a < a$$

which is a contradiction.† ∎

The symbol $\Rightarrow$ takes very little less space than the word "implies" or "if ···, then ···," but its use seems to make some arguments easier to follow. The double-headed double arrow $\Leftrightarrow$ will also be used for "if and only if" which may be contracted to "iff." Thus

$$a + c < b + c \Leftrightarrow a < b, \text{ and}$$

$$\text{given } c > 0, \text{ then } ac < bc \Leftrightarrow a < b.$$

1–5 COMPLETION

A set A of real numbers is **bounded above** if there is a number b such that

(1) $$a \in A \Rightarrow a \leqslant b;$$

otherwise A is **unbounded above**. An upper bound b of A is said to be a **least upper bound** of A (or **supremum** of A, written $b = \sup A$) if in addition

(2) $$c < b \Rightarrow \exists\, a \in A \quad \text{such that‡ } a > c.$$

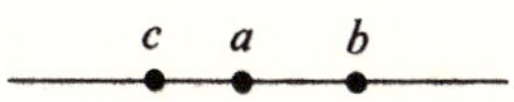

The rational numbers are not sufficient for all desirable arithmetic operations. For example, there is no rational square root of 2 (Theorem 1–4.1).§ In this informal orientation, the interstices in the rational number system are filled by accepting:

The Completion Principle. *Any set of real numbers which is nonempty and bounded above has a supremum.*

Theorem 1. *A nonempty set of real numbers which is bounded above has a unique supremum.*

Proof. Suppose a set $A \neq \varnothing$ has two different suprema b_1 and b_2 which we consider named so that $b_1 < b_2$. Since b_2 is a supremum of A, then $\exists\, a \in A$ such that $b_1 < a$ (from (2), since $b_1 < b_2$), thus showing that b_1 is not even an upper bound of A. ∎

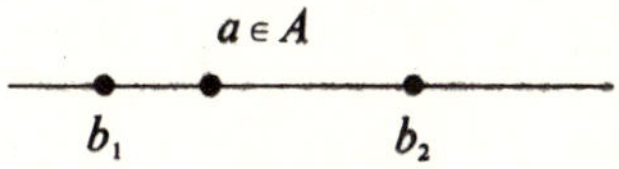

FIG. 1

† Writing stops here, but thought continues: "Hence $a > b$ is wrong and the only remaining possibilities are $a \leqslant b$."

‡ Meaning "corresponding to each number c such that $c < b$ there is at least one $a \in A$ such that $a > c$."

§ This means Section 1-4, Theorem 1.

A special case of the next theorem shows that there is a positive number whose square is 2. The proof will use the binomial expansion (with a and b real and n a positive integer)

$$(a + b)^n = \binom{n}{0}a^n + \binom{n}{1}a^{n-1}b + \binom{n}{2}a^{n-2}b^2 + \cdots + \binom{n}{n}b^n$$

where

$$\binom{n}{k} = \frac{n!}{k!(n-k)!}, \qquad k = 0, 1, 2, \cdots, n,$$

it being understood that $k! = 1 \cdot 2 \cdots k$ if $k > 0$ and $0! = 1$. This expansion is established by using the induction principle.

Theorem 2. *Given a real number $x > 0$ and a positive integer n, there is one and only one real number $y > 0$ such that*

$$y^n = x. \tag{3}$$

Proof. Let

$$A = \{a \mid a > 0 \text{ and } a^n < x\}.$$

Since $x > 0$, then $a_0 = x/(1 + x)$ is such that $0 < a_0 < 1$ and

$$a_0{}^n < a_0 < x$$

so that $a_0 \in A$, showing that $A \neq \varnothing$. The number $x + 1$ is an upper bound of A, since

$$(1 + x)^n \geqslant 1 + nx > x.$$

Hence let $b = \sup A$. Note that $b \geqslant a_0 > 0$ and hence that $(b + 1)^n > b^n$.

Assume $b^n < x$. Choose a number h such that

$$0 < h < \frac{x - b^n}{(b + 1)^n - b^n} \qquad \text{and also } h < 1. \tag{4}$$

Then

$$\begin{aligned}(b + h)^n &= b^n + \binom{n}{1}b^{n-1}h + \binom{n}{2}b^{n-2}h^2 + \cdots + \binom{n}{n}h^n \\ &\leqslant b^n + h\left[\binom{n}{1}b^{n-1} + \binom{n}{2}b^{n-2} + \cdots + \binom{n}{n}\right] \qquad \text{since } 0 < h < 1 \\ &= b^n + h[(b + 1)^n - b^n] \\ &< b^n + (x - b^n) \quad \text{by (4)} \\ &= x.\end{aligned}$$

Therefore $b + h \in A$, contradicting $b = \sup A$, since $0 < h$. Hence

$$b^n \geqslant x. \tag{5}$$

Assume $b^n > x$. This time choose h such that

$$(6) \qquad 0 < h < \frac{b^n - x}{(b+1)^n - b^n} \qquad \text{and also } h < b.$$

Then $0 < b - h$ and

$$\begin{aligned}(b-h)^n &= b^n - \binom{n}{1}b^{n-1}h + \binom{n}{2}b^{n-2}h^2 + \cdots + (-1)^n\binom{n}{n}h^n \\ &= b^n - h\left[\binom{n}{1}b^{n-1} - \binom{n}{2}b^{n-2}h + \cdots + (-1)^{n-1}\binom{n}{n}h^{n-1}\right] \\ &\geqslant b^n - h\left[\binom{n}{1}b^{n-1} + \binom{n}{2}b^{n-2} + \cdots + \binom{n}{n}\right] \\ &= b^n - h[(b+1)^n - b^n] > b^n - (b^n - x) = x.\end{aligned}$$

Thus $b - h$ is an upper bound of A, contradicting $b = \sup A$. Hence $b^n \leqslant x$, which with (5) shows that $b^n = x$.

If $c > 0$ is different from b, then either $c < b$ and $c^n < b^n = x$ or else $c > b$ and $c^n > b^n = x$. Hence b is the only positive number whose nth power is x. ∎

Definition 1. *For $x > 0$ and $n > 0$ an integer, the positive number whose nth power is x will be written as*

$$\sqrt[n]{x} \quad \text{or as} \quad x^{1/n};$$

that is,

$$y = \sqrt[n]{x} \Leftrightarrow (y > 0 \text{ and } y^n = x).$$

Theorem 3. *Let $x > 0$ and positive integers m, n, p, q be given such that $m/n = p/q$. Then*

$$(7) \qquad (x^{1/n})^m = (x^{1/q})^p.$$

Proof. Set $y = x^{1/nq}$. Then $y > 0$ and $y^{nq} = x$. Since q and n are positive integers, then by Theorem 1–3.2,

$$(y^q)^n = (y^n)^q = y^{nq} = x.$$

Thus, by Definition 1, both

$$y^q = x^{1/n} \qquad \text{and} \qquad y^n = x^{1/q}.$$

Consequently, by Theorem 1–3.2 again,

$$y^{qm} = (y^q)^m = (x^{1/n})^m \qquad \text{and} \qquad y^{np} = (y^n)^p = (x^{1/q})^p.$$

But $qm = np$ is a positive integer so that $y^{qm} = y^{nq}$ and hence (7) follows. ∎

Definition 2. *For $x > 0$ and $r > 0$ rational, define*

$$x^r = (x^{1/n})^m$$

where m and n are any positive integers such that $r = m/n$. Also define

$$0^r = 0 \qquad \textit{and} \qquad x^{-r} = \frac{1}{x^r}.$$

Theorem 4. *For $x > 0$, and r and s positive rationals, then*

$$x^r \cdot x^s = x^{r+s}. \tag{8}$$

Proof. Set $r = m/n$ and $s = p/q$ with m, n, p, q positive integers. Also set

$$y = x^{1/(nq)}.$$

Then $y > 0$ and $x = y^{nq} = (y^n)^q = (y^q)^n$. Hence $x^{1/n} = y^q$, $x^{1/q} = y^n$ so that

$$x^r = (x^{1/n})^m = y({}^q)^m = y^{qm} \qquad \text{and} \qquad x^s = (x^{1/q})^p = (y^n)^p = y^{np}.$$

Thus

$$\begin{aligned} x^r \cdot x^s &= y^{qm} \cdot y^{np} = y^{qm+np} \\ &= (x^{1/(nq)})^{qm+np} \\ &= x^{(qm+np)/(nq)} = x^{r+s} \end{aligned}$$

since one representation of $r + s$ as the ratio of positive integers is

$$r + s = \frac{qm + np}{nq}. \quad \blacksquare$$

PROBLEM 1. If $x > 0$ and $y > 0$, and if r and s are rational, then

$$x^r x^s = x^{r+s},$$

$$\frac{x^r}{x^s} = x^{r-s},$$

$$(x^r)^s = x^{rs},$$

$$(xy)^r = x^r y^r,$$

$$\left(\frac{x}{y}\right)^r = \frac{x^r}{y^r}.$$

PROBLEM 2. If $r > s > 0$ are rational, then

$$\begin{aligned} x^r &> x^s \qquad \text{for } x > 1, \\ x^r &< x^s \qquad \text{for } 0 < x < 1. \end{aligned}$$

PROBLEM 3. If $0 < x < y$ and r is rational, then

$$\begin{aligned} 0 < x^r < y^r \qquad &\text{for } r > 0, \\ 0 < y^r < x^r \qquad &\text{for } r < 0. \end{aligned}$$

1–6 DEDEKIND CUTS

A set may have sets as its elements and on several occasions it will be necessary to consider sets of sets. If $\mathscr{A}$ is a set, one of whose elements is a set A, which in turn has a as an element, then we write

$$a \in A \in \mathscr{A}.$$

For example, N.Y. is an element of U.S.A., which in turn is an element (member) of U.N. Note that N.Y. $\notin$ U.N.

The letter $\mathfrak{C}$ will be reserved until near the end of this section, but not thereafter, to denote the set of sets defined below.

Definition 1. *Let $\mathfrak{C}$ be the set, each element α of which is a set of rationals satisfying:*

(i) *α is not empty;*
(ii) *There is at least one rational not in α;*
(iii) *If $p \in \alpha$, q is rational and $q < p$, then $q \in \alpha$; and*
(iv) *α has no greatest element; that is, if $p \in \alpha$, there is an $r \in \alpha$ such that $r > p$.*

Any $\alpha \in \mathfrak{C}$ is said to be a **cut** *(in the rationals).*

Note that $p \in \alpha \in \mathfrak{C}$ means (among other things) that p is rational.

The remainder of this section provides further experience with set operations and some basic abstract algebra, and in addition develops one of the standard methods of proceeding from the rationals to a definition of real numbers. Hence, until the definition of a real number is given (near the end of the section), a mental attitude to take is that we have been whisked back in time to a hypothetical era in the dawn of civilization. The integers† and their ratios are called rationals and are the only numerical quantities "being they sufficient unto all practical purposes." For the first time in any brain anywhere, that of the Sage of the day beheld, however, that none of these quantities expresses the length of the hypotenuse of an isosceles right triangle having edges one cubit long.

Thus the sum, product, and quotient (denominator not zero) of two rationals are rationals and if r and s are any rationals, then one and only one of

$$r < s, \qquad r = s, \qquad r > s$$

holds. In particular, any integer (even if zero or negative) is a rational and the rationals have the Archimedean property.

Incidentally, our shortening "rational numbers" to "rationals" is not due to indolence or ignorance of adjective versus noun, but is a deliberate sop to the delusion that nothing is known about real numbers, and nothing will be known about them, until they are defined. All we know is that there is a set (the rationals) with a pair of binary operations (called addition and multiplication) on its

† With digits written ٠ ١٢٣٤٥٦٧٨٩, as still used in Arab countries today. We shall, however, replace ٠ by 0 and ١ by 1.

elements and the elements are ordered as stated above. A couple of special rationals, and we know their specialities, are represented by the symbols 0 and 1. The result of the operation symbolized by $1 + 1$ is represented by 2. Thus the theorems, definitions, etc., of this section are not numbered; we merely use some symbols for reference purposes—should you happen to remember the common names of these symbols no harm will be done.

Theorem 1. *If r is a rational, then the set r^* defined by*

$$r^* = \{p \mid p \text{ is rational and } p < r\}$$

is a cut (called the **rational cut** *determined by r).*

Proof. Since $r - 1$ and $r + 1$ are rationals and $r - 1 < r < r + 1$, then $r - 1 \in r^*$ and $r + 1 \notin r^*$ so that (i) and (ii) hold.

If $p \in r^*$, q is rational and $q < p$, then $q < p < r$ so that $q \in r^*$ and (iii) holds, but also the rational $\frac{1}{2}(p + r) < r$ is in r^* and is greater than p, thus showing that (iv) holds. ∎

Since 0 is a rational, the cut 0^* is the set of negative rationals. Hence 0 is not an element of 0^*, but is the smallest rational not in 0^*.

Theorem 2. *If $p \in \alpha \in \mathfrak{C}$, q is rational and $q \notin \alpha$, then $q > p$.*

Proof. The assumption $q \leqslant p$ leads, via (iii), to the contradiction $q \in \alpha$. ∎

If $\alpha \in \mathfrak{C}$ and $\beta \in \mathfrak{C}$, then α and β are sets (satisfying (i)–(iv)), and $\alpha = \beta$ means that $\alpha \subset \beta$ and $\beta \subset \alpha$.

Theorem 3. *If α and β are cuts, then the set $\alpha + \beta$ defined by*

$$\alpha + \beta = \{r \mid r = p + q \text{ for some } p \in \alpha \text{ and } q \in \beta\}$$

is a cut (called the **sum** *of α and β). Also*

$$\alpha + \beta = \beta + \alpha.$$

Proof. We shall establish that $\alpha + \beta$ is a cut by checking (i)–(iv) in (a)–(d) below.

(a) Since $\alpha \neq \varnothing$ and $\beta \neq \varnothing$, select $p \in \alpha$ and $q \in \beta$. Thus $p + q \in \alpha + \beta$ and hence $\alpha + \beta \neq \varnothing$.

(b) Select r_1 and r_2 rationals, $r_1 \notin \alpha$ and $r_2 \notin \beta$. Hence

$$r_1 > p \quad \text{for each } p \in \alpha \qquad \text{and} \qquad r_2 > q \quad \text{for each } q \in \beta$$

by Theorem 2. Thus $r_1 + r_2$ is a rational not equal to (in fact greater than) each element $p + q$ of $\alpha + \beta$ and hence $r_1 + r_2 \notin \alpha + \beta$.

(c) Take $r \in \alpha + \beta$, s rational and $s < r$. Write $r = p + q$ with $p \in \alpha$ and $q \in \beta$. Thus $p - (r - s)$ is a rational less than p and hence is an element of α by (iii). Consequently

$$s = [p - (r - s)] + q \in \alpha + \beta.$$

(d) Again take any $r \in \alpha + \beta$ and write $r = p + q$ with $p \in \alpha$ and $q \in \beta$. From (iv) take $s \in \alpha$ and $s > p$. Then

$$s + q \in \alpha + \beta$$

and $s + q > p + q = r$ so that $\alpha + \beta$ has no largest element.

Thus $\alpha + \beta$ is a cut. Also

$$\alpha + \beta = \{p + q \mid p \in \alpha, q \in \beta\} = \{q + p \mid q \in \beta, p \in \alpha\} = \beta + \alpha. \blacksquare$$

PROBLEM 1. For any rationals r and s,

$$(r + s)^* = r^* + s^*.$$

Theorem 4. *If α, β, and γ are cuts, then*

$$(\alpha + \beta) + \gamma = \alpha + (\beta + \gamma)$$

(and hence $\alpha + \beta + \gamma$ is unambiguous). Also

$$\alpha + 0^* = \alpha.$$

Proof. From Theorem 3, $\alpha + \beta$ is a cut and then $(\alpha + \beta) + \gamma$ is a cut and likewise for $\alpha + (\beta + \gamma)$.

To obtain the stated associativity, select any $s \in (\alpha + \beta) + \gamma$. Hence there is a $t \in (\alpha + \beta)$ and an $r \in \gamma$ such that $s = t + r$. Now write $t = p + q$ with $p \in \alpha$ and $q \in \beta$. Thus

$$s = (p + q) + r = p + (q + r) \in \alpha + (\beta + \gamma).$$

Hence $(\alpha + \beta) + \gamma \subset \alpha + (\beta + \gamma)$. The reverse inclusion follows in a similar way and thus the first equality holds.

Take $s \in \alpha + 0^*$ and write $s = p + q$ with $p \in \alpha$ and $q \in 0^*$. Hence q is a negative rational and $s = p + q < p$ so that $s \in \alpha$ by (iii). Thus

$$\alpha + 0^* \subset \alpha.$$

Next take any $p \in \alpha$. Use (iv) to select an $r \in \alpha$ with $r > p$. Hence $p - r \in 0^*$. Thus

$$p = r + (p - r) \in \alpha + 0^*$$

so that $\alpha \subset \alpha + 0^*$. Consequently $\alpha + 0^* = \alpha$. $\blacksquare$

The adjective "upper" in the next definition is suggested by Theorem 2.

Definition 2. *Given a cut α, any rational q such that $q \notin \alpha$ is said to be an* **upper rational** *for α. A* **proper upper rational** *for α is any upper rational for α such that there is a smaller upper rational for α.*

Note that if q_0 is an upper rational for a cut α but not a proper upper rational for α, then any rational $r < q_0$ is in α and any rational $s > q_0$ is a proper upper rational for α.

PROBLEM 2. Let $\alpha = 0^* \cup \{p \mid p$ is rational, $p \geqslant 0$ and $p^2 < 2\}$. Then α is a cut. Also the set of upper rationals for α has no smallest element.

[*Hint for checking* (iii): Take $p > 0$ rational and $p^2 < 2$. Let h be a rational such that both $0 < h < 1$ and

$$h < \frac{2 - p^2}{2p + 1}.$$

Then $(p + h)^2 = p^2 + (2p + h)h < p^2 + (2p + 1)h <$ etc.

[*Hint for upper rational property:* Take $r > 0$ rational such that $2 < r^2$. Let

$$s = r - \frac{r^2 - 2}{2r}.$$

Check that $0 < s < r$ and $2 < s^2$.]

This problem shows that there is a nonempty set of rationals which is bounded above but has no rational least upper bound.

Lemma. *If α is a cut and $r > 0$ is rational, then there is a $p \in \alpha$ and a proper upper rational q for α such that*

$$q - p = r.$$

Proof. Choose $s \in \alpha$ and t an upper rational for α. Then $t - s > 0$ and $r > 0$ so that, by the Archimedean property of the rationals, there is a positive integer n such that

$$nr > t - s; \qquad \text{that is, } t < s + nr.$$

Hence there is a positive integer $m \leqslant n$ such that

$$s + (m - 1)r \in \alpha \qquad \text{and} \qquad s + mr \qquad \text{is an upper rational for } \alpha.$$

If $s + mr$ is a proper upper rational for α, we set

$$p = s + (m - 1)r \qquad \text{and} \qquad q = s + mr,$$

but if $s + mr$ is not a proper upper rational for α, we set

$$p = s + (m - \tfrac{1}{2})r \qquad \text{and} \qquad q = s + (m + \tfrac{1}{2})r.$$

In either case $p \in \alpha$, q is a proper upper rational for α and $q - p = r$. ∎

Theorem 5. *If α is a cut, then the set $(-\alpha)$ defined by*

$$(-\alpha) = \{-p \mid p \text{ is a proper upper rational for } \alpha\}$$

is a cut and $\alpha + (-\alpha) = 0^*$. Also, if β is a cut and $\alpha + \beta = 0^*$, then $\beta = (-\alpha)$.

Proof. Take q an upper rational for α and set $s = \max(q, 0)$. Then $s + 1$ is a positive rational which (being $>q$) is a proper upper rational for α. Thus $-(s + 1) \in (-\alpha)$ so that $(-\alpha) \neq \varnothing$.

Any negative $q \in \alpha$ (and there are some) is such that $-q$ is rational but $-(-q)$ is not a proper upper rational for α so that $-q \notin (-\alpha)$. Hence (ii) holds.

To check (iii), take $r \in (-\alpha)$, s rational and $s < r$. Then $-r$ is a proper upper rational for α. Also $-s > -r$ so that $-s$ is also a proper upper rational for α; that is, $s \in (-\alpha)$.

Take any $r \in (-\alpha)$. Then $-r$ is a proper upper rational for α and hence there is a smaller upper rational p for α; that is,

$$p < -r \qquad \text{and} \qquad p \notin \alpha.$$

Hence $r < -p$ and $s = \frac{1}{2}[r + (-p)]$ is a rational such that

$$r < s < -p \qquad \text{and} \qquad p < -s < -r.$$

From $p < -s$ and $p \notin \alpha$ we have that $-s$ is a proper upper rational for α and therefore $s = -(-s) \in (-\alpha)$. Then $r < s$ shows that $(-\alpha)$ has no greatest element. Thus (iv) holds.

Hence $(-\alpha)$ is a cut and we next show that $\alpha + (-\alpha) = 0^*$.

Take any $r \in \alpha + (-\alpha)$. Write $r = s + (-p)$ with $s \in \alpha$ and $-p \in (-\alpha)$. Hence $p \notin \alpha$ (in fact p is a proper upper rational for α) and therefore $s < p$ so that $r = s - p < 0$. Consequently $r \in 0^*$ and we have $\alpha + (-\alpha) \subset 0^*$.

On the other hand, take any $r \in 0^*$. Hence $-r > 0$. We use $-r$ in place of the r in the lemma to obtain a $p \in \alpha$ and a proper upper rational q for α such that

$$q - p = -r.$$

Hence $-q \in (-\alpha)$ and

$$r = p - q = p + (-q) \in \alpha + (-\alpha).$$

Therefore $0^* \subset \alpha + (-\alpha)$ and we have $\alpha + (-\alpha) = 0^*$.

Finally, let β be any cut such that $\alpha + \beta = 0^*$. Then

$$\begin{aligned}\beta = 0^* + \beta &= [\alpha + (-\alpha)] + \beta = (-\alpha) + (\alpha + \beta)\\ &= (-\alpha) + 0^* = (-\alpha). \ \blacksquare\end{aligned}$$

Theorem 6. *If α and β are cuts, then there is one and only one cut γ such that*

$$\alpha + \gamma = \beta.$$

Proof. Set $\gamma = \beta + (-\alpha)$. Then

$$\alpha + \gamma = \alpha + [\beta + (-\alpha)] = \alpha + (-\alpha) + \beta = 0^* + \beta = \beta.$$

Let δ be any cut such that $\alpha + \delta = \beta$. Then uniqueness follows from

$$\delta = 0^* + \delta = (-\alpha) + (\alpha + \delta) = (-\alpha) + \beta = \gamma. \ \blacksquare$$

We shall interpret $\beta - \alpha$ and $-\alpha + \beta$ to mean $\beta + (-\alpha)$.

PROBLEM 3. For any cuts α and β,

$$-(-\alpha) = \alpha \qquad \text{and} \qquad -(\alpha - \beta) = -\alpha + \beta.$$

[*Hint for the second equality:*

$$-\alpha + \beta = -\alpha + \beta + 0^* = -\alpha + \beta + [-(\alpha - \beta) + (\alpha - \beta)] = etc.]$$

Theorem 7. *If α and β are cuts, then they are either the same or one is a proper subset of the other.*

Proof. Since α and β are sets, either $\alpha = \beta$ (meaning $\alpha \subset \beta$ and $\beta \subset \alpha$) or else $\alpha \neq \beta$, in which case there is an element r in one of them but not in the other. Consider that $r \in \beta$ but $r \notin \alpha$. Then r is an upper rational for α. Hence

$$\begin{aligned} q \in \alpha &\Rightarrow q < r \\ &\Rightarrow q \in \beta \qquad \text{(from } r \in \beta \text{ and (iii))}. \end{aligned}$$

Thus $\alpha \subset \beta$ and, moreover, is a proper subset of β.

Similarly, if α contains an element not in β, then β is a proper subset of α. ∎

Hence, if $\alpha \in \mathfrak{C}$ and $\beta \in \mathfrak{C}$, then as sets either $\alpha \subset \beta$ or $\beta \subset \alpha$ and it is natural to think of the elements of $\mathfrak{C}$ as being simply ordered. The next definition introduces a new role for the symbols $\leqslant$ and $<$ in expressing an ordering of the elements of $\mathfrak{C}$.

Definition 3. *For α and β cuts we write*

$$\alpha \leqslant \beta \qquad (\textit{equivalently } \beta \geqslant \alpha)$$

whenever $\alpha \subset \beta$, and the sharper inequality

$$\alpha < \beta \qquad (\textit{equivalently } \beta > \alpha)$$

whenever α is a proper subset of β.

It then follows from Theorem 7 that one and only one of

$$\alpha < \beta, \qquad \alpha = \beta, \qquad \alpha > \beta$$

holds for any cuts α and β.

We shall, of course, continue to use $\leqslant$ and $<$ between rationals.

Theorem 8. *If α, β, γ are cuts with $\alpha \leqslant \beta \leqslant \gamma$, then*

$$\alpha \leqslant \gamma.$$

If $\alpha < \beta \leqslant \gamma$, or if $\alpha \leqslant \beta < \gamma$, then

$$\alpha < \gamma.$$

Proof. If $\alpha \leqslant \beta$ and $\beta \leqslant \gamma$, then $\alpha \subset \beta \subset \gamma$ so that $\alpha \subset \gamma$ and $\alpha \leqslant \gamma$.

Consider $\alpha < \beta \leqslant \gamma$. Take p such that

$$p \in \beta \qquad \text{but} \qquad p \notin \alpha.$$

Since $\beta \subset \gamma$, then $p \in \gamma$, but $p \notin \alpha$ which says that $\alpha < \gamma$. Similarly, if $\alpha \leqslant \beta < \gamma$ there is a $q \in \gamma$, $q \notin \beta$, and thus $q \notin \alpha$ since $\alpha \subset \beta$. Hence $q \in \gamma$ and $q \notin \alpha$ so that $\alpha < \gamma$. ∎

PROBLEM 4. Given a cut α, then $\alpha < 0^*$ iff $-\alpha > 0^*$.

Theorem 9. *For any cuts α, β, and γ,*

$$\alpha + \gamma = \beta + \gamma \qquad \textit{iff } \alpha = \beta$$

and

$$\alpha + \gamma < \beta + \gamma \qquad \textit{iff } \alpha < \beta.$$

Proof. If $\alpha = \beta$, then $\alpha + \gamma$ means the same as $\beta + \gamma$. On the other hand, if $\alpha + \gamma = \beta + \gamma$, then

$$\alpha = 0^* + \alpha = \gamma - \gamma + \alpha = -\gamma + (\alpha + \gamma) = -\gamma + (\beta + \gamma) = \beta.$$

Thus both theorems about equality hold.

Consider $\alpha < \beta$. Then $\alpha \subset \beta$,

$$\alpha + \gamma = \{p + q \mid p \in \alpha \text{ and } q \in \gamma\} \subset \{p + q \mid p \in \beta \text{ and } q \in \gamma\} = \beta + \gamma$$

and hence at least $\alpha + \gamma \leqslant \beta + \gamma$. But equality here would force the contradiction $\alpha = \beta$ and hence $\alpha + \gamma < \beta + \gamma$.

Finally consider $\alpha + \gamma < \beta + \gamma$. Add $-\gamma$ to both sides and obtain (by the part just proved)

$$(\alpha + \gamma) + (-\gamma) < (\beta + \gamma) + (-\gamma);$$

that is, $\alpha < \beta$. ∎

PROBLEM 5. (a) With r and s rationals

$$r^* < s^* \qquad \text{iff } r < s.$$

(b) If α is a cut, then $p \in \alpha$ iff $p^* < \alpha$.

PROBLEM 6. If α and β are cuts such that $\alpha < \beta$, then there is a rational r such that

$$\alpha < r^* < \beta.$$

[*Hint:* Take $p \in \beta$, $p \notin \alpha$; then an $r \in \beta$ such that $r > p$.]

This problem will be used in the proof of the next theorem.

We now have the portion of arithmetic involving addition (and subtraction) of elements of $\mathfrak{C}$ and have ordered the elements of $\mathfrak{C}$. Also there is a subset of $\mathfrak{C}$ which is indistinguishable from the rationals as far as addition and ordering are concerned. By this is meant that corresponding to each rational r we have assigned one and only one element of $\mathfrak{C}$ (namely, r^*) in such a way that for any rationals r and s,

$$(r + s)^* = r^* + s^* \qquad \text{and}$$
$$r^* < s^* \qquad \text{iff } r < s.$$

$\mathfrak{C}$ has, moreover, a desirable completeness property not possessed by the set of rationals (or the set of rational cuts). We establish this property in the next two theorems, after which a full arithmetic for $\mathfrak{C}$ will be developed by introducing multiplication of elements of $\mathfrak{C}$.

Theorem 01 (Dedekind). *Let L and U be two sets of cuts satisfying*

I. $L \neq \varnothing, U \neq \varnothing$;
II. $L \cup U = \mathfrak{C}$;
III. $\alpha \in L$ *and* $\beta \in U \Rightarrow \alpha < \beta$.

Then there is one and only one cut γ *such that*

(1) $$\alpha \leqslant \gamma \leqslant \beta \quad \textit{for all } \alpha \in L \textit{ and all } \beta \in U.$$

Also, this cut γ *is either the largest element of L or else is the smallest element of U.*

Proof. An immediate consequence of III is that

$$L \cap U = \varnothing.$$

We shall show that the specific set γ defined by

(2) $$\gamma = \{p \mid p \in \alpha \text{ for some } \alpha \in L\}$$

is a cut, then that this cut satisfies (1), and follow this by establishing the stated uniqueness and maximum-minimum properties.

(a) L is not empty, each cut α is not empty, and thus γ is not empty.

(b) Take any $\beta \in U$ and then any rational $q \notin \beta$. By III, $\alpha \subset \beta$ for all $\alpha \in L$ so that $q \notin \alpha$ for all $\alpha \in L$ and therefore $q \notin \gamma$.

(c) Take any $r \in \gamma$, s rational, and $s < r$. Since $r \in \gamma$, there is a cut α such that $r \in \alpha \in L$, and for this α we have $s \in \alpha$ by (iii). Therefore $s \in \gamma$.

(d) Again take any $r \in \gamma$ and choose a cut α such that $r \in \alpha \in L$. Since no cut has a largest element, there is an $s \in \alpha$ such that $s > r$. From $s \in \alpha$ this s is an element of γ but is also greater than the arbitrarily chosen $r \in \gamma$. Hence γ has no largest element.

Thus γ is a cut and we next show that this γ satisfies (1).

Take any $\alpha \in L$. Then $p \in \alpha \Rightarrow p \in \gamma$ and hence $\alpha \subset \gamma$. Thus $\alpha \leqslant \gamma$, and this is true of each $\alpha \in L$.

Take any $p \in \gamma$. Select a cut α such that $p \in \alpha \in L$. Then $p \in \alpha \subset \beta$ for every $\beta \in U$, by III. Thus $p \in \beta$ for every $\beta \in U$. Since p was any element of γ, we have $\gamma \subset \beta$; that is, $\gamma \leqslant \beta$ for every $\beta \in U$.

Thus the cut γ defined by (2) satisfies (1) and we next assume there are two cuts γ_1 and γ_2, so named that $\gamma_1 < \gamma_2$, both satisfying

$$\alpha \leqslant \gamma_1 \leqslant \beta \quad \text{and} \quad \alpha \leqslant \gamma_2 \leqslant \beta \quad \text{for all } \alpha \in L \text{ and all } \beta \in U.$$

There is then a rational r such that the cut r^* satisfies

$$\gamma_1 < r^* < \gamma_2,$$

by Problem 6. Then $\alpha \leqslant \gamma_2$ for all $\alpha \in L$ puts $r^* \in L$, whereas $\gamma_1 \leqslant \beta$ for all $\beta \in U$ puts $r^* \in U$. This contradiction to $L \cap U = \varnothing$ establishes uniqueness.

Since $L \cup U = \mathfrak{C}$ and $L \cap U = \varnothing$, we know that γ is in one of L or U but not in both. Hence, from (1), if $\gamma \in L$, then γ is the largest member of L, but if $\gamma \in U$ then γ is the smallest member of U. ∎

To avoid any misunderstanding, we say that a nonempty set $A \subset \mathfrak{C}$ is bounded above if there is a $\beta \in \mathfrak{C}$ such that

$$\alpha \in A \Rightarrow \alpha \leqslant \beta.$$

Then upper bound, a (the) least upper bound (if it exists), bounded below, etc., should be evident in this context.

Theorem 11. *If $A \subset \mathfrak{C}$ is nonempty and bounded above, then A has a least upper bound in $\mathfrak{C}$.*

Proof. Define $L = \{\delta \mid \delta \in \mathfrak{C}$ and $\delta < \alpha$ for at least one $\alpha \in A\}$ and $U = \{\beta \mid \beta \in \mathfrak{C}$ and $\beta \notin L\}$.

We check that L and U satisfy I, II, and III in (a), (b), and (c) below.

(a) Take any $\alpha \in A$ (possible since $A \neq \varnothing$), then any $r \in \alpha$. Hence $r^* < \alpha$ by Problem 5(b). Therefore $r^* \in L$ and hence $L \neq \varnothing$.

Let $\beta \in \mathfrak{C}$ be an upper bound of A, which is possible since A is bounded above. Take any proper upper rational r for β and note that

$$\alpha \in A \Rightarrow \alpha \leqslant \beta < r^*.$$

Hence $r^* \notin L$, $r^* \in U$, and thus $U \neq \varnothing$.

(b) $L \cup U = \mathfrak{C}$ by the definition of U.

(c) First note that $L \cap U = \varnothing$ by the definition of U. Assume there is a $\delta \in L$ and a $\beta \in U$ with $\beta \leqslant \delta$. Since $\delta \in L$, pick an $\alpha \in A$ such that $\delta < \alpha$. Then

$$\beta \leqslant \delta < \alpha \Rightarrow \beta < \alpha \qquad \text{(Theorem 8)}$$
$$\Rightarrow \beta \in L \Rightarrow \beta \notin U.$$

L and U, therefore, satisfy the conditions of Theorem 01 and hence there is a cut γ such that

$$\delta \leqslant \gamma \leqslant \beta \qquad \text{for all } \delta \in L \text{ and all } \beta \in U.$$

Assume $\gamma < \alpha$ for some $\alpha \in A$. Pick a rational r such that $\gamma < r^* < \alpha$. Then

$$r^* < \alpha \Rightarrow r^* \in L \quad \text{but} \quad \gamma < r^* \Rightarrow r^* \in U \Rightarrow r^* \notin L.$$

Thus γ is an upper bound of A.

Assume A has an upper bound $\beta_0 \in \mathfrak{C}$ such that $\beta_0 < \gamma$. Then $\beta_0 \in L$ and hence there is an $\alpha \in A$ such that $\beta_0 < \alpha$, contradicting β_0 being an upper bound of A. Thus γ is the least upper bound of A. ∎

We now introduce another binary operation (called multiplication) on $\mathfrak{C}$.

Theorem 21. *If α and β are cuts such that $\alpha \geqslant 0^*$ and $\beta \geqslant 0^*$, then the set $\alpha\beta$ defined by*

$$\alpha\beta = 0^* \cup \{pq \mid p \in \alpha,\ q \in \beta,\ p \geqslant 0 \text{ and } q \geqslant 0\}$$

is a cut (called the **product** *of α and β). Also $\alpha\beta = \beta\alpha$.*

Proof. If either $\alpha = 0^*$ or $\beta = 0^*$, then $\alpha\beta = 0^*$ since the second set in the definition is empty.

We thus consider $\alpha > 0^*$ and $\beta > 0^*$. Then (i) and (ii) are easily checked. Also it is easy to establish (iii) and (iv) for the set

$$\{pq \mid p \in \alpha,\ q \in \alpha,\ p > 0 \text{ and } q > 0\}$$

and to see that this suffices. ∎

With $\alpha \in \mathfrak{C}$ such that $\alpha > 0^*$, we temporarily associate the set α^+ defined by

$$\alpha^+ = \{p \mid p \in \alpha \text{ and } p \geqslant 0\},$$

call this set the **plus-part** of α, and note that

$$\alpha = 0^* \cup \alpha^+ \qquad \text{and}$$
$$(p \in \alpha^+,\ q \text{ rational, and } 0 \leqslant q \leqslant p) \Rightarrow q \in \alpha^+.$$

If $0^* < \alpha \in \mathfrak{C}$ and $0^* < \beta \in \mathfrak{C}$, then a check will show that the sets defined by

$$\alpha^+ + \beta^+ = \{p + q \mid p \in \alpha^+,\ q \in \beta^+\} = \beta^+ + \alpha^+ \qquad \text{and}$$
$$\alpha^+\beta^+ = \{pq \mid p \in \alpha^+,\ q \in \beta^+\} = \beta^+\alpha^+$$

are the plus-sets of $\alpha + \beta$ and $\alpha\beta$, respectively.

Theorem 31.† *For cuts* $\alpha > 0^*$, $\beta > 0^*$, *and* $\gamma > 0^*$,

$$\alpha^+(\beta^+ + \gamma^+) = \alpha^+\beta^+ + \alpha^+\gamma^+,$$
$$\alpha(\beta + \gamma) = \alpha\beta + \alpha\gamma,$$
$$(\alpha\beta)\gamma = \alpha(\beta\gamma).$$

Proof. Take any $s \in \alpha^+(\beta^+ + \gamma^+)$. Choose p, q, r from appropriate sets to obtain

$$s = p(q + r) = pq + pr \in \alpha^+\beta^+ + \alpha^+\gamma^+.$$

Hence $\alpha^+(\beta^+ + \gamma^+) \subset \alpha^+\beta^+ + \alpha^+\gamma^+$.

Next take any $s \in \alpha^+\beta^+ + \alpha^+\gamma^+$. Choose

$$p_1 \in \alpha^+, \quad q \in \beta^+; \qquad p_2 \in \alpha^+, \quad r \in \gamma^+$$

such that $s = p_1 q + p_2 r$. Set $\max(p_1, p_2) = p$. Thus $p \in \alpha^+$ and

$$s \leqslant pq + pr = p(q + r) \in \alpha^+(\beta^+ + \gamma^+).$$

Thus, as noted above, $s \in \alpha^+(\beta^+ + \gamma^+)$ and we are able to assert that the first equality of the theorem holds. Then

$$\begin{aligned}
\alpha(\beta + \gamma) &= 0^* \cup [\alpha(\beta + \gamma)]^+ \\
&= 0^* \cup [\alpha^+(\beta + \gamma)^+] \\
&= 0^* \cup [\alpha^+(\beta^+ + \gamma^+)] \\
&= 0^* \cup [\alpha^+\beta^+ + \alpha^+\gamma^+] \\
&= 0^* \cup [(\alpha\beta)^+ + (\alpha\gamma)^+] \\
&= 0^* \cup [\alpha\beta + \alpha\gamma]^+ \\
&= \alpha\beta + \alpha\gamma.
\end{aligned}$$

The last equation of the theorem follows in a similar way. ∎

† For a name of the symbol 31 see the problem labeled 13 at the end of this chapter.

The next definition formalizes what we are accustomed to in practice.†

Definition 4. *For any cuts α and β:*

If $\alpha \geqslant 0^*$ *and* $\beta \geqslant 0^*$, *then* $\alpha\beta$ *is the cut defined in Theorem* 21.
If $\alpha < 0^*$ *and* $\beta \geqslant 0^*$, *then* $\alpha\beta = -[(-\alpha)\beta]$.
If $\alpha < 0^*$ *and* $\beta < 0^*$, *then* $\alpha\beta = (-\alpha)(-\beta)$.

The first three parts of the following problem may be proved by using the fact that the equation holds if the cuts involved are $\geqslant 0^*$ and considering possible cases in which some or all of the cuts are $<0^*$.

PROBLEM 7. For any cuts α, β, and γ:

(a) $$\alpha\beta = \beta\alpha.$$
(b) $$\alpha(\beta\gamma) = (\alpha\beta)\gamma.$$
(c) $$\alpha(\beta + \gamma) = \alpha\beta + \alpha\gamma.$$
(d) $$\alpha 1^* = \alpha.$$
(e) If $\alpha \neq 0^*$, then there is one and only one cut α^{-1} such that

$$\alpha\alpha^{-1} = 1^*.$$

(f) If $\alpha \neq 0^*$, then there is one and only one cut δ such that

$$\alpha\delta = \beta.$$

(g) If $\alpha < \beta$ and $0^* < \gamma$, then $\alpha\gamma < \beta\gamma$.
[*Hint for* (e): Show that if $\alpha > 0^*$, then the set defined by

$$\alpha^{-1} = 0^* \cup \{0\} \cup \left\{\frac{1}{p} \,\middle|\, 0 < p \in \alpha\right\}$$

is a cut, etc. With this "inverse" defined for all cuts $>0^*$, define $\alpha^{-1} = -(-\alpha)^{-1}$ if $\alpha < 0^*$.]

The following definition introduces a mere change in terminology.

Definition 5. *From now on we will use* **real number** *instead of "cut,"* **rational number** *instead of "rational cut," and* E^1 *instead of* $\mathfrak{C}$.

To be pedantic, a real number *is* a cut, a cut *is* a real number, and E^1 denotes the set of real numbers.

The frightening prospect of being forced forever to visualize a whole set of rationals satisfying (i)–(iv) every time a real number is mentioned will now be dispelled. We want real numbers for what can be done with them. Thus we must know that they exist and have the properties we want, but other than this

† For example, $(-2.3)(4.5) = -10.35$ is obtained by finding $(2.3)(4.5) = 10.35$ and then inserting a minus sign.

we can be utterly indifferent about what they may look like. The whole point is that we now know:

There exists a set having the properties traditionally expected of the real numbers

whereas we were previously intimidated (on not too good authority) into accepting this on faith. At any rate, we now have confidence that nothing will go wrong if we adhere to the following collection of workable rules.

There is a set E^1 *and two binary operations, denoted by* $+$ *and* $\times$, *such that if* $a \in E^1$ *and* $b \in E^1$, *then* $a + b$ *and* $a \times b$ *are also elements of* E^1. *These operations are* (*with also* $c \in E^1$)

F₁. Commutative: $a + b = b + a, \qquad a \times b = b \times a;$

F₂. Associative: $(a + b) + c = a + (b + c), \qquad (a \times b) \times c = a \times (b \times c);$

F₃. Distributive: $a \times (b + c) = (a \times b) + (a \times c).$

The set E^1 *has two unique distinct elements denoted by* 0 *and* 1 *such that*

F₄. $a + 0 = a \qquad a \times 1 = a.$

Corresponding to each element a *there is a unique element* $(-a)$ *and correspondingly to each element* $a \neq 0$ *there is a unique element* a^{-1} *such that*

F₅. $a + (-a) = 0 \qquad a \times a^{-1} = 1.$

The elements of E^1 *are ordered by a binary relation* $<$ *in such a way that*

O₁. *One and only one of* $a < b$, $a = b$, $b < a$ *holds.*
O₂. *If* $a < b$ *and* $b < c$, *then* $a < c$ (*called* **transitivity**).
O₃. *If* $a < b$, *then* $a + c < b + c$.
O₄. *If* $a < b$ *and* $0 < c$, *then* $ac < bc$.

The set E^1 *has the completeness property:*

C. *If* A *is a nonempty subset of* E^1 *and* A *is bounded above, then there is an element of* E^1 *which is the least upper bound of* A.

A set and two binary operations having the **F**-properties is said to be a **field** which is **ordered** if it has the **O**-properties, and **complete** if it has the property **C**. Thus a curt truism is:

The set of real numbers is a complete ordered field.

The set of real numbers contains the set of rational numbers which in turn contains the set of integers. Since we started with the rationals there may be some little doubt about the existence of the set of real numbers. Anyone interested in beginning without even "God given" integers should consult the excellent little book by E. Landau listed in the Bibliography.

Definition 6. *Any real number which is not a rational number is said to be an* **irrational** *number.*

PROBLEM 8. Let $\mathscr{F}$ be a complete ordered field (not necessarily the field of real numbers). Copy Theorem 01, but replace $\mathfrak{C}$ by $\mathscr{F}$ and "cut" by "element of $\mathscr{F}$." Prove this theorem.

1–7 GEOMETRY

Experience with the number line to represent real numbers is taken for granted and geometric language will be used. With $a < b$ given real numbers (or points),

$$(a, b) = \{x \mid a < x < b\}$$

is an **open interval** and

$$[a, b] = \{x \mid a \leqslant x \leqslant b\}$$

is a **closed interval** (even if $a = b$). Closed-open and open-closed intervals are

$$[a, b) = \{x \mid a \leqslant x < b\} \qquad \text{and}$$

$$(a, b] = \{x \mid a < x \leqslant b\}.$$

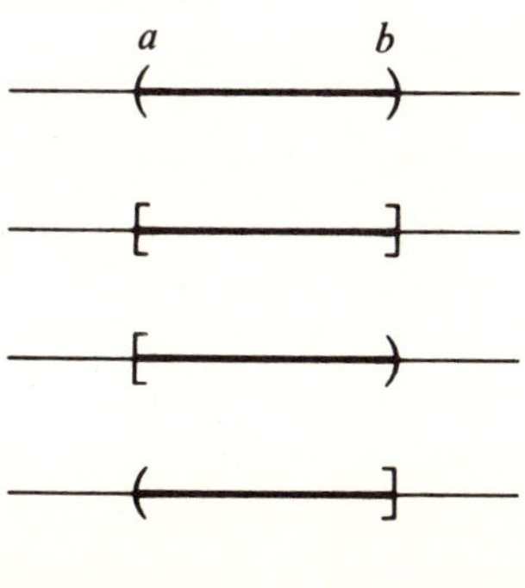

All of these intervals have **length** $b - a$.

Whether (a, b) represents an open interval, or (later on) the plane rectangular coordinates of a point, or an ordered pair of numbers, or a complex number, or a vector must be judged by the context in which it appears.

Geometric intuition may be used, and even cultivated, provided it is never relied upon as "proof" of an analytic statement. For example, Fig. 1, page 11 illustrates the impossibility of b_1 being a supremum of the set A in question, but this figure alone does not constitute a proof that the set does not have a second supremum.

There is a famous set, called the **Cantor set**, some of whose properties defy geometric intuition. This set is on the closed interval $[0, 1]$ and is defined in geometric terms as follows:

1. Erase the open interval $(\frac{1}{3}, \frac{2}{3})$ of length $\frac{1}{3}$.
2. Of the two remaining closed intervals erase the middle open third of each; that is, erase $(\frac{1}{9}, \frac{2}{9})$ and $(\frac{7}{9}, \frac{8}{9})$ the sum of whose lengths is $2 \cdot \frac{1}{9} = 2/3^2$.

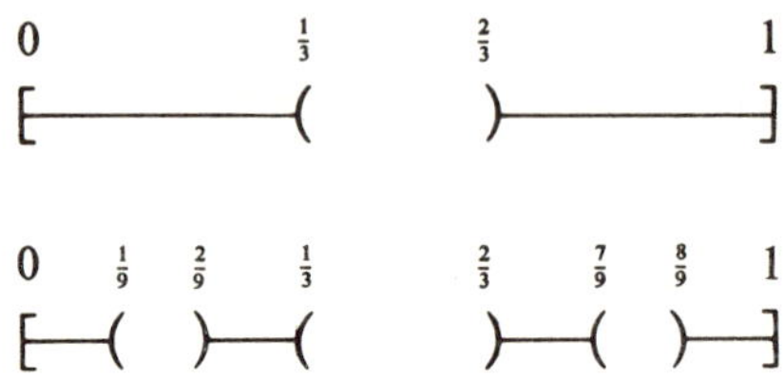

3. Of the four $= 2^2$ remaining closed intervals, each of length $1/3^2$, erase the middle open third of each so that the sum of lengths of intervals erased at this stage is $2^2/3^3$.

. . .

n. At the *n*th stage there are 2^{n-1} closed intervals remaining on [0, 1] each of length $1/3^{n-1}$; now erase the middle open third of each. The sum of the lengths of intervals erased at this stage is $2^{n-1}/3^n$.

. . .

The process is to be continued inductively.

The points of [0, 1] which are never erased constitute the **Cantor set**.

Note: We shall reserve the Greek letter Γ to represent the Cantor set.

Facts borrowed from previous courses are

$$s_n = 1 + r + r^2 + \cdots + r^{n-1} = \frac{1 - r^n}{1 - r} \qquad \text{if } r \neq 1,$$

$$\sum_{k=0}^{\infty} r^k = \lim_{n \to \infty} s_n = \frac{1}{1 - r} \qquad \text{if } |r| < 1,$$

and a few other results (which will also be established later) about series.

In forming the Cantor set the sum of the lengths of intervals erased at the 1st, 2nd, $\cdots$, *n*th stages is

$$s_n = \frac{1}{3} + \frac{2}{3^2} + \frac{2^2}{3^3} + \cdots + \frac{2^{n-1}}{3^n} = 1 - \left(\frac{2}{3}\right)^n.$$

Hence

$$\lim_{n \to \infty} s_n = \lim_{n \to \infty} \left[1 - \left(\frac{2}{3}\right)^n\right] = 1;$$

that is, the sum of lengths of all intervals erased is the length of the original interval [0, 1]. The set remaining on [0, 1] is the Cantor set and may seem so sparse as to be insignificant.

Points surely in the Cantor set are

$$0, 1, \tfrac{1}{3}, \tfrac{2}{3}, \tfrac{1}{9}, \tfrac{2}{9}, \tfrac{7}{9}, \tfrac{8}{9}, \text{etc.},$$

for all end points of those open intervals which were erased. But are there any points of the Cantor set other than these end points? There are (as shown later). And this Cantor set has other remarkably unintuitive properties which will be given at various places throughout the book to illustrate and amplify basic concepts.

It should be emphasized that the Cantor set and its properties are not given to destroy geometric intuition or to instill fear of using geometric intuition, but rather to guide the use of geometric intuition in analysis to a supporting role in discovering what to suspect and how then to prove or disprove the suspected. Hence the construction of the Cantor set should be placed, and kept, clearly in mind.

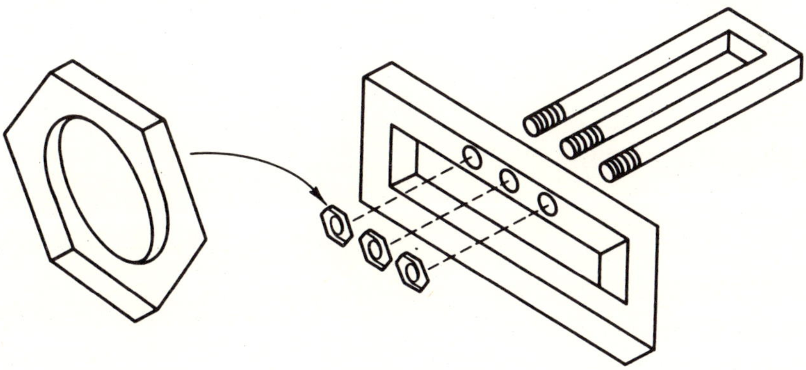

1-8 DECIMAL, TERNARY, AND BINARY REPRESENTATIONS

The digits 0, 1, 2, ⋯, 9 are called the **decimal digits**. If

$$\sum_{n=1}^{\infty} \frac{\alpha_n}{10^n} \text{ converges to } x,$$

with each α_n a decimal digit, then a **decimal representation** of x is

$$x = 0.\alpha_1\alpha_2\alpha_3 \cdots.$$

Lemma 1. *Let* $0 < x < 1$ *be given. Then for each* $n = 1, 2, 3, \cdots$ *there are decimal digits* $\alpha_1, \alpha_2, \cdots, \alpha_n$ *such that*

$$\frac{\alpha_1}{10} + \frac{\alpha_2}{10^2} + \cdots + \frac{\alpha_n}{10^n} \leqslant x < \frac{\alpha_1}{10} + \frac{\alpha_2}{10^2} + \cdots + \frac{\alpha_n}{10^n} + \frac{1}{10^n}. \tag{1}$$

Proof. (by induction). Let α_1 be defined, using the greatest integer notation, by

$$\alpha_1 = [10x].$$

Since $0 < x < 1$, then $0 < 10x < 10$ so that α_1 is a decimal digit and $\alpha_1 \leqslant 10x < \alpha_1 + 1$. Thus

$$\frac{\alpha_1}{10} \leqslant x < \frac{\alpha_1}{10} + \frac{1}{10}$$

and (1) holds for $n = 1$.

Now let n be a positive integer for which decimal digits $\alpha_1, \alpha_2, \cdots, \alpha_n$ have already been determined such that (1) holds. To save writing, set

$$a_n = \frac{\alpha_1}{10} + \frac{\alpha_2}{10^2} + \cdots + \frac{\alpha_n}{10^n}. \tag{2}$$

Then (1) becomes $a_n \leqslant x < a_n + (1/10^n)$ so that

$$0 \leqslant 10^{n+1}(x - a_n) < 10.$$

Thus upon setting

$$\alpha_{n+1} = [10^{n+1}(x - a_n)]$$

α_{n+1} is one of the decimal digits,

$$\alpha_{n+1} \leqslant 10^{n+1}(x - a_n) < \alpha_{n+1} + 1, \qquad \text{and}$$

$$a_n + \frac{\alpha_{n+1}}{10^{n+1}} \leqslant x < a_n + \frac{\alpha_{n+1}}{10^{n+1}} + \frac{1}{10^{n+1}},$$

which, in view of the definition (2) of a_n, is (1) with n replaced by $n + 1$. ∎

Thus Lemma 1 is proved and it then follows that

Theorem 1. *Any number x, such that $0 < x < 1$, has at least one decimal representation*

$$x = 0.\alpha_1\alpha_2\alpha_3 \cdots.$$

The number $\frac{1}{4} = 0.25$ has a second decimal representation

$$\tfrac{1}{4} = 0.24999 \cdots,$$

since

$$\frac{2}{10} + \frac{4}{10^2} + \frac{9}{10^3}\sum_{k=0}^{\infty} \frac{1}{10^k} \text{ converges to } \frac{1}{4}.$$

Theorem 2. *Any number $0 < x < 1$ has at most two decimal representations.*

Proof. Consider an x, $0 < x < 1$, which has at least two different decimal representations

$$\begin{aligned} x &= 0.\alpha_1\alpha_2\alpha_3 \cdots \\ &= 0.\beta_1\beta_2\beta_3 \cdots. \end{aligned}$$

Let m be the first value of n for which $\alpha_n \neq \beta_n$ so that

(3) $\qquad \alpha_1 = \beta_1, \qquad \alpha_2 = \beta_2, \cdots, \alpha_{m-1} = \beta_{m-1}, \qquad$ but $\qquad \alpha_m \neq \beta_m.$

Let these representations be named so that $\alpha_m < \beta_m$. Since α_m and β_m are decimal digits, then

$$1 \leqslant \alpha_m + 1 \leqslant \beta_m \leqslant 9.$$

Hence

$$\begin{aligned}
x &= \sum_{n=1}^{m-1} \frac{\alpha_n}{10^n} + \frac{\alpha_m}{10^m} + \sum_{k=0}^{\infty} \frac{\alpha_{m+k+1}}{10^{m+k+1}} \\
&\leqslant \sum_{n=1}^{m-1} \frac{\alpha_n}{10^n} + \frac{\alpha_m}{10^m} + \frac{9}{10^{m+1}} \sum_{k=0}^{\infty} \frac{1}{10^k} \\
&= \sum_{n=1}^{m-1} \frac{\beta_n}{10^n} + \frac{\alpha_m}{10^m} + \frac{1}{10^m} = \sum_{n=1}^{m-1} \frac{\beta_n}{10^n} + \frac{\alpha_m + 1}{10^m} \\
&\leqslant \sum_{n=1}^{m-1} \frac{\beta_n}{10^n} + \frac{\beta_m}{10^m} = \sum_{n=1}^{m-1} \frac{\beta_n}{10^n} + \frac{\beta_m}{10^m} + \sum_{k=0}^{\infty} \frac{0}{10^{m+k+1}} \\
&\leqslant \sum_{n=1}^{m-1} \frac{\beta_n}{10^n} + \frac{\beta_m}{10^m} + \sum_{k=0}^{\infty} \frac{\beta_{m+k+1}}{10^{m+k+1}} \\
&= \sum_{n=1}^{\infty} \frac{\beta_n}{10^n} = x.
\end{aligned}$$

Thus all equalities hold so that

$$\alpha_m + 1 = \beta_m$$

and also

$$\alpha_n = 9 \quad \text{and} \quad \beta_n = 0 \quad \text{for } n = m+1, m+2, \cdots.$$

This definite relation between the α's and β's shows that x does not have still another different decimal representation. ∎

By definition, 0, 1, and 2 are called the **ternary digits**. If

$$\sum_{n=1}^{\infty} \frac{\alpha_n}{3^n} \text{ converges to } x,$$

with each α_n a ternary digit, then a **ternary representation** of x is

$$x = 0 \underset{3}{.} \alpha_1 \alpha_2 \alpha_3 \cdots,$$

where the small 3 is put under the ternary point to emphasize that it is the ternary (and not the decimal) point.

PROBLEM 1. If $0 < x < 1$, then x has at least one and no more than two ternary representations.

Note that $\frac{1}{3}$ has the two ternary representations,

$$\tfrac{1}{3} = \underset{3}{0.}1000\cdots \qquad \text{and} \qquad \tfrac{1}{3} = \underset{3}{0.}0222\cdots$$

and also

$$\tfrac{2}{3} = \underset{3}{0.}2000\cdots \qquad \text{and} \qquad \tfrac{2}{3} = \underset{3}{0.}1222\cdots.$$

Each of $\frac{1}{3}$ and $\frac{2}{3}$ is seen to have a ternary representation with the digit 1 in the first place and another ternary representation not using the digit 1 in any place.

Lemma 2. *If* $\frac{1}{3} < x < \frac{2}{3}$ *and a ternary representation of* x *is*

$$x = \underset{3}{0.}\alpha_1\alpha_2\alpha_3\cdots$$

then $\alpha_1 = 1$ *and among* $\alpha_2, \alpha_3, \cdots$ *there is at least one* $\neq 0$ *and at least one* $\neq 2$.

Proof. Suppose $\alpha_1 = 0$. Then

$$x = \frac{0}{3} + \sum_{n=2}^{\infty} \frac{\alpha_n}{3^n} \leqslant \sum_{n=2}^{\infty} \frac{2}{3^n}$$

$$= \frac{2}{3^2}\sum_{k=0}^{\infty}\frac{1}{3^k} = \frac{2}{3^2}\frac{1}{1-\frac{1}{3}} = \frac{1}{3}$$

contradicting $\frac{1}{3} < x$. Suppose $\alpha_1 = 2$. Then $x \geqslant \frac{2}{3}$, contradicting $x < \frac{2}{3}$. Thus $\alpha_1 = 1$.

Now the supposition $\alpha_2 = \alpha_3 = \cdots = 0$ gives $x = \frac{1}{3}$, contradicting $\frac{1}{3} < x$ while the supposition $\alpha_2 = \alpha_3 = \cdots = 2$ gives $x = \frac{2}{3}$, contradicting $x < \frac{2}{3}$. ∎

It should be seen, conversely, that

Lemma 3. *If* $x = \underset{3}{0.}1\,\alpha_2\alpha_3\cdots$ *with at least one* $\alpha_n \neq 0$ *and at least one* $\alpha_n \neq 2$, *then the point* x *is on the open interval* $(\frac{1}{3}, \frac{2}{3})$ *and hence is not a point of the Cantor set.*

A way of indicating "x has a ternary representation not involving the digit 1" is "x has a ternary representation of the form

$$(4) \qquad x = \underset{3}{0.}(2\gamma_1)(2\gamma_2)(2\gamma_3)\cdots \qquad \text{where } \gamma_n = \begin{Bmatrix}0\\1\end{Bmatrix}\text{''}.$$

All of this is motivation for

Theorem 3. *A point* x *is in the Cantor set* Γ *if and only if at least one ternary representation of* x *does not have the digit* 1 *in any place, that is,*

$$\Gamma = \{x \mid x \textit{ has a ternary representation of the form } (4)\}.$$

A proof of this theorem will be given after the next lemma.

Whenever a ternary representation has only zeros from some place on, these zeros will not be indicated. For example,

$$\tfrac{1}{9} = \underset{3}{0.01}, \qquad \tfrac{2}{9} = \underset{3}{0.02}, \qquad \tfrac{7}{9} = \underset{3}{0.21}, \qquad \tfrac{8}{9} = \underset{3}{0.22}.$$

In defining the Cantor set, the open intervals erased at the second stage were $(\frac{1}{9}, \frac{2}{9})$ and $(\frac{7}{9}, \frac{8}{9})$.

From the above comments it will be seen that the following lemma holds for $n = 1$ and $n = 2$.

Lemma 4. *In defining the Cantor set Γ, the open interval $(\underline{x}_n, \bar{x}_n)$ was erased at the nth stage if and only if $\underline{x}_n$ and $\bar{x}_n$ can be written as*

$$\left.\begin{aligned} &(5) \qquad \underline{x}_n = \underset{3}{0.}(2\gamma_1)(2\gamma_2)\cdots(2\gamma_{n-1})1 \\ &(6) \qquad \bar{x}_n = \underset{3}{0.}(2\gamma_1)(2\gamma_2)\cdots(2\gamma_{n-1})2 \end{aligned}\right\} \textit{with only zeros and ones among the } \gamma_k\textit{'s.}$$

Proof. To make the induction step, let $n \geqslant 2$ be an integer for which the lemma is true. Then the open interval $(\underline{x}_{n+1}, \bar{x}_{n+1})$ was erased at the $(n + 1)$st stage iff $\underline{x}_{n+1} = 0 + 1/3^{n+1}$ or $\underline{x}_{n+1} = \bar{x}_n + 1/3^{n+1}$ with $\bar{x}_n$ in the form (6); i.e., respectively,

$$\underline{x}_{n+1} = \frac{1}{3^{n+1}} = \underset{3}{0.}\underbrace{00\cdots0}_{n}1 \qquad \text{where } \gamma_n = 0 \qquad \text{or else}$$

$$\begin{aligned} \underline{x}_{n+1} &= \underset{3}{0.}(2\gamma_1)(2\gamma_2)\cdots(2\gamma_{n-1})2 + \frac{1}{3^{n+1}} \\ &= \underset{3}{0.}(2\gamma_1)(2\gamma_2)\cdots(2\gamma_{n-1})(2\gamma_n)1 \qquad \text{where } \gamma_n = 1. \end{aligned}$$

With γ_n so determined, then $\bar{x}_{n+1} = \underline{x}_{n+1} + 1/3^{n+1}$ so that

$$\bar{x}_{n+1} = \underset{3}{0.}(2\gamma_1)(2\gamma_2)\cdots(2\gamma_{n-1})(2\gamma_n)2.$$

Since these representations are in the forms (5) and (6) with n replaced by $n + 1$ the lemma is proved. ∎

Then, as in the proof of Lemma 3, a number x is such that

$$\underline{x}_n < x < \bar{x}_n,$$

with $\underline{x}_n$ and $\bar{x}_n$ in the forms (5) and (6), iff

$$x = \underset{3}{0.}(2\gamma_1)(2\gamma_2)\cdots(2\gamma_{n-1})1\alpha_{n+1}\alpha_{n+2}\cdots$$

with at least one $\alpha_{n+k} \neq 0$ and at least one $\alpha_{n+k} \neq 2$. Hence a point x is erased at some stage iff either (in case there are two) ternary representation of x has a 1 in some place. Hence a point of $[0, 1]$ is not erased at any stage iff it can be written in ternary form without using the digit 1. Thus Theorem 3 is proved.

Incidentally, points of the Cantor set Γ other than end points of erased open intervals may now be shown. One such point is

$$0.\underset{3}{}202020\cdots = \tfrac{3}{4}.$$

The digits 0 and 1 are called the **binary digits**. If

$$\sum_{n=1}^{\infty} \frac{\alpha_n}{2^n} \text{ converges to } x,$$

with each α_n a binary digit, then a binary representation of x is

$$x = \underset{2}{0.}\alpha_1\alpha_2\alpha_3\cdots.$$

PROBLEM 2. If $0 < x < 1$, then x has at least one and no more than two binary representations.

The only binary representation of the number zero is $\underset{2}{0.}0$ but 1 may be represented by $1 = \underset{2}{0.}111\cdots$.

If $x = \underset{2}{0.}\,\alpha_1\alpha_2\cdots\alpha_n 1$, with each α_k a binary digit, then the second binary representation of this x is

$$x = \underset{2}{0.}\alpha_1\alpha_2\cdots\alpha_n 0111\cdots$$

and is called the **infinite binary representation** of x. It should now be seen that

Corollary. *If $0 < x < 1$, then x has one and only one infinite binary representation.*

1–9 ABSOLUTE VALUE AND INEQUALITIES

The **absolute value** of a real number x is denoted by $|x|$, where

$$|x| = \begin{cases} x & \text{if } x \geqslant 0 \\ -x & \text{if } x < 0. \end{cases} \tag{1}$$

From the definition of the square root, $y = \sqrt{x^2} \Leftrightarrow (y \geqslant 0 \text{ and } y^2 = x^2)$. Hence, since

$$|x|^2 = \begin{Bmatrix} x^2 & \text{if } x \geqslant 0 \\ (-x)(-x) = x^2 & \text{if } x < 0 \end{Bmatrix}, \qquad \text{then } \sqrt{x^2} = |x|.$$

From (1),

$$-|x| = \begin{Bmatrix} -x & \text{if } x \geqslant 0 \\ x & \text{if } x < 0 \end{Bmatrix} \leqslant 0 \leqslant \begin{Bmatrix} x & \text{if } x \geqslant 0 \\ -x & \text{if } x < 0 \end{Bmatrix} = |x|,$$

and in either case

$$-|x| \leqslant x \leqslant |x| \qquad \text{and} \qquad -|x| \leqslant -x \leqslant |x|. \tag{2}$$

Theorem 1. *Given a number $a \geqslant 0$, then*

(3) $$|x| \leqslant a \quad \text{if and only if} \quad -a \leqslant x \leqslant a.$$

Proof. Given $|x| \leqslant a$, then $-a \leqslant -|x|$ which with (2) gives $-a \leqslant -|x| \leqslant x \leqslant |x| \leqslant a$; that is, $-a \leqslant x \leqslant a$.

Conversely, given $-a \leqslant x \leqslant a$, then $0 \leqslant a + x$ and $0 \leqslant a - x$, so that $0 \leqslant (a + x)(a - x) = a^2 - x^2$, $x^2 \leqslant a^2$, $\sqrt{x^2} \leqslant \sqrt{a^2}$; and hence $|x| \leqslant a$. ∎

From (2), $-|y| \leqslant y \leqslant |y|$ which with (2) by addition gives

$$-(|x| + |y|) \leqslant x + y \leqslant |x| + |y|.$$

Now from (3)

(4) $$|x + y| \leqslant |x| + |y|.$$

Writing (4) as $|X + y| \leqslant |X| + |y|$ and then setting $X = x - y$ we obtain $|x| \leqslant |x - y| + |y|$ so that $|x| - |y| \leqslant |x - y|$. Then $|y| - |x| \leqslant |y - x| = |x - y|$ and thus by Theorem 1,

(5) $$\big||x| - |y|\big| \leqslant |x - y|.$$

Little more than "the square of a real number is nonnegative and the sum of nonnegative numbers is nonnegative" leads to a result which is used in several places later on.

Theorem 2. (Schwarz Inequality).† *If $a_1, a_2, \cdots, a_n$ and $b_1, b_2, \cdots, b_n$ are real, then*

(6) $$\left(\sum_{k=1}^{n} a_k b_k\right)^2 \leqslant \sum_{k=1}^{n} a_k{}^2 \sum_{k=1}^{n} b_k{}^2$$

with equality if there is a number c such that $ca_k = b_k$, $k = 1, 2, \cdots, n$. Conversely, if the equality holds and at least on $a_k \neq 0$, then there is a constant c such that $b_k = ca_k$, $k = 1, 2, \cdots, n$.

Proof. For x real,

(7) $$0 \leqslant \sum_{k=1}^{n} (a_k x - b_k)^2 = \sum_{k=1}^{n} (a_k{}^2 x^2 - 2a_k b_k x + b_k{}^2)$$

$$= x^2 \sum_{k=1}^{n} a_k{}^2 - 2x \sum_{k=1}^{n} (a_k b_k) + \sum_{k=1}^{n} b_k{}^2 = Ax^2 - 2Bx + C$$

upon setting

$$A = \sum_{k=1}^{n} a_k{}^2, \qquad B = \sum_{k=1}^{n} a_k b_k, \qquad \text{and } C = \sum_{k=1}^{n} b_k{}^2.$$

Hence $A \geqslant 0$. If $A = 0$, then $a_1 = a_2 = \cdots = a_k = 0$ and both sides of (6) are

† Sometimes called the Cauchy-Bunyakovsky-Schwarz inequality.

zero, so consider $A > 0$. Then

$$0 \leqslant A(Ax^2 - 2Bx + C) = (Ax - B)^2 + AC - B^2$$

holds. In particular for $x = B/A$ we obtain $0 \leqslant AC - B^2$ so that $B^2 \leqslant AC$, which is (6).

If $b_k = ca_k$ for some constant c, $k = 1, 2, \cdots, n$, then both sides of (6) are equal to c^2A^2.

Conversely, consider that the equality in (6) holds and at least one $a_k \neq 0$. Then $AC - B^2 = 0$, $A > 0$ and for $c = B/A$

$$0 = (Ac - B)^2 = (Ac - B)^2 + AC - B^2 = A\sum_{k=1}^{n}(a_kc - b_k)^2$$

so that $b_k = ca_k$, $k = 1, 2, \cdots, n$. ∎

Since $B \leqslant |B| = \sqrt{B^2} \leqslant \sqrt{AC}$, the Schwarz inequality leads to

Corollary (Minkowski's Inequality)

$$\sqrt{\sum_{k=1}^{n}(a_k + b_k)^2} \leqslant \sqrt{\sum_{k=1}^{n}a_k{}^2} + \sqrt{\sum_{k=1}^{n}b_k{}^2}. \tag{8}$$

In case at least one $a_k \neq 0$ and at least one $b_k \neq 0$, then the equality holds iff there is a positive constant c such that $b_k = ca_k$, $k = 1, 2, \cdots, n$.

Proof.

$$\sum_{k=1}^{n}(a_k + b_k)^2 = A + 2B + C \leqslant A + 2\sqrt{AC} + C = (\sqrt{A} + \sqrt{C})^2 \tag{9}$$

and (8) follows by taking square roots. If $b_k = ca_k$, $c > 0$, $k = 1, 2, \cdots, n$, then (even if $A = 0$) both sides of (8) are equal to $(1 + c)\sqrt{A}$.

Conversely, let the equality in (8) hold with $A > 0$ and $C > 0$. Then the equality in (9) holds and $B = \sqrt{AC} > 0$. Thus there is a constant c such that $b_k = ca_k$ by the theorem. But then $0 < B = cA$ so that $c > 0$. ∎

1-10 COMPLEX NUMBERS

Need motivated the introduction of complex numbers, but the unfortunate adjective "imaginary" hindered their ready acceptance. A nonimaginary method by which complex numbers could have been introduced is by means of ordered pairs of real numbers.

Definition 1. *An ordered pair (x, y) of real numbers is defined to be a* **complex number** *z. Equality, addition, and multiplication are extended to complex numbers $z_1 = (x_1, y_1)$ and $z_2 = (x_2, y_2)$ by*

(i) $z_1 = z_2$ iff $x_1 = x_2$ and $y_1 = y_2$.
(ii) $z_1 + z_2 = (x_1 + x_2, y_1 + y_2)$.
(iii) $z_1z_2 = (x_1x_2 - y_1y_2, x_1y_2 + x_2y_1)$.

Commutativity of the addition and multiplication operations for complex numbers then follows from the same laws for real numbers:

$$z_1 + z_2 = (x_1 + x_2, y_1 + y_2) = (x_2 + x_1, y_2 + y_1) = z_2 + z_1,$$

and

$$\begin{aligned} z_1 z_2 &= (x_1 x_2 - y_1 y_2, x_1 y_2 + x_2 y_1) \\ &= (x_2 x_1 - y_2 y_1, y_2 x_1 + y_1 x_2) = z_2 z_1. \end{aligned}$$

For a third complex number $z_3 = (x_3, y_3)$,

$$\begin{aligned} z_1 + (z_2 + z_3) &= (x_1, y_1) + (x_2 + x_3, y_2 + y_3) \\ &= (x_1 + (x_2 + x_3), y_1 + (y_2 + y_3)) \\ &= ((x_1 + x_2) + x_3, (y_1 + y_2) + y_3) \\ &= (z_1 + z_2) + z_3. \end{aligned}$$

Proof of multiplicative associativity of complex numbers uses distributivity, as well as commutativity and associativity, of addition and multiplication for real numbers:

$$\begin{aligned} z_1(z_2 z_3) &= (x_1, y_1)(x_2 x_3 - y_2 y_3, x_2 y_3 + x_3 y_2) \\ &= (x_1 x_2 x_3 - x_1 y_2 y_3 - x_2 y_1 y_3 - x_3 y_1 y_2, x_1 x_2 y_3 + x_1 x_3 y_2 \\ &\qquad + x_2 x_3 y_1 - y_1 y_2 y_3) \end{aligned}$$

and it should be checked that $(z_1 z_2)z_3$ may be displayed in the same form so that

$$z_1(z_2 z_3) = (z_1 z_2)z_3.$$

PROBLEM 1. The distributive property

$$z_1(z_2 + z_3) = z_1 z_2 + z_1 z_3$$

holds for any complex numbers z_1, z_2, z_3.

The facts

$$(x, y) = (0, 0) \qquad \text{iff } x = 0 \text{ and } y = 0,$$
$$(x, y) + (0, 0) = (x + 0, y + 0) = (x, y)$$

and

$$(x, y)(0, 0) = (x \cdot 0 - y \cdot 0, x \cdot 0 + y \cdot 0) = (0, 0)$$

show that (0, 0) acts as the "zero" of the complex number system. The symbol 0 is then given double duty by setting

$$0 = (0, 0).$$

Then $(x, y) \neq 0$ means that either $x \neq 0$ or $y \neq 0$; that is,

$$(x, y) \neq 0 \qquad \text{iff } x^2 + y^2 \neq 0.$$

Note that

$$(x, y)(1, 0) = (x \cdot 1 - y \cdot 0, x \cdot 0 + y \cdot 1) = (x, y).$$

Also, for $z_1 = (x_1, y_1) \neq 0$, then $(x_1, y_1)(x, y) = (x_1, y_1)$ if and only if

$$x_1 x - y_1 y = x_1 \qquad \text{and}$$
$$x_1 y + y_1 x = y_1$$

whose unique solution (since $x_1{}^2 + y_1{}^2 \neq 0$) is $x = 1, y = 0$. Thus (1, 0) acts as the "unit" in the complex number system. The symbol 1 is also given double duty by setting

$$1 = (1, 0).$$

In fact, for a any real number, it is customary to set

(1) $$a = (a, 0)$$

and thus to "embed" the real number system in the complex number system.

The complex number (0, 1) is observed to be such that

$$(0, 1)(0, 1) = (0 \cdot 0 - 1 \cdot 1, \quad 0 \cdot 1 + 1 \cdot 0) = (-1, 0) = -1,$$

wherein (1) justifies the last equality. Hence, by extending the use of positive integer exponents to complex numbers, the equality

(2) $$(0, 1)^2 = -1$$

is obtained. There is now nothing "imaginary" in agreeing to use i for the complex number (0, 1) by setting

(3) $$i = (0, 1).$$

Consequently, (2) becomes

(4) $$i^2 = -1.$$

Hence for a and b real, this definition of i together with the agreement (1) yields

$$(a, 0) + (0, 1)(b, 0) = a + ib,$$

the familiar way of writing a complex number.

With a and b real, the complex number $z = a + ib$ is said to have

(i) Real part (or component) a;
(ii) Imaginary part ib; and
(iii) Imaginary component b.

Given a complex number $z = (a, b)$, then

$$(a, b) + (x, y) = 0 \qquad \text{iff } x = -a, \qquad \text{and } y = -b.$$

It is therefore natural to define $-z$ by setting

$$-z = (-a, -b); \qquad \text{that is, } -(a + ib) = -a - ib.$$

Also for $(c, d) \neq 0$, then $(c, d)(x, y) = (a, b)$ holds iff

$$cx - dy = a$$
$$dx + cy = b$$

whose unique solution is

$$x = \frac{ac + bd}{c^2 + d^2}, \qquad y = \frac{bc - ad}{c^2 + d^2}, \quad \text{provided } (c, d) \neq 0.$$

Thus, by definition,

$$\frac{a + ib}{c + id} = \frac{ac + bd}{c^2 + d^2} + i\,\frac{bc - ad}{c^2 + d^2}, \qquad c + id \neq 0,$$

validates the multiplication of both numerator and denominator by $c - id$.

The **absolute value** of a complex number $z = x + iy$ is defined by

(5) $$|z| = \sqrt{x^2 + y^2}.$$

Hence $|-z| = \sqrt{(-x)^2 + (-y)^2} = \sqrt{x^2 + y^2} = |z|$.

With $z_1 = x_1 + iy_1$ and $z_2 = x_2 + iy_2$, then Minkowski's inequality (with $n = 2$) yields

$$|z_1 + z_2| = |(x_1 + x_2) + i(y_1 + y_2)| = \sqrt{(x_1 + x_2)^2 + (y_1 + y_2)^2}$$
$$\leqslant \sqrt{x_1{}^2 + y_1{}^2} + \sqrt{x_2{}^2 + y_2{}^2}, \qquad \text{see (8) of Sec. 1–9,}$$

so that

(6) $$|z_1 + z_2| \leqslant |z_1| + |z_2|.$$

Since this holds for any two complex numbers, then

$$|z_1| = |(z_1 - z_2) + z_2| \leqslant |z_1 - z_2| + |z_2|$$

and

(7) $$|z_1| - |z_2| \leqslant |z_1 - z_2| = \sqrt{(x_1 - x_2)^2 + (y_1 - y_2)^2}$$

from which both

(8) $$\big||z_1| - |z_2|\big| \leqslant |z_1 - z_2| \qquad \text{and}$$

(9) $$\big||z - z_1| - |z_2 - z_1|\big| \leqslant |z - z_2|$$

should be seen to follow.

With x and y real, then the complex numbers $z_1 = x + i\cdot 0$ and $z_2 = 0 - iy$ are such that, from (7) and then (6),

$$|x| - |y| \leqslant |x + iy| \leqslant |x| + |y|.$$

From $z = x + iy$ and $w = u + iv$ and

$$|zw| = |(x + iy)(u + iv)| = |xu - yv + i(xv + yu)|$$
$$= \{(x^2u^2 - 2xyuv + y^2v^2) + (x^2v^2 + 2xyuv + y^2u^2)\}^{1/2}$$
$$= \sqrt{(x^2 + y^2)(u^2 + v^2)}$$

it follows that

$$|zw| = |z|\,|w|. \tag{10}$$

The **conjugate** $\bar{z}$ of $z = x + iy$ is defined by

$$\bar{z} = x - iy. \tag{11}$$

Consequently,

$$\begin{aligned} |\bar{z}| &= \sqrt{x^2 + (-y)^2} = |z|, \\ \bar{\bar{z}} &= \overline{x - iy} = x + iy = z, \\ z\bar{z} &= (x + iy)(x - iy) = x^2 + y^2 = |z|^2 = |\bar{z}|^2, \end{aligned}$$

and it should also be checked that

$$\begin{aligned} z + \bar{z} &= 2x \qquad \text{(real)}, \\ z - \bar{z} &= 2iy \qquad \text{(imaginary if } y \neq 0), \end{aligned}$$

$$z \text{ is real iff } z = \bar{z},$$

and if w is also complex, then

$$\overline{z \pm w} = \bar{z} \pm \bar{w} \qquad \text{and} \qquad \overline{zw} = \bar{z}\bar{w}.$$

Notice that the complex number system lacks the order property of the real number system; that is, the symbol $<$ is never used between complex numbers unless they are known to be real.

The Schwarz inequality involving complex numbers is given in

Theorem 1. *If $z_1, z_2, \cdots, z_n$ and $w_1, w_2, \cdots, w_n$ are complex numbers, then*

$$\left| \sum_{k=1}^{n} z_k \bar{w}_k \right|^2 \leqslant \sum_{k=1}^{n} |z_k|^2 \sum_{k=1}^{n} |w_k|^2 = \sum_{k=1}^{n} z_k \bar{z}_k \sum_{k=1}^{n} w_k \bar{w}_k. \tag{12}$$

Proof. With summations from $k = 1$ to n, set

$$A = \Sigma\, |z_k|^2, \qquad B = \Sigma\, z_k \bar{w}_k, \qquad C = \Sigma\, |w_k|^2.$$

Then A and C are real; in fact $A \geqslant 0$ and $C \geqslant 0$, but B might not be real. Thus $\bar{A} = A$ but $\bar{B} = \Sigma\, \bar{z}_k w_k$. Then

$$\begin{aligned} 0 \leqslant \Sigma\, |\bar{B}z_k - Aw_k|^2 &= \Sigma\, (\bar{B}z_k - Aw_k)\overline{(\bar{B}z_k - Aw_k)} \\ &= \Sigma\, (\bar{B}z_k - Aw_k)(B\bar{z}_k - A\bar{w}_k) \\ &= \Sigma\, (B\bar{B}z_k\bar{z}_k - A\bar{B}z_k\bar{w}_k - AB\bar{z}_k w_k + A^2 w_k \bar{w}_k) \\ &= B\bar{B}A - A\bar{B}B - AB\bar{B} + A^2C = -A\,|B|^2 + A^2C \\ &= A(AC - |B|^2), \end{aligned}$$

which shows that if $A > 0$ then $|B|^2 \leqslant AC$ which is (12). If, however, $A = 0$ then for each k, $|z_k| = 0$, $z_k = 0$, and (12) holds since all parts are zero. ∎

The complex number $z = x + iy$ (as the ordered pair (x, y) of real numbers) and the point labeled (x, y) of the Euclidean plane E^2 are naturals to identify in what is called the Argand diagram (Fig. 1). Another identification is with the vector

$$\mathbf{v} = \mathbf{i}x + \mathbf{j}y$$

where **i** and **j** are the unit vectors from the origin to the points (1, 0) and (0, 1), respectively. It would be more suggestive of this identification to write

$$z = x + jy$$

as the engineers do, but mathematicians prefer $x + iy$. The form†

$$z = |z|(\cos\theta + i\sin\theta)$$

thus appears for many **amplitudes** θ with the one such that $-\pi < \theta \leqslant \pi$, called the **principal** amplitude.

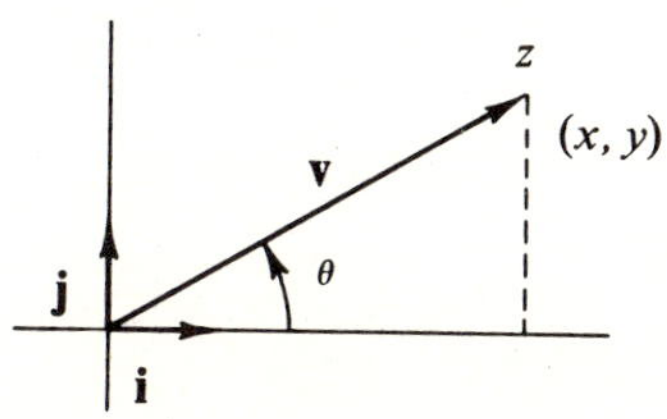

FIG. 1

Alternative ways of thinking about complex numbers and their properties are thus available. Whether complex numbers are "identified" with points or are "represented" by points, or "are" points is of so little concern that a metaphoric statement such as

"the Euclidean plane E^2 is the set of all complex numbers z with $|z_1 - z_2|$ the distance between z_1 and z_2"

is usually not frowned upon. It is mentally more economical to think of

$$\{z \,|\, |z - z_0| = r > 0\}$$

as the circle, and of

$$\{z \,|\, |z - z_0| < r\}$$

as the open disk with center z_0 and radius r, rather than of these expressions representing, or being identified with, the geometric figure.

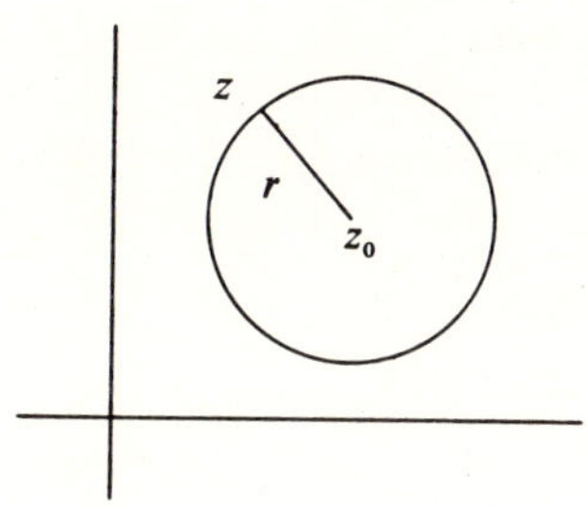

FIG. 2

However, Fig. 3 alone is not considered a proof (but certainly suggests a proof) of the

† This form will not be used, however, until the sine and cosine functions are defined in Section 7–3.

Theorem 2. *If z_1 and z_2 are complex numbers and $r_1 > 0$ are such that*

(13) $$z_2 \in \{z \,||z - z_1| < r_1\}$$

then there exists a number $r_2 > 0$ such that

(14) $$\{z \,||z - z_2| < r_2\} \subset \{z \,||z - z_1| < r_1\}.$$

Proof. From (13) it follows that $|z_2 - z_1| < r_1$. Set

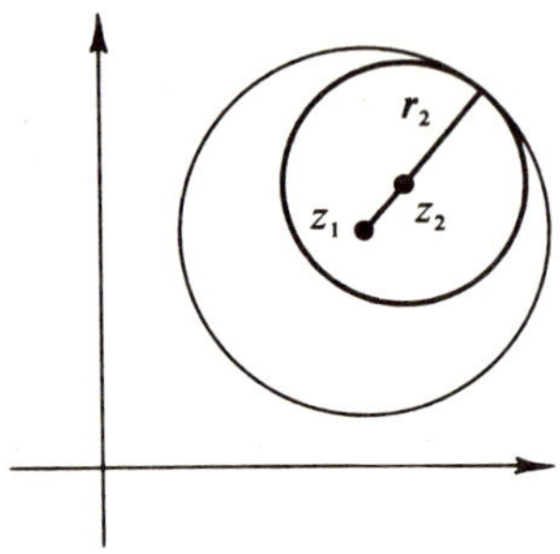

FIG. 3

$$r_2 = r_1 - |z_2 - z_1|$$

so that $r_2 > 0$. Then

$$|z - z_2| < r_2 \Rightarrow |z - z_1| - |z_2 - z_1| < r_2 \qquad \text{from (9)}$$
$$\Rightarrow |z - z_1| < |z_2 - z_1| + (r_1 - |z_2 - z_1|) = r_1$$

which in turn implies that (14) holds for this number r_2. ∎

The formal manipulations

$$\frac{d}{d\theta}(\cos\theta + i\sin\theta) = -\sin\theta + i\cos\theta$$
$$= i^2\sin\theta + i\cos\theta$$
$$= i(\cos\theta + i\sin\theta)$$

could have inspired the definition (and it is only a definition) that $e^{i\theta}$ is the complex number given by the Euler formula

(15) $$e^{i\theta} = \cos\theta + i\sin\theta.$$

Then

$$\frac{d}{d\theta}e^{i\theta} = ie^{i\theta} \quad \text{resembles} \quad \frac{d}{dx}e^{cx} = ce^{x}$$

for c a real constant and x a real variable.

For y real, $e^{iy} = \cos y + i\sin y$ and for x real, a further extension of the laws of exponents leads to defining e^z, as a complex number, by setting

$$e^z = e^{x+iy} = e^x e^{iy} = e^x(\cos y + i\sin y).$$

Then for $z_1 = x_1 + iy_1$ and $z_2 = x_2 + iy_2$,

$$\begin{aligned}
e^{z_1+z_2} &= e^{(x_1+x_2)+i(y_1+y_2)} = e^{x_1+x_2}e^{i(y_1+y_2)} \\
&= e^{(x_1+x_2)}\{\cos(y_1+y_2) + i\sin(y_1+y_2)\} \\
&= e^{x_1}e^{x_2}\{\cos y_1 \cos y_2 - \sin y_1 \sin y_2 \\
&\quad + i(\sin y_1 \cos y_2 + \cos y_1 \sin y_2)\} \\
&= e^{x_1}e^{x_2}\{\cos y_1(\cos y_2 + i\sin y_2) \\
&\quad + i\sin y_1(\cos y_2 + i\sin y_2)\} \\
&= e^{x_1}(\cos y_1 + i\sin y_1)e^{x_2}(\cos y_2 + i\sin y_2) \\
&= e^{z_1}e^{z_2}.
\end{aligned}$$

For the sake of theoretical niceties, this process will be almost exactly reversed later on in Section 7–3.

PROBLEMS

1. The sum, difference, product, and quotient (denominator $\neq 0$) of two rational numbers are rational numbers.
2. The sum of two irrational numbers may be rational.
3. If $r \neq 0$ is rational and x is irrational, then $r + x$ is irrational.
4. Between any two real numbers there is at least one irrational number.
5. If $r > 0$ is rational and $x > 0$, then

$$x^r = \sqrt[n]{x^m}$$

where $r = m/n$ with m and n positive integers.
6. If $a < b + 100\varepsilon$ for each positive number ε, then $a \leqslant b$.
7. $(a \leqslant b + \varepsilon$ for each $\varepsilon > 0) \Rightarrow a \leqslant b$.
8. If $a < b + \varepsilon$ for each number ε such that $0 < \varepsilon < 1$, then $a \leqslant b$.
9. If $a + \varepsilon < b$ for each number ε such that $0 < \varepsilon < 1$, then $a + 1 \leqslant b$.
10. There is no rational number whose square is 5.
11. There is no rational number whose cube is 2.
12. $2^n > n$ for each positive integer n.
13. Read the following passage from right to left:

stigid ruo setirw barA na suhT .tfel ot thgir morf seog gnitirw dna gnidaer barA
10 gniees dna gninnigeb eht ot kcab seog neht eH .9 8 7 6 5 4 3 2 1 0 redro eht ni
nehT .net rof siht sesu eh
,nevele = 11
,evlewt = 12
,net dna eerht = neetriht = 13
.cte

Since our reading and writing goes from left to right, we write the digits in order as 0 1 2 3 4 5 6 7 8 9. We then go back to the beginning and seeing 01 we should use this for ten. Then

11 = eleven,
21 = twelve,
31 = three and ten = three'n ten = thirteen.
etc. (Continued on p. 44.)

Hence 02 = twenty, 12 = one and twenty (as the Germans say), etc.

$$\begin{array}{rl} 6\;5\;8 & = 6,\ 5 \text{ tens, and } 8 \text{ hundreds.} \\ +7\;0\;4 & = 7,\ 0 \text{ tens, and } 4 \text{ hundreds.} \\ \hline 3\;6\;21 & = 3,\ 6 \text{ tens, } 2 \text{ hundreds, and } 1 \text{ thousand.} \end{array}$$

$$\begin{array}{r} 3\;9 \\ -9\;1 \\ \hline 4\;7 \end{array} \qquad\qquad \begin{array}{r} 5\;6 \\ \times 3\;2 \\ \hline 5\;9\;1 \\ 0\;3\;1 \\ \hline 5\;9\;4\;1 \end{array}$$

Try a few additions, subtractions, and multiplications in "forward" arithmetic and then work out the long-division algorithm.

14. With n a positive integer, then
(a) $2^n \leqslant n!$ iff $n \geqslant 4$.
(b) $(\frac{4}{3})^n > n$ iff $n = 1$ or $n \geqslant 7$.
(c) $n^3 + 5n$ is divisible by 6.
(d) $3^{2n+2} - 8n - 9$ is divisible by 64.
(e) $1 + na + \dfrac{n(n-1)}{2}a^2$ is $\begin{cases} \geqslant (1+a)^n & \text{if } -1 < a \leqslant 0 \\ \leqslant (1+a)^n & \text{if } a > 0. \end{cases}$
(f) If $a > 0$ and $a \neq 1$, then

$$\frac{a^{2n+2} - 1}{a(a^{2n} - 1)} > \frac{n+1}{n}.$$

[*Hint:* With α_n the ratio on the left,

$$\alpha_{n+1} = \alpha_1 - \frac{1}{\alpha_n}.\Big]$$

15. Use the result of Problem 1-3.2 to establish that for each positive integer n:
(a) $n^3 - n + 1$ is not divisible by 3.
(b) $2n^3 + 3n^2 + n + 5$ is not divisible by 6.
(c) n has some interesting property (if you agree that "being the only positive integer whose square is itself" is an interesting property and "being the smallest positive integer with no interesting property" is an interesting property).

16. Set $s_1 = \sqrt{2}$ and for $n > 1$, $s_n = \sqrt{2 + \sqrt{s_{n-1}}}$. Then
(a) $s_n < 2$;
(b) $s_n < s_{n+1}$.

17. Let $x > 0$ be given. Then there is one and only one way of expressing the greatest integer in x as

$$[x] = a_m 2^m + a_{m-1} 2^{m-1} + \cdots + a_1 2 + a_0$$

with each a_k either 0 or 1.

18. If a and b are real numbers, then

$$\frac{a + b + |a - b|}{2} = \text{the larger of } a \text{ and } b,$$

$$\frac{a + b - |a - b|}{2} = \text{the smaller of } a \text{ and } b.$$

19. Give an alternative proof of Schwarz inequality (6) of Section 1–9 by starting with
(a) $0 \leqslant \sum_{k=1}^{n} (Ba_k - Ab_k)^2$ where A, B, C are the sums defined in the proof of Theorem 1–9.2.
(b) $0 \leqslant \sum_{j=1}^{n} \sum_{k=1}^{n} (a_k b_j - a_j b_k)^2$.
(c) Prove Theorem 1–10.1 by the technique of part (b).

20. Given $a_1 \geqslant a_2 \geqslant \cdots \geqslant a_n$ and $b_1 \geqslant b_2 \geqslant \cdots \geqslant b_n$, show that

$$\sum_{k=1}^{n} a_k \sum_{k=1}^{n} b_k \leqslant n \sum_{k=1}^{n} a_k b_k .$$

[*Hint:* $$0 \leqslant \sum_{k=1}^{n} \sum_{j=1}^{n} (a_k - a_j)(b_k - b_j).]$$

21. If x and y are real, then $|x+y|^2 = x^2 + 2xy + y^2$. What is a similar formula for $|z+w|^2$ with z and w complex? Show that

$$|z+w|^2 + |z-w|^2 = 2|z|^2 + 2|w|^2$$

and give a geometric interpretation of this equation.

22. (a) If $a > 0$, $b > 0$, and $a^2 = 2b^2$, then

$$0 < a - b < b \qquad \text{and} \qquad \frac{a}{b} = \frac{2b-a}{a-b}.$$

[*Hint:* $b^2 < a^2 < (2b)^2$.]

(b) Use part (a) to prove there is no rational number whose square is 2.

23. (a) Let n be a positive integer expressed in ternary form. Show that the rightmost nonzero digit in

$$n^2 \text{ is } 1, \qquad \text{but in } 2n^2 \text{ is } 2.$$

(b) Use part (a) to prove that no rational number has its square equal to 2.

24. If x is any number (positive, negative, or zero) and n is a positive integer, then in terms of the greatest integer notation,

$$[x] + \left[x + \frac{1}{n}\right] + \cdots + \left[x + \frac{n-1}{n}\right] = [nx].$$

[*Hint:* Let $f(y) = [y] + [y + 1/n] + \cdots + [y + (n-1)/n] - [ny]$. Then $f(y + 1/n) = f(y)$ and hence $f(x) = f(x - k/n)$ for any positive or negative integer k. Take $k = [nx]$ and show that $f(x - k/n) = 0$.]

25. (a) Let $-1 < x < 1$ and $-1 < y < 1$ be given. Then

(i) $-1 < xy < 1$.

(ii) $0 < 1 + x + y + xy$ and $0 < 1 - x - y + xy$.

(iii) $-1 < \dfrac{x+y}{1+xy} < 1$.

(b) Given a number $c > 0$, let $V = \{v \mid -c < v < c\}$. Then

$$(u \in V,\ v \in V) \Rightarrow \frac{u+v}{1 + (uv/c^2)} \in V.$$

26. If a and b are real or complex numbers, then

$$\frac{|a+b|}{1+|a+b|} \leqslant \frac{|a|}{1+|a|} + \frac{|b|}{1+|b|}.$$

[*Hint:* Contrapositive proof or use elementary calculus to show that φ is an increasing function and (with $|b| > 0$) f is a decreasing function where

$$\varphi(t) = 1 - \frac{1}{1+t} \qquad \text{and} \qquad f(t) = \varphi(t + |b|) - \varphi(t)].$$

27. (a) If p is a positive integer, then there is at least one pair of nonnegative integers m and n such that

$$(*) \qquad p = 1 + 2 + \cdots + (m + n) + m.$$

[*Hint:* Let q be such that $1 + 2 + \cdots + q \leqslant p < 1 + 2 + \cdots + q + (q + 1)$. Set $m = p - (1 + \cdots + q)$ and $n = q - m$.]

(b) Given a positive integer p, let m and n be any pair of nonnegative integers satisfying (*). Then (check each of these)

$$2p = (m + n)^2 + 3m + n,$$

$$8p + 1 = (2m + 2n + 1)^2 + 8m < (2m + 2n + 3)^2,$$

$$(2m + 2n^2 + 1) \leqslant 8p + 1 < (2m + 2n + 3)^2,$$

$$m + n + 1 \leqslant \frac{\sqrt{8p + 1} + 1}{2} < m + n + 2,$$

$$m + n = \left[\frac{\sqrt{8p + 1} + 1}{2}\right] - 1,$$

$$3m + n = 2p - \left\{\left[\frac{\sqrt{8p + 1} + 1}{2}\right] - 1\right\}^2.$$

In the last two equations the right sides depend only on p and, since their simultaneous solution is unique, there is one and only one pair of nonnegative integers m and n satisfying (*).

28. A set P is said to be **partially ordered** by a binary relation $\prec$ defined for some (not necessarily all) pairs $a, b \in P$ such that

(i) $a \prec a$ for all $a \in P$;

(ii) Whenever $a \prec b$ and $b \prec c$, then $a \prec c$; and

(iii) Whenever $a \prec b$ and $b \prec a$, then $a = b$.

In each of the following show that P is partially ordered by $\prec$ as defined.

(a) Let P be the family of all subsets of a set X and define $A \prec B$ iff $A \subset B \subset X$.

(b) Let P be the set of all complex numbers and define $x + iy \prec u + iv$ iff $x \leqslant u$ and $y \leqslant v$.

(c) P is the set of docks on the left bank (facing downstream) of the Mississippi River, its tributaries, or its delta. Define $a \prec a$ for each $a \in P$, and if any of the water which flows by a later flows by $b \in P$, then set $a \prec b$. [Note that there is at least one dock a for which there is no dock b satisfying both $a \prec b$ and $b \neq a$, and there may be several such **maximal** docks.]

[*Note:* If P is partially ordered and in addition either $a \prec b$ or $b \prec a$ for every pair $a, b \in P$, then P is said to be **simply** (or completely, or linearly, or totally) ordered.]

Problems on St. Andrews Crosses

We first introduce an iconic notation for representing set operations.

The plane is identified with the universal set. A 45° line divides the plane into two sets and the points above the line are identified with a set A; those below the line with the complement A^c. Rather than shading the upper half-plane to indicate A, a dot is placed above the line to indicate A, but a dot below the line indicates A^c:

$$\cdot\!\!/ = A, \qquad /\!\!. = A^c.$$

If a second set B is involved, then a 135° stroke is used:

$$\backslash\!\!\cdot = B, \qquad .\!\!\backslash = B^c.$$

When the two strokes are placed together they form a St. Andrews cross . Then the single dot on  indicates the set common to both A and B; that is.

 $= A \cap B.$

Also  $=$ A $\cap B^c$ since the dot is above the A-line and below the B-line. Similarly,

 $= A^c \cap B$ and  $= A^c \cap B^c$.

The union and intersection symbols will be used between St. Andrews crosses. Thus

 $\cup$  $=$ 

is a representation of $(A \cap B^c) \cup (A \cap B) = A$. Note that  $= A$, since both quadrants above the A-line are filled. Also 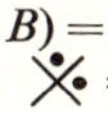 $= B$. Hence

 $\cup$  $=$  $= A \cup B.$

The dots on crosses thus act as point sets in forming unions and intersections of crosses. For example,

 $\cap$  $=$ 

illustrates $(A \cup B) \cap B^c = A \cap B^c$.

To form the complement of a cross, erase all dots and dot all spaces that were blank. Hence

 $=$ 

is a schematic way of illustrating the de Morgan law

$$(A \cup B)^c = A^c \cap B^c.$$

The other de Morgan law is

$(A \cap B)^c =$  $=$ 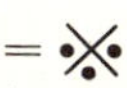 $=$  $\cup$  $= A^c \cup B^c.$

The universal set is represented by  and  $= \varnothing$.

One set A divides the universal set into two mutually exclusive sets A and A^c. Two sets A and B divide the universal set into four mutually exclusive sets $A \cap B$, $A \cap B^c$, $A^c \cap B$, and $A^c \cap B^c$. With three sets A, B, C, choose A or A^c, then B or B^c, then C or C^c, and thus see that there are $2^3 = 8$ mutually exclusive sets whose union is the universal set. To represent these possibilities on a cross, divide each of the four spaces of an AB-cross by a 45° stroke. Thus the single dot on

is above the A-line, below the B-line, and above a C-line, and thus represents $A \cap B^c \cap C$.

With A primary, B secondary, and C tertiary, it now takes four dots to represent C alone:

 $= C$ while  $= C^c$.

Since  $= A \cup B$, to represent $(A \cup B) \cap C$, it is only necessary to put a stroke under each of these three dots:

$(A \cup B) \cap C =$  $=$ 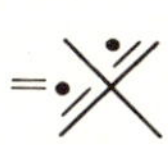 $\cup$  $= (A \cap C) \cup (B \cap C).$

Also

$(A \cup B) \cap (A \cup C) =$  $\cap$  $=$  $=$  $\cup$  $= A \cup (B \cap C).$

The expression $(A \cap B^c \cap C) \cup (A \cap B) \cup (A^c \cap B \cap C^c)$ may be represented by

The upper dot may be replaced by  to give

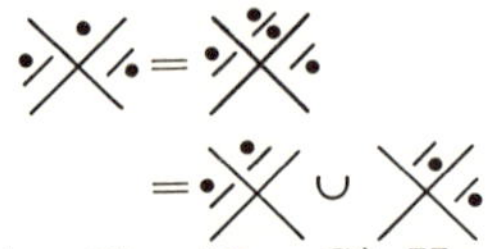

and then this union reads as $(A \cap C) \cup (B \cap C^c)$. Hence it is seen that

$$(A \cap B^c \cap C) \cup (A \cap B) \cup (A^c \cap B \cap C^c) = (A \cap C) \cup (B \cap C^c).$$

PROBLEM 1.

(a) $A \cup A^cB = A \cup B$.
(b) $(ABC) \cup (A^cBC) \cup B^c \cup C^c =$ universal set.
(c) $(A^c \cup B^c)(B \cup C)(A^c \cup B) \subset A^c$.
(d) $(BC^c) \cup (A^cB^cC) \subset A^c \cup BC^c$.
(e) $(AB) \cup (AC^c) \cup (BC) = (AC^c) \cup (BC)$.
(f) $(A \cup B)^c \cup [B(A^c \cup C)] = A^c \cup BC$.
(g) $(A \cup B) \cap (B \cup C) \cap (C \cup A) = (A \cap B) \cup (B \cap C) \cup (C \cap A)$.

The **symmetric difference** of two sets A and B is represented by $A \triangle B$ and is defined by

$$A \triangle B = (A - B) \cup (B - A) = AB^c \cup A^cB =$$ 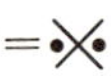 .

Hence $A \triangle B = B \triangle A$ and

$$(A \triangle B)^c =$$  .

If C is a third set, then

$$(A \triangle B) \triangle C = (A \triangle B)C^c \cup (A \triangle B)^cC$$

Also $B \triangle C =$ 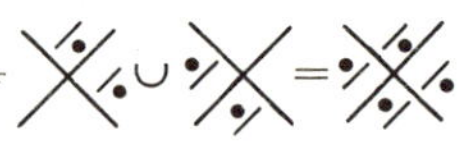 and $(B \triangle C)^c =$  . Therefore

$$A \triangle (B \triangle C) =$$  .

Hence, by comparing the crosses on the right in the last two displays, $(A \triangle B) \triangle C = A \triangle (B \triangle C)$ and it is unambiguous to write

$$A \triangle B \triangle C =$$  .

Then in forming the cross representation of $A \triangle B \triangle C^c$, merely move each dot across its C-line to obtain

$$A \triangle B \triangle C^c =$$  .

But this erases each dot and dots each blank space, and we thus see that

$$(A \triangle B \triangle C)^c = A \triangle B \triangle C^c.$$

If four sets A_1, B_1, A_2, B_2 are involved, then draw a large $A_1 B_1$-cross and in each quadrant put a small $A_2 B_2$-cross. For example, to represent A_2 alone, we have

 $= A_2$.

As examples, check how each set is represented by the cross below it in

$A_1 \cup A_2$ $\quad$ $B_1 \cup B_2$ $\quad$ $(A_1 \cup A_2) \triangle (B_1 \cup B_2)$

and

$A_1 \triangle B_1$ $\quad$ $A_2 \triangle B_2$ $\quad$ $(A_1 \triangle B_1) \cup (A_2 \triangle B_2)$

   = 

It therefore follows that

 $\subset$ 

since each dot on the left is matched by a dot on the right. This indicates that

$$(A_1 \cup A_2) \triangle (B_1 \cup B_2) \subset (A_1 \triangle B_1) \cup (A_2 \triangle B_2).$$

Some of the parts in the following problem have already been obtained, but we include them all because they will be needed for some of the problems at the end of Chapter 5.

PROBLEM 2.

(a) $A \triangle B = B \triangle A$.
(b) $A \triangle A = \varnothing$.
(c) $A \triangle B \subset (A \triangle C) \cup (B \triangle C)$.
(d) $$\left.\begin{array}{l}(A_1 \cup A_2) \triangle (B_1 \cup B_2)\\(A_1 \cap A_2) \triangle (B_1 \cap B_2)\\(A_1 - A_2) \triangle (B_1 - B_2)\end{array}\right\} \subset (A_1 \triangle B_1) \cup (A_2 \triangle B_2).$$

PROBLEM 3.

(a) $(ABCD) \cup (A^cBC) \cup (B^cC) \cup (CD^c) = C$.
(b) $\{ABC \cup [(A \triangle B)C^c]\}^c(A^cC \cup C^c)(B \cup AC) = B(A \triangle C)$.

Chapter 2

SETS AND SPACES

2–1 EQUIVALENT SETS

The letter J will be reserved for the set of all *positive* integers and J_n, with $n \in J$, for

$$J_n = \{1, 2, 3, \cdots, n\}.$$

Hence, if $A \subset J$ and $A \neq \varnothing$, then $\exists\, n_0 \in A$ such that $n \in A \Rightarrow n_0 \leqslant n$; that is, any nonempty set of positive integers has a smallest element (Problem 1-3.2).

The letter E^1 will be used for the set of all real numbers, including the usual number-line representation in which the distance between points $x \in E^1$ and $y \in E^1$ is $|x - y|$. Then E^1, with distance so defined, is called **Euclidean one-dimensional space** (or Euclidean 1-space).

Euclidean 2-space is denoted by E^2, where

$$E^2 = \{(x, y) \mid x \in E^1 \text{ and } y \in E^1\},$$

with the distance between $(x_1, y_1) \in E^2$ and $(x_2, y_2) \in E^2$ defined by

$$\sqrt{(x_2 - x_1)^2 + (y_2 - y_1)^2}.$$

Also $z = (x, y) \in E^2$, with components $x \in E^1$ and $y \in E^1$, will be called a point, a complex number, or a vector according to the context in which it occurs.

If A and B are sets, then $A \times B$ is defined by

$$A \times B = \{(a, b) \mid a \in A \text{ and } b \in B\}$$

and is called the **Cartesian product** of A and B in this order. Also $A^2 = A \times A$ is used. For example, J^2 is the set of all ordered pairs of positive integers, and $E^2 = E^1 \times E^1$.

If to each $x \in A$ is associated one and only one $y \in B$, denoted by $y = f(x)$, then f is said to be a **function** (or **mapping**) on A into B and

$$f: A \to B$$

is written. The set A is called the **domain** of f. With $C \subset A$, then $f(C)$ will be used for the **image** of the set C by f where

$$f(C) = \{b \mid b = f(c) \quad \text{for some } c \in C\}.$$

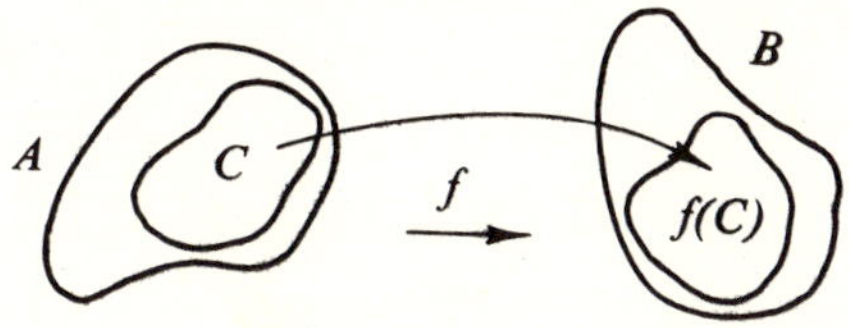

FIG. 1

Hence $f(C) \subset B$. The set $f(A)$ is called the **range** of f. Since $\varnothing$ is a subset of every set, then $f(\varnothing) = \varnothing$. Clearly:

(1) $$\text{If } C_1 \subset C_2 \subset A, \quad \text{then } f(C_1) \subset f(C_2).$$

The distinction between an "into" mapping as given above and the following definition of an "onto" mapping should be carefully observed.

A mapping $f: A \to B$ is said to be on A **onto** B (or briefly "is onto") if $f(A) = B$.

In some situations there is a preference for considering a function f, with domain A and range in B, to be a subset of $A \times B$ having the additional property:

If $(a, b) \in f$ and $(a, c) \in f$, then $b = c$.

Theorem 1. *Let $X_1 \subset A$, $X_2 \subset A$ and $f: A \to B$ be given. Then*

(2) $$f(X_1 \cap X_2) \subset f(X_1) \cap f(X_2)$$

and this inclusion may be proper, but

(3) $$f(X_1 \cup X_2) = f(X_1) \cup f(X_2).$$

Proof. Since $X_1 \cap X_2$ is a subset of both X_1 and X_2, then by (1),

$$f(X_1 \cap X_2) \subset f(X_1) \quad \text{and} \quad f(X_1 \cap X_2) \subset f(X_2)$$

so that (2) follows.

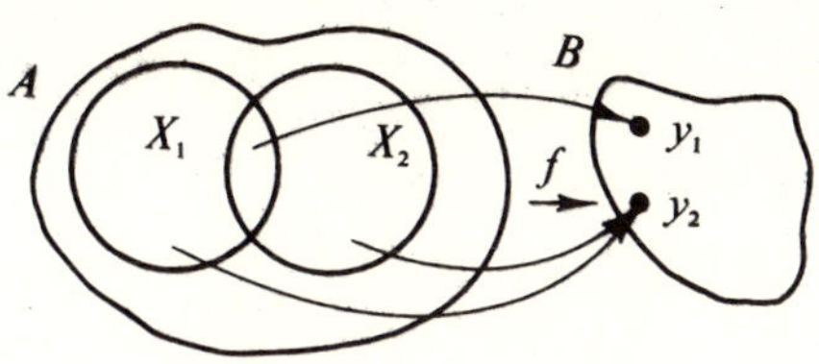

FIG. 2

Figure 2 illustrates a situation in which

$$f(X_1) = f(X_2) = \{y_1, y_2\} \quad \text{and} \quad f(X_1 \cap X_2) = \{y_1\}$$

thus showing that the inclusion may be proper.

Since X_1 and X_2 are both subsets of $X_1 \cup X_2$, it follows that $f(X_1) \subset f(X_1 \cup X_2)$, $f(X_2) \subset f(X_1 \cup X_2)$ and hence

$$f(X_1) \cup f(X_2) \subset f(X_1 \cup X_2). \tag{4}$$

To obtain the opposite inclusion, let $y \in f(X_1 \cup X_2)$ and select an $x \in X_1 \cup X_2$ such that $y = f(x)$. If $x \in X_1$, then $y \in f(X_1)$; if $x \in X_2$, then $y \in f(X_2)$; and in either case $y \in f(X_1) \cup f(X_2)$. Thus $y \in f(X \cup X_2) \Rightarrow y \in f(X_1) \cup f(X_2)$ so that

$$f(X_1 \cup X_2) \subset f(X_1) \cup f(X_2)$$

which with (4) establishes the equality in (3). ∎

Given $f\colon A \to B$ is into and $Y \subset B$, then the inverse image of Y by f is denoted by $f^{-1}(Y)$ and is defined by

$$f^{-1}(Y) = \{x \mid x \in A \text{ and } f(x) \in Y\}.$$

Hence $f^{-1}(Y) \subset A$ and $f(f^{-1}(Y)) \subset Y$. If $Y \subset B - f(A)$, then $f^{-1}(Y) = \varnothing$, and in particular $f^{-1}(\varnothing) = \varnothing$.

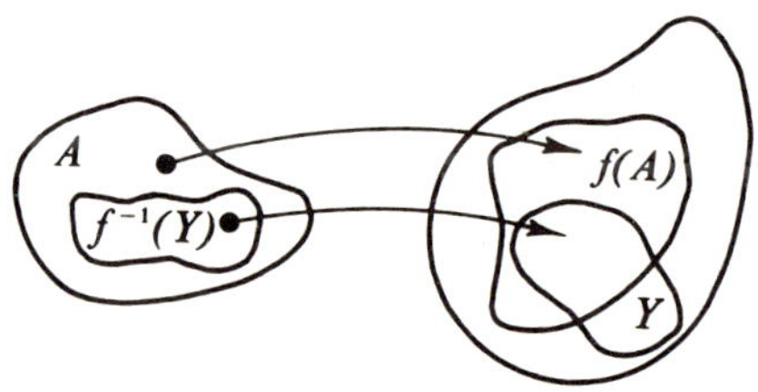

FIG. 3

Even though $f(f^{-1}(Y)) \subset Y$, note that:

$$\text{If } X \subset A, \text{ then } f^{-1}(f(X)) \supset X$$

and X may be a proper subset of $f^{-1}(f(X))$. For example, let f be defined by $f(x) = x^2$ for $x \in E^1$. Then for this function

$$f([-2, -1]) = [1, 4] = f([1, 2]),$$
$$f^{-1}(f([1, 2])) = f^{-1}([1, 4]) = [-2, 1] \cup [1, 2],$$

and hence $[1, 2]$ is a proper subset of $f^{-1}(f([1, 2]))$.

PROBLEM 1. Let $f\colon A \to B$ be onto. Given $Y_1 \subset B$ and $Y_2 \subset B$, then

$$f^{-1}(Y_1 \cap Y_2) = f^{-1}(Y_1) \cap f^{-1}(Y_2)$$

and

$$f^{-1}(Y_1 \cup Y_2) = f^{-1}(Y_1) \cup f^{-1}(Y_2).$$

Notice that $f\colon A \to B$ will have the additional property

$$(x_1 \in A,\ x_2 \in A,\ x_1 \neq x_2) \Rightarrow f(x_1) \neq f(x_2)$$

iff for each $y \in f(A)$ the set $f^{-1}(\{y\})$ has exactly one element (of A).

Even though $f^{-1}(D)$ has been used with D a set, neither f^{-1} by itself nor $f^{-1}(y)$ with y an element has been defined as yet.

Definition 1.

(i) *If $f: A \to B$ is such that $f^{-1}(\{y\})$ has exactly one element for each $y \in f(A)$, then f is said to be* **one to ore** *(briefly 1–1) on A into B.*†

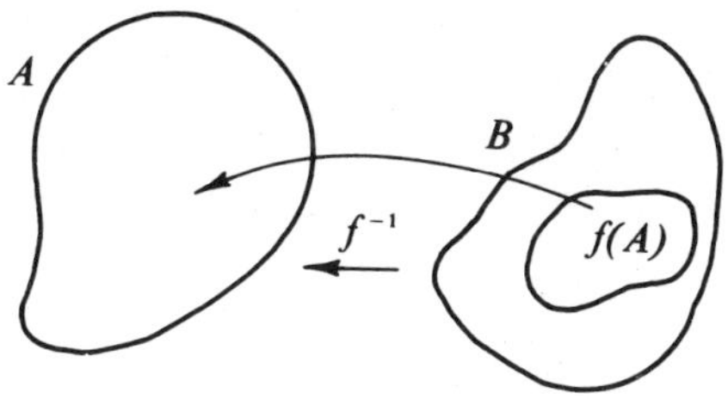

FIG. 4

(ii) *If $f: A \to B$ is 1–1 on A into B, then the function f^{-1}, defined for each $y \in f(A)$ by setting $f^{-1}(y)$ = the element of $f^{-1}(\{y\})$, is called the* **inverse** *of f.*

Hence if $f: A \to B$ has an inverse, then $f^{-1}: f(A) \to A$ is an onto mapping.

Definition 2. *If sets A and B are such that there is an*

$$f: A \to B \qquad \text{which is both 1–1 and onto,}$$

then A is said to be **equivalent** *to B, or A can be put into 1–1 correspondence with B, or A has the same* **cardinal** *as B, and we write $A \sim B$.*

The notation

$$f: A \leftrightarrow B$$

will be used to say "$f: A \to B$ is 1–1 and onto." Hence $f: A \leftrightarrow B$ may be used if both $B = f(A)$ and $(x_1 \in A,\ x_2 \in A,\ x_1 \neq x_2) \Rightarrow f(x_1) \neq f(x_2)$. Thus

$$(A \sim B) \Leftrightarrow (\exists f: A \leftrightarrow B).$$

Theorem 2. *The relation $\sim$ is*

Reflexive: $A \sim A$.

Symmetric: $A \sim B \Rightarrow B \sim A$.

Transitive: $(A \sim B \text{ and } B \sim C) \Rightarrow A \sim C$.

† Some authors use **surjective** instead of onto, **injective** instead of one-to-one, and say that a function is **bijective** if it is both surjective and injective.

Proof. The reflexive property follows by using the identity function; that is, $f(a) = a$ for each $a \in A$.

If f: $A \leftrightarrow B$, then f^{-1}:$B \leftrightarrow A$ and hence the symmetric property holds. [We may thus read $A \sim B$ as "A and B are equivalent."]

To prove the transitive property, take $A \sim B$ and $B \sim C$, then an f and g such that

$$f: A \leftrightarrow B \quad \text{and} \quad g: B \leftrightarrow C,$$

and define the function h by setting $h(x) = g(f(x))$ for each $x \in A$. Then

$$\begin{aligned}(x_1 \in A,\ x_2 \in A,\ x_1 \neq x_2) &\Rightarrow f(x_1) \neq f(x_2)\\ &\Rightarrow g(f(x_1)) \neq g(f(x_2))\\ &\Rightarrow h(x_1) \neq h(x_2).\end{aligned}$$

Also, $h(A) = g(f(A)) = g(B) = C$. Thus h: $A \leftrightarrow C$ and hence $A \sim C$. ∎

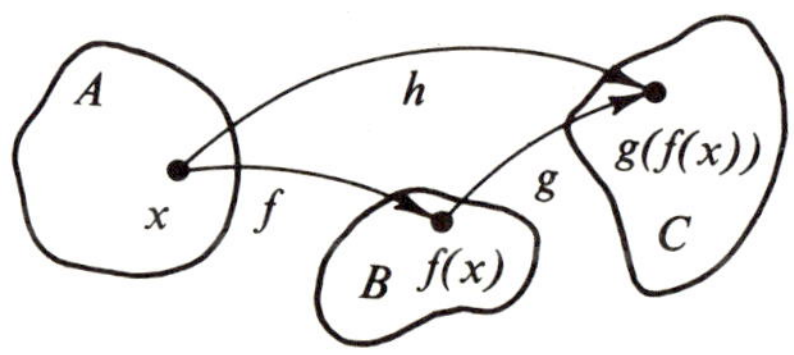

FIG. 5

Theorem 3. *The set of real numbers is equivalent to*

$$\{x \mid -1 < x < 1\}; \text{ that is } E^1 \sim (-1, 1).$$

Proof. The function f defined by

$$f(x) = \frac{x}{1 - x^2}, \qquad -1 < x < 1,$$

is clearly such that f: $(-1, 1) \leftrightarrow E^1$. ∎

Even though E^1 has no largest or smallest element, the following theorem and corollaries include the fact that $E^1 \sim [-1, 1]$.

Theorem 4. *The open-closed interval* $(0, 1]$ *and the open interval* $(0, 1)$ *are equivalent.*

Proof. This will be proved by defining a function f such that

(5) $$f: (0, 1] \leftrightarrow (0, 1).$$

[Intuitively, $(\frac{1}{2}, 1]$, $(\frac{1}{3}, \frac{1}{2}]$, $(\frac{1}{4}, \frac{1}{3}]$ $\cdots$ are successively relocated as illustrated in Fig. 6. Also the ordinary analytic geometry graph of f is indicated in Fig. 7.]

The function f is defined for each x, $0 < x \leqslant 1$, as follows:

Determine $n \in J$ such that $\dfrac{1}{n+1} < x \leqslant \dfrac{1}{n}$; then set

$$f(x) = x + 1 - \frac{1}{n} - \frac{1}{n+1}.$$

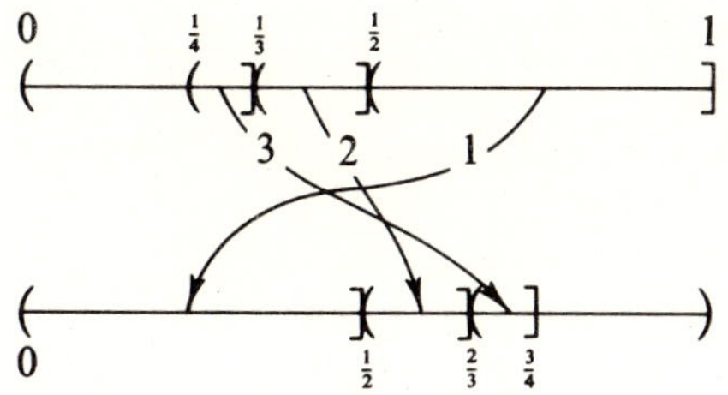

FIG. 6

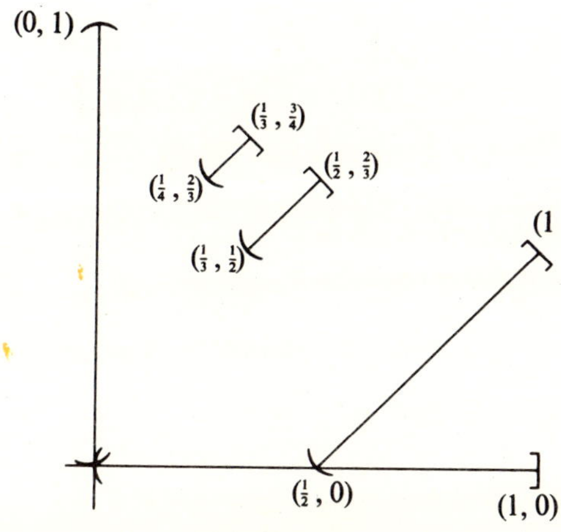

FIG. 7

Hence f has a different definition on each of the intervals

$$(\tfrac{1}{2}, 1], (\tfrac{1}{3}, \tfrac{1}{2}], (\tfrac{1}{4}, \tfrac{1}{3}], \cdots.$$

If $(n+1)^{-1} < x \leqslant n^{-1}$, then

$$f(x) > \frac{1}{n+1} + 1 - \frac{1}{n} - \frac{1}{n+1} = \frac{n-1}{n} \qquad \text{and} \qquad f(x) \leqslant \frac{n}{n+1}.$$

Also, given a number y, $0 < y < 1$, there is a unique $n \in J$ for which

$$\frac{1}{n+1} \leqslant 1 - y < \frac{1}{n} \quad \text{and hence} \quad \frac{1}{n+1} < y - \left(1 - \frac{1}{n} - \frac{1}{n+1}\right) \leqslant \frac{1}{n}$$

so that

$$f\left(y - 1 + \frac{1}{n} + \frac{1}{n+1}\right) = y.$$

It thus follows that (5) holds. ∎

Corollary 1. $[-1, 1] \sim (-1, 1)$.

Proof. By using the function f of the above proof and defining the function g by

$$g(x) = \begin{cases} f(x) & \text{if } 0 < x \leqslant 1 \\ -f(-x) & \text{if } -1 \leqslant x < 0 \\ 0 & \text{if } x = 0 \end{cases}$$

then g: $[-1, 1] \leftrightarrow (-1, 1)$. ∎

The following results are seen by using the transitivity of $\sim$ and the easily established equivalence $(-1, 1) \sim (0, 1)$.

Corollary 2. $E^1 \sim (-1, 1) \sim [-1, 1]$

$$\sim (0, 1) \sim (0, 1] \sim [0, 1).$$

Corollary 3. *If $a \in E^1$, $b \in E^1$, and $a < b$, then*

$$E^1 \sim (a, b) \sim [a, b] \sim (a, b] \sim [a, b).$$

2–2 INFINITE SETS

It may seem that there are "twice as many" positive integers as positive even integers, but

$$\{1, 2, 3, 4, 5, 6, \cdots\} \sim \{2, 4, 6, \cdots\}$$

as $f(n) = 2n$ for each $n \in J$ shows. Any preconceived notion as to the "size" of sets (even in E^1 or E^2) should not influence an opinion about whether the sets are equivalent or not, as the corollaries of Theorem 2-1.4 show. Also, sets of different "dimension" may be equivalent; Theorem 9 shows that the edge of a square is equivalent to the whole square, including its interior. On the other hand, the set of rational numbers and the set of irrational numbers are not equivalent (Problem 3 below).

Although a set "being finite" seems easier intuitively than a set "being infinite," the following definition gives the second notion first.

Definition 1. *A set is said to be* **infinite** *if it is equivalent to at least one of its nonempty proper subsets. A set is* **finite** *if it is not infinite.* [In particular $\varnothing$ is finite.]

The technique of inductive selection used in the next proof will be employed in several following proofs.

Theorem 1. *Any nonempty set is either infinite or is equivalent to J_n for some $n \in J$.*

Proof. Let $A \neq \varnothing$ be given. Select an element of A and label it a_1. Set $A_1 = A - \{a_1\}$.

Either $A_1 = \varnothing$, in which case $A = \{a_1\}$ so that $A \sim J_1$, or $A_1 \neq \varnothing$, in which case select $a_2 \in A_1$.

Consider that elements $a_1, a_2, \cdots, a_n$ of A have been selected, no two the same, and set

$$A_n = A - \{a_1, a_2, \cdots, a_n\}.$$

Either $A_n = \varnothing$, in which case $A \sim J_n$, or $A_n \neq \varnothing$, in which case the process of selection continues.

If the process terminates, it does so at some stage n with $n \in J$ and $A \sim J_n$.

Consider the situation in which the process does not terminate. Then elements $a_1, a_2, a_3, \cdots$ of A are selected, no two the same. Let

$$B = A - \{a_1, a_2, a_3, \cdots\}.$$

If $B = \varnothing$, then f defined by $f(a_n) = a_{n+1}$, sets up a 1–1 correspondence between $A = \{a_1, a_2, a_3, \cdots\}$ and the proper subset $\{a_2, a_3, a_4, \cdots\}$ of A and hence A is infinite. If $B \neq \varnothing$, then f defined by

$$f(a) = \begin{cases} a & \text{if } a \in B \\ a_{n+1} & \text{if } a = a_n, \quad n = 1, 2, 3, \cdots, \end{cases}$$

is a 1–1 correspondence between A and the proper subset $A - \{a_1\}$. Hence A is infinite in this case also. ∎

The above proof shows that

Corollary. *Every infinite set has a subset which is equivalent to J.*

Theorem 2. *Every superset of an infinite set is infinite.*

Proof. Let A be infinite and $B \supset A$. Let $C \sim A$ be a proper subset of A. Hence there is a mapping $f\colon C \leftrightarrow A$. Then $D = C \cup (B - A)$ is a proper subset of B and

$$g\colon D \leftrightarrow B \qquad \text{where } g(d) = \begin{cases} f(d) & \text{if } d \in C \\ d & \text{if } d \in B - A. \end{cases}$$

Hence B is infinite. ∎

Definition 2. *A set is said to be* **countable** *if it is finite (possibly empty) or is equivalent to J, but* **uncountable** *(or* **noncountable***) if it is not countable.*

If a set A is both infinite and $A \sim J$, then A may be said to be† **countably infinite.**

The corollary of Theorem 1 may now be restated.

† In some books "A is countable (or denumerable)" is used to mean "$A \sim J$," and then "A is at most countable" to mean "A is finite or countable."

Corollary. *Every infinite set has a countably infinite subset.*

Theorem 3. *J^2 is countably infinite; that is, the set of ordered pairs of positive integers can be put into 1–1 correspondence with the set of positive integers.*

15	20	26		
(1, 4) 10	(2, 4) 14	(3, 4) 19	(4,4) 25	
(1, 3) 6	(2, 3) 9	(3, 3) 13	(4, 3) 18	24
(1, 2) 3	(2, 2) 5	(3, 2) 8	(4, 2) 12	17
(1, 1) 1	(2, 1) 2	(3, 1) 4	(4, 1) 7	11

FIG. 1

Proof. The infinite set $\{(1, 1), (1, 2), (1, 3), \cdots\}$ is a subset of J^2 and hence J^2 is infinite. The accompanying diagram is an aid in seeing one way of intuitively obtaining a function $f: J^2 \leftrightarrow J$. For example,

$$f(1, 1) = 1, f(2, 1) = 2, f(1, 2) = 3, f(3, 1) = 4, f(2, 2) = 5, \cdots.$$

In this scheme if $(m, n) \in J^2$, then

$$f(m, n) = \tfrac{1}{2}(m + n - 2)(m + n - 1) + n. \tag{1}$$

To see this, start at (m, n) and go down the line of slope -1. Below (m, n) on this line there are $n - 1$ of the labeled lattice points, the lowest of which is $(m + n - 1, 1)$ and to its left is $(m + n - 2, 1)$. Thus the number of points counted before reaching $(m + n - 1, 1)$ is

$$1 + 2 + 3 + \cdots + (m + n - 2) = \tfrac{1}{2}(m + n - 2)(m + n - 1).$$

Hence this integer plus n is the positive integer associated with (m, n).

Had we defined f by (1), without the geometry, there might have been some question about f being 1–1. To see that it is, let $(m', n') \in J^2$ be such that

$$f(m, n) = f(m', n') = \tfrac{1}{2}(m' + n' - 2)(m' + n' - 1) + n'.$$

Then

$$n - n' = \tfrac{1}{2}[(m' + n' - 2)(m' + n' - 1) - (m + n - 2)(m + n - 1)]. \tag{2}$$

Assume $m' + n' \neq m + n$ and so named that $m' + n' > m + n$. Then, since all are positive integers, $m' + n' - 1 \geqslant m + n \geqslant 2$. Therefore

$$n - n' \geqslant \tfrac{1}{2}[(m + n - 1)(m + n) - (m + n - 2)(m + n - 1)]$$
$$= m + n - 1.$$

Thus $1 \geqslant m + n'$, which is a contradiction to $m + n' \geqslant 2$. Hence

$$m' + n' = m + n.$$

This, with (2), gives $n' = n$ and then $m' = m$. ∎

[In this connection see Problem 27, Chapter 1.]

The following theorem uses the fact that a nonempty set of positive integers has a smallest element. The point of this theorem is: "To establish the countability of a set, it is sufficient to obtain a mapping on J onto the set without the 1–1 requirement."

Theorem 4. *If $A \neq \varnothing$ and if there is a mapping $f: J \to A$ which is onto, then A is countable.*

Proof. Define $g(1) = f(1)$. With $g(n)$ defined, $n \geqslant 1$, let

$$A_n = A - \{g(1), g(2), \cdots, g(n)\}.$$

If $A_n = \varnothing$, the process stops.

If $A_n \neq \varnothing$, set $g(n + 1) = f(m_n)$, where m_n is the smallest integer in the nonempty set of positive integers

$$\{k \mid f(k) \in A_n\}.$$

Hence if $n > 1$ and $g(n)$ is defined, then $g(i) \neq g(j)$ for $1 \leqslant i < j \leqslant n$. Also, $m_n > n$ whenever m_n is defined.

In case the process stops at some stage n, then

$$g: J_n \leftrightarrow A$$

so that A is finite.

In case the process does not stop, set

$$B = A - \{g(1), g(2), g(3), \cdots\}.$$

Assume $B \neq \varnothing$ and let $b \in B$. There is then a $p \in J$ such that $f(p) = b$. For each $n \in J$, $g(n) \neq f(p)$. Hence

$$f(p) \in A_1, \qquad f(p) \in A_2, \cdots, \qquad f(p) \in A_n, \cdots,$$

so that $p > m_n > n$ for every $n \in J$, contradicting $p \in J$. Thus $B = \varnothing$ and $g: J \leftrightarrow A$. Hence A is countable. ∎

Theorem 5. *Every subset of a countable set is countable.*

Proof. Let A be countable and $B \subset A$. If B is finite, then B is countable so consider B infinite. Hence A is countably infinite by Theorem 2. Consider a mapping

$$f: J \to A \qquad \text{which is onto.}$$

Select a $b_0 \in B$. Define a mapping g on J into B by

$$g(n) = \begin{cases} f(n) & \text{if } f(n) \in B \\ b_0 & \text{if } f(n) \notin B. \end{cases}$$

But g is onto, since if $b \in B$, then $b \in A$ and $\exists n \in J$ such that $g(n) = f(n) = b$. Thus B is countable by Theorem 4. ∎

PROBLEM 1. Every superset of a noncountable set is noncountable.

Theorem 6. *The union of a countable collection of countable sets is countable.*

Proof. Let $A_1, A_2, A_3, \cdots$ be a finite or countably infinite collection of sets such that each A_k in the collection is countable. Then the sets displayed as

$$\begin{aligned} A_1 &= \{a_{11}, a_{12}, a_{13}, \cdots\} \\ A_2 &= \{a_{21}, a_{22}, a_{23}, \cdots\} \\ A_3 &= \{a_{31}, a_{32}, a_{33}, \cdots\} \\ &\cdots \end{aligned}$$

show the union $A = A_1 \cup A_2 \cup A_3 \cup \cdots$ to be equivalent to J^2 or to a subset of J^2 if there are only finitely many sets $A_1, A_2, \cdots, A_n$, or if some set in the collection is finite, or if an element of one set also occurs in another. In any case, A is equivalent to a countable set so is itself countable. ∎

Theorem 7. *The set of rational numbers is countably infinite.*

Proof. The set is infinite, since J is an infinite subset.

The set of positive rationals, denoted by R^+, is countable as seen by associating each positive rational m/n in lowest terms with the ordered pair (m, n), thus showing that R^+ is equivalent to a subset of J^2. Clearly, $R^- \sim R^+$, where R^- is the set of negative rationals. Then the set of all rationals is $R^- \cup R^+ \cup \{0\}$ and is thus countable. ∎

PROBLEM 2. The set of rationals on $[0, 1]$ is countably infinite.

The **diagonal process** used in the proof of the next theorem is a technique which will be employed on several later occasions.

Theorem 8. *The set* $[0, 1]$ *is noncountable.*

Proof. Suppose $[0, 1]$ is countable and let $x_1, x_2, \cdots$ be such that

$$[0, 1) = \{x_1, x_2, x_3, \cdots\},$$

with $x_i \neq x_j$, $i \neq j$. For each $m \in J$ select *the* decimal representation of x_m which *does not have all 9's* from some place on; that is,

$$x_m = 0.\alpha_{m1}\alpha_{m2}\alpha_{m3}\cdots,$$

where for each $N \in J$ there is an $n > N$ for which $\alpha_{mn} \neq 9$. Display the totality of numbers of $[0, 1)$ as

$$\begin{aligned} x_1 &= 0.\alpha_{11}\alpha_{12}\alpha_{13}\cdots \\ x_2 &= 0.\alpha_{21}\alpha_{22}\alpha_{23}\cdots \\ x_3 &= 0.\alpha_{31}\alpha_{32}\alpha_{33}\cdots \\ &\cdots. \end{aligned}$$

Now define $x_0 = 0.\beta_1\beta_2\beta_3\cdots$ by setting

$$\beta_n = \begin{cases} 0 & \text{if } \alpha_{nn} \neq 0 \\ 1 & \text{if } \alpha_{nn} = 0. \end{cases}$$

Then $0 \leqslant x_0 \leqslant \frac{1}{9} < 1$. Since neither x_0 nor x_n, $n \in J$, has all 9's from any place on and $\beta_n \neq \alpha_{nn}$, then $x_0 \neq x_n$. Thus the above display does not contain all numbers of $[0, 1)$, which is a contradiction. Thus $[0, 1)$ is noncountable and hence the superset $[0, 1]$ is noncountable. ∎

PROBLEM 3. The set of irrational numbers on $[0, 1]$ is noncountable.

PROBLEM 4. If $a \in E^1$, $b \in E^1$, and $a < b$, then each of the sets (a, b), $[a, b]$, $(a, b]$, and $[a, b)$ is noncountable.

With n a positive integer, the **König-binary** symbol $(n]$ stands for $n - 1$ zeros followed by a 1. Thus

$$(2] \text{ means } 01, \qquad (4] \text{ means } 0001, \qquad \text{and } (1] \text{ means } 1.$$

In connection with binary representations of numbers, for example,

$$\begin{aligned} \frac{5}{64} = \frac{1}{2^4} + \frac{1}{2^{4+2}} &= 0\underset{2}{.}000101 = 0\underset{2}{.}(4](2] \\ &= 0\underset{2}{.}000100111\cdots = 0\underset{2}{.}(4](3](1](1](1]\cdots. \end{aligned}$$

Given a number t, $0 < t \leqslant 1$, write t in its unique binary form *using infinitely many 1's.* Then there is a unique infinite sequence $n_1, n_2, n_3, \cdots$ of positive integers such that

$$\begin{aligned} t &= \frac{1}{2^{n_1}} + \frac{1}{2^{n_1+n_2}} + \frac{1}{2^{n_1+n_2+n_3}} + \cdots \\ &= 0\underset{2}{.}(n_1](n_2](n_3]\cdots. \end{aligned} \tag{1}$$

This is called the **König-binary representation** of t. Thus

Any number $0 < t \leqslant 1$ has one and only one König-binary representation.

For example, although $0.(4](2] = \frac{5}{16}$, the König-binary representation of $\frac{5}{16}$ is $0\underset{2}{.}(4](3](1](1](1]\cdots$.

Theorem 9. $[0, 1] \sim [0, 1] \times [0, 1]$.

Proof. We first prove that $(0, 1] \sim (0, 1] \times (0, 1]$. Given a number t, $0 < t \leqslant 1$, represent t in its unique König-binary form (1). Set

$$(2)\qquad \begin{aligned} f(t) &= 0_{\overset{.}{2}}(n_1](n_3](n_5]\cdots(n_{2k-1}]\cdots,\\ g(t) &= 0_{\overset{.}{2}}(n_2](n_4](n_6]\cdots(n_{2k}]\cdots, \end{aligned}$$

and then associate the point $(f(t), g(t))$ with t by setting

$$(3)\qquad F(t) = (f(t), g(t)).$$

For example,

$$F(\tfrac{5}{64}) = (0_{\overset{.}{2}}000111\cdots, 0_{\overset{.}{2}}00111\cdots) = (\tfrac{1}{8}, \tfrac{1}{4}).$$

Also, $F(0)$ is not defined, but $F(1)$ is the point $(1, 1)$.

Since $0 < f(t) \leqslant 1$ and $0 < g(t) \leqslant 1$, then F is on $(0, 1]$ into $(0, 1] \times (0, 1]$. But F is also onto. For let $0 < x \leqslant 1$ and $0 < y \leqslant 1$ be given and represent each in its König-binary form:

$$\begin{aligned} x &= 0_{\overset{.}{2}}(m_1](m_2](m_3]\cdots,\\ y &= 0_{\overset{.}{2}}(p_1](p_2](p_3]\cdots. \end{aligned}$$

Then the number t defined by

$$t = 0_{\overset{.}{2}}(m_1](p_1](m_2](p_2](m_3](p_3]\cdots$$

is such that $0 < t \leqslant 1$ and $F(t) = (x, y)$. Also if $0 < s \leqslant 1$ and $s \neq t$, then $F(s) \neq F(t)$ because of the uniqueness of the König-binary representation. Thus

$$F\colon (0, 1] \leftrightarrow (0, 1] \times (0, 1].$$

Finally the function G defined for $0 \leqslant t \leqslant 1$ by

$$G(t) = \begin{cases} (4t, 0) & \text{if } 0 \leqslant t \leqslant \frac{1}{4}\\ (0, 4t - 1) & \text{if } \frac{1}{4} < t \leqslant \frac{1}{2}\\ F(2t - 1) & \text{if } \frac{1}{2} < t \leqslant 1 \end{cases}$$

is on $[0, 1]$ onto $[0, 1] \times [0, 1]$ and is 1–1. ∎

PROBLEM 5. With $0 < t \leqslant 1$, let $t = 0_{\overset{.}{2}}\alpha_1\alpha_2\alpha_3\cdots$ be the unique binary representation of t using infinitely many 1's. Set

$$H(t) = (0_{\overset{.}{2}}\alpha_1\alpha_3\alpha_5\cdots, 0_{\overset{.}{2}}\alpha_2\alpha_4\alpha_6\cdots).$$

Show that the resulting function

$$H\colon (0, 1] \to (0, 1] \times (0, 1]$$

is onto, but is not 1–1.

[*Hint:* Compute, for example, $H(\tfrac{7}{16})$, $H(\tfrac{23}{48})$, $H(\tfrac{37}{48})$.]

As sparse on [0, 1] as the Cantor set seems to be, the following theorem shows that the Cantor set is as "numerous" as the interval [0, 1] itself.

Theorem 10. *There is a function with domain the Cantor set* Γ *and range the interval* [0, 1].

Proof. One such function f is defined in the following way. With $x \in \Gamma$, represent x in its unique ternary form *without using the digit* 1:

$$x = \underset{3}{0.}(2\gamma_1)(2\gamma_2) \cdots (2\gamma_k) \cdots, \qquad \gamma_k = \begin{Bmatrix} 0 \\ 1 \end{Bmatrix}.$$

Then use binary notation and set

$$f(x) = \underset{2}{0.}\gamma_1\gamma_2 \cdots \gamma_k \cdots. \tag{4}$$

[As examples:

$$f(\tfrac{1}{3}) = f(\underset{3}{0.}0222 \cdots) = \underset{2}{0.}0111 \cdots = \tfrac{1}{2} \qquad \text{and}$$

$$f(\tfrac{3}{4}) = f(\underset{3}{0.}202020 \cdots) = \underset{2}{0.}101010 \cdots = \tfrac{2}{3}.]$$

In (4), $0 \leqslant \underset{2}{0.}\gamma_1\gamma_2\gamma_3 \cdots \leqslant 1$ and hence f is on [0, 1] into [0, 1]. To show that f is onto [0, 1], that is, that [0, 1] is the range of f, select $0 \leqslant y \leqslant 1$ arbitrarily. Represent y in binary form:

$$y = \underset{2}{0.}\delta_1\delta_2 \cdots \delta_k \cdots, \qquad \delta_k = \begin{Bmatrix} 0 \\ 1 \end{Bmatrix}.$$

If y has two binary representations, either may be used. Then x defined by $x = \underset{3}{0.}(2\delta_1)(2\delta_2)(2\delta_3) \cdots$ is in the Cantor set and $f(x) = y$.

Hence the theorem is proved. ∎

It was shown above that $f(\frac{1}{3}) = \frac{1}{2}$. Also

$$f(\tfrac{2}{3}) = f(\underset{3}{0.}2000 \cdots) = \underset{2}{0.}1000 \cdots = \tfrac{1}{2}.$$

In fact if $(\underline{x}_n, \bar{x}_n)$ is one of the open intervals erased at the nth stage, then $f(\underline{x}_n) = f(\bar{x}_n)$. To see this, recall ((5) and (6) of Section 1–8) that $\underline{x}_n$ and $\bar{x}_n$ have the specific forms

$$\begin{aligned} \underline{x}_n &= \underset{3}{0.}(2\gamma_1)(2\gamma_2) \cdots (2\gamma_{n-1})\,1 \\ &= \underset{3}{0.}(2\gamma_1)(2\gamma_2) \cdots (2\gamma_{n-1})\,0\,2\,2\,2 \cdots \qquad \text{and} \\ \bar{x}_n &= \underset{3}{0.}(2\gamma_1)(2\gamma_2) \cdots (2\gamma_{n-1})\,2. \end{aligned}$$

Hence

$$\begin{aligned} f(\underline{x}_n) &= \underset{2}{0.}\gamma_1\gamma_2 \cdots \gamma_{n-1}\,0\,1\,1\,1 \cdots \\ &= \underset{2}{0.}\gamma_1\gamma_2 \cdots \gamma_{n-1}\,1 \qquad \text{and also} \\ f(\bar{x}_n) &= \underset{2}{0.}\gamma_1\gamma_2 \cdots \gamma_{n-1}\,1. \end{aligned}$$

Conversely, let $0 \leqslant y \leqslant 1$ be such that there is an $x_1 \in \Gamma$ and an $x_2 \in \Gamma$ with $x_1 \neq x_2$ such that

$$f(x_1) = f(x_2) = y.$$

From the way f was defined, this means that y has two binary representations; that is, for some n,

$$\begin{aligned} y &= 0\underset{2}{.}\delta_1\delta_2 \cdots \delta_n 0\,1\,1\,1 \cdots \\ &= 0\underset{2}{.}\delta_1\delta_2 \cdots \delta_n 1. \end{aligned}$$

The first representation gives

$$\begin{aligned} x_1 &= 0\underset{3}{.}(2\delta_1)(2\delta_2) \cdots (2\delta_n)\, 0\,2\,2\,2 \cdots \\ &= 0\underset{3}{.}(2\delta_1)(2\delta_2) \cdots (2\delta_n)\, 1 \end{aligned}$$

while the second gives

$$x_2 = 0\underset{3}{.}(2\delta_1)(2\delta_2) \cdots (2\delta_n)\, 2$$

such that $f(x_1) = f(x_2) = y$. From these specific forms it follows that x_1 is the lower end and x_2 is the upper end of an open interval erased in forming the Cantor set.

PROBLEM 6. For the function f defined in the proof of Theorem 10:

$$\text{If } c_1 \in \Gamma \text{ and } c_2 \in \Gamma, \text{ with } c_1 < c_2, \quad \text{then } f(c_1) \leqslant f(c_2).$$

The function f: $\Gamma \to [0, 1]$, in the proof of Theorem 10, and its properties given above are used in defining a function on $[0, 1]$ onto $[0, 1]$.

Definition 3. *The* **Cantor function** ψ: $[0, 1] \to [0, 1]$ *is defined by setting*

$$\begin{aligned} \psi(x) &= f(x) && \text{if } x \in \Gamma \\ &= f(\underline{x}) = f(\bar{x}) && \text{if } \underline{x} < x < \bar{x} \end{aligned}$$

for each open interval $(\underline{x}, \bar{x}) \subset [0, 1] - \Gamma$ *with* $\underline{x} \in \Gamma$ *and* $\bar{x} \in \Gamma$.

The Cantor function will be used so many times for illustrative, or counterexample, purposes that we reserve ψ for it. Also, Γ will represent the Cantor set.

2–3 SEQUENCES OF SETS

A sequence is a function whose domain is equivalent to the set J of positive integers. Whether the domain is J or some set equivalent to J (such as $\{0, 1, 2, \cdots\}$) is usually of little concern. The familiar subscript notation will be used. A sequence whose members are $a_1, a_2, a_3, \cdots$ will be denoted by $\{a_n\}$ (not to be confused with $\{a\}$ for the set whose only element is a).

The members of a sequence may be sets. With $\{A_n\}$ a sequence of sets, the union and intersection of all members are defined and denoted by

$$\bigcup_{n=1}^{\infty} A_n = \{x \mid x \in \text{of at least one } A_n\},$$

$$\bigcap_{n=1}^{\infty} A_n = \{x \mid x \in A_n, \qquad n = 1, 2, 3, \cdots\}.$$

$\bigcup A_n$ and $\bigcap A_n$ may be used if it is understood that the domain is J. The interpretations of

$$\bigcup_{k=n}^{\infty} A_k = \bigcup_{k \geqslant n} A_k \qquad \text{and} \qquad \bigcup_{k=m}^{n} A_k \qquad \text{with } m \leqslant n$$

should be clear.

PROBLEM 1. Let $\{A_n\}$ be a sequence of sets in a universal set. Then de Morgan's rules hold; that is,

$$(\bigcup A_n)^c = \bigcap (A_n)^c \qquad \text{and} \qquad (\bigcap A_n)^c = \bigcup (A_n)^c.$$

Definition 1. *A sequence of sets $\{A_n\}$ is said to be* **monotonically decreasing** *if*

$$A_1 \supset A_2 \supset A_3 \supset \cdots$$

and **monotonically increasing** *if*

$$A_1 \subset A_2 \subset A_3 \subset \cdots.$$

For example, the sequence $\{A_n\}$ defined by

$$A_n = \left\{x \,\middle|\, 0 \leqslant x < 1 + \frac{1}{n}\right\}$$

is monotonically decreasing,

$$\bigcap A_n = \bigcap_{k=n}^{\infty} A_k = \{x \mid 0 \leqslant x \leqslant 1\} \qquad \text{and} \qquad \bigcup_{k=n}^{\infty} A_k = A_n.$$

Definition 2. *Given a sequence of sets $\{A_n\}$, form the sequences $\{\underline{A}_n\}$ and $\{\bar{A}_n\}$ by setting*

$$\underline{A}_n = \bigcap_{k \geqslant n} A_k \qquad \text{and} \qquad \bar{A}_n = \bigcup_{k \geqslant n} A_k \qquad \text{for } n = 1, 2, 3, \cdots$$

and then set

$$\underline{A} = \bigcup_{n \geqslant 1} \underline{A}_n \qquad \text{and} \qquad \bar{A} = \bigcap_{n \geqslant 1} \bar{A}_n.$$

By definition, $\underline{A}$ is the **lower limit** *(or limit inferior) and $\bar{A}$ is the* **upper limit** *(or limit superior) of $\{A_n\}$ and the notations*

$$\underline{A} = \varliminf_{n\to\infty} A_n = \liminf_{n\to\infty} A_n \qquad \textit{and}$$

$$\bar{A} = \varlimsup_{n\to\infty} A_n = \limsup_{n\to\infty} A_n$$

are used. [$n \to \infty$ will usually be deleted.]

EXAMPLE. Let $\{A_n\}$ be defined by

$$A_n = \begin{cases} \left\{(x, y) \,\middle|\, 0 \leqslant x \leqslant 1 - \dfrac{1}{n}, 0 \leqslant y \leqslant 2 - \dfrac{1}{n}\right\} & \text{for } n \text{ odd} \\[2ex] \left\{(x, y) \,\middle|\, 0 \leqslant x \leqslant 2 + \dfrac{1}{n}, 0 \leqslant y \leqslant 1 + \dfrac{1}{n}\right\} & \text{for } n \text{ even.} \end{cases}$$

Then

$$\varliminf A_n = \{(x, y) \mid 0 \leqslant x < 1, 0 \leqslant y \leqslant 1\} \qquad \text{and}$$
$$\varlimsup A_n = \{(x, y) \mid 0 \leqslant x \leqslant 2, 0 \leqslant y < 2\}.$$

Theorem 1. *If m and n are any positive integers, then*

$$\underline{A}_n \subset \bar{A}_m. \tag{1}$$

Proof. If $n \leqslant m$, then

$$\underline{A}_n = \bigcap_{k \geqslant n} A_k \subset \bigcap_{k \geqslant m} A_k \subset \bigcup_{k \geqslant m} A_k = \bar{A}_m, \tag{2}$$

but if $m \leqslant n$, then

$$\bar{A}_m = \bigcup_{k \geqslant m} A_k \supset \bigcup_{k \geqslant n} A_k \supset \bigcap_{k \geqslant n} A_k = \underline{A}_n. \;\blacksquare \tag{3}$$

Theorem 2. $\varliminf A_n \subset \varlimsup A_n$.

Proof. Let $n \in J$. Then from (1)

$$\underline{A}_n \subset \bigcap \bar{A}_m = \bar{A}.$$

Thus $\underline{A} = \bigcup \underline{A}_n \subset \bar{A}$. ∎

The results in the following corollary are interpretations of the first three terms of (2) and (3), respectively.

Corollary. *$\{\underline{A}_n\}$ is monotonically increasing and $\{\bar{A}_n\}$ is monotonically decreasing.*

Definition 3. *In the notation of Definition 1, if $\underline{A} = \bar{A}$, then this common set is denoted by A, the sequence $\{A_n\}$ is said to have* **limit** *A, and this is denoted by*

$$A = \lim A_n.$$

Theorem 3. *Any monotonically decreasing sequence $\{A_n\}$ has*

$$\lim A_n = \bigcap A_n .$$

Proof. Given that $\{A_n\}$ is monotonically decreasing, then

$$\underline{A}_n = \bigcap_{k \geqslant n} A_k = \bigcap_{k \geqslant n-1} A_k = \cdots = \bigcap_{k \geqslant 1} A_k$$

and hence

$$\underline{A} = \bigcup \underline{A}_n = \bigcap A_k ,$$

but also

$$\bar{A}_n = \bigcup_{k \geqslant n} A_k = A_n \qquad \text{so that} \qquad \bar{A} = \bigcap \bar{A}_n = \bigcap A_n . \blacksquare$$

PROBLEM 2. Any monotonically increasing sequence $\{A_n\}$ has

$$\lim A_n = \bigcup A_n .$$

Theorem 4. *For every sequence of sets $\{A_n\}$*

$$\underline{\lim}\, A_n = \{x \mid x \in A_n \textit{ for all except finitely many } n\};$$

that is, $x \in \underline{A}$ iff $\exists N$ such that $n \geqslant N \Rightarrow x \in A_n$.

Proof. Let $x \in \underline{A} = \bigcup \underline{A}_n$. Hence there is an N such that $x \in \underline{A}_N = \bigcap_{n \geqslant N} A_n$. Therefore $n \geqslant N \Rightarrow x \in A_n$.

Let x and N be such that $n \geqslant N \Rightarrow x \in A_n$. Then

$$x \in \bigcap_{n \geqslant N} A_n = \underline{A}_N \subset \bigcup_{n \geqslant N} \underline{A}_n \subset \bigcup \underline{A}_n = \underline{A}. \blacksquare$$

PROBLEM 3. For every sequence of sets $\{A_n\}$,

$$\overline{\lim}\, A_n = \{x \mid x \in A_n \text{ for infinitely many } n\};$$

that is, $x \in \bar{A}$ iff for each N there is an $n \geqslant N$ for which $x \in A_n$.

Theorem 5. *By considering complements with respect to a universal set,*

$$\underline{\lim}\, A_n^c = (\overline{\lim}\, A_n)^c \qquad \textit{and} \qquad \overline{\lim}\, A_n^c = (\underline{\lim}\, A_n)^c .$$

Proof. Let $B_n = A_n^c$. Then

$$\underline{B}_n = \bigcap_{k \geqslant n} B_k = \bigcap_{k \geqslant n} A_k^c = \left(\bigcup_{k \geqslant n} A_k \right)^c = (\bar{A}_n)^c \qquad \text{and}$$

$$\underline{B} = \bigcup \underline{B}_n = \bigcup (\bar{A}_n)^c = (\bigcap \bar{A}_n)^c = (\bar{A})^c,$$

which is the first equation in different notation. The second equation follows by applying the first to $\{A_n^c\}$. $\blacksquare$

Theorem 6. *Given two sequences* $\{A_n\}$ *and* $\{B_n\}$, *form the sequence* $\{C_n\}$ *by setting* $C_n = A_n \cup B_n$. *Then*

$$\underline{A} \cup \underline{B} \subset \underline{C} \subset \left\{\begin{matrix} \underline{A} \cup \bar{B} \\ \bar{A} \cup \underline{B} \end{matrix}\right\} \subset \bar{C} = \bar{A} \cup \bar{B}. \tag{4}$$

Proof. The following proof depends heavily upon Theorem 4.

Let $x \in \underline{A}$. Choose N such that $n \geqslant N \Rightarrow x \in A_n$. Then $n \geqslant N \Rightarrow x \in A_n \cup B_n = C_n$. Thus $x \in \underline{C}$. Hence $\underline{A} \subset \underline{C}$ and similarly $\underline{B} \subset \underline{C}$ so that $\underline{A} \cup \underline{B} \subset \underline{C}$.

Let $x \in \underline{C}$ and choose N_1 such that $n \geqslant N_1 \Rightarrow x \in C_n = A_n \cup B_n$. If $x \in B_n$ for only finitely many n, there is an N_2 such that $n \geqslant N_2 \Rightarrow x \in A_n$ and hence, with N the larger of N_1 and N_2, $n \geqslant N \Rightarrow x \in A_n$ so that $x \in \underline{A}$. If, however, $x \in B_n$ for infinitely many n, then $x \in \bar{B}$. In either case $x \in \underline{A} \cup \bar{B}$. Hence $\underline{C} \subset \underline{A} \cup \bar{B}$. By interchanging the roles of A_n and B_n, then $\underline{C} \subset \bar{A} \cup \underline{B}$.

The remaining inclusions follow in a similar way.

The first inclusion in (4) may be written

$$\underline{\lim}\, A_n \cup \underline{\lim}\, B_n \subset \underline{\lim}(A_n \cup B_n).$$

PROBLEM 4

(a) Establish the equality in (4). ∎

(b) Show by examples that each inclusion in (4) may be proper.

(c)
$$\overline{\lim}\, A_n \cap \overline{\lim}\, B_n \supset \overline{\lim}(A_n \cap B_n) \supset \left\{\begin{matrix} \overline{\lim}\, A_n \cap \underline{\lim}\, B_n \\ \underline{\lim}\, A_n \cap \overline{\lim}\, B_n \end{matrix}\right\}$$
$$\supset \underline{\lim}(A_n \cap B_n) = \underline{\lim}\, A_n \cap \underline{\lim}\, B_n.$$

[*Hint for* (*b*). $A_{2n-1} = B_{2n} = [0, 1]$, $A_{2n} = B_{2n-1} = [1, 2]$, $n = 1, 2, 3, \ldots$. Check that $\underline{C} = [0, 2]$, but $\underline{A} = \underline{B} = \{1\}$.]

A function with domain a set I may have sets as "values." [A sequence of sets is such a function (by definition) in which $I \sim J$.] The domain I need not be countable. For example, let $\mathscr{Y}$ be the family of all subsets of E^1. Then the function f defined for each x such that $0 < x \leqslant 1$ by

$$f(x) = \{t \mid 0 < t \leqslant x\}$$

is on (0, 1] into $\mathscr{Y}$.

With

$$f\colon I \to \mathscr{B}$$

where I is a set and $\mathscr{B}$ is a family of sets, it is sometimes convenient to use subscript notation. When this is done the domain I is referred to as the **index** set and the function is denoted by $\{B_x\}$. Also, the union and intersection symbols are used by defining, for $A \subset I$,

$$\bigcup_{x \in A} B_x = \{t \mid t \in B_x \quad \text{for at least one } x \in A\}$$

$$\bigcap_{x \in A} B_x = \{t \mid t \in B_x \quad \text{for all } x \in A\}.$$

For example, with $B_x = \{t \mid 0 < t < x\}$ for each $x \in I = (0, 2]$, then

$$\bigcup_{x \in I} B_x = B_2, \qquad \bigcap_{x \in I} B_x = \varnothing, \qquad \bigcup_{x \in (0, \sqrt{2})} B_x = \bigcup_{x \in (0, \sqrt{2}]} B_x = B_{\sqrt{2}}$$

and $B_x \subset B_y$ iff $0 < x \leqslant y \leqslant 2$. Also, with R the set of rational numbers on (0, 2),

$$\bigcup_{x \in R} B_x = B_2 = \{t \mid 0 < t < 2\}. \tag{5}$$

To see this last statement, first note that the union is a subset of B_2. Next select any $t_0 \in B_2$. Then $0 < t_0 < 2$ and a rational number r may be selected such that $t_0 < r < 2$. For such an r,

$$t_0 \in \{t \mid 0 < t < r\} = B_r \subset \bigcup_{x \in R} B_x.$$

Hence

$$\left(t \in B_2 \Rightarrow t \in \bigcup_{x \in R} B_x\right) \Rightarrow B_2 \subset \bigcup_{x \in R} B_x.$$

B_2 is therefore a subset and a superset of the union and hence (5) holds.

If $\mathscr{B}$ is any family of sets (whether indexed or not) the union and intersection symbols are used as follows:

$$\bigcup_{B \in \mathscr{B}} B = \{t \mid t \in B \quad \text{for some } B \in \mathscr{B}\},$$

$$\bigcap_{B \in \mathscr{B}} B = \{t \mid t \in B \quad \text{for every } B \in \mathscr{B}\}.$$

The de Morgan rules also hold for families of sets:

$$\left(\bigcup_{B \in \mathscr{B}} B\right)^c = \bigcap_{B \in \mathscr{B}} B^c \quad \text{and} \quad \left(\bigcap_{B \in \mathscr{B}} B\right)^c = \bigcup_{B \in \mathscr{B}} B^c.$$

2–4 METRIC SPACES

With n a positive integer, **Euclidean n-space** E^n is the set of all ordered n-tuples of real numbers with the **distance** between

$$\mathbf{x} = (x^1, x^2, \cdots, x^n) \in E^n \quad \text{and} \quad \mathbf{y} = (y^1, y^2, \cdots, y^n) \in E^n,$$

defined to be

$$\sqrt{(x^1 - y^1)^2 + (x^2 - y^2)^2 + \cdots + (x^n - y^n)^2}.$$

In particular, $E^2 = E^1 \times E^1, \cdots, E^n = E^{n-1} \times E^1$ with the appropriate distance functions defined.

The purpose of this section is to give some of the topological properties of E^n. It is, however, usually as easy (and more often, easier) to study these properties in the less encumbered environment of metric spaces.

Definition 1. *A* **pseudometric space** *S is any set for which the* **distance** *$d(s, t)$ between any pair of* **points** *$s \in S$ and $t \in S$ is defined, satisfying*

(i) $d(s, t) = d(t, s) \geqslant 0$;
(ii) $d(s, s) = 0$; *and*
(iii) $d(s, t) \leqslant d(s, u) + d(u, t), u \in S$.

S is said to be a **metric space** *if in addition*

(iv) $s \neq t \Rightarrow d(s, t) > 0$.

Thus E^n is a pseudometric space since (i) and (ii) clearly hold and (iii) follows from Minkowski's inequality (for $n \geqslant 2$). If E^n is under consideration, then a point will be designated by

$$\mathbf{x} = (x^1, x^2, \cdots, x^n)$$

in which x^k, $k = 1, 2, \cdots, n$ is a real number called the kth coordinate (or component) of $\mathbf{x}$ (and does not mean x raised to the kth power). The notation and "component" terminology anticipates that E^n will be "made into a vector space."† Hence $\mathbf{x} \neq \mathbf{y}$ means that $x^k \neq y^k$ for at least one k, $1 \leqslant k \leqslant n$. Thus if $\mathbf{x} \neq \mathbf{y}$ then $d(\mathbf{x}, \mathbf{y}) > 0$ and hence E^n is a metric space.

Any set A may be made into a pseudometric space by defining $d(a, b) = 0$ for each $a \in A$ and $b \in A$, but such a triviality will be shunned. There are, however, significant pseudometric spaces for which $d(s, t)$ is zero for some "points" s and t which are not identical.

For the time being we shall consider only metric spaces and therefore

(iv) $$d(s, t) > 0 \qquad \text{whenever } s \neq t$$

holds in addition to (i), (ii), and (iii).

Definition 2. *With S a metric space and $s \in S$, a* **neighborhood** *of s is a set of the form*

$$\{t \mid t \in S \text{ and } d(s, t) < r\}$$

for some **radius** *$r > 0$. Such an r-neighborhood, with* **center** *$s \in S$, will be denoted by $N(s, r)$. Also*

$$N_0(s, r) = \{t \mid t \in S \text{ and } 0 < d(s, t) < r\}$$

is called a **deleted neighborhood** *of s.*

In the following theorems and definitions, S is a metric space and reminders that $s, t, u, \cdots$ are elements of S will be kept to a minimum. Also, S is the universal set so that $A \subset S$ for any set A. It is expected that visualization will be done by considering $S = E^1$, or E^2, and occasionally E^3. In fact, if $S = E^n$, then $N(x, r)$ is sometimes called an **n-ball**, or more specifically an **open n-ball.**

† Vector spaces will be defined in Section 2–12.

The statements "N is a neighborhood" and "N_0 is a deleted neighborhood" mean that there is an s and an $r > 0$ such that, respectively,

$$N = N(s, r) \qquad \text{and} \qquad N_0 = N_0(s, r).$$

Also, $N(s)$ will be used for a neighborhood of s without commitment as to its radius (except that it be positive).

Theorem 1. *The family of neighborhoods in S have the properties:*

(a) *Every $s \in S$ has a neighborhood and for each $N(s)$, $s \in N(s)$.*
(b) *If $t \in N(s)$, then $\exists\, N(t) \subset N(s)$.*
(c) *If $s \neq t$, then $\exists\, N(s)$ and $N(t)$ such that $N(s) \cap N(t) = \varnothing$.*

Proof. (a) is obvious [but was included because in abstract space theory any set having a family of subsets satisfying (a), (b), and (c) is called a **Hausdorff space**].

(b) is tautological if $t = s$ so consider $t \neq s$. Then $d(s, t) > 0$ by (iv). Since $N(s)$ is a neighborhood, let $r > 0$ be its radius. Hence $d(s, t) < r$, since $t \in N(s)$, and therefore

$$r_1 = r - d(s, t) > 0.$$

Take $N(t) = N(t, r_1)$. Now

$$u \in N(t) \Rightarrow d(s, u) \leqslant d(s, t) + d(t, u) < d(s, t) + [r - d(s, t)] = r;$$

that is,

$$(u \in N(t) \Rightarrow d(s, u) < r) \Rightarrow N(t) \subset N(s).$$

(c) Take $r = \frac{1}{2}d(s, t)$. Then $r > 0$ and $N(s, r) \cap N(t, r) = \varnothing$, for otherwise there would be a u in the intersection and

$$r = d(s, t) \leqslant d(s, u) + d(u, t) < \tfrac{1}{2}r + \tfrac{1}{2}r = r. \quad \blacksquare$$

Definition 3. *Let A be a set.* [Recall that this means $A \subset S$.]

(i) *A point s is said to be a* **cluster point** (or limit point) *of A if for each $N_0(s)$*

$$N_0(s) \cap A \neq \varnothing$$

that is, for each $r > 0$ there is a $t \in A$ such that $0 < d(s, t) < r$; so in particular $t \neq s$.

(ii) *Any point (of S) which is not a cluster point of A is called an* **isolated point** *of A.*

(iii) *$A' = \{s \mid s$ is a cluster point of $A\}$ is called the* **derived set** *of A and $\bar{A} = A \cup A'$ the* **closure** *of A.*

(iv) *A is said to be* **closed** *if $A' \subset A$; that is, if every cluster point of A is a point of A.*

(v) *A is said to be* **dense in itself** *if $A \subset A'$,* **dense in a set** *B if $\bar{A} \supset B$, and* **everywhere dense** *if $\bar{A} = S$.*

(vi) *A is said to be* **perfect** *if it is both closed and dense in itself; that is, if $A = A'$.*

(vii) *A is said to be* **nowhere dense** *if for each point s and each $N(s)$ there is a neighborhood $N(t) \subset N(s)$ such that $N(t) \cap A = \varnothing$.*

Note that the empty set $\varnothing$ has no cluster point. This may be stated as $\varnothing' = \varnothing$. Thus $\varnothing$ is perfect and in particular $\varnothing$ is closed.

Since we are considering S the universal space and S' is a set, then $S' \subset S$ so that S is closed.

Theorem 2. *$s \in A'$ if and only if $N(s) \cap A$ is infinite for every neighborhood $N(s)$ of s.*

Proof. If $N(s) \cap A$ is infinite for every $N(s)$, then $N_0(s) \cap A \neq \varnothing$ for every $N_0(s)$ so that $s \in A'$.

Assume there is a set A having a point $s \in A'$ which has $N(s)$ such that $N(s) \cap A$ is a finite set $\{a_1, a_2, \cdots, a_n\}$ with $a_i \neq s$ for $i = 1, 2, \cdots, n$. Let r be the smallest of the n numbers.

$$d(s, a_i), \qquad i = 1, 2, \cdots, n,$$

all of which are positive (by iv of Definition 1). Then $r > 0$,

$$N(s, r) \cap A \subset \{s\} \qquad \text{and} \qquad N_0(s, r) \cap A = \varnothing.$$

Hence s is not a cluster point of A, contradicting the assumption $s \in A'$. ∎

Corollary 1. *A finite set has no cluster point.*

Corollary 2. *Every finite set is closed.*

PROBLEM 1. Given that $S = E^1$.

(a) Any closed interval $[a, b]$ is a closed set and is perfect if $a < b$.

(b) The set R of rational numbers on $[0, 1]$ is not closed, but $R' = [0, 1]$ and R is dense in $[0, 1]$.

(c) The interval $(0, 1]$ is not a closed set.

(d) The Cantor set Γ is perfect.

PROBLEM 2. With R the set of all rational numbers, then $\overline{R^2} = E^2$; that is, in Euclidean 2-space the set of points having both coordinates rational is a countable set which is everywhere dense.

PROBLEM 3. In a metric space S, if $C \subset A$, then $C' \subset A'$.

Theorem 3. *For any two sets A and B*

$$(A \cup B)' = A' \cup B' \qquad \text{and} \qquad \overline{(A \cup B)} = \bar{A} \cup \bar{B}.$$

Proof. Since A and B are both subsets of $A \cup B$, then we have $A' \subset (A \cup B)'$, $B' \subset (A \cup B)'$, and

$$A' \cup B' \subset (A \cup B)'.$$

Let $s \in (A \cup B)'$ so that $N(s) \cap (A \cup B)$ is infinite for every $N(s)$.

Case 1. $N(s) \cap A$ is infinite for every $N(s)$. Then $s \in A' \subset A' \cup B'$. Hence $(A \cup B)' \subset A' \cup B'$.

Case 2. There is an $N_1(s)$ such that $N_1(s) \cap A$ is finite. Then $N(s) \cap B$ is infinite for every $N(s)$ and $s \in B'$. In this case $(A \cup B)' \subset B' \subset A' \cup B'$.

Therefore $A' \cup B' = (A \cup B)'$ and then

$$\begin{aligned}\bar{A} \cup \bar{B} &= (A \cup A') \cup (B \cup B') \\ &= (A \cup B) \cup (A' \cup B') \\ &= (A \cup B) \cup (A \cup B)' = \overline{A \cup B}. \blacksquare\end{aligned}$$

PROBLEM 4. For any two sets A and B,

$$(A \cap B)' \subset A' \cap B', \qquad \overline{A \cap B} \subset \bar{A} \cap \bar{B}$$

and the inclusions may be proper.

Theorem 4. *A set is closed if and only if it is equal to its closure.*

Proof. Let A be a closed set. Then $A' \subset A$ so that $\bar{A} = A \cup A' = A$.

Let A be such that $A = \bar{A}$. Then $A' - A = A' - \bar{A} = \varnothing$ since $A' \subset \bar{A}$. Thus $A' \subset A$ and A is closed. $\blacksquare$

PROBLEM 5. For any set A, the sets A' and $\bar{A}$ are closed. [*Hint:* Let $s \in (A')'$ and take any $N(s)$. Select a $t \in N_0(s) \cap A'$, then an $N(t) \subset N(s)$, and show that $N(t) \cap A$ is infinite, etc.]

Theorem 5. *If A_1 and A_2 are closed sets, then $A_1 \cup A_2$ is a closed set.*

Proof. From Theorems 3 and 4

$$A_1 \cup A_2 = \bar{A}_1 \cup \bar{A}_2 = \overline{A_1 \cup A_2} \quad \text{is closed. } \blacksquare$$

Corollary. *The union of a finite number of closed sets is closed.*

EXAMPLE ($S = E^1$). With $A_n = [1/n, 1]$, $n = 1, 2, 3, \cdots$,

$$\cup A_n = (0, 1]$$

is not closed. Hence the union of a countably infinite collection of closed sets need not be closed.

Theorem 6. *If p is a cluster point of a set A, then A has a countably infinite subset having p as its only cluster point.*

Proof. For each $n \in J$ let

$$A_n = A \cap N\left(p, \frac{1}{n}\right)$$

and note that each of these sets is infinite. Select $a_1 \in A_1$, then $a_2 \in A_2 - \{a_1\}, \cdots$, $a_{n+1} \in A_{n+1} - \{a_1, a_2, \cdots, a_n\}, \cdots$. This is possible for each $n \in J$ since after a_1,

$a_2, \cdots, a_n$ have been selected, then $A_{n+1} - \{a_1, a_2, \cdots, a_n\}$ is infinite (and hence $\neq \varnothing$). Then

$$B = \{a_1, a_2, a_3, \cdots\} \subset A$$

is countably infinite since $a_i \neq a_j$ if $i \neq j$.

First $p \in B'$, since for any number $r > 0$ the set $B \cap N(p, r)$ is infinite; in fact it contains $\{a_n, a_{n+1}, \cdots\}$ whenever $n > 1/r$.

Next p is the only cluster point of B. For if $q \neq p$, then with $r = \frac{1}{2}d(p, q) > 0$, the set $B \cap N(q, r)$ does not contain a_n for $n > 1/r$ and is hence finite so that $q \notin B'$. ∎

Definition 4. *A set is said to be* **bounded** *if there is some $s \in S$ and a $N(s)$ such that $A \subset N(s)$.*

The union of a finite number of neighborhoods $N(s_1, r_1)$, $N(s_2, r_2), \cdots$, $N(s_n, r_n)$ is bounded. To see this, let

$$r = \max[d(s_1, s_k) + r_k, \ k = 1, 2, \cdots, n].$$

Then, with $p \in N(s_k, r_k)$,

$$d(s_1, p) \leqslant d(s_1, s_k) + d(s_k, p) < d(s_1, s_k) + r_k \leqslant r$$

for $k = 1, 2, \cdots, n$ and hence $p \in N(s_1, r)$. Hence

$$\bigcup_{k=1}^{n} N(s_k, r_k) \subset N(s_1, r).$$

Clearly, if $B \subset A$ and A is bounded, then B is bounded.

Theorem 7. *Any unbounded set has a countably infinite subset having no cluster point.*

Proof. Let A be an unbounded set, take a point s, and set

$$A_n = A \cap [N(s, n)]^c, \qquad n = 1, 2, 3, \cdots.$$

A_n is infinite for each $n \in J$, since A is unbounded. The inductive selection process then yields a countably infinite set

$$B = \{a_1, a_2, \cdots\} \subset A$$

having no two elements the same and $a_n \in A_n$ for $n \in J$. Also, for any point q, $N(q, 1) \subset N(s, d(s, q) + 1)$ and thus $N(q, 1)$ does not include a_n whenever $n > d(s, q) + 1$. Thus $A \cap N(q, 1)$ is finite, q is not a cluster point of B, and hence B has no cluster point. ∎

Theorem 8. *Let A be a set having the property: Every infinite subset of A has at least one cluster point in A.*

Then A is closed and bounded.

Proof. Let p be any cluster point of A and select a countably infinite set $B \subset A$ having p as its only cluster point (possible by Theorem 6). By the given property of A, B has at least one cluster point in A. Since p is the only cluster point of B, then $p \in A$. Thus $p \in A' \Rightarrow p \in A$; that is, A is closed.

Next A is bounded. Assume A is not bounded and on the basis of this select a countably infinite set $B \subset A$ having no cluster point (by Theorem 7). Since B has no cluster point at all, it has none in A, thus contradicting the given property of A. ∎

Theorem 9. *Let F be a nonempty closed subset of a metric space S and $a \in S$ but $a \notin F$. Then there is a number $\delta > 0$ such that $b \in F \Rightarrow d(a, b) > \delta$.*

Proof. Assume this is not so. Then for $n = 1, 2, 3, \cdots$, there is a $b_n \in F$ such that $d(a, b_n) < 1/n$. Let $r > 0$ be arbitrary. Then

$$b_n \in N_0(a, r) \qquad \text{whenever } 1/n < r$$

so that $a \in \{b_1, b_2, b_3, \cdots\}' \subset F' \subset F$, contrary to $a \notin F$. ∎

2–5 OPEN SETS

The setting remains in a metric space S.

Definition 1. *Let A be a set.*

(i) *A point s is said to be an* **interior point** *of A if there is a $N(s)$ such that $N(s) \subset A$.*
(ii) *$A^\circ = \{s \mid s \text{ is an interior point of } A\}$ is called the* **interior** *of A.* [Hence $A^\circ \subset A$].
(iii) *A is said to be* **open** *if $A^\circ = A$.*

Thus $\varnothing$ has no interior point; $\varnothing^\circ = \varnothing$; $\varnothing$ is open.
S is open since $N \subset S$ for every neighborhood N.
Hence both $\varnothing$ and S are open and they are both closed (as stated after Definition 2-4.2).

Theorem 1. *Any neighborhood is an open set.*

Proof. Let $N = N(s)$ be a neighborhood. If $t \in N(s)$, then $\exists N(t) \subset N(s)$ by Theorem 2-4.1(b), that is,

$$(t \in N \Rightarrow t \in N^\circ) \Rightarrow N \subset N^\circ.$$

Thus $N = N^\circ$, since any set is a superset of its interior. ∎

In E^1, for example, the open interval $(0, 1)$ is an open set. However, $(0, 1]$ is not open (because 1 is not an interior point) and is not closed (because $(0, 1]' = [0, 1] \not\subset (0, 1]$). Hence there are sets which are neither open nor closed.

Theorem 2. *A set is open if and only if its complement is closed.*

Proof. Let A be any open set. Assume that the complement A^c is not closed. Then $(A^c)' \not\subset A^c$ so that $(A^c)' - A^c \neq \varnothing$, and this may be written $(A^c)' \cap A \neq \varnothing$. Select a point

$$p \in (A^c)' \cap A = (A^c)' \cap \overset{\circ}{A}.$$

But

$$p \in (A^c)' \Rightarrow N_0(p, r) \cap A^c \neq \varnothing \qquad \text{for all } r > 0, \qquad \text{while}$$
$$p \in \overset{\circ}{A} \Rightarrow \exists r > 0 \quad \text{such that } N(p, r) \subset A \quad \text{so } N(p, r) \cap A^c = \varnothing.$$

This contradiction shows that A^c is closed.

Let A be a set whose complement A^c is closed. If $A = \varnothing$ (hence open), then $A^c = S$ is closed so consider $A \neq \varnothing$. Select $a \in A$. Then there is an $r_0 > 0$ such that $N(a, r_0) \subset A$; for otherwise $N_0(a, r) \cap A^c \neq \varnothing$ for all $r > 0$ thus making $a \in (A^c)' \subset A^c$ contrary to $a \in A$. Since a was any element of A, A is open. ∎

Theorem 3. *The intersection of two open sets is open.*

Proof. Let A and B be open sets. Then A^c and B^c are closed (by Theorem 2), $A^c \cup B^c$ is closed (by Theorem 2-4.5), and hence $(A^c \cup B^c)^c = A \cap B$ is open (by Theorem 2). ∎

Corollary. *The intersection of a finite number of open sets is open.*

EXAMPLE (in E^1). The countably infinite family $\{A_n\}$ of open sets $A_n = (-1/n, 1/n)$, $n = 1, 2, 3, \cdots$ has

$$\bigcap A_n = \{0\}$$

which is not open.

Theorem 4. (a) *The union of any family of open sets is open.*
(b) *The intersection of any family of closed sets is closed.*

Proof. Take any family $\mathscr{G}$ of open sets whose union

$$\bigcup_{G \in \mathscr{G}} G$$

is $\neq \varnothing$ (to avoid triviality) and let p be any point of this union. Let G_p be a member of the family such that $p \in G_p$. Since G_p is open, take a neighborhood $N(p) \subset G_p$. Then

$$N(p) \subset G_p \subset \bigcup_{\mathscr{G}} G.$$

Hence every point of the union is an interior point of the union and the union is open.

Then (b) follows by using Theorem 2 and de Morgan's rule. ∎

PROBLEM 1. The Cantor set is a perfect, nowhere-dense set.

Lemma (for $S = E^1$). *Let $G \neq \varnothing$ be an open subset of* $[0, 1]$ *and take an $x \in G$. Then there are two numbers a_x and b_x such that $a_x \notin G$, $b_x \notin G$ but the open interval*

$$(a_x, b_x) \subset G.$$

Proof. First, since $x \in G = G^\circ$, let $r > 0$ be such that

$$(x - r, x + r) \subset G.$$

Next define the set A_x by

$$A_x = \{t \mid (t, x + r) \subset G\}.$$

Then $A_x \neq \varnothing$ (for example, $x - r \in A_x$) and A_x is bounded below by zero, since $G \subset [0, 1]$. Hence, inf A_x exists and we set

$$a_x = \inf A_x.$$

Then $a_x \notin G$. For suppose $a_x \in G = G^\circ$. Let h be such that $0 < h < r$ and $(a_x - h, a_x + h) \subset G$. Then $a_x + \frac{1}{2}h \in A_x$ so that

$$(a_x - h, x + r) = (a_x - h, a_x + h) \cup (a_x + \tfrac{1}{2}h, x + r) \subset G,$$

making $a_x - h \in A_x$ with $a_x - h < \inf A_x$. This contradiction shows that $a_x \notin G$.

Since $x - r \in A_x$, then $a_x \leqslant x - r$.

Also $(a_x, x + r) \subset G$. There could be doubt about this only in case $a_x < x - r$. In this case let $r_1, r_2, r_3, \cdots$ be all the rational numbers on $(a_x, x - r)$. Then each $r_n \in A_x$ so that $(r_n, x + r) \subset G$ and

$$(a_x, x + r) = \bigcup_{n=1}^{\infty} (r_n, x + r) \subset G.$$

In a similar way, b_x may be determined such that

$$b_x \notin G \qquad \text{but} \qquad (x - r, b_x) \subset G.$$

Then $(a_x, b_x) = (a_x, x + r) \cup (x - r, b_x) \subset G$ show that the lemma is true. ∎

Definition 2. *If $G \subset E^1$ and $(a, b) \subset G$ with $a \notin G$ and $b \notin G$, then (a, b) is called an* **open component** *of G.*

Hence if (a, b) and (c, d) are open components of the same set G, then either

$$(a, b) = (c, d) \qquad \text{or else} \qquad (a, b) \cap (c, d) = \varnothing. \tag{1}$$

With Γ the Cantor set on $[0, 1]$, note that the open components of $[0, 1] - \Gamma$ are countably infinite.

Theorem 5 (for $S = E^1$). *Let $G \neq \varnothing$ be an open subset of* $[0, 1]$. *Then there is a countable collection of open intervals $I_1, I_2, I_3, \cdots$ having the properties*

(a) $I_i \cap I_j = \varnothing$ for $i \neq j$.
(b) $\cup I_i = G$.

Proof. Consider a component (a, b) of G. This open interval contains rational numbers; pick one and only one of them to represent (a, b). Do this for each component of G. Then different components have different representatives because of (1). The set of representatives is countable (being a subset of the rational numbers) and hence the family of components is countable. The components may then be ordered as $I_1, I_2, \cdots$ satisfying (a).

Let x be an element of G. Then, by the above lemma, x is in a component of G and hence $x \in I_i$ for some $i \in J$. Thus $x \in \cup I_n$. Hence $G \subset \cup I_n$. But each $I_n \subset G$ so $\cup I_n \subset G$ and thus (b) holds. ∎

[A definite procedure for picking representatives and ordering the components is to first display all rationals (each in simplest form) as a sequence $r_1, r_2, r_3, \cdots$. Then $\{n \mid r_n \in G\}$ is a nonempty set of positive integers and has a smallest integer. Let I_1 be the component containing the r with subscript this smallest integer. If $G - I_1 \neq \varnothing$, then $\{n \mid r_n \in G - I_1\}$ is nonempty, etc.]

PROBLEM 2. If G and H are nonempty open subsets of $(0, 1)$ such that $G \cup H = (0, 1)$, then $G \cap H \neq \varnothing$.

The following problem involves a special case of results to be given in the next section. The more general setting will be easier to understand if the problem is thoroughly understood and an airtight proof written out.

PROBLEM 3 (in E^1). Let $\mathscr{G}$ be a family of open sets having the property

$$[0, 1] \subset \bigcup_{G \in \mathscr{G}} G.$$

Then there is a finite number $G_1, G_2, \cdots, G_n$ of elements of $\mathscr{G}$ such that

$$[0, 1] \subset \bigcup_{k=1}^{n} G_k.$$

[*Hint:* Let A be the set defined by

$$A = \{x \mid 0 < x < 2 \text{ and } [0, x] \subset \text{union of a finite number of sets in } \mathscr{G}\}.$$

Careful reasoning will show that $\sup A > 1$ and hence $1 \in A$.]

The above lemma, Theorem 5, and Problem 2 could, of course, be generalized by replacing $[0, 1]$ by any closed interval $[a, b]$ with $a < b$.

The following definition gives another situation in which the notion of sets of sets is used.

Definition 3. *We use F_σ to denote the set of sets defined by the stipulation that $A \in F_\sigma$ iff A is a set expressible as a countable union of closed sets.*

Also G_δ is the set (family) of sets, each expressible as a countable intersection of open sets.

For example,

$$(0, 1] = \bigcup_n \left\{x \,\middle|\, \frac{1}{n} \leqslant x \leqslant 1\right\}$$
$$= \bigcap_n \left\{x \,\middle|\, 0 < x < 1 + \frac{1}{n}\right\}$$

shows that the open-closed interval (0, 1] is an element of F_σ and also is an element of G_δ.

We shall usually use the terminology "A is an F_σ set" rather than the notation $A \in F_\sigma$.

PROBLEM 4. In any metric space:

(a) If A is closed, then A is a G_δ set.
[*Hint:* Set $H_n = \bigcup_{x \in A} N(x, 1/n)$ and $\cap H_n$.]

(b) A set A is a G_δ set iff A^c is an F_σ set.

(c) Every closed set and every open set is both an F_σ set and a G_δ set. (In particular the whole space is both an F_σ set and a G_δ set.)

(d) Every F_σ set may be expressed as the countable union of a monotonically increasing sequence of closed sets.

(e) The countable union of F_σ sets is an F_σ set.

Later on (Problem 2-8.1) it will be seen that (in E^1) the set of rationals is an F_σ set, but not a G_δ set. There are, moreover, sets which are neither F_σ nor G_δ sets—this will be an easy consequence of results in Chapter 5.

2-6 COMPACT SETS

Unless otherwise restricted, the discussion is for a metric space S.

Definition 1. *With A a set and $\mathscr{F}$ a family of sets:*

(i) *$\mathscr{F}$ is said to* **cover** *A (or be a covering of A) if*

$$A \subset \bigcup_{F \in \mathscr{F}} F;$$

that is, *if each point of A is a point of at least one of the sets of $\mathscr{F}$.*

(ii) *$\mathscr{F}$ is said to be an* **open cover** *of A if each member of $\mathscr{F}$ is an open set and $\mathscr{F}$ covers A.*

(iii) *If $\mathscr{F}$ covers A, then any subfamily of $\mathscr{F}$ which also covers A is called a* **subcover** *of A.*

In this terminology, Problem 2-5.3 (generalized to an interval $[a, b]$) could be stated:

Any open cover of a closed interval $[a, b]$ has a finite subcover.†

† This is the essence of the classical Heine-Borel theorem.

COMPACT

Definition 2. *A set K is said to be* **compact** *if every open cover of K has a finite subcover.*

Now the above generalization of Problem 2-5.3 may be stated as:

Every closed interval in E^1 is a compact set.

EXAMPLE (in E^1). The set $A = (0, 1]$ is not compact. For let $\mathscr{G}$ be the family defined by $G_n \in \mathscr{G}$ if

$$G_n = \left(\frac{1}{n}, 2\right), \qquad n = 1, 2, 3, \cdots.$$

Then $A \subset \cup G_n$. But consider a finite subfamily of $\mathscr{G}$ and among the members let G_{n_0} be the one with largest subscript. This finite subfamily therefore does not cover any point of $(0, 1/n_0] \subset A$ and hence does not cover A.

Theorem 1. *Every compact set in a metric space is bounded and closed.*

Proof. Let K be a compact set.

Take $\mathscr{F}$ the family of neighborhoods of the form $N(p, 1)$, $p \in K$. This is an open cover of K and thus has a finite subcover. With p_k, $k = 1, 2, \cdots, n$ the centers of the members of this subcover, then

$$K \subset \bigcup_{k=1}^{n} N(p_k, 1).$$

This union is a bounded set and thus K is bounded.

If the complement $K^c = \varnothing$, then $K = S$ is closed so consider $K^c \neq \varnothing$. Select any point $q \in K^c$. It will be shown that q is an interior point of K^c so that K^c is open and K closed. Around each point $p \in K$ put the neighborhood

$$N(p, \rho(p)) \qquad \text{where} \qquad \rho(p) = \tfrac{1}{2}d(p, q) > 0 \qquad (\text{since } p \neq q).$$

The family of all such neighborhoods is an open cover of K and thus has a finite subcover; that is, there are points $p_1, p_2, \cdots, p_n$ of K such that

$$K \subset \bigcup_{k=1}^{n} N(p_k, \rho(p_k)).$$

Let $r = \min[\rho(p_k), k = 1, 2, \cdots, n]$ and take the neighborhood $N(q, r)$. Then

$$N(q, r) \cap K \subset N(q, r) \cap \left[\bigcup_{k=1}^{n} N(p_k, \rho(p_k))\right] = \varnothing,$$

since $s \in N(q, r)$ implies, for $k = 1, 2, \cdots, n$, that

$$2\rho(p_k) = d(q, p_k) \leqslant d(q, s) + d(s, p_k) < r + d(s, p_k)$$
$$\leqslant \rho(p_k) + d(s, p_k),$$

which implies $d(s, p_k) > \rho(p_k)$; that is, $s \notin N(p_k, \rho(p_k))$. Hence $N(q, r) \subset K^c$, and therefore K^c is open so that K is closed. ∎

Theorem 2. *In a metric space every closed subset of a compact set K is compact.*

Proof. Let $A \subset K$ with A closed and take any open covering $\mathcal{G}$ of A. Then the family $\mathcal{G}$ augmented by the single open set A^c is an open covering of K (even of S). This open covering has a finite subcover of K and, after removing A^c, if it were used, the resulting finite subfamily of $\mathcal{G}$ covers A. Hence A is compact. ∎

PROBLEM 1. If K is compact and F is closed, then $K \cap F$ is compact.

PROBLEM 2. The union of a finite number of compact sets is compact.

PROBLEM 3. Give an example of a noncompact set which is the union of an infinite sequence of compact sets.

Theorem 3. *In a metric space every infinite subset of a compact set K has a cluster point (which is a point of K, since K is closed).*

Proof. Let $A \subset K$ be such that $A' = \varnothing$. Around each point $p \in K$ put a neighborhood $N(p)$ such that $A \cap N(p)$ is finite. The family of all such neighborhoods is an open cover of K. Since K is compact, let $p_1, p_2, \cdots, p_n$ be such that

$$K \subset \bigcup_{i=1}^{n} N(p_i) \qquad \text{and hence} \qquad A = A \cap K \subset \bigcup_{i=1}^{n} [A \cap N(p_i)].$$

Thus A is finite.

Consequently, any infinite set $A \subset K$ has $A' \neq \varnothing$. ∎

Any monotonically decreasing sequence $A_1 \supset A_2 \supset A_3 \supset \cdots$ of nonempty sets is such that any two, any three, $\cdots$, any $n \in J$ of them, selected anywhere from the sequence, have at least one point in common. Figure 1 illustrates a family of three sets such that any two have nonempty intersection, but the intersection of all three is empty.

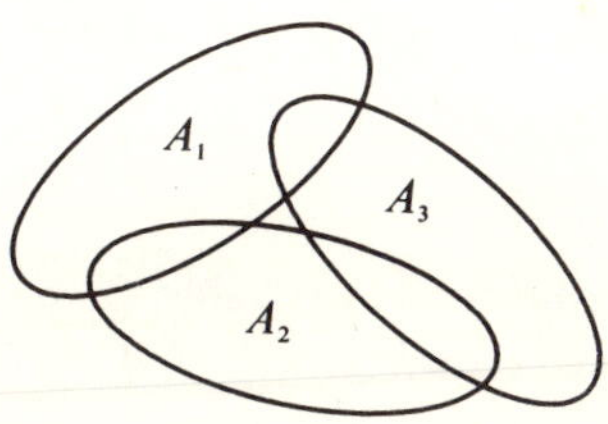

FIG. 1

Definition 3. *A family of sets is said to have the* **finite intersection property** *if every subfamily of any $n \in J$ members has nonempty intersection; that is, every finite nonempty subfamily has nonempty intersection.*

EXAMPLE (in E^1). Given $A_n = \{x \mid 0 < x \leqslant 1/n\}$, $n = 1, 2, 3, \cdots$, then the family $\{A_n\}$ has the finite intersection property, but not an "infinite intersection" property; that is,

$$\bigcap_{n=1}^{\infty} A_n = \varnothing.$$

Theorem 4. *In the metric space S, let $\mathscr{K}$ be a family of compact sets having the finite intersection property. Then*

$$\bigcap_{K \in \mathscr{K}} K \neq \varnothing.$$

Proof. (For convenience $K \in \mathscr{K}$ will be deleted from $\bigcap_{K \in \mathscr{K}}$).

Assume $\bigcap K = \varnothing$. Then $(\bigcap K)^c = S$. Choose any member of $\mathscr{K}$ and label it K_0. Then $K_0 \subset S = (\bigcap K)^c = \bigcup K^c$; that is, the family of complements of members of $\mathscr{K}$ is an open covering of the compact set K_0 ("open covering" since each $K \in \mathscr{K}$ is closed). Hence there is a finite number $K_1, K_2, \cdots, K_n$ of members of $\mathscr{K}$ such that

$$K_0 \subset \bigcup_{k=1}^{n} K_k^c = \left(\bigcap_{k=1}^{n} K_k\right)^c.$$

Thus

$$\varnothing = K_0 \cap \left[\bigcap_{k=1}^{n} K_k\right] = \bigcap_{k=0}^{n} K_k,$$

which violates the finite intersection property of $\mathscr{K}$. ∎

Corollary 1. *If $K_1 \supset K_2 \supset K_3 \supset \cdots$ are all nonempty compact sets, then*

$$\bigcap_{n=1}^{\infty} K_n \neq \varnothing.$$

This corollary will be used many times in the following work.

The next corollary follows easily from the fact that closed intervals in E^1 are compact.

Corollary 2 (in E^1). *For $n = 1, 2, 3, \cdots$, let $I_n = [a_n, b_n]$ be such that $I_1 \supset I_2 \supset I_3 \supset \cdots$. Then $\cap I_n \neq \varnothing$.*

PROBLEM 4. Prove the above corollary without relying upon the compactness of closed intervals in E^1.

Definition 4. *A set A is said to be* **separable** *if there is a countable set $D \subset A$ such that $\bar{D} \supset A$; that is, a countable subset of A which is dense in A.*

Thus a metric space S is **separable** if there is (in S) an everywhere-dense countable set. See Definition 2-4.3 (v).

PROBLEM 5. E^n is separable.

Theorem 5. *Every compact subset of a metric space is separable.*

Proof. Let K be a compact set. With $n \in J$, let $\mathscr{G}_n$ be the family of neighborhoods

$$N\left(x, \frac{1}{n}\right) \qquad \text{for all } x \in K.$$

Then $\mathscr{G}_n$ is an open covering of K and hence there is a finite subset $D_n = \{x_{n1}, x_{n2}, \cdots, x_{nm_n}\}$ of K such that

$$K \subset \bigcup_{i=1}^{m_n} N\left(x_{ni}, \frac{1}{n}\right).$$

Since D_n is a finite subset of K for each $n \in J$, then

$$D = \bigcup D_n$$

is a countable subset of K. Thus, since K is closed,

$$\bar{D} = D \cup D' \subset K. \tag{1}$$

To see that D is dense in K, take any $y \in K$ and any neighborhood $N(y, r)$ of y. Select $n \in J$ such that $1/n < r$. Then there is an i among $1, 2, \cdots, m_n$ such that

$$y \in N\left(x_{ni}, \frac{1}{n}\right).$$

Hence $y \in D$ or $y \in D'$. Thus $y \in K \Rightarrow y \in D \cup D' = \bar{D}$; that is, $K \subset \bar{D}$. Therefore, from (1), $K = \bar{D}$. ∎

Theorem 6. *Given a nonempty compact set $F \subset E^1$, then F has a smallest element; that is, there is an $a \in F$ such that $b \in F \Rightarrow a \leqslant b$.*

Proof. Assume F has no smallest element. For each $a \in F$ let $H_a = \{x \mid x > a\}$. Then the family $\{H_a\}$ is an open covering of F, since for $b \in F$ there is an $a \in F$ with $a < b$ and $b \in H_a$. Thus there are a finite number of points $a_1, a_2, \cdots, a_n$ of F such that

$$\bigcup_{i=1}^{n} H_{a_i} \supset F. \tag{2}$$

Choose $a_0 \in F$ with $a_0 < \min(a_1, a_2, \cdots, a_n)$, which is possible by assumption. Then a_0 is not an element of the union in (2), which contradicts (2). ∎

Corollary. *Any nonempty closed subset of E^1 which is bounded below (or above) has a smallest (respectively largest) element.*

Theorem 7. *Let G and H be open sets in E^1 such that $G \cap H = \varnothing$ and $G \cup H$ includes a closed interval $[a, b]$ with $a < b$. Then either $[a, b] \subset G$ or $[a, b] \subset H$.*

Proof. The end point a is in one of the sets, say $a \in G$. Then there is an $a_1 > a$ such that $(a, a_1) \subset G$.

Assume $[a, b] \cap H \neq \varnothing$. Then also $(a, b) \cap H$ is a nonempty open set and $[a, b] \cap \bar{H}$ is a nonempty compact set. Let c be the smallest element of $[a, b] \cap \bar{H}$. Then c is not less than a_1 and is not greater than any element of $(a, b) \cap H$.

Hence $a < c < b$ so that $c \in G \cup H$ and also $c \in G' \cup H'$. But these two are incompatible, since

$$(G \cup H) \cap (G' \cup H') \subset (G \cap H') \cup (H \cap G') = \varnothing$$

because G and H are open with $G \cap H = \varnothing$. Thus $[a, b] \cap H = \varnothing$ and hence $[a, b] \subset G$. ▮

2–7 PROPERTIES IN E^k

In this section primary concern is in Euclidean k-space E^k. Hence a point $\mathbf{x}$ may be expressed as an ordered k-tuple,

$$\mathbf{x} = (x_1, x_2, \cdots, x_k),$$

of real numbers. The distance between $\mathbf{x}$ and $\mathbf{y} = (y_1, y_2, \cdots, y_k)$ is therefore such that, for $i = 1, 2, 3, \cdots, k$,

$$|x_i - y_i| \leqslant \left[\sum_{j=1}^{k} (x_j - y_j)^2\right]^{1/2} = d(\mathbf{x}, \mathbf{y}).$$

The following definition generalizes the notion of intervals in E^1 and rectangles in E^2.

Definition 1. *In E^k a set R is called a* **k-cell** *if there are intervals $I_1, I_2, \cdots, I_k$ in E^1 such that*

$$R = I_1 \times I_2 \times \cdots \times I_k$$

and each I_i, $i = 1, 2, \cdots, k$ is then called an **edge** *of R. R is said to be a* **closed k–cell** *or an* **open k–cell** *if all its edges are closed intervals or all are open intervals, respectively. If a_i and b_i ($a_i \leqslant b_i$) are the end points of the ith edge, $i = 1, 2, \cdots, n$, then R is said to have* **diagonal length**

$$\left[\sum_{i=1}^{n} (a_i - b_i)^2\right]^{1/2}.$$

In the notation of the definition, if $\mathbf{x} \in R$ and $\mathbf{y} \in R$, then $a_i \leqslant x_i \leqslant b_i$, $a_i \leqslant y_i \leqslant b_i$, and $|x_i - y_i| \leqslant b_i - a_i$, $i = 1, 2, 3, \cdots, k$, so that

$$d(\mathbf{x}, \mathbf{y}) = \left[\sum_{i=1}^{k} (x_i - y_i)^2\right]^{1/2} \leqslant \text{diagonal length of } R.$$

PROBLEM 1. A closed k-cell is a closed set and an open k-cell is an open set.

A neighborhood $N(\mathbf{c}, r)$ will be referred to as a ***k*-ball** (with center $\mathbf{c}$ and radius r). Note that a k-ball is an open set, but a k-cell may be open, or closed, or neither. A set may be empty, but there is no empty k-ball and no empty k-cell. Also

$$N(\mathbf{c}, r) \subset R$$

if R has edges $(c_i - r, c_i + r)$, $i = 1, 2, \cdots, k$. This follows, since

$$\begin{aligned} \mathbf{x} \in N(\mathbf{c}, r) &\Rightarrow d(\mathbf{c}, \mathbf{x}) < r \\ &\Rightarrow |x_i - c_i| \leqslant d(\mathbf{c}, \mathbf{x}) < r, \qquad i = 1, 2, \cdots, k \\ &\Rightarrow c_i - r < x_i < c_i + r, \qquad i = 1, 2, \cdots, k. \end{aligned}$$

On the other hand, if R has edges $[a_i, b_i]$ for $i = 1, 2, \cdots, k$, then

$$R \subset N(\mathbf{a}, r), \qquad \text{where } r = d(\mathbf{a}, \mathbf{b}) + 1.$$

Hence:

A set A is bounded iff there is a k-cell $R \supset A$.

Theorem 1. *If $R_1 \supset R_2 \supset R_3 \supset \cdots$ are all closed k-cells, then*

$$\bigcap_{n=1}^{\infty} R_n \neq \varnothing.$$

Proof. For $n = 1, 2, 3, \cdots$, let $I_{n1}, I_{n2}, \cdots, I_{nk}$ be closed intervals in E^1 such that

$$R_n = I_{n1} \times I_{n2} \times \cdots \times I_{nk}.$$

Then for each $i = 1, 2, \cdots, k$ it follows that $I_{1i}, I_{2i}, I_{3i}, \cdots$ is a monotonically decreasing sequence of closed intervals which therefore has nonempty intersection (by Corollary 2, p. 82) and we let x_i^* be a number such that

$$x_i^* \in \bigcap_{n=1}^{\infty} I_{ni}.$$

Hence the point $\mathbf{x}^* = (x_1^*, x_2^*, \cdots, x_k^*)$ is such that

$$\mathbf{x}^* \in \bigcap_{n=1}^{\infty} [I_{n1} \times I_{n2} \times \cdots \times I_{nk}] = \bigcap_{n=1}^{\infty} R_n. \blacksquare$$

Theorem 2. *Every closed k-cell is compact.*

Proof. Assume there is a closed k-cell R which is not compact. Let $\mathscr{G}$ be an open cover of R which has no finite subcover.

For $i = 1, 2, \cdots, k$, let $[a_i, b_i]$ be the edges of R and set $c_i = \frac{1}{2}(b_i + a_i)$. Form sub k-cells of R by choosing k edges, one from each of the pairs

$$[a_1, c_1], [c_1, b_1]; \quad [a_2, c_2], [c_2, b_2]; \cdots; \quad [a_k, c_k], [c_k, b_k]$$

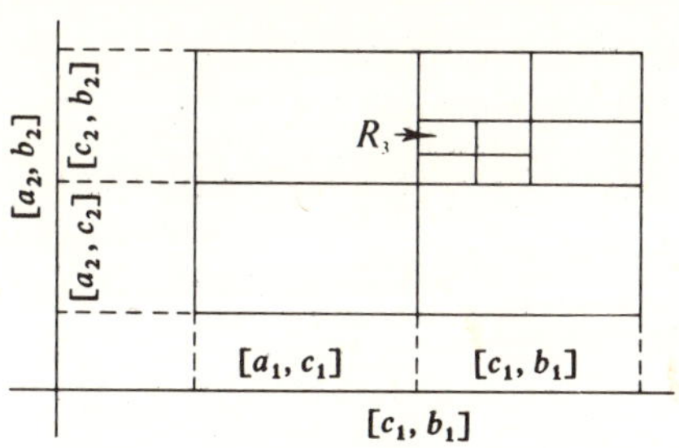

FIG. 1

and do so in all possible combinations. This yields 2^k k-cells whose union is R. At least one of these sub k-cells has no finite cover (in $\mathscr{G}$), for if each could be finitely covered, then the union of all 2^k of them (namely R) could be finitely covered. Let R_1 be one of these nonfinitely covered sub k-cells of R. With δ the diagonal length of R, then R_1 has

$$\text{diagonal length} \leqslant \left[\sum_{i=1}^{n} \left(\frac{b_i - a_i}{2}\right)^2\right]^{1/2} = \frac{1}{2}\delta.$$

Next perform the same process on R_1 that was performed on R, and let R_2 be the resulting closed k-cell.

By continuing the process, a sequence

$$R \supset R_1 \supset R_2 \supset R_3 \supset \cdots \tag{1}$$

of closed k-cells is obtained such that, for $n = 1, 2, 3, \cdots$,

$$R_n \text{ is not covered by any finite subfamily of } \mathscr{G}, \text{ and} \tag{2}$$

$$(\mathbf{x} \in R_n \text{ and } \mathbf{y} \in R_n) \Rightarrow d(\mathbf{x}, \mathbf{y}) \leqslant 2^{-n}\delta. \tag{3}$$

Property (1) and Theorem 1 assure a point $\mathbf{x}^* \in \cap R_n \subset R$. Since $\mathscr{G}$ covers R, $\mathscr{G}$ has at least one member containing $\mathbf{x}^*$. Select such a member and label it G^* so that

$$\mathbf{x}^* \in G^*.$$

Since G^* is open, let $r > 0$ be such that $N(\mathbf{x}^*, r) \subset G^*$. Now choose $n^* \in J$ such that $2^{-n^*}\delta < r$. Since $\mathbf{x}^* \in R_n$ for every $n \in J$, then $\mathbf{x}^* \in R_{n^*}$. From (3),

$$\mathbf{y} \in R_{n^*} \Rightarrow d(\mathbf{x}^*, \mathbf{y}) \leqslant 2^{-n^*}\delta < r; \qquad \text{that is, } R_{n^*} \subset N(\mathbf{x}^*, r) \subset G^*.$$

Thus R_{n^*} is covered by the subfamily of $\mathscr{G}$ consisting of the single member G^*, in contradiction to (2). ∎

Theorem 3 (Weierstrass-Bolzano). *Every bounded infinite set in E^k has at least one cluster point.*

Proof. Let A be a bounded infinite set. From boundedness alone, take a closed k-cell $R \supset A$. Then R is compact (Theorem 2) and now A is an infinite subset of a compact set so that $A' \neq \varnothing$ by Theorem 2-6.3. ∎

Theorem 4 (Heine-Borel). *Every bounded closed set $A \subset E^k$ is compact.*

Proof. Take a closed, hence compact, k-cell $R \supset A$. Then A is a closed subset of a compact set and is thus compact by Theorem 2-6.2. ∎

The following theorem correlates some previously proved results.

Theorem 5. *If a set $A \subset E^k$ has any one of the properties:*

(a) *A is closed and bounded, or*
(b) *A is compact, or*
(c) *Every infinite subset of A has a cluster point in A,*
then it has the other two as well.

Proof. Each of the following implications is a restatement of the theorem cited:

$$\begin{aligned} &(a) \Rightarrow (b) \qquad \text{(Theorem 4).} \\ &(b) \Rightarrow (c) \qquad \text{(Theorem 2-6.3).} \\ &(c) \Rightarrow (a) \qquad \text{(Theorem 2-4.8).} \end{aligned}$$

Thus (a) ⇒ (b) ⇒ (c) ⇒ (a) and the theorem is proved. ▮

[Note that the link (a) ⇒ (b) is the only one in this chain requiring† $A \subset E^k$.]

Theorem 6. *The Cantor set* Γ *contains all distances from* 0 *to* 1; *that is, given a number* $0 < b \leqslant 1$, *there are points* $c_1 \in \Gamma$ *and* $c_2 \in \Gamma$ *such that*

(4) $$c_2 - c_1 = b.$$

Proof. (Intuitive‡). Visualize everything in the ordinary analytic geometry way. Start with the closed square [0, 1] × [0, 1], remove all except the closed squares of edge lengths $\frac{1}{3}$ in each corner, and let A_1 be the union of these four squares. Then the line $y = x + b$ intersects A_1 in a nonempty bounded closed set K_1, which is thus compact. In each of the four squares of A_1 take the closed

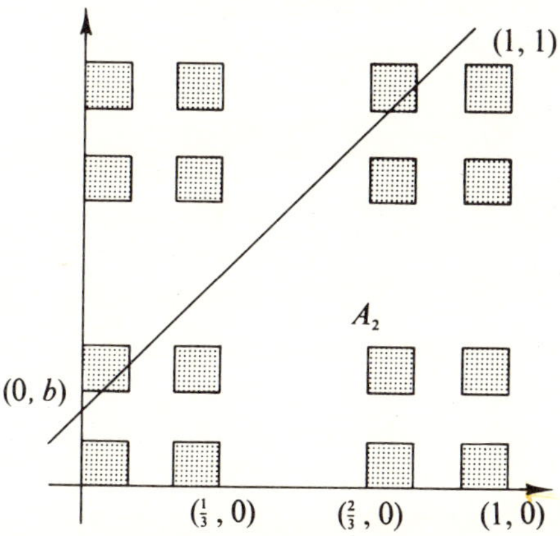

FIG. 2

squares of edge length $\frac{1}{9}$ at each corner and let A_2 be the union of these 16 squares. Then the line $y = x + b$ intersects A_2 in a nonempty compact set $K_2 \subset K_1$. Continuing in this way, sets

$$A_1 \supset A_2 \supset A_3 \supset \cdots \qquad \text{and} \qquad K_1 \supset K_2 \supset K_3 \supset \cdots$$

are obtained with $A_n \supset K_n \neq \varnothing$ and K_n compact for each $n \in J$. Thus, by using Corollary 1 of Theorem 2-6.4,

$$\varnothing \neq \bigcap K_n \subset \bigcap A_n = \Gamma \times \Gamma.$$

Let $(c_1, c_2) \in \bigcap K_n$. Then $x = c_1 \in \bigcap \Gamma$, and $y = c_2 \in \Gamma$ satisfy $y = x + b$; that is, $c_2 = c_1 + b$ from which (4) follows. ▮

† Even (a) ⇒ (b) in any separable metric space, but the proof is harder than in E^k.

‡ A proof is given in Section 2-9.

PROBLEM 2. Every point of (0, 1) is midway between two points of the Cantor set.

The following covering theorem is similar to the Heine-Borel theorem (Theorem 4), but has no restriction on the set $A \subset E^k$.

Theorem 7 (Lindelöf). *Given any set $A \subset E^k$ and an open covering $\mathscr{H}$ of A, then $\mathscr{H}$ contains a countable subcovering of A.*

Proof. Let $P = \{\mathbf{p} | \mathbf{p} = (p_1, p_2, \cdots, p_k)$, each p_i rational$\}$. Then P is countable and dense in E^k.

With $a \in A$, any $H \in \mathscr{H}$ such that $a \in H$ and let $h > 0$ be such that $N(a, h) \subset H$. Select a point $\mathbf{p}(a) \in P$ such that $d(a, \mathbf{p}(a)) < h/4$. Let $r(a)$ be a rational number such that $\frac{1}{2}h < r(a) < \frac{3}{4}h$. Then

$$N(\mathbf{p}(a), r(a)) \subset N(a, h) \subset H.$$

The family $\mathscr{G}$ of all such $N(\mathbf{p}(a), r(a))$ for all $a \in A$ is an open covering of A but moreover is countable, since each coordinate of each $\mathbf{p}(a)$ and each $r(a)$ is rational. Now, with each $N \in \mathscr{G}$, pick one and only one $H \in \mathscr{H}$ such that $H \supset N$. The collection of all such selected H is then a countable subfamily of $\mathscr{H}$ which covers A. ∎

PROBLEM 3. Given any nonempty open set $A \subset E^k$, then A is the union of a countable family of neighborhoods.

2–8 PERFECT SETS IN E^k

The empty set $\varnothing$, any closed interval $[a, b]$ with $a < b$, and the Cantor set are examples of perfect sets in E^1. The following result shows how "extensive" a nonempty perfect set must be in any space E^k, $k = 1, 2, \cdots$.

Theorem 1. *Any nonempty perfect set in E^k is noncountable.*†

Both a contradictive and a constructive proof will be given. Points of E^k will be denoted by $x, y, \cdots$ rather than by $\mathbf{x}, \mathbf{y}, \cdots$.

Proof. (Contradictive). Assume there is a nonempty perfect set P which is countable:

$$(1) \qquad P = \{x_1, x_2, x_3, \cdots\}, \qquad x_i \neq x_j \text{ for } i \neq j.$$

We first show how to select points $y_1, y_2, y_3, \cdots$ of P and positive numbers $r_1, r_2, r_3, \cdots$ such that

$$(2) \qquad (Ny_1, r_1) \supset N(y_2, r_2) \supset N(y_3, r_3) \supset \cdots$$

and such that for each $n \in J$,

$$(3) \qquad x_n \notin \overline{N}(y_n, r_n) = \{x \mid d(y_n, x) \leqslant r_n\}.$$

† In particular, any closed interval $[a, b]$ with $a < b$ is noncountable, and we have an alternate proof that *the set of real numbers is noncountable.*

Take $y_1 = x_2$ and $r_1 = \frac{1}{2}d(y_1, x_1)$. Then $r_1 > 0$, since $y_1 \neq x_1$. Also $x_1 \notin \bar{N}(y_1, r_1)$, but $y_1 \in P = P'$ so that

$$N(y_1, r_1) \cap P \text{ is infinite.}$$

Thus $\{i \mid x_i \in N(y_1, r_1) \cap P \text{ and } i \geqslant 3\}$ is a nonempty set of positive integers and we designate the smallest of these by m_2. Set

$$y_2 = x_{m_2}.$$

Then $y_2 \in P$, and since $m_2 \geqslant 3$, $y_2 \neq x_2 = y_1$ and $0 < d(y_2, y_1) < r_1$. Choose

$$r_2 = \min[\tfrac{1}{2}d(y_2, x_2) \text{ and } r_1 - d(y_2, y_1)].$$

Then $x_2 \notin \bar{N}(y_2, r_2)$ while $N(y_2, r_2) \subset N(y_1, r_1)$.

For $n \geqslant 3$, and with $y_{n-1} \in P$ and $r_{n-1} > 0$ already selected such that

$$x_{n-1} \notin \bar{N}(y_{n-1}, r_{n-1}) \subset N(y_{n-2}, r_{n-2}),$$

then $\{i \mid x_i \in P \cap N(y_{n-1}, r_{n-1}) \text{ and } i \geqslant n+1\}$ is nonempty, and we let m_n be its smallest integer. Set

$$y_n = x_{m_n}.$$

Then $y_n \in P$, $y_n \neq x_n$, since $m_n \geqslant n+1$, and $0 < d(y_n, y_{n-1}) < r_{n-1}$. Set

$$r_n = \min[\tfrac{1}{2}d(y_n, x_n), r_{n-1} - d(y_n, y_{n-1})].$$

Since $x \in N(y_n, r_n) \Rightarrow$

$$d(y_{n-1}, x) \leqslant d(y_{n-1}, y_n) + d(y_n, x) < (r_{n-1} - r_n) + r_n = r_{n-1},$$

then

$$N(y_{n-1}, r_{n-1}) \supset N(y_n, r_n).$$

If $x \in \bar{N}(y_n, r_n)$, then $d(x, y_n) \leqslant r_n$ so that

$$2r_n \leqslant d(x_n, y_n) \leqslant d(x_n, x) + d(x, y_n) \leqslant d(x_n, x) + r_n$$

and hence $d(x_n, x) \geqslant r_n > 0$, showing that $x_n \neq x$ and thus

$$x_n \notin \bar{N}(y_n, r_n).$$

Hence, by induction, this process yields points $y_1, y_2, y_3, \cdots$ of P and positive numbers $r_1, r_2, r_3, \cdots$ for which (2) and (3) hold. Thus

$$P \cap \bar{N}(y_n, r_n)$$

is closed and bounded, hence compact (since the space is E^k), and is nonempty (has at least y_n). Moreover

$$P \cap \bar{N}(y_1, r_1) \supset P \cap \bar{N}(y_2, r_2) \supset P \cap \bar{N}(y_3, r_3) \supset \cdots.$$

Thus

$$\bigcap_{n=1}^{\infty} [P \cap \bar{N}(y_n, r_n)]$$

is a nonempty subset of P (by Corollary 1 of Theorem 2-6.4), but is empty since it contains none of the points $x_1, x_2, x_3, \cdots$ which make up all of P by assumption. This finishes the contradictive proof. ∎

Let A be the set of all binary sequences. One way of thinking of this is to consider $\boldsymbol{\alpha} \in A$ as a "vector with countably many components" each either 0 or 1:

$$\boldsymbol{\alpha} = (\alpha_1, \alpha_2, \alpha_3, \cdots).$$

Let f: $A \to [0, 1]$ be defined by $f(\boldsymbol{\alpha}) = \underset{2}{0.}\alpha_1\alpha_2\alpha_3\cdots$. Since each x, $0 \leqslant x \leqslant 1$ has at least one binary representation, then f: $A \to [0, 1]$ is an onto mapping. Hence there is a subset of A which is equivalent to the noncountable set [0, 1]. The second (constructive) proof of Theorem 1 will use the result:

The set of all binary sequences is noncountable.

Let $P \neq \varnothing$ and $P \subset E^k$ be a perfect set, let $Q = \{p_1, p_2, \cdots, p_n\}$ be a set of $n \geqslant 2$ distinct points of P, and let Δ be such that

(4) $$0 < \Delta < \tfrac{1}{3} \min d(p_i, p_j), \qquad 1 \leqslant i < j \leqslant n.$$

We shall show how to pick out $2n$ more points in a particular way that will be used in the constructive proof we shall give of Theorem 1. [Figure 1 with $n = 2$ may help to visualize the situation.]

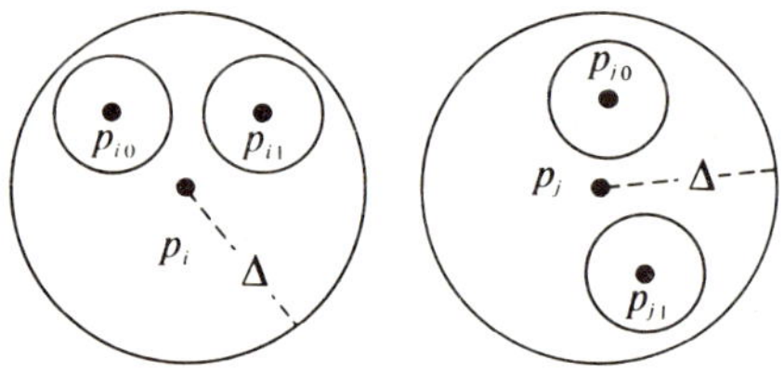

FIG. 1

First note from (4) that

(5) $$\overline{N}(p_i, \Delta) \cap \overline{N}(p_j, \Delta) = \varnothing, \qquad 1 \leqslant i < j \leqslant n;$$

(6) $$N_0(p_i, \Delta) \cap Q = \varnothing;$$

(7) $$N_0(p_i, \Delta) \cap P \text{ is infinite};$$

where (7) follows, since $p_i \in P = P'$. For $i = 1, 2, \cdots, n$, pick two points of $N_0(p_i, \Delta) \cap P$ and label them p_{io} and p_{i1}. Then

$$Q_1 = \{p_{10}, p_{11}, p_{20}, p_{21}, \cdots, p_{no}, p_{n1}\}$$

consist of $2n$ distinct points of P and $Q_1 \cap Q = \varnothing$.

Also, for $i = 1, 2, \cdots, n$, let

$$\delta_i = \tfrac{1}{3} \min[d(p_{io}, p_i), d(p_{i1}, p_i), \Delta - d(p_{io}, p_i), \Delta - d(p_{i1}, p_i)]$$

and then set $\delta = \min[\delta_1, \delta_2, \cdots, \delta_n]$. Then $\delta > 0$ and, in addition to (5), (6), and (7), the properties to be used later are

(8) $$\overline{N}(p_{io}, \delta) \subset N(p_i \Delta,), \overline{N}(p_{i1}, \delta) \subset N(p_i, \Delta),$$

(9) $$\overline{N}(p_{io}, \delta) \cap \overline{N}(p_{i1}, \delta) = \varnothing,$$

(10) $$\delta \leqslant \tfrac{1}{3}\Delta.$$

Proof. (Constructive). Select two points of P, label them

(11) $$x_0 \text{ and } x_1, \quad \text{and let } r_1 = \tfrac{1}{3}d(x_0, x_1).$$

With the set $\{x_0, x_1\}$ in place of Q and r_1 in place of Δ, select four points

(12) $$\{x_{00}, x_{01}, x_{10}, x_{11}\},$$

the first two in $N_0(x_0, r_1)$, the second two in $N_0(x_1, r_1)$, and use r_2 in place of δ above.

The third stage is to use the set with the points (12) as Q, and r_2 in place of Δ. The result is a selection of 2^3 new distinct points of P labeled systematically as

$$x_{000}, x_{001}, x_{010}, x_{011}, x_{100}, x_{101}, x_{110}, x_{111}.$$

Also, r_3 is obtained where

$$0 < r_3 \leqslant \frac{1}{3} r_2 \leqslant \frac{1}{3^2} r_1 \leqslant \frac{1}{3^3} d(x_0, x_1).$$

The process is to be continued by induction. At the nth stage, 2^n new distinct points of P are selected and labeled with the 2^n binary sequences of n digits. Then with $\alpha_1, \alpha_2, \cdots, \alpha_n$ and $\beta_1, \beta_2, \cdots, \beta_n$ two different sequences of n binary digits,

(13) $$\overline{N}(x_{\alpha_1\alpha_2\cdots\alpha_n}, r_n) \cap \overline{N}(x_{\beta_1\beta_2\cdots\beta_n}, r_n) = \varnothing;$$

but

$$\overline{N}(x_{\alpha_1\alpha_2\cdots\alpha_{n-1}0}, r_n) \qquad \text{and} \qquad \overline{N}(x_{\alpha_1\alpha_2\cdots\alpha_{n-1}1}, r_n)$$

also have empty intersection, although both are subsets of $N_0(x_{\alpha_1\alpha_2\cdots\alpha_{n-1}}, r_{n-1})$. Moreover,

$$0 < r_n \leqslant \frac{1}{3^n} d(x_0, x_1).$$

Now, with $\alpha_1, \alpha_2, \alpha_3, \cdots$ an infinite sequence of binary digits,

$$\overline{N}(x_{\alpha_1}, r_1) \supset \overline{N}(x_{\alpha_1\alpha_2}, r_2) \cap \overline{N}(x_{\alpha_1\alpha_2\alpha_3}, r_3) \cap \cdots.$$

Hence the intersection of these nonempty compact sets is nonempty (by Corollary 1 of Theorem 2-6.4, since the space is E^k), but consists of exactly one point (since $r_n \to 0$) and we label this point

$$x_{\alpha_1\alpha_2\alpha_3\cdots}.$$

Also this point is a point of $P = P'$ since it is a cluster point of the set of distinct points,

$$\{x_{\alpha_1}, x_{\alpha_1\alpha_2}, x_{\alpha_1\alpha_2\alpha_3}, \cdots\} \subset P.$$

Moreover, if $\beta_1, \beta_2, \beta_3, \cdots$ is any sequence of binary digits having $\alpha_n \neq \beta_n$ for at least one n, then (because of (13))

$$x_{\beta_1\beta_2\beta_3\cdots} \neq x_{\alpha_1\alpha_2\alpha_3\cdots}.$$

Hence a 1–1 correspondence is established between the (noncountable) set of infinite binary sequences and a subset of P. Thus P is noncountable. ∎

PROBLEM 1. The set of rational points on [0, 1] is an F_σ set, but not a G_δ set. In fact any G_δ set containing the rationals on [0, 1] is noncountable.

[*Hint:* Let $G_1 \supset G_2 \supset \cdots$ be open sets each containing all rationals on [0, 1]. With r_0 and r_1 rationals on (0, 1), take nondegenerate, nonabutting closed intervals I_0 and I_1 centered at r_0 and r_1, and contained in $G_1 \cap (0, 1)$. Continue much as in the above constructive proof.]

2–9 CARDINALS

In Definition 2-1.2 one of the alternative expressions for $A \sim B$ was "A has the same cardinal as B." Since the relation $\sim$ is known to be symmetric (Theorem 2-1.2) we may read $A \sim B$ as "A and B have the same cardinal." For $n \in J$, the set J_n and any set $A \sim J_n$ is said to have cardinal n. The empty set $\varnothing$ has cardinal 0. Thus 0, 1, 2, 3, $\cdots$ are the finite cardinals.

The symbol $\aleph_0$ (read "Aleph naught") is used for the cardinal of J. Hence any set $A \sim J$ also has cardinal $\aleph_0$ and we use "A has $\aleph_0$ elements" and "A is countably infinite" as synonymous, both meaning that $A \sim J$. For example, the set of all rational numbers has $\aleph_0$ members and the same is true of the subset of points on [0, 1] having finite binary expansions. Since every set is either finite or infinite and every infinite set has a countably infinite subset (corollary of Theorem 2-2.1), $\aleph_0$ is considered to be the first (or smallest) **transfinite** cardinal.

The fact that the union of a finite set of countably infinite sets is countably infinite may be expressed by $n\aleph_0 = \aleph_0$ for each $n \in J$. Also, $\aleph_0\aleph_0 = \aleph_0$ is interpreted to mean that the union of $\aleph_0$ sets each having $\aleph_0$ members has $\aleph_0$ members (Theorem 2-2.6).

The cardinal of the set E^1 of all real numbers is denoted by $\mathfrak{c}$ (a small German script c) and is called the **cardinal of the continuum**. Hence, if $a \in E^1$, $b \in E^1$, and $a < b$, then each of the intervals

$$(a, b), (a, b], [a, b), \text{ and } [a, b]$$

has cardinal $\mathfrak{c}$ (Theorem 2-1.4 and corollaries). Hence, if $d \in E^1$, then each of the sets

$$\{x \mid x > d\}, \{x \mid x \geqslant d\}, \{x \mid x < d\}, \text{ and } \{x \mid x \leqslant d\}$$

has cardinal $\mathfrak{c}$. Moreover, Theorem 2-2.9 may be restated as $[0, 1] \times [0, 1]$ has cardinal $\mathfrak{c}$; thus

$$[0, 1] \times [0, 1] = \bigcup_{0 \leqslant x \leqslant 1} \{(x, y) \mid 0 \leqslant y \leqslant 1\}$$

expresses a set of $\mathfrak{c}$ elements as the union of $\mathfrak{c}$ sets each having $\mathfrak{c}$ elements and this is denoted by

$$\mathfrak{c} \cdot \mathfrak{c} = \mathfrak{c}. \tag{1}$$

Theorem 1. *Given a set A of cardinal $\mathfrak{c}$, then A may be expressed as the union of $\aleph_0$ subsets $A_1, A_2, A_2, \cdots$, each of cardinal $\mathfrak{c}$ and such that $A_i \cap A_j = \varnothing$ for $1 \leqslant i < j$.*

Proof. Since both A and $(0, 1]$ have cardinal $\mathfrak{c}$, there is a mapping f: $(0, 1] \leftrightarrow A$. Then the sequence of sets defined for $n = 1, 2, 3, \cdots$ by

$$f\left(\left(\frac{1}{n+1}, \frac{1}{n}\right]\right) = A_n$$

has the stated properties. ∎

The result in Theorem 1 may be expressed by

$$\aleph_0 \mathfrak{c} = \mathfrak{c}. \tag{2}$$

Theorem 2. *The set B of infinite binary sequences has cardinal $\mathfrak{c}$.*

Proof. Let $B_1 = \{b \mid b \in B \text{ and } b \text{ has } \aleph_0 \text{ 1's}\}$ and set

$$B_2 = B - B_1.$$

The fact that each number x, $0 < x \leqslant 1$, has a unique binary representation using $\aleph_0$ 1's may be interpreted as

"B_1 has cardinal $\mathfrak{c}$."

Express $B_1 = \bigcup_{n=1}^{\infty} A_n$, where each A_n has cardinal $\mathfrak{c}$ and $A_i \cap A_j = \varnothing$ for $1 \leqslant i < j$. Now for $n = 1, 2, 3, \cdots$, select a mapping

$$f_n\colon \left(\frac{1}{n+1}, \frac{1}{n}\right) \leftrightarrow A_n.$$

By associating each member $(\alpha_1, \alpha_2, \cdots, \alpha_n, 0, 0, 0, \cdots)$ of B_2 with the number $0\underset{2}{.}\alpha_1\alpha_2\cdots\alpha_n$ it is seen that B_2 is equivalent to an infinite subset of the rational numbers; that is, B_2 has cardinal $\aleph_0$. Express

$$B_2 \text{ as } \{r_1, r_2, r_3, \cdots\}, \qquad r_i \neq r_j, \quad 1 \leqslant i < j.$$

Then the mapping f defined by

$$f(x) = \left\{\begin{matrix} f_n(x) \text{ if } \dfrac{1}{n+1} < x < \dfrac{1}{n} \\ \\ r_n \text{ if } x = \dfrac{1}{n} \end{matrix}\right\}, \qquad n = 1, 2, 3, \cdots$$

is seen to be f: $(0, 1] \leftrightarrow B$. Hence B has cardinal $\mathfrak{c}$. ∎

The set $A = \{a_1, a_2, a_3\}$ of three distinct elements has

$$\varnothing, \{a_1\}, \{a_2\}, \{a_3\}, \{a_1, a_2\}, \{a_1, a_3\}, \{a_2, a_3\}, \{a_1, a_2, a_3\}$$

as all of its subsets. These $2^3 = 8$ subsets may be put into 1–1 correspondence with the three-digit binary numbers

$$000, 100, 010, 001, 110, 101, 011, 111$$

by the

RULE. Put a 1 or a 0 in the ith place according as a_i is present in or absent from the subset.

Similarly, a set A having $n \in J$ distinct elements has 2^n subsets (including $\varnothing$ and A), since for each of the n-elements there is a "yes-no" decision of including that element in a subset. Hence the 2^n subsets can be put into 1–1 correspondence with the 2^n distinct n-digit binary numbers.

It is therefore natural to let $2^{\aleph_0}$ denote the cardinal of the set of all subsets of a set A having $\aleph_0$ distinct elements. Moreover, the subsets of A may be put into 1–1 correspondence with the infinite binary sequences by the above rule. But the set of infinite binary sequences has cardinal $\mathfrak{c}$ (Theorem 2) and hence

$$2^{\aleph_0} = \mathfrak{c}. \tag{3}$$

Stated another way:

If there are $\aleph_0$ yes-no decisions, then the set of all combinations of choices has cardinal $2^{\aleph_0} = \mathfrak{c}$.

The following result is one application of the theory of cardinals.†

Theorem 3. *Let Γ be the Cantor set and define, for $-1 \leqslant d \leqslant 1$,*

$$D_d = \{(x, y) \mid x \in \Gamma, y \in \Gamma \text{ and } y - x = d\}.$$

Then $D_d \neq \varnothing$ for any d, $-1 \leqslant d \leqslant 1$, and

(a) *D_d has cardinal $\mathfrak{c}$ for some d.*
(b) *D_d is finite for some d.*
(c) *There is no d for which D_d has cardinal $\aleph_0$.*

The following lemma will be used in a proof of this theorem.

Lemma 1. *Let d be a number such that $-1 \leqslant d \leqslant 1$. Then there is at least one sequence $\{\gamma_n\}$ such that*

$$d = \sum_{n=1}^{\infty} \frac{2\gamma_n}{3^n}, \qquad \textit{with each } \gamma_n = -1, 0, \text{ or } 1. \tag{4}$$

If there is a second sequence $\{\gamma_n'\}$ such that

$$d = \sum_{n=1}^{\infty} \frac{2\gamma_n'}{3^n}, \qquad \text{with each } \gamma_n' = -1, 0, \textit{or } 1,$$

then one of the sequences has only -1's *and the other only* 1's *from some place on, and there is no other such representation. If the representation of d in the form* (4) *is unique, then $\aleph_0$ of the γ_n are zero.*

† N. C. Bose Majumder, On the distance set of the Cantor middle third set, III. *The American Mathematical Monthly*, **72** (1965), pp. 725–729.

Proof. Represent the number $0 \leqslant \frac{1}{2}(d+1) \leqslant 1$ in ternary form

$$\frac{1}{2}(d+1) = \sum_{n=1}^{\infty} \frac{\delta_n}{3^n}, \qquad \text{with each } \delta_n = 0, 1, \text{ or } 2.$$

Hence

$$\frac{1}{2}d = \sum_{n=1}^{\infty} \frac{\delta_n}{3^n} - \frac{1}{2} = \sum_{n=1}^{\infty} \frac{\delta_n}{3^n} - \sum_{n=1}^{\infty} \frac{1}{3^n} = \sum_{n=1}^{\infty} \frac{\delta_n - 1}{3^n}.$$

Upon setting $\gamma_n = \delta_n - 1$, the representation (4) is obtained.

The rest of the lemma follows from known facts about alternate ternary representations of numbers on [0, 1]. ∎

Proof of Theorem 3. Given a number d, $-1 \leqslant d \leqslant 1$, express d in the form (4), thus determining a sequence $\{\gamma_n\}$ with each $\gamma_n = -1, 0,$ or 1. Two numbers

$$x = \sum_{n=1}^{\infty} \frac{2\alpha_n}{3^n} \quad \text{and} \quad y = \sum_{n=1}^{\infty} \frac{2\beta_n}{3^n}, \tag{5}$$

with the sequences $\{\alpha_n\}$ and $\{\beta_n\}$ chosen by setting

(i) $\alpha_n = 1$ and $\beta_n = 0$ if $\gamma_n = -1$;
(ii) $\alpha_n = \beta_n = 0$ or $\alpha_n = \beta_n = 1$ if $\gamma_n = 0$;
(iii) $\alpha_n = 0$ and $\beta_n = 1$ if $\gamma_n = 1$;

are such that

$$y - x = \sum_{n=1}^{\infty} \frac{2(\beta_n - \alpha_n)}{3^n} = \sum_{n=1}^{\infty} \frac{2\gamma_n}{3^n} = d.$$

Also, $x \in \Gamma$ and $y \in \Gamma$, since each α_n and each β_n in (5) is 0 or 1. (See Theorem 1-8.3.)

In (ii) either choice may be made whenever $\gamma_n = 0$. Thus if $\aleph_0$ of the γ_n are 0, then there are $2^{\aleph_0} = \mathfrak{c}$ pairs (x, y) in Γ^2 with $y - x = d$; and this is true if d has a unique representation in the form (4). On the other hand, if d has an alternate representation, then both representations have a finite number of zero terms and hence only a finite number of pairs (x, y) in Γ^2 with $y - x = d$.

Conversely, given any $x_1 \in \Gamma$ and $y_1 \in \Gamma$, then $y_1 - x_1$ has a representation in the form (4). Hence there is no d, $-1 \leqslant d \leqslant 1$, for which D_d has cardinal $\aleph_0$. ∎

Visualize several functions having domain [0, 1] and ranges in E^1; for example, the function f defined by $f(t) = t^2$ for $0 \leqslant t \leqslant 1$. The following theorem shows that the set of all such functions does not have cardinal $\mathfrak{c}$.

Theorem 4. *Let $F = \{f \mid f\colon [0, 1] \to E^1\}$. Then there is a subset of F having cardinal $\mathfrak{c}$, but F does not have cardinal $\mathfrak{c}$.*

Proof. Let C be the subset of F defined by $f \in C$ iff for some $y \in E^1$, $f(t) = y$ for all $t \in [0, 1]$; that is, C consists of the constant real-valued functions with domain [0, 1]. The very definition of C shows that $C \sim E^1$, and hence C has cardinal $\mathfrak{c}$.

Assume F has cardinal $\mathfrak{c}$. Then there is a 1–1 correspondence between [0, 1] and F. A convenient way of thinking of this is to consider $I = [0, 1]$ as an index set for F and for each $x \in I$ to let f_x be the element (function) of F corresponding to x. Hence, for $x \in I$, it follows that $f_x(0) \in E^1$, $f_x(\frac{1}{2}) \in E^1$, $f_x(1) \in E^1$, and if $0 \leqslant t \leqslant 1$, then $f_x(t) \in E^1$. In particular

$$\text{for } x \in I, \qquad \text{then } f_x(x) \in E^1.$$

We now define a function g: $[0, 1] \to E^1$ by setting

$$g(x) = f_x(x) + 1 \qquad \text{for each } x \in I = [0, 1]. \tag{6}$$

Thus $g \in F$ and hence, by assumption, there is an $x_0 \in I$ such that

$$g = f_{x_0}.$$

For each $t \in [0, 1]$, $g(t) = f_{x_0}(t)$ and in particular

$$g(x_0) = f_{x_0}(x_0).$$

But from the definition (6) of g,

$$g(x_0) = f_{x_0}(x_0) + 1.$$

This contradiction shows that there is no 1–1 correspondence between [0, 1] and F; that is, F does not have cardinal $\mathfrak{c}$. ∎

Definition 1. *Let α and β be cardinals. If a set A of cardinal α has a subset $B \subset A$ of cardinal β, then*

$$\beta \leqslant \alpha$$

and if in addition A does not have cardinal β, then

$$\beta < \alpha.$$

Upon letting $\mathfrak{f}$ denote the cardinal of the set of all real-valued functions, each with domain [0, 1], then from Theorem 4,

$$\mathfrak{c} < \mathfrak{f}.$$

PROBLEM 1. Given a set A, let α denote its cardinal and let 2^α denote the cardinal of the set $\mathscr{A}$ of all subsets of A. Then

$$\alpha < 2^\alpha. \tag{7}$$

[*Hint:* After proving $\alpha \leqslant 2^\alpha$, assume $\mathscr{A}$ has cardinal α. Then A may be used as an index set for $\mathscr{A}$; that is, $A_a \in \mathscr{A}$ for each $a \in A$, etc. Let $B = \{a \mid a \in A$ but $a \notin A_a\}$. Then $B \subset A$ so there is an index $a_0 \in A$ such that $B = A_{a_0}$. Show that $a_0 \in B$ and $a_0 \notin B$ both lead to contradictions.]

Note that (7) holds for finite, as well as transfinite, cardinals. Even if $A = \varnothing$, then $\alpha = 0$, but the family of subsets of $\varnothing$ has one element (namely $\{\varnothing\}$) and $0 < 1 = 2^0$.

2–10 CONNECTED SETS

A convenient terminology is to say that two nonempty sets U and V (in any space whatever) are **disjoint** if $U \cap V = \varnothing$. Thus in E^1 the intervals (0, 1] and (1, 2) are disjoint, but they are not "separate."

Definition 1. *In a metric space S:*

(i) *Two nonempty sets* U and V *are* **separate** *if*

$$U \cap \bar{V} = \bar{U} \cap V = \varnothing.$$

(ii) *A set A is* **disconnected** *if there are separate sets U and V such that $A = U \cup V$.*
(iii) *A nonempty set A is* **connected** *if it is not disconnected.*

Thus a set A is disconnected if there are nonempty sets U and V such that

$$A = U \cup V \qquad \text{and} \qquad U \cap V = U \cap V' = U' \cap V = \varnothing.$$

A nonempty set A is connected iff

$$(A = U \cup V, U \cap V = U \cap V' = U' \cap V = \varnothing) \Rightarrow (U = \varnothing \text{ or } V = \varnothing).$$

The notion of connectedness does not apply to the empty set. The whole metric space S may, however, either be connected or disconnected.

Theorem 1. *A metric space S is disconnected iff it contains a set other than $\varnothing$ or S which is both closed and open.*

Proof. Let S be disconnected. Then there are nonempty sets U and V such that

$$S = U \cup V, \qquad U \cap V = U \cap V' = U' \cap V = \varnothing.$$

Hence $U = V^c$. Since $U' \cap V = \varnothing$, any point of V has a neighborhood free of points of U; that is, every point of V is an interior point of V so that V is open. In the same way, U is open. Thus $U = V^c$ is both open and closed, is nonempty (given), and is not all of S, since its complement $V \neq \varnothing$.

Conversely, let S contain a set $U \neq \varnothing$, $U \neq S$, and U both open and closed. Then U^c is both open and closed. Also $U' \subset U$ and $(U^c)' \subset U^c$, since both U and U^c are closed so that $U \cap (U^c)' \subset U \cap U^c = \varnothing$ and $U^c \cap U' \subset U^c \cap U = \varnothing$. Moreover, $U^c \neq \varnothing$, since $U \neq S$. Thus U and U^c are nonempty sets such that

$$S = U \cup U^c \qquad \text{and} \qquad U \cap \overline{U^c} = \bar{U} \cap U^c = \varnothing.$$

Hence S is disconnected. ∎

Theorem 2. *A nonempty set A in a metric space S is disconnected iff there are open sets G and H such that*

$$A \subset G \cup H, A \cap G \neq \varnothing, A \cap H \neq \varnothing, \text{and } G \cap H = \varnothing.$$

Equivalently, A is connected iff G open, H open, $A \subset G \cup H$, and $G \cap H = \varnothing$ imply that either $A \subset G$ or $A \subset H$.

Proof. Let A be a set for which there are such sets G and H. Since G and H are open and $G \cap H = \varnothing$, then also $G \cap \bar{H} = \bar{G} \cap H = \varnothing$. The sets $U = G \cap A$ and $V = H \cap A$ are nonempty, $A = U \cup V$,

$$U \cap \bar{V} = (G \cap A) \cap \overline{H \cap A} \subset G \cap \bar{H} = \varnothing,$$

and in the same way, $\bar{U} \cap V = \varnothing$ so that A is disconnected.

Conversely, let A be disconnected and select nonempty sets U and V such that $A = U \cup V$ and $U \cap \bar{V} = \bar{U} \cap V = \varnothing$. Take any $a \in U$. Then $a \notin \bar{V}$. Since $\bar{V}$ is a nonempty closed set, let $v(a) > 0$ be such that $v \in \bar{V} \Rightarrow d(a, v) > v(a)$, which is possible by Theorem 2-4.9. Then the set G defined by

$$G = \bigcup_{a \in U} N(a, \tfrac{1}{2}v(a)) \supset \bigcup_{a \in U} \{a\} = U$$

and also G is open. In the same way, H is open and $H \supset V$, where

$$H = \bigcup_{a \in V} N(a, \tfrac{1}{2}\mu(a)),$$

with $\mu(a)$ selected for each $a \in V$ such that $0 < \mu(a) < d(a, u)$ whenever $u \in U$. Then $G \cup H \supset U \cup V = A$, $A \cap G \supset U \neq \varnothing$, and $A \cap H \supset V \neq \varnothing$.

Assume that $G \cap H \neq \varnothing$ and let $c \in G \cap H$. Then there is an $a_1 \in U$ and an $a_2 \in V$ such that

$$c \in N(a_1, \tfrac{1}{2}v(a_1)) \cap N(a_2, \tfrac{1}{2}\mu(a_2)).$$

Hence $d(a_1, c) < \frac{1}{2}v(a_1) < \frac{1}{2}d(a_1, a_2)$ and $d(a_2, c) < \frac{1}{2}\mu(a_2) < \frac{1}{2}d(a_1, a_2)$. Thus

$$d(a_1, a_2) \leqslant d(a_1, c) + d(c, a_2) < d(a_1, a_2)$$

so that $G \cap H = \varnothing$ and the theorem is proved. ∎

Trivially, one-element sets are connected.

Theorem 3. *A set $A \subset E^1$ of at least two points is connected iff*

$$(1) \qquad (a_1 \in A, a_2 \in A, a_1 < a_2) \Rightarrow [a_1, a_2] \subset A.$$

Proof. Let $A \subset E^1$ have two points $a_1 < a_2$ such that $[a_1, a_2] \not\subset A$. Then there is a $c \notin A$ and $a_1 < c < a_2$. Hence $G = \{x \mid x > c\}$ and $H = \{x \mid x < c\}$ are nonempty open sets, $G \cap H = \varnothing$ and $G \cup H \supset A$. Thus A is disconnected.

On the other hand, let $A \subset E^1$ be a set of at least two points for which (1) holds. Assume A is disconnected and take open sets G and H such that $G \cup H \supset A$, $G \cap A \neq \varnothing$, $H \cap A \neq \varnothing$, and $G \cap H = \varnothing$. Select a point in $G \cap A$ and another (necessarily different) in $H \cap A$ and name these a_1 and a_2 with $a_1 < a_2$. Then, from (1) $[a_1, a_2] \subset A \subset G \cup H$. Thus, by Theorem 2-6.7, either $[a_1, a_2] \subset G$ or $[a_1, a_2] \subset H$, contradicting the choice of one end point in G and the other in H. Hence A is connected. ∎

PROBLEM 1. The set

$$A = \{(0, y) \mid -1 < y < 1\} \cup \{(x, 0) \mid 0 < x < 1\}$$

is connected.

PROBLEM 2. Figure 1 represents a set $A \subset E^2$ consisting of the point $(0, \frac{1}{2})$ and all points on the graph of $y = f(x)$ where f is the function defined on the interval $(0, 1]$ by

$$f(x) = \begin{cases} 2n(2n-1)x - 2n + 1 \text{ if } \dfrac{1}{2n} \leqslant x \leqslant \dfrac{1}{2n-1} \\ -2n(2n+1)x + 2n + 1 \text{ if } \dfrac{1}{2n+1} \leqslant x \leqslant \dfrac{1}{2n} \end{cases} \qquad n = 1, 2, 3, \cdots.$$

Show that A is a connected set.

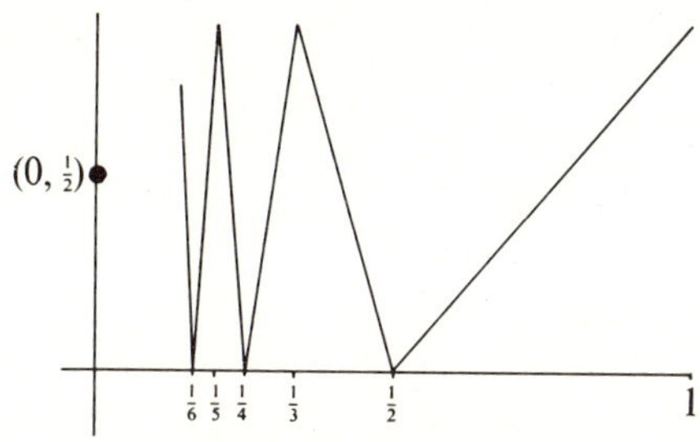

FIG. 1

2–11 DISTANCE AND RELATIVE PROPERTIES

Definition 1. *In a metric space S, the* **distance** *$d(b, A)$ between a point b and a nonempty set A is defined by*

$$d(b, A) = \inf\{d(b, a) \mid a \in A\}.$$

Theorem 1. *If a and b are points and $C \neq \varnothing$ is a set in a metric space, then*

$$d(a, C) \leqslant d(a, b) + d(b, C).$$

Proof. $d(a, C) \leqslant d(a, c) \leqslant d(a, b) + d(b, c)$ for all $c \in C$. Since $0 \leqslant d(a, b) < \infty$, we have

$$d(a, C) - d(a, b) \leqslant \inf\{d(b, c) \mid c \in C\} = d(b, C). \blacksquare$$

If $\varnothing \neq A \subset B$, then for any point b,

$$\{d(b, a) \mid a \in A\} \subset \{d(b, a) \mid a \in B\},$$

and hence $\inf\{d(b, a) \mid a \in A\} \geqslant \inf\{d(b, a) \mid a \in B\}$. Thus

$$\varnothing \neq A \subset B \Rightarrow d(b, A) \geqslant d(b, B) \qquad \text{for any point } b.$$

Theorem 2. *If b is a point and A is a nonempty set in a metric space, then*

$$d(b, \bar{A}) = d(b, A).$$

Proof. Since $A \subset \bar{A}$, then $d(b, \bar{A}) \leqslant d(b, A)$. Assume there is a point b such that

$$d(b, \bar{A}) < d(b, A).$$

Select an $\bar{a} \in \bar{A} = A \cup A'$ such that $d(b, \bar{a}) < d(b, A)$. Hence

$$d(b, \bar{a}) < d(b, A) \leqslant d(b, a) \qquad \text{for all } a \in A$$

and therefore $\bar{a} \notin A$. Thus $\bar{a} \in A'$, and we next select $a \in A$ such that

$$d(a, \bar{a}) < d(b, A) - d(b, \bar{a}).$$

This says that $d(b, a) \leqslant d(b, \bar{a}) + d(\bar{a}, a) < d(b, A)$, which is impossible, since $a \in A$. Thus

$$d(b, \bar{A}) \geqslant d(b, A)$$

and hence the equality holds. ▮

Theorem 3. *In a metric space, if A is a nonempty compact set and b is a point, then there is an $a_0 \in A$ such that*

$$d(b, a_0) = d(b, A).$$

Proof. If $b \in A$, then with $a_0 = b$, $d(b, a_0) = 0 = d(b, A)$ and thus consider $b \notin A$. Then $a \in A \Rightarrow d(b, a) > 0$. Select

$$a_n \in A \qquad \text{with } d(b, a_n) < d(b, A) + \frac{1}{n}, \qquad n = 1, 2, 3, \cdots.$$

Then $B = \{a_1, a_2, a_3, \cdots\}$ is an infinite subset of A and thus has a cluster point $a_0 \in A$, by Theorem 2-6.3, since A is compact. Hence $a_0 \neq b$ and

$$d(b, a_0) \geqslant d(b, A)$$

by the definition of $d(b, A)$. Let $\varepsilon > 0$ be given and fix an n such that

$$\frac{1}{n} < \varepsilon \qquad \text{and} \qquad a_n \in N_0(a_0, \varepsilon) \cap B,$$

which is possible, since this set is infinite. Hence

$$d(b, a_0) \leqslant d(b, a_n) + d(a_n, a_0) < d(b, A) + 2\varepsilon.$$

Consequently, $d(b, a_0) \leqslant d(b, A)$ and thus the equality holds. ▮

Theorem 4. *Let G be an open nonempty set in a metric space such that also $G^c \neq \varnothing$ and set*

$$B_n = \left\{x \,\middle|\, d(x, G^c) > \frac{1}{n}\right\}, \qquad n = 1, 2, 3, \cdots.$$

Then $G = \cup B_n$, $\{B_n\}$ is monotonically increasing, and each B_n is open.

Proof. Select any point $g \in G$ and then $r > 0$ such that $N(g, r) \subset G^o$. Hence $b \in G^c \Rightarrow d(b, g) \geqslant r$ so that $d(g, G^c) \geqslant r$. Thus, for $1/n < r$, we have $g \in B_n$. Hence $G \subset \cup B_n$, since g was any point of G. But there is no point of G^c in any B_n so that $G = \cup B_n$.

Since

$$x \in B_n \Rightarrow d(x, G^c) > \frac{1}{n} > \frac{1}{n+1} \Rightarrow x \in B_{n+1}$$

the sequence $\{B_n\}$ is *monotonically increasing.*

Fix an n such that $B_n \neq \varnothing$ and take any point $g \in B_n$. Then $d(g, G^c) > 1/n$ and we take r such that

$$0 < r < d(g, G^c) - \frac{1}{n}.$$

Then for any $x \in N(g, r)$ we have

$$\begin{aligned} d(g, G^c) &\leqslant d(g, x) + d(x, G^c) \qquad \text{by Theorem 1} \\ &< d(g, G^c) - \frac{1}{n} + d(x, G^c). \end{aligned}$$

Thus $d(x, G^c) > 1/n$, $N(g, r) \subset B_n$, and hence g is an interior point of B_n. Consequently B_n is open. ∎

Definition 2. *The distance $d(A, B)$ between nonempty sets A and B in a metric space is defined by*

$$d(A, B) = \inf\{d(a, b) \mid a \in A, b \in B\}.$$

Even though $d(A, B)$ is called a distance, the triangular inequality does not hold. Figure 1 represents three sets A, B, C with

$$d(A, C) > d(A, B) + d(B, C) = 0.$$

Also $d(A, B)$ may be zero and $A \neq B$.

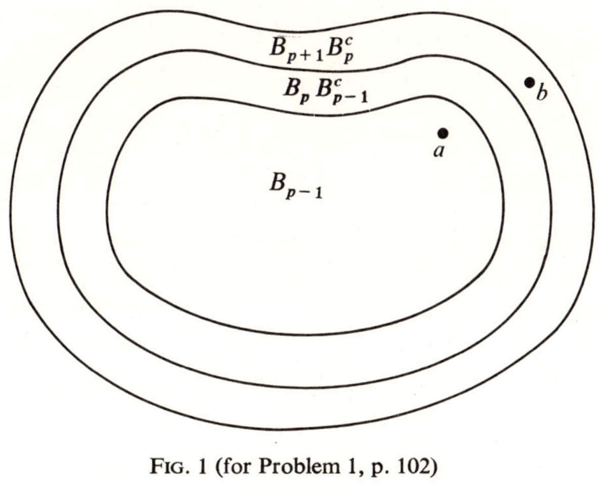

FIG. 1 (for Problem 1, p. 102)

PROBLEM 1. Let G and B_n be as in Theorem 4. If $p > 1$ is an integer and $B_{p-1} \neq \varnothing$, then

$$d(B_{p-1}, B_{p+1}B_p^c) \geqslant \frac{1}{p(p-1)} > 0.$$

The results in Theorem 4 and Problem 1 play a central role in proving the important Theorem 5-1.4.

Definition 3. *The* **diameter** $\delta(A)$ *of a set A in a metric space is defined by* $\delta(\varnothing) = 0$, $\delta(A) = \infty$ *if A is unbounded, and*

$$\delta(A) = \sup\{d(a, b) \mid a \in A \text{ and } b \in A\},$$

if $A \neq \varnothing$ and bounded.

PROBLEM 2.

(a) $\delta(A) = \delta(\bar{A})$.
(b) If $A \cap B \neq \varnothing$, then $\delta(A \cup B) \leqslant \delta(A) + \delta(B)$.
(c) $\delta(A \cup B) \geqslant \max[\delta(A), \delta(B)]$.
(d) $\delta(A \cup B) \leqslant \delta(A) + \delta(B) + d(A, B)$.

The following notions are used in discussing continuous functions in Chapter 6.

By identifying each $x \in E^1$ with the ordered pair $(x, 0) \in E^2$, it is customary to consider $E^1 \subset E^2$. Then both $\mathbf{x} = (x, 0)$ and $\mathbf{y} = (y, 0)$ are points of E^2 and

$$d(\mathbf{x}, \mathbf{y}) = \sqrt{(x-y)^2 + (0-0)^2} = |x - y|,$$

which is the distance between the points $x \in E^1$ and $y \in E^1$. Then, for example, the interval $(0, 1)$ is an open subset of E^1 but is not open when considered as a subset of E^2.

Definition 4. *Given a metric space S with distance function d, then a* **submetric space** *of S is any set $T \subset S$ with distance function d_T defined by*

$$d_T(t_1, t_2) = d(t_1, t_2) \qquad \text{for each } (t_1, t_2) \in T^2.$$

[Since d (defined on S^2) satisfies Definition 2-4.1, (i)–(iv), it is clear that d_T also satisfies these conditions.]

Hereafter we shall consider that any subset of a metric space S is "made into a submetric space" and shall not write the subscript on its distance function.

Definition 5. *Given a subset A of a metric space S, then a set $B \subset A$ is said to be* **open relative** *to A if B is open as a set in the submetric space A. In a similar way,* **closed relative** *to A,* **perfect relative** *to A, etc., are defined. Saying that a set is open means that it is open relative to S.*

The above definition presupposes the notion that a **neighborhood relative** to A of any $a \in A$ is a set of the form

$$\{x \mid x \in A \text{ and } d(a, x) < r\} = N(a, r) \cap A \qquad \text{for some } r > 0.$$

Theorem 5. *With S a metric space and $A \subset S$, then a set $B \subset A$ is open relative to A iff there is an open set H such that $B = H \cap A$.*

Proof. Let B be open relative to A. This means that $B \subset A$ and each $b \in B$ is an interior, relative to A, point of B; that is, there is an $r > 0$ such that

$$N(b, r) \cap A \subset B.$$

Let H be the union of all such $N(b, r)$ for all $b \in B$. Then $H \cap A \subset B$ and also H is open (Theorem 2-5.4). But

$$B = \bigcup_{b \in B} \{b\} \subset H.$$

Thus $B = B \cap A \subset H \cap A$ so that $H \cap A = B$.

Conversely, let H be an open set. If $H \cap A = \varnothing$, then $H \cap A$ is open relative to any metric space, so consider $H \cap A \neq \varnothing$ and take any $b \in H \cap A$. Let $r > 0$ be such that $N(b, r) \subset H = H^{\circ}$. Then $N(b, r) \cap A \subset H \cap A$; that is, b is an interior relative to A point of $H \cap A$ and hence $H \cap A$ is open relative to A. ∎

Note that in E^1 the open-closed interval $B = (0, 1]$ is closed relative to $A = (0, 2)$. Even though B is bounded and closed relative to A, it is not "compact relative to A" and has no "smallest element relative to A." Also, B is not closed relative to $T = (-1, 2)$. This is a warning not to apply theorems proved for E^k to submetric spaces of E^k without checking their validity there.

PROBLEM 3. Given a subset A of a metric space S, then:

(a) $B \subset A$ is closed relative to A iff $B' \cap A \subset B$; that is, iff every cluster point of B which is a point of A is also a point of B.

(b) $B \subset A$ is open relative to A iff $A - B$ is closed relative to A.

(c) Any set $A \subset S$ is both open relative to itself and closed relative to itself.

(d) A nonempty set $A \subset S$ is disconnected iff there is a $B \subset A$ with $B \neq \varnothing$, $B \neq A$, and B both open and closed relative to A.

(e) Given a set $A \subset S$, then a set $B \subset A$ is closed relative to A iff there is a closed set F such that $B = F \cap A$.

2–12 E^k AS A VECTOR SPACE

Z is used in the following definition to represent the field of complex numbers (instead of E^2); we do this to emphasize the "field," rather than the "space," properties of complex numbers in this setting. We shall, however, continue to use E^1 for Euclidean one-dimensional space as well as the real number field.

Definition 1. *Let V be a set on which a binary operation, called addition, is defined so that $\mathbf{u} + \mathbf{v} \in V$ for each $\mathbf{u}, \mathbf{v} \in V$. Also let an operation, called multiplication of each $a \in Z$ by each $\mathbf{u} \in V$, be defined for which $a\mathbf{u} \in V$. Then V, together with these two operations, is said to be a* **vector space over the complex number**

field (*or a vector space with complex numbers as* **scalars**) *provided the following axioms hold for all* $\mathbf{u}$, $\mathbf{v}$, $\mathbf{w} \in V$ *and* a, $b \in Z$:

$$\mathbf{u} + \mathbf{v} = \mathbf{v} + \mathbf{u}$$
$$\mathbf{u} + (\mathbf{v} + \mathbf{w}) = (\mathbf{u} + \mathbf{v}) + \mathbf{w}$$
$$a(\mathbf{u} + \mathbf{v}) = a\mathbf{u} + a\mathbf{v}$$
$$(a + b)\mathbf{u} = a\mathbf{u} + b\mathbf{u}$$
$$1\mathbf{u} = \mathbf{u}$$

there is a unique element $\mathbf{0} \in V$, *called the* **origin** *or* **null** *vector, such that*

$$\mathbf{u} + \mathbf{0} = \mathbf{u} \qquad \text{and} \qquad 0\mathbf{u} = \mathbf{0}$$

and each $\mathbf{u} \in V$ *has a unique additive inverse* $(-\mathbf{u}) \in V$ *such that*

$$\mathbf{u} + (-\mathbf{u}) = \mathbf{0}.$$

We denote this vector space, and its field, by (V, Z).

These axioms are not independent, nor the least that could be imposed and still have them all. They are, moreover, more general than we need for some time (as is most of the further development in this section), but this generality avoids virtual repetition later on. Since the real field E^1 may be considered as embedded in Z, the definition of a **vector space** over the **real field** is included. A vector space in which the field is Z is usually called a **complex vector space** and is called a **real vector space** if restriction is to the real field (regardless, in both cases, of the nature of the vectors $\mathbf{v}$ themselves).

Definition 2. *A complex vector space* (V, Z) *is said to be a* **normed vector space** *if a real-valued function* $\| \;\|$ *is defined with domain* V *such that for all* $\mathbf{u}$, $\mathbf{v} \in V$ *and* $c \in Z$,

$$\|\mathbf{v}\| \geqslant 0 \qquad \text{and} \qquad = 0 \qquad \text{iff } \mathbf{v} = \mathbf{0},$$
$$\|c\mathbf{v}\| = |c| \; \|\mathbf{v}\|, \qquad \text{and}$$
$$\|\mathbf{u} + \mathbf{v}\| \leqslant \|\mathbf{u}\| + \|\mathbf{v}\|.$$

Then $\|\mathbf{v}\|$ *is called the* **norm** *of* $\mathbf{v} \in V$.

The use of vector notation $\mathbf{x} = (x_1, x_2, \cdots, x_k) \in E^k$ anticipated making E^k into a real vector space by defining vector addition of $\mathbf{x}$ and $\mathbf{y} = (y_1, y_2, \cdots, y_k)$ as

$$\mathbf{x} + \mathbf{y} = (x_1 + y_1, x_2 + y_2, \cdots, x_k + y_k)$$

and scalar multiplication of a real number a and $\mathbf{x}$ as

$$a\mathbf{x} = (ax_1, ax_2, \cdots, ax_k).$$

Also $\mathbf{0} = (0, 0, \cdots, 0)$ has been used for the null vector. By checking, via components, it is seen that E^k with these operations is a real vector space, (E^k, E^1).

Similarly, with Z^k the k-factor cartesian product of Z, we use (Z^k, Z) for the vector space in which each vector **z** is an ordered k-tuple of complex numbers

$$\mathbf{z} = (z_1, z_2, \cdots, z_k) \in Z^k,$$

$\mathbf{z} + \mathbf{w}$ is obtained by component addition,

$$c\mathbf{z} = (cz_1, cz_2, \cdots, cz_k), \qquad c \in Z,$$

and $\mathbf{0} = (0, 0, \cdots, 0)$.

In the following definition $\bar{z}$ is the complex conjugate of z.

Definition 3. *In* (Z^k, Z) *the* **inner** (*or scalar, or dot*) *product of vectors* **z** *and* **w** *is defined by*

$$\mathbf{z} \cdot \mathbf{w} = z_1 \bar{w}_1 + z_2 \bar{w}_2 + \cdots + z_k \bar{w}_k. \tag{1}$$

[Recall that if **u** and **v** are ordinary geometric vectors in the plane, then, by definition,

$$\mathbf{u} \cdot \mathbf{v} = |\mathbf{u}|\ |\mathbf{v}| \cos\theta,$$

where θ is the smallest undirected angle between the vectors. With **i** and **j** the unit vectors on the axes, it was shown in the elementary course that if $\mathbf{u} = a_1\mathbf{i} + a_2\mathbf{j}$ and $\mathbf{v} = b_1\mathbf{i} + b_2\mathbf{j}$, then

$$\mathbf{u} \cdot \mathbf{v} = a_1 b_1 + a_2 b_2 .$$

If $\mathbf{u} \neq \mathbf{0}$ and $\mathbf{v} \neq \mathbf{0}$, then

$$\cos\theta = \frac{\mathbf{u} \cdot \mathbf{v}}{|\mathbf{u}|\ |\mathbf{v}|} = \frac{a_1 b_1 + a_2 b_2}{\sqrt{a_1{}^2 + a_2{}^2}\,\sqrt{b_1{}^2 + b_2{}^2}}$$

and thus $\theta = 90°$ iff $a_1 b_1 + a_2 b_2 = 0$.

An explanation of why conjugates appear in (1) is given in Section 4-7].

We make the following observations (in the notation of Definition 3), which are easily checked by using known properties of complex conjugates:

$$\mathbf{z} \cdot \mathbf{z} \geq 0 \quad \text{and} \quad = 0 \qquad \text{iff } \mathbf{z} = \mathbf{0},$$

$$\mathbf{w} \cdot \mathbf{z} = \overline{\mathbf{z} \cdot \mathbf{w}},$$

$$(c\mathbf{z}) \cdot \mathbf{w} = c(\mathbf{z} \cdot \mathbf{w}) \qquad \text{but} \qquad \mathbf{z} \cdot (c\mathbf{w}) = \bar{c}(\mathbf{z} \cdot \mathbf{w}), \qquad c \in Z.$$

Also, if $\mathbf{u} = (u_1, u_2, \cdots, u_k) \in Z^k$, then

$$\mathbf{z} \cdot (\mathbf{u} \pm \mathbf{w}) = \mathbf{z} \cdot \mathbf{u} \pm \mathbf{z} \cdot \mathbf{w} \qquad \text{and} \qquad (\mathbf{u} \pm \mathbf{w}) \cdot \mathbf{z} = \mathbf{u} \cdot \mathbf{z} + \mathbf{w} \cdot \mathbf{z}.$$

We shall say that **z** and **w** are **orthogonal**, denoted by

$$\mathbf{z} \perp \mathbf{w}, \qquad \text{whenever } \mathbf{z} \cdot \mathbf{w} = 0,$$

even if one or both are **0**. The real number $\sqrt{\mathbf{z} \cdot \mathbf{z}}$ is called the **norm** of **z** and is denoted by

$$\|\mathbf{z}\| = \sqrt{\mathbf{z} \cdot \mathbf{z}}. \tag{2}$$

Thus $\|\mathbf{z}\| \geqslant 0$ and $= 0$ iff $\mathbf{z} = \mathbf{0}$, $\|c\mathbf{z}\|^2 = (c\mathbf{z})\cdot(c\mathbf{z}) = c\bar{c}(\mathbf{z}\cdot\mathbf{z}) = |c|^2 \|z\|^2$ which with the following theorem reconciles this use of "norm" with that of Definition 2. [See inequality (5) below.]

A set $\{\mathbf{v}_1, \mathbf{v}_2, \cdots, \mathbf{v}_n\}$ of vectors in (Z^k, Z) is said to be an **orthonormal** set if

$$\mathbf{v}_i \cdot \mathbf{v}_j = \begin{cases} 0 & \text{if } i \neq j \\ 1 & \text{if } i = j. \end{cases}$$

For example, the k-vectors of (Z^k, Z)

$$\begin{aligned} \mathbf{e}_1 &= (1, 0, 0, \cdots, 0) \\ \mathbf{e}_2 &= (0, 1, 0, \cdots, 0) \\ &\cdots \\ \mathbf{e}_k &= (0, 0, \cdots, 0, 1) \end{aligned}$$

is an orthonormal set of vectors which has the further property that

$$\mathbf{z} = (z_1, z_2, \cdots, z_k) = \sum_{i=1}^{k} z_i \mathbf{e}_i .$$

The set $\{\mathbf{e}_1, \mathbf{e}_2, \cdots, \mathbf{e}_k\}$ is called the **standard basis** of (Z^k, Z).

In connection with this type of work, the **Kronecker delta**, defined by

$$\delta_{ij} = \begin{cases} 0 & \text{if } i \neq j \\ 1 & \text{if } i = j \end{cases}$$

is a convenient notation. Thus $\mathbf{e}_i \cdot \mathbf{e}_j = \delta_{ij}$.

In the proof of the following theorem it will be seen that (4) and then (5) follow easily from (3). If the only object were to prove (4) and (5), then a simpler overall method could be (actually has been) used. Much later (Section 7-4) it will be seen, however, that (3) is useful in its own right and it is well to see how various facts are interrelated.

Theorem 1. *In (Z^k, Z) if $\mathbf{v}_1, \mathbf{v}_2, \cdots, \mathbf{v}_n$, $n \geqslant 1$, is any orthonormal set of vectors and $\mathbf{z} = (z_1, z_2, \cdots, z_k) \in Z^k$, then*

(3) $$\sum_{i=1}^{n} |\mathbf{z} \cdot \mathbf{v}_i|^2 \leqslant \|\mathbf{z}\|^2 \qquad \textit{(called Bessel's inequality)}.$$

Also, if $\mathbf{w} = (w_1, w_2, \cdots, w_k) \in Z_k$, then

(4) $$|\mathbf{z}\cdot\mathbf{w}| \leq \|\mathbf{z}\| \, \|\mathbf{w}\| \qquad \text{(Schwarz inequality)} \quad \text{and}$$

(5) $$\|\mathbf{z} + \mathbf{w}\| \leq \|\mathbf{z}\| + \|\mathbf{w}\| \qquad \text{(Minkowski's inequality).}$$

Proof. For convenience we set $c_i = \mathbf{z}\cdot\mathbf{v}_i$. Thus $c_i \in Z$ and

$$\begin{aligned} (c_i \mathbf{z}) \cdot \mathbf{v}_i &= c_i(\mathbf{z} \cdot \mathbf{v}_i) = c_i c_i = c_i^2 \in Z, && \text{but} \\ \mathbf{z} \cdot (c_i \mathbf{v}_i) &= \bar{c}_i(\mathbf{z} \cdot \mathbf{v}_i) = \bar{c}_i c_i = |c_i|^2 \geqslant 0 && \text{is real.} \end{aligned}$$

Let $\mathbf{z}'$ be the vector defined by

$$\mathbf{z}' = \mathbf{z} - \sum_{i=1}^{n} c_i \mathbf{v}_i .$$

Then (check each step by some property obtained above)

$$0 \leqslant \|\mathbf{z}'\|^2 = \mathbf{z}' \cdot \mathbf{z}' = \left(\mathbf{z} - \sum_{i=1}^{n} c_i \mathbf{v}_i\right) \cdot \left(\mathbf{z} - \sum_{j=1}^{n} c_j \cdot \mathbf{v}_j\right)$$

$$= \mathbf{z} \cdot \mathbf{z} - \mathbf{z} \cdot \left(\sum_j c_j \mathbf{v}_j\right) - \left(\sum_i c_i \mathbf{v}_i\right) \cdot \mathbf{z} + \left(\sum_i c_i \mathbf{v}_i\right) \cdot \left(\sum_j c_j \mathbf{v}_j\right)$$

$$= \|\mathbf{z}\|^2 - \sum_j \mathbf{z} \cdot (c_j \mathbf{v}_j) - \sum_i (c_i \mathbf{v}_i) \cdot \mathbf{z} + \sum_i \sum_j (c_i \mathbf{v}_i) \cdot (c_j \mathbf{v}_j)$$

$$= \|\mathbf{z}\|^2 - \sum_j \bar{c}_j (\mathbf{z} \cdot \mathbf{v}_j) - \sum_i c_i (\mathbf{v}_i \cdot \mathbf{z}) + \sum_i \sum_j c_i \bar{c}_j (\mathbf{v}_i \cdot \mathbf{v}_j)$$

$$= \|\mathbf{z}\|^2 - \sum_j \bar{c}_j c_j - \sum_i c_i \bar{c}_i + \sum_i \sum_j c_i \bar{c}_j \, \delta_{ij}$$

$$= \|\mathbf{z}\|^2 - \sum_j |c_j|^2 - \sum_i |c_i|^2 + \sum_i |c_i|^2 = \|\mathbf{z}\|^2 - \sum_j |c_j|^2.$$

Thus (3) is seen to hold.

The inequality (4) holds if $\mathbf{w} = \mathbf{0}$ and thus consider $\mathbf{w} \neq \mathbf{0}$. Then $\|\mathbf{w}\| > 0$ and the vector $\mathbf{v}_1$ defined by

$$\mathbf{v}_1 = \frac{\mathbf{w}}{\|\mathbf{w}\|} \qquad \text{has} \qquad \mathbf{v}_1 \cdot \mathbf{v}_1 = \frac{1}{\|\mathbf{w}\|^2} (\mathbf{w} \cdot \mathbf{w}) = \frac{\|\mathbf{w}\|^2}{\|\mathbf{w}\|^2} = 1.$$

Hence the single vector $\mathbf{v}_1$ is an orthonormal system of $n = 1$ vectors (the first condition of orthonormality holding trivially). Thus, from (3), with $n = 1$ and taking the square root,

$$|\mathbf{z} \cdot \mathbf{v}_1| \leqslant \|\mathbf{z}\|; \qquad \text{that is,} \qquad \left| \mathbf{z} \cdot \frac{\mathbf{w}}{\|\mathbf{w}\|} \right| = \frac{1}{\|\mathbf{w}\|} |\mathbf{z} \cdot \mathbf{w}| \leqslant \|\mathbf{z}\|$$

from which (4) follows.

[Before establishing (5) we note from (4) that $|\mathbf{z} \cdot \mathbf{w}|^2 \leqslant \|\mathbf{z}\|^2 \, \|\mathbf{w}\|^2$. By using the definitions of the inner product and norm we have

$$\left| \sum_{i=1}^{k} z_i \bar{w}_i \right|^2 \leqslant \sum_{i=1}^{k} |z_i|^2 \sum_{i=1}^{k} |w_i|^2 = \sum_{i=1}^{k} z_i \bar{z}_i \sum_{i=1}^{k} w_i \bar{w}_i ,$$

which is the first form of Schwarz inequality given in Section 1-10, (12)]

Note that $\mathbf{z} \cdot \mathbf{w} + \mathbf{w} \cdot \mathbf{z}$ has typical element $z_i \bar{w}_i + w_i \bar{z}_i$, which is real. From this reality we have

$$\mathbf{z} \cdot \mathbf{w} + \mathbf{w} \cdot \mathbf{z} \leqslant |\mathbf{z} \cdot \mathbf{w} + \mathbf{w} \cdot \mathbf{z}| \leqslant |\mathbf{z} \cdot \mathbf{w}| + |\mathbf{w} \cdot \mathbf{z}| \leqslant 2\|\mathbf{z}\| \, \|\mathbf{w}\|,$$

by using (4). Hence

$$\begin{aligned}\|\mathbf{z}+\mathbf{w}\|^2 &= (\mathbf{z}+\mathbf{w})\cdot(\mathbf{z}+\mathbf{w})\\ &= \mathbf{z}\cdot\mathbf{z}+\mathbf{z}\cdot\mathbf{w}+\mathbf{w}\cdot\mathbf{z}+\mathbf{w}\cdot\mathbf{w}\\ &\leqslant \|\mathbf{z}\|^2+2\|\mathbf{z}\|\,\|\mathbf{w}\|+\|\mathbf{w}\|^2\\ &= (\|\mathbf{z}\|+\|\mathbf{w}\|)^2\end{aligned}$$

from which (5) follows. ▮

PROBLEM 1. The vector $\mathbf{z}'$ defined in the above proof is orthogonal to each $\mathbf{v}_i$, $i = 1, 2, \cdots, n$.

PROBLEM 2.

(a) In (V, Z), $\|\mathbf{z}+\mathbf{w}\|^2 + \|\mathbf{z}-\mathbf{w}\|^2 = 2\|\mathbf{z}\|^2 + 2\|\mathbf{w}\|^2$.

[This is called the **Parallelogram law**, since in (E^2, E^1) it says that the sum of the squares of the diagonals of a parallelogram is equal to the sum of the squares of the sides.]

(b). In (Z^k, Z), $\|\mathbf{z}+\mathbf{w}\|\,\|\mathbf{z}-\mathbf{w}\| \leq \|\mathbf{z}\|^2 + \|\mathbf{w}\|^2$ with equality iff $\mathbf{z}\perp\mathbf{w}$. Give an interpretation in (E^2, E^1).

PROBLEM 3. For each $\mathbf{x} = (x_1, x_2, \cdots, x_k) \in E^k$, set

$$\|\mathbf{x}\| = \max_{1\leq i\leq k} |x_i|.$$

Show that the function $\|\ \|$ so defined makes E^k into a normed vector space.

With $\mathbf{x}_0 \in E^k$ and $s > 0$, call $\{\mathbf{x} \mid \|\mathbf{x}-\mathbf{x}_0\| < s\}$ a **cubic neighborhood** of $\mathbf{x}_0$. By using cubic neighborhood to define cluster point of a set, show that a set $A \in E^k$ has $\mathbf{y}$ as a cluster point iff $\mathbf{y}$ is a cluster point as originally defined.

A norm for a vector $\mathbf{z} \in (Z^k, Z)$ was defined specifically by (2), and if all imaginary parts are zero, this gives a norm

(6) $$\|\mathbf{x}\| = \sqrt{\mathbf{x}\cdot\mathbf{x}} = (x_1{}^2 + x_2{}^2 + \cdots + x_k{}^2)^{1/2} \qquad \text{for } \mathbf{x} \in E^k.$$

A different norm for $\mathbf{x} \in E^k$ is the one of Problem 3. Thus a vector space (V, Z) may have various norms all satisfying Definition 2. Also a space may have various inner products. The following definition is then another step in abstraction for use in studying properties of inner products per se.

Definition 4. *A vector space (V, Z) such that a function $(\cdot)$ is defined on $V \times V$ into Z, for which*

$$\begin{gathered}(\mathbf{u}\cdot\mathbf{v}) = \overline{(\mathbf{v}\cdot\mathbf{u})}\\ (\mathbf{u}+\mathbf{v})\cdot\mathbf{w} = \mathbf{u}\cdot\mathbf{w}+\mathbf{v}\cdot\mathbf{w}\\ (c\mathbf{u})\cdot\mathbf{v} = c(\mathbf{u}\cdot\mathbf{v}), \qquad c \in Z, \qquad \text{and}\\ \mathbf{u}\cdot\mathbf{u} \geqslant 0 \quad \textit{and} \quad = 0 \quad \text{iff } \mathbf{u} \neq \mathbf{0},\end{gathered}$$

is said to be an **inner-product** *space with* **inner product** $(\cdot)$.

The given inner product of an inner-product space is used to define the **standard norm** *of each* $\mathbf{v} \in V$ *by*

$$\|\mathbf{v}\| = \sqrt{\mathbf{v} \cdot \mathbf{v}}.$$

In an inner-product space the **distance** *between vectors* $\mathbf{u}, \mathbf{v} \in V$ *is defined to be*

$$\|\mathbf{u} - \mathbf{v}\|.$$

Whenever an inner-product space is given and a norm is mentioned without qualifications, it is to be understood that the standard norm is intended. On the other hand, a "normed vector space" means that Definition 2 is satisfied, but the norm need not have come from an inner product. [In this connection see Problems 22 and 23 at the end of the chapter.]

PROBLEM 4. In an inner-product space:

(a) The Schwarz and Minkowski inequalities

$$|\mathbf{x} \cdot \mathbf{y}| \leqslant \|\mathbf{x}\| \, \|\mathbf{y}\| \qquad \text{and}$$
$$\|\mathbf{x} + \mathbf{y}\| \leqslant \|\mathbf{x}\| + \|\mathbf{y}\|$$

hold. [Thus any inner-product space is at least a pseudo metric space.]

(b). The parallelogram law holds.

[*Hint for Schwarz:* With t real and c complex, expand $0 \leqslant \|t\mathbf{x} + c\mathbf{y}\|^2$ and, when applicable, see that $c = (\mathbf{x} \cdot \mathbf{y})/|\mathbf{x} \cdot \mathbf{y}|$ simplifies this to

$$t^2\|\mathbf{x}\|^2 + 2t|\mathbf{x} \cdot \mathbf{y}| + \|\mathbf{y}\|^2 \geqslant 0.]$$

The Definition 4 defines a norm in terms of an inner product which (Problem 4) satisfies the parallelogram law. For the possibility of reversing this process, see Problems 22 and 23. Problem 19 at the end of Chapter 6 shows how an inner product may be defined in terms of any real-valued nonnegative continuous function which satisfies the parallelogram law.

We return now to the inner-product vector space E^k having inner product as in Definition 3 with all imaginary parts zero (hence norm given by (6)) and the distance between $\mathbf{x} \in E^k$ and $\mathbf{y} \in E^k$ defined by

$$d(\mathbf{x}, \mathbf{y}) = \|\mathbf{x} - \mathbf{y}\|.$$

This, of course, is an involved way of saying that we consider E^k as the ordinary, garden variety, Euclidean space of k dimensions. Some thought should be given, however, to how much (maybe all) of the following discussion could be for any inner-product vector space.

The following definition extends the notion of a line and line segment in E^3 (and parametric representations of them).

Definition 5. *Given* $\mathbf{x}_1, \mathbf{x}_2 \in E^k$, *then*

$$\{\mathbf{x} \mid \mathbf{x} = t\mathbf{x}_1 + (1 - t)\mathbf{x}_2 \textit{ for some } t \in E^1\} \qquad \textit{with } \mathbf{x}_1 \neq \mathbf{x}_2$$

is said to be the **line** *containing* $\mathbf{x}_1$ and $\mathbf{x}_2$, *and*

$$\{\mathbf{x} \mid \mathbf{x} = t\mathbf{x}_1 + (1-t)\mathbf{x}_2, \qquad 0 \leqslant t \leqslant 1\},$$

is called the **line segment** *joining* $\mathbf{x}_1$ *and* $\mathbf{x}_2$, *even if* $\mathbf{x}_1 = \mathbf{x}_2$. *A set* $A \subset E^k$ *is said to be* **convex** *if whenever* $\mathbf{x}_1 \in A$ *and* $\mathbf{x}_2 \in A$, *the line segment joining* $\mathbf{x}_1$ *and* $\mathbf{x}_2$ *lies in* A.

The empty set and one-element sets are (logically) convex sets. Clearly E^k is convex.

Theorem 2. *Every neighborhood (k-ball) of* E^k *is a convex set.*

Proof. Take any $N = N(\mathbf{x}_0, r)$; $\mathbf{x}_0 \in E^k$ and $r > 0$. Select any $\mathbf{x}_1 \in N$ and $\mathbf{x}_2 \in N$. Then

$$\|\mathbf{x}_1 - \mathbf{x}_0\| < r \qquad \text{and} \qquad \|\mathbf{x}_2 - \mathbf{x}_0\| < r.$$

With t such that $0 \leqslant t \leqslant 1$ and $\mathbf{x} = t\mathbf{x}_1 + (1-t)\mathbf{x}_2$, then

$$\begin{aligned}\|\mathbf{x} - \mathbf{x}_0\| &= \|t\mathbf{x}_1 + (1-t)\mathbf{x}_2 - \mathbf{x}_0\| \\ &= \|t(\mathbf{x}_1 - \mathbf{x}_0) + (1-t)(\mathbf{x}_2 - \mathbf{x}_0)\| \\ &\leqslant t\|\mathbf{x}_1 - \mathbf{x}_0\| + (1-t)\|\mathbf{x}_2 - \mathbf{x}_0\| \\ &< tr + (1-t)r = r.\end{aligned}$$

Thus $\mathbf{x} \in N$ and hence N is convex. ∎

Theorem 3. *If* $A \subset E^k$ *is convex, then its closure* $\bar{A}$ *is convex.*

Proof. Assume A is convex but $\bar{A}$ is not convex. Then $\bar{A}$ contains at least two points and $\bar{A} \neq E^k$. Also, there are points $\mathbf{x}_1, \mathbf{x}_2 \in \bar{A}$ and a $t_0 \in [0, 1]$ such that

$$\mathbf{x}_0 = t_0\mathbf{x}_1 + (1-t)\mathbf{x}_2 \notin \bar{A}.$$

Since $\bar{A}^c$ is a nonempty open set, let $r > 0$ be any number such that

$$N(\mathbf{x}_0, r) \subset \bar{A}^c.$$

Since $\mathbf{x}_1, \mathbf{x}_2 \in \bar{A} = A \cup A'$, let $\mathbf{y}_1, \mathbf{y}_2 \in A$ such that

$$0 < \|\mathbf{y}_1 - \mathbf{x}_1\| < r \qquad \text{and} \qquad 0 < \|\mathbf{y}_2 - \mathbf{x}_2\| < r.$$

Then $\mathbf{y} = t_0\mathbf{y}_1 + (1-t_0)\mathbf{y}_2 \in A$, since A is convex, but

$$\|\mathbf{y} - \mathbf{x}_0\| \leqslant t_0\|\mathbf{y}_1 - \mathbf{x}_1\| + (1-t_0)\|\mathbf{y}_2 - \mathbf{x}_2\| < r,$$

so that $\mathbf{y} \in \bar{A}^c \subset A^c$. This contradiction shows that $\bar{A}$ is also convex.

PROBLEM 5. Any family of convex sets has its intersection a convex set.

Definition 6. *Take any* $\mathbf{a} \in E^k$ *such that* $\mathbf{a} \neq \mathbf{0}$ *and any* $c \in E^1$. *Then we define*

(i) $\{\mathbf{x} \in E^k | \mathbf{a} \cdot \mathbf{x} = c\}$ is a **hyperplane**.
(ii) $\{\mathbf{x} \in E^k | \mathbf{a} \cdot \mathbf{x} < c\}$ and $\{\mathbf{x} \in E^k | \mathbf{a} \cdot \mathbf{x} > c\}$ are **half-spaces**.

PROBLEM 6. (a) A half-space is a convex set. (b) The whole line containing any two distinct points of a hyperplane lies in the hyperplane.

PROBLEM 7. Let P be a hyperplane having equation $\mathbf{a}\cdot\mathbf{x}=c$, take any $\mathbf{x}_0\in E^k$ and define

$$\mathbf{y}_1=\mathbf{x}_0-\frac{\mathbf{a}\cdot\mathbf{x}_0-c}{\|\mathbf{a}\|^2}\mathbf{a}.$$

Then

(a) $\mathbf{y}_1\in P$.

(b) $\|\mathbf{y}_1-\mathbf{x}_0\|=\dfrac{|\mathbf{a}\cdot\mathbf{x}_0-c|}{\|\mathbf{a}\|}$.

(c) If $\mathbf{y}\in P$, then $(\mathbf{y}-\mathbf{y}_1)\perp(\mathbf{x}_0-\mathbf{y}_1)$.

PROBLEM 8. Any half-space is an open set.

PROBLEMS

1. Given $A=\{1, 2,\cdots,n\}$ and $B=\{0, 1\}$, how many functions $f\colon A\to B$ are there which map:

(a) A into B? (b) A onto B?

2. Describe a set $A\subset E^1$ such that (a) $A'=\{\frac{1}{4},\frac{1}{3},\frac{1}{2}\}$; (b) A' is countably infinite.

3. If ξ is irrational, then $\{x\mid x=m+n\xi,\ m \text{ and } n \text{ integers}\}$ is countable and dense in E^1.

4. A function $f\colon[0, 1]\to E^1$ has the property that there is a number $\lambda>0$ such that

$$|f(x_1)+f(x_2)+\cdots+f(x_n)|\leqslant\lambda$$

for any choice of any finite number of distinct points $x_1, x_2,\cdots,x_n$ of $[0, 1]$. Show that $\{x\mid 0\leqslant x\leqslant 1 \text{ and } f(x)\neq 0\}$ is countable.

5. (a) Given a set $A\subset E^1$ and a number $k\neq 0$, let $B=\{ka\mid a\in A\}$. Then $x\in A'$ iff $kx\in B'$.

(b) Given a set $A\subset E^1$ and a number k, let $B=\{a-k\mid a\in A\}$. Then $x\in B'$ iff $x+k\in A'$.

6. Any nonempty bounded closed set $A\subset E^1$ is either a closed interval or may be obtained from a closed interval by removing a countable disjoint family of open intervals each having both end points in A.

7. If $0\leqslant\lambda\leqslant 1$ there are points $c_1, c_2\in\Gamma$ such that

$$\frac{c_1+3c_2}{3}=\lambda.$$

8. Let R be the space of all rational numbers with the distance between $p, q\in R$ defined by

$$d(p, q)=|p-q|.$$

Then

(a) R is a metric space.

(b) The set $A=\{p\mid p \text{ rational and } 2<p^2<3\}$ is closed and bounded, but is not compact.

9. If $A \subset E^k$ has the property that every $a \in A$ has a neighborhood $N(a)$ such that $N(a) \cap A$ is countable, then A is countable.

10. A number is said to be **algebraic** if it is the root of some polynomial with integer coefficients. Prove that the set of algebraic numbers is countable.

11. The family $\mathscr{F}$ of intervals $\{x \mid 1/n < x < 2/n\}$, $n = 2, 3, 4, \cdots$ is an open covering of the open interval $(0, 1)$, but no finite subfamily covers $(0, 1)$.

12. Theorem. If X is a separable metric space and F is any closed set in X, then $F = P \cup C$ with P perfect, C countable, and $P \cap C = \varnothing$. [Remember that $\varnothing$ is both perfect and countable. Extend Problem 9 to X.]

Corollary. Every nonempty countable closed set in E^k has isolated points.

13. Theorem 1. If $\{G_n\}$ is a sequence of open sets in E^k such that each G_n is dense in E^k, then $\cap G_n$ is not empty.

Theorem 2. If $\{F_n\}$ is a sequence of closed sets in E^k such that $E^k = \cup F_n$, then at least one F_n has a nonempty open subset.

Prove either theorem [*Hint:* Method of Theorem 2-8.1] and show that the theorems are equivalent.

14. Let S be a metric space in which every infinite subset has a cluster point. Then
(a) S is separable.
(b) S is compact.

15. The family of open circular disks in E^2, each having center (r, r) and radius r for some rational $r > 0$, is a countable covering of $\{(x, y) \mid x > 0, y > 0\}$.

16. Definition. A family $\mathscr{G}$ of open sets in a metric space X is said to be a **base** for X if whenever H is an open set in X and $x \in H$, there is a $G \in \mathscr{G}$ such that $x \in G \subset H$.

Theorem 1. If X has a base $\mathscr{G}$, then every open set $H \subset X$ is the union of sets in $\mathscr{G}$.

Theorem 2. If X is separable, then
(a) X has a countable base.
(b) Every open covering of X has a countable subcover.

17. Theorem 1. If X is a metric space in which every infinite subset has a cluster point (in X), then X is separable and compact.

Theorem 2. If X is a metric space in which every bounded infinite subset has a cluster point, then X is separable and is the countable union of compact sets.

[*Hint for Theorem* 1: With n a positive integer choose $x_1 \in X$ and let $N_1 = N(x_1, 1/n)$. Choose $x_2 \in XN_1{}^c$, if there is one, and let $N_2 = N(x_2, 1/n)$. Choose $x_3 \in X(N_1 \cup N_2)^c$, if there is one, etc. Show that the process stops in a finite number of steps. Do this for $n = 1, 2, 3, \cdots$ and consider all points selected to obtain separability.

Then use Problem 16 to see that any open cover of X has a countable subcover, say $G_1, G_2, G_3, \cdots$. This countable subcover has a finite subcover, for otherwise we could find $x_1, x_2, x_3, \cdots$ in X such that

$$x_n \notin (G_1 \cup G_2 \cup \cdots \cup G_n)$$

and any cluster point of $\{x_1, x_2, x_3, \cdots\}$ leads to a contradiction.]

18. Let f: $E^1 \to E^1$ be such that

$$f(x+y)f(x-y) \leqslant f^2(x) - f^2(y) \qquad \text{for all } x, y \in E^1.$$

Then $f(x+y)f(x-y) = f^2(x) - f^2(y)$ for all $x, y \in E^1$.

[*Hint:* Show that $f(u)f(v) \leqslant f^2[(u+v)/2] - f^2[(u-v)/2]$. Put $u = v = 0$ and find $f(0) = 0$, put $u = -v = x$ and find $f(x)f(-x) \leqslant 0$, put $u = 0$, $v = -2x$ and find $f^2(x) \leqslant f^2(-x) \leqslant f^2(-(-x))$. Deduce that $f(-x) = -f(x)$ and then show that the given inequality may be reversed.]

19. (a) If f: $E^1 \to E^1$ is such that $f(-x) = -f(x)$ and $f(x+y) \leqslant f(x) + f(y)$ for all $x, y \in E^1$, then $f(x+y) = f(x) + f(y)$ for all $x, y \in E^1$ and $f(0) = 0$.

(b) If $\varphi: E^1 \to E^1$ is such that $\varphi(x+y) = \varphi(x) + \varphi(y)$ for all $x, y \in E^1$, then $\varphi(0) = 0$, $\varphi(-x) = -\varphi(x)$ and

$$\varphi(x+y)\varphi(x-y) = \varphi^2(x) - \varphi^2(y) \qquad \text{for all } x,\ y \in E^1.$$

20. In a vector space V, if a function $p: V \to [0, \infty)$ satisfies the parallelogram law,

$$p^2(\mathbf{x}+\mathbf{y}) + p^2(\mathbf{x}-\mathbf{y}) = 2p^2(\mathbf{x}) + 2p^2(\mathbf{y}),$$

then p also satisfies the triangular inequality

$$p(\mathbf{x}+\mathbf{y}) \leqslant p(\mathbf{x}) + p(\mathbf{y}).$$

[*Hint:* First show that $p(\mathbf{0}) = 0$ and $p(2\mathbf{x}) = 2p(\mathbf{x})$. Assume the triangular inequality does not hold. Take vectors $\mathbf{a}$ and $\mathbf{b}$ and a number $\delta > 0$ such that

$$(*) \qquad p(\mathbf{a}) + p(\mathbf{b}) + \delta < p(\mathbf{a}+\mathbf{b})$$

so named that $p(\mathbf{a}) \geqslant p(\mathbf{b})$. Hence, from the parallelogram law,

$$\begin{aligned} p^2(\mathbf{a}-\mathbf{b}) &= 2p^2(\mathbf{a}) + 2p^2(\mathbf{b}) - p^2(\mathbf{a}+\mathbf{b}) \\ &< p^2(\mathbf{a}) - 2p(\mathbf{a})p(\mathbf{b}) + p^2(\mathbf{b}) \\ &\quad -2\delta p(\mathbf{a}) - 2\delta p(\mathbf{b}) - \delta^2 \\ &\leqslant \begin{cases} [p(\mathbf{a}) - p(\mathbf{b})]^2 - \delta^2 \\ [p(\mathbf{a}) - p(\mathbf{b}) - \delta]^2. \end{cases} \end{aligned}$$

The upper line gives $p(\mathbf{a}) - p(\mathbf{b}) - \delta > 0$ and then from the lower line $p(\mathbf{a}-\mathbf{b}) \leqslant p(\mathbf{a}) - p(\mathbf{b}) - \delta$. This, the inequality (*), and $2p(\mathbf{b}) = p(2\mathbf{b})$ yield

$$p(\mathbf{a}-\mathbf{b}) + p(2\mathbf{b}) + 2\delta < p(\mathbf{a}+\mathbf{b}).$$

Set $\mathbf{a}-\mathbf{b} = \mathbf{a}_1$ and $2\mathbf{b} = \mathbf{b}_1$. Then $\mathbf{a}_1 + \mathbf{b}_1 = \mathbf{a}+\mathbf{b}$ and

$$p(\mathbf{a}_1) + p(\mathbf{b}_1) + 2\delta < p(\mathbf{a}_1 + \mathbf{b}_1)$$

which is in the form of (*) with δ replaced by 2δ. Hence for $n = 1, 2, 3, \cdots$, there are $\mathbf{a}_n$ and $\mathbf{b}_n$ such that

$$p(\mathbf{a}_n) + p(\mathbf{b}_n) + 2^n\delta < p(\mathbf{a}_n + \mathbf{b}_n) \qquad \text{and} \qquad \mathbf{a}_n + \mathbf{b}_n = \mathbf{a}+\mathbf{b}.$$

But $2^n\delta < p(\mathbf{a}+\mathbf{b})$ leads to a contradiction.]

21. For a real vector space V, if a "circle product" is defined on $V \times V$ into E^1 and satisfies the distributive law,

$$(\mathbf{x}+\mathbf{y}) \circ \mathbf{z} = \mathbf{x} \circ \mathbf{z} + \mathbf{y} \circ \mathbf{z},$$

then $(r\mathbf{x}) \circ \mathbf{y} = r(\mathbf{x} \circ \mathbf{y})$ for each rational number r.

22. Let V be a real normed vector space whose norm satisfies the parallelogram law. Define a "circle product" on $V \times V$ by

$$\mathbf{x} \circ \mathbf{y} = \tfrac{1}{4}[\,\|\mathbf{x}+\mathbf{y}\|^2 - \|\mathbf{x}-\mathbf{y}\|^2].$$

Show that this circle product satisfies the conditions of a dot product except that $(c\mathbf{x}) \circ \mathbf{y} = c(\mathbf{x} \circ \mathbf{y})$ is known only for c rational.
[*Hint: for the distributive law*

$$\begin{aligned} 4[(\mathbf{x}+\mathbf{z}) \circ \mathbf{y} - \mathbf{x} \circ \mathbf{y} - \mathbf{z} \circ \mathbf{y}] = {} & \|(\mathbf{x}+\mathbf{y}) + \mathbf{z}\|^2 - \|(\mathbf{z}-\mathbf{y}) + \mathbf{x}\|^2 \\ & - \|\mathbf{x}+\mathbf{y}\|^2 + \|\mathbf{x}-\mathbf{y}\|^2 \\ & - \|\mathbf{z}+\mathbf{y}\|^2 + \|\mathbf{z}-\mathbf{y}\|^2; \end{aligned}$$

then use the parallelogram law on the first, second, fourth, and fifth terms.]

23. In a normed complex vector space (V, Z) whose norm satisfies the parallelogram law, the "circle product" defined by

$$\mathbf{x} \circ \mathbf{y} = \tfrac{1}{4}[\|\mathbf{x} + \mathbf{y}\|^2 - \|\mathbf{x} - \mathbf{y}\|^2 + i\|\mathbf{x} + i\mathbf{y}\|^2 - i\|\mathbf{x} - i\mathbf{y}\|^2]$$

has the distributive property

$$(\mathbf{x} + \mathbf{y}) \circ \mathbf{z} = \mathbf{x} \circ \mathbf{z} + \mathbf{y} \circ \mathbf{z}.$$

24. A set $A \subset E^k$ is said to be **polygonally connected** if for each pair $a, b \in A$ there is a finite number of points $a = a_0, a_1, a_2, \cdots, a_n = b$ such that the whole segment joining a_{i-1} and a_i is in A, $i = 1, 2, \cdots, n$.

(a) Any open, nonempty, connected set $A \subset E^k$ is polygonally connected. [*Hint:* For each $a \in A$ the set P_a of all points polygonally connectable in A to a is open and so is $A - P_a$.]

(b) The complement in E^k, $k > 1$, of the set of all points with rational coordinates is polygonally connected.

Problems on Relations

For the explanations and problems of this section we use X, Y, Z, and W as nonempty sets (or spaces).

Definition 1. A subset $R \subset X \times Y$ is called a **relation** in $X \times Y$, or a relation from X into Y. The terminology "y is in the relation R to x" is used if $(x, y) \in R$ and may also be expressed by yRx.

Definition 2. Given a relation $R \subset X \times Y$, the inverse of R is denoted by R^{-1} and is the relation in $Y \times X$ defined by

$$R^{-1} = \{(y, x) \mid (x, y) \in R\}.$$

Expressed another way $xR^{-1}y \Leftrightarrow yRx$.

It then follows that

$$\begin{aligned}(R^{-1})^{-1} &= \{(a, b) \mid (b, a) \in R^{-1}\} \\ &= \{(a, b) \mid (a, b) \in R\} = R.\end{aligned}$$

If R_1 and R_2 are both relations in $X \times Y$, then $R_1 \cup R_2$ and $R_1 \cap R_2$ are also relations in $X \times Y$ such that

$$(R_1 \cup R_2)^{-1} = R_1^{-1} \cup R_2^{-1} \qquad \text{and} \qquad (R_1 \cap R_2)^{-1} = R_1^{-1} \cap R_2^{-1}.$$

The first of these is seen as follows:

$$\begin{aligned}(R_1 \cup R_2)^{-1} &= \{(a, b) \mid (b, a) \in R_1 \cup R_2\} \\ &= \{(a, b) \mid (b, a) \in R_1 \quad \text{or} \quad (b, a) \in R_2\} \\ &= \{(a, b) \mid (b, a) \in R_1\} \cup \{(a, b) \mid (b, a) \in R_2\} \\ &= R_1^{-1} \cup R_2^{-1},\end{aligned}$$

and the second follows in a similar way.

Definition 3. *Given $R \subset X \times Y$ and $S \subset Y \times Z$, we define $S \circ R$* by

$$S \circ R = \{(x, z) \mid \exists\, y \in Y \text{ such that } (x, y) \in R \text{ and } (y, z) \in S\}.$$

Thus $S \circ R \subset X \times Z$ and this relation from X into Z is called the **composition** *of S and R (in this order).*

Note that if $S \circ R$ can be formed, then the second space for R must be in the first space for S. For example, if $R \subset X \times Y$ and $S \subset W \times Z$, then $S \circ R$ is undefined unless $Y \subset W$.

With $R \subset X \times Y$ and $S \subset Y \times Z$, then

$$R^{-1} \subset Y \times X \qquad \text{and} \qquad S^{-1} \subset Z \times Y.$$

Thus the common space Y is in the "wrong" position for forming $S^{-1} \circ R^{-1}$, but is in the "right" position for forming $R^{-1} \circ S^{-1}$.

PROBLEM 1. If $R \subset X \times Y$ and $S \subset Y \times Z$, then

$$(S \circ R)^{-1} = R^{-1} \circ S^{-1}.$$

We display $R \subset X \times Y$ and its inverse R^{-1} as

$$R \subset X \times Y, \qquad R^{-1} \subset Y \times X$$

to show that Y is in the right position for forming

$$R^{-1} \circ R \subset X \times X.$$

Note also that $R \circ R^{-1}$ may be formed and

$$R \circ R^{-1} \subset Y \times Y.$$

Then, upon replacing S in Problem 1 by R^{-1}, we have

$$(R^{-1} \circ R)^{-1} = R^{-1} \circ (R^{-1})^{-1} = R^{-1} \circ R.$$

This fact expressed another way is

$$(x, s) \in R^{-1} \circ R \Leftrightarrow (s, x) \in R^{-1} \circ R.$$

Similarly,

$$(R \circ R^{-1})^{-1} = R \circ R^{-1} \qquad \text{and} \qquad (y, t) \in R \circ R^{-1} \Leftrightarrow (t, y) \in R \circ R^{-1}.$$

Theorem 1. *If $R \subset X \times Y$, $S \subset Y \times Z$, and $T \subset Z \times W$, then*

$$(T \circ S) \circ R = T \circ (S \circ R)$$

[*and it is unambiguous to write $T \circ S \circ R$*].

Proof. Take $(a, b) \in (T \circ S) \circ R$. Then there is a c such that $(a, c) \in R$ and $(c, b) \in T \circ S$. Hence there is a d such that $(c, d) \in S$ and $(d, b) \in T$. We thus have

$$(a, c) \in R, \qquad (c, d) \in S, \qquad \text{and} \qquad (d, b) \in T.$$

The first two of these say that $(a, d) \in S \circ R$ and this with $(d, b) \in T$ says that $(a, b) \in T \circ (S \circ R)$. Thus

$$(a, b) \in (T \circ S) \circ R \Rightarrow (a, b) \in T \circ (S \circ R).$$

A similar argument replaces $\Rightarrow$ by $\Leftarrow$. ∎

If $R \subset X \times Y$, then $R^{-1} \circ R \subset X \times X$ and $R \circ R^{-1} \subset Y \times Y$ so that both $R \circ (R^{-1} \circ R)$ and $(R \circ R^{-1}) \circ R$ may be formed, they are the same subset of $X \times Y$, and are denoted by

$$R \circ R^{-1} \circ R.$$

The next example shows that this double composition need not be equal to R.

EXAMPLE. With $X = \{1, 2, 3\}$ and $y = \{2, 4\}$ let

$$R = \{(1, 2), (1, 4), (2, 2)\}.$$

Then $R^{-1} = \{(2, 1), (4, 1), (2, 2)\}$. Thus

$$R^{-1} \circ R = \{(1, 1), (1, 2), (2, 1), (2, 2)\}$$

and

$$R \circ (R^{-1} \circ R) = \{(1, 2), (1, 4), (2, 2), (2, 4)\}$$
$$= R \cup \{(2, 4)\}.$$

PROBLEM 2. Take X and Y as in the example.

(a) Let $R = \{(3, 2), (3, 4)\}$ and show that

$$R \circ R^{-1} \circ R = R.$$

(b) Find $R \circ R^{-1} \circ R$ if $R = \{(1, 2), (2, 4), (3, 4)\}$.

Theorem 2. *If R is any relation from any X into any Y, then*

$$R \subset R \circ R^{-1} \circ R.$$

Proof. Take any $(a, b) \in R$. Then $(b, a) \in R^{-1}$ and hence (a, a) is among the elements of $R^{-1} \circ R$; that is,

$$(a, b) \in R \qquad \text{and} \qquad (a, a) \in R^{-1} \circ R.$$

Then $(a, b) \in R \circ (R^{-1} \circ R)$, since we know the existence of a c such that $(a, c) \in R^{-1} \circ R$ and $(c, b) \in R$; namely, $c = a$. Thus

$$(a, b) \in R \Rightarrow (a, b) \in R \circ R^{-1} \circ R. \ \blacksquare$$

The following definition of a function is a rephrasing of the "ordered pair" definition given in Section 2-1.

Definition 4. A relation $f \subset X \times Y$ is said to be a **function** from X into Y if

$$[(x, y_1) \in f \text{ and } (x, y_2) \in f] \Rightarrow y_1 = y_2.$$

Theorem 3. *If $f \subset X \times Y$ is a function, then*

$$f \circ f^{-1} \circ f = f.$$

Proof. Since f is a relation we already know that

$$f \subset f \circ f^{-1} \circ f.$$

Take any $(a, b) \in f \circ f^{-1} \circ f = (f \circ f^{-1}) \circ f$. Then there is a c such that

$$(a, c) \in f \qquad \text{and} \qquad (c, b) \in f \circ f^{-1}.$$

From $(c, b) \in f \circ f^{-1}$ there is a d such that

$$(c, d) \in f^{-1} \qquad \text{and} \qquad (d, b) \in f; \qquad \text{that is}$$
$$(d, c) \in f \qquad \text{and} \qquad (d, b) \in f.$$

This last pair means that $c = b$, by the above definition of a function. Thus the above expression $(a, c) \in f$ becomes $(a, b) \in f$. Hence

$$(a, b) \in f \circ f^{-1} \circ f \Rightarrow (a, b) \in f,$$

which gives the opposite inclusion to the one already known. $\blacksquare$

PROBLEM 3. If $R \subset X \times Y$ is such that R^{-1} is a function (from Y into X), then $R \circ R^{-1} \circ R = R$.

Problems on Topological Spaces

As the following definition is read, keep in mind known properties of the family of open sets in a metric space.

Definition 1. *Let X be a nonempty set and let $\mathscr{T}$ be a family of subsets of X having the properties:*

1. *$\varnothing$ and X are members of $\mathscr{T}$;*
2. *The union of any collection of members of $\mathscr{T}$ is a member of $\mathscr{T}$; and*
3. *The intersection of any finite number of members of $\mathscr{T}$ is a member of $\mathscr{T}$.*

Then $\mathscr{T}$ is said to be a **topology** *for X. Also, X together with $\mathscr{T}$ is said to be a* **topological space,** *which we denote by $(X, \mathscr{T})$. The members of $\mathscr{T}$ are called the* **open** *sets for this topology.*

The space X need not be metric, and if $\mathscr{T}$ is a topology for X, then "open" in this new topological sense conveys no implication about interior points. In fact, any nonempty set A may be made into a topological space by choosing for $\mathscr{T}$ *all* subsets of A. In this so-called **discrete** topology every subset of A is open, even the one-element subsets.

EXAMPLE 1. Let X be a nonempty subset of a metric space S and let $\mathscr{T}$ be the family of subsets of X each of which is of the form $G \cap X$ for some open (in the metric sense) set G of S. Then $\mathscr{T}$ is a topology for X. [This is merely the topological form of notation of relative openness. See Theorem 2-11.5.] This $\mathscr{T}$ is said to be the topology **induced** on X by the **standard topology** of S.

EXAMPLE 2. $(E^2, \mathscr{T})$ is a topological space if $\mathscr{T}$ is defined by saying that $T \in \mathscr{T}$ iff there is an open (in the standard topology of E^1) set $G \subset E^1$ such that $T = G \times E^1$. In particular, $\varnothing = \varnothing \times E^1$, where the first $\varnothing$ is the empty set of E^2 and the second of E^1.

PROBLEM 1. Let $(X, \mathscr{T})$ be a topological space. If $A \subset X$ is a set having the property that corresponding to each $a \in A$ there is a $T \in \mathscr{T}$ such that

$$a \in T \subset A,$$

then A is an open set (meaning $A \in \mathscr{T}$).

[*Hint:* Let G be the union of all members of $\mathscr{T}$ which are subsets of A.]

Definition 2. *Given a topological space $(X, \mathscr{T})$, then*

(i) *The* **closed** *sets of X are those sets which are complements of open sets;*
(ii) *A point $x \in X$ is a* **cluster point** *of a set $A \subset X$ if whenever $x \in T \in \mathscr{T}$ there is a $y \neq x$ such that $y \in A \cap T$;*
(iii) *The* **derived** *set A' of A is the set of all cluster points of A and $\bar{A} = A \cup A'$ is the* **closure** *of A;*
(iv) *A set A is said to be* **dense** *in a set B if $B \subset \bar{A}$. (Hence A is dense in X if $\bar{A} = X$.)*

PROBLEM 2. A set A is closed iff $A' \subset A$.

In a topological space $(X, \mathscr{T})$, let A be a set and $\mathscr{F}$ a family of members of $\mathscr{T}$. Then $\mathscr{F}$ is said to be an **open covering** of A if each point of A is in at least one member of $\mathscr{F}$. The definition of A being **compact** also reads as before.

PROBLEM 3. In a topological space, if A is a compact set, then

(a) Every closed subset of A is compact.
(b) Every infinite subset of A has a cluster point in A.

Definition 3. *With X a nonempty set, corresponding to each $x \in X$, let $\mathscr{N}_x$ be a nonempty family of nonempty subsets of X and let $\mathscr{N}$ be the family of all such sets* (that is, $N \in \mathscr{N}$ iff there is an x such that $N \in \mathscr{N}_x$). *Then $(X, \mathscr{N})$ is said to be a* **neighborhood space**, *provided*

(i) *For each $x \in X$, if $N_1 \in \mathscr{N}_x$ and $N_2 \in \mathscr{N}_x$, then there is an $N_3 \in \mathscr{N}_x$ such that $N_3 \subset N_1 \cap N_2$.*
(ii) *For each $x \in X$ and each $N \in \mathscr{N}_x$, if $y \in N$ then there is an $N_1 \in \mathscr{N}_y$ such that $N_1 \subset N$.*

Each $N \in \mathscr{N}_x$ is said to be a **neighborbood** *of x.*

Although a neighborhood space is an abstraction of $\{y \mid d(x, y) < r\}$ being a neighborhood of x in a metric space, the above definition does not require that x be in any of its neighborhoods. For example (in descriptive terms), in E^2 if we specify that $\mathscr{N}_x$ is the family of all open circular disks each having x as the lowest point of its rim, then $(E^2, \mathscr{N})$ is a neighborhood space, but $x \notin N$ for all $N \in \mathscr{N}_x$.

For the following problems, forget that "open" had any previous connotation.

PROBLEM 4. Given a neighborhood space $(X, \mathscr{N})$, a set T is said to be open if corresponding to each $x \in T$ there is an $N \in \mathscr{N}_x$ such that $N \subset T$. Let $\mathscr{T}$ be the family of all open sets in X together with $\varnothing$. Then $(X, \mathscr{T})$ is a topological space.

PROBLEM 5. If $(X, \mathscr{T})$ is a topological space and we define $\mathscr{N}_x$ as the family of all $T \in \mathscr{T}$ for which $x \in T$, then $(X, \mathscr{N})$ is a neighborhood space.

The following definition may clarify the parenthetic statement of the first sentence in the proof of Theorem 2-4.1.

Definition 4. *A neighborhood space $(X, \mathscr{N})$ is said to be a* **Hausdorff** *space* (or a T_2 *space*) *if*

(i) *$x \in N$ for each $N \in \mathscr{N}_x$; and*
(ii) *if $x, y \in X$ and $x \neq y$, then there is an $N_1 \in \mathscr{N}_x$ and an $N_2 \in \mathscr{N}_y$ such that $N_1 \cap N_2 = \varnothing$.*

PROBLEM 6. In a Hausdorff space:

(a) Any compact set is closed.
(b) If $A \cap B = \varnothing$ with A and B compact, then there are open sets $U \supset A$ and $V \supset B$ such that $U \cap V = \varnothing$.

PROBLEM 7. In E^1 let $\mathscr{T}$ be the family of sets defined by $T \in \mathscr{T}$ iff either $T = \varnothing$ or T^c is countable.

(a) Then $(E^1, \mathscr{T})$ is a topological space.
(b) Make $(E^1, \mathscr{T})$ into the neighborhood space $(E^1, \mathscr{N})$ as in Problem 5. Then $(E^1, \mathscr{N})$ is not a Hausdorff space.

Chapter 3

SEQUENCES AND SERIES

3–1 THE EXTENDED REAL NUMBER SYSTEM

Definition 1. *The* **extended real number system** *is denoted by* E^* *and consists of* E^1 *together with the symbols* ∞ *and* $-\infty$ *satisfying*

(i) $\infty + \infty = \infty \cdot \infty = (-\infty)(-\infty) = \infty,$
$-\infty + (-\infty) = \infty(-\infty) = -\infty < \infty.$

(ii) *For each* $x \in E^1$,
$-\infty < x < \infty,$
$x + \infty = \infty, \qquad x - (-\infty) = \infty,$
$x + (-\infty) = -\infty, \qquad x - \infty = -\infty$
$\dfrac{x}{\infty} = \dfrac{x}{-\infty} = 0,$
$0(\infty) = 0(-\infty) = 0.$

(iii) *If* $0 < x < \infty$, *then* $x(\infty) = \infty$ *and* $x(-\infty) = -\infty$.
If $-\infty < x < 0$, *then* $x(\infty) = -\infty$ *and* $x(-\infty) = \infty$.

(iv) *If* $x \in E^1$ *and* $y \in E^*$, *then*
$x + y = y + x \qquad \text{and} \qquad xy = yx.$

Notice that none of the following are defined

$$\infty + (-\infty), \qquad -\infty + \infty, \qquad \frac{x}{0} \quad \text{for } x \in E^*.$$

In particular, E^* is not a field.

The following theorem extends the fact: If $a, b \in E^1$ and $a < b + \varepsilon$ for each $\varepsilon > 0$, then $a \leqslant b$.

Theorem 1. *If* $a, b \in E^*$ *and* $a < c$ *for each* $c > b$, *then* $a \leqslant b$.

Proof. Assume that a and b satisfy the hypotheses but not the conclusion. Then $a > b$ so that $a \neq -\infty$, $b \neq \infty$, and there is a $c \in E^1$ such that $a > c > b$. Now $c > b \Rightarrow a < c$ in contradiction to $a > c$. ∎

The use of inf A and sup A *for* $A \subset E^1$ is extended by the following definition.

Definition 2. *If a nonempty set* $A \subset E^*$ *is*

(i) *Unbounded below, then* $\inf A = -\infty$.

(ii) *Unbounded above, then* $\sup A = \infty$.

Also

(iii) $\inf \varnothing = \infty$ *and* $\sup \varnothing = -\infty$.

The definition (iii) is given on the grounds that neither $x < \infty$ nor $-\infty < x$ holds for any $x \in \varnothing$.

PROBLEM 1. Given that $A \subset E^*$ and $A \neq \varnothing$,

(a) If $b \leqslant a$ for all $a \in A$, then $b \leqslant \inf A$.

(b) If $b \geqslant a$ for all $a \in A$, then $b \geqslant \sup A$.

For $A \subset E^*$ both inf A and sup A are uniquely defined and $\inf A \leqslant \sup A$ iff $A \neq \varnothing$. The phraseology "inf A and sup A exist* for any $A \subset E^*$" will be used. Then the statement "inf A or sup A exists," for a certain set $A \subset E^*$, means that $A \neq \varnothing$ and $-\infty < \inf A \leqslant \sup A < \infty$ for this set A. Thus, if

$$A = \{1, 2, 3, \cdots\},$$

then inf A and sup A exist*, but also $\inf A = 1$ exists and $\sup A = \infty$.

Also if $A = \{\infty, \infty, \infty, \cdots\} = \{\infty\}$, then $\sup A = \inf A = \infty$.

The proof of the following theorem gives practice in thinking about supremums and infimums which may be infinite.

Theorem 2. *Given* $A \subset E^*$ *let* $A^- = \{x \mid -x \in A\}$. *Then*

$$\sup A^- = -\inf A \quad \text{and} \quad \inf A^- = -\sup A. \tag{1}$$

Proof. If $A = \varnothing$, then $A^- = \varnothing$ and the equalities hold by Definition 2 (iii). Hence consider $A \neq \varnothing$.

Case 1. $\inf A = \infty$. Then $A = \{\infty\}$, $A^- = \{-\infty\}$ and $\sup A^- = -\infty = -\inf A$.

Case 2. $\inf A < \infty$. Take any $c > \inf A$ and $c \in E^1$. Then there is an $a \in A$ such that $a < c$, $-a > -c$, and hence $\sup A^- > -c$, since $-a \in A^-$. Thus $-\sup A^- < c$ for each $c > \inf A$, and hence $-\sup A^- \leqslant \inf A$ by Theorem 1. Therefore

$$\sup A^- \geqslant -\inf A.$$

However, $b \in A^- \Rightarrow -b \in A \Rightarrow -b \geqslant \inf A \Rightarrow b \leqslant -\inf A$, so that

$$\sup A^- \leqslant -\inf A.$$

Thus the first equality in (1) holds, and then from this equality with A replaced by A^-

$$\sup(A^-)^- = -\inf A^-; \qquad \text{that is} \qquad \sup A = -\inf A^-$$

and hence the second equality in (1) also holds. ∎

The supremum of a set of real numbers is used in the following definition to extend the notion of the diameter of a circle.

Definition 3. *A nonempty set A in a metric space S is said to have* **diameter** $\operatorname{di}(A) \in E^*$, *where*

$$\operatorname{di}(A) = \sup\{d(x, y) \mid x \in A, \quad y \in A\}.$$

Hence the diameter of a set is a nonnegative number or is ∞.

Theorem 3. *A nonempty set A in a metric space is bounded iff its diameter is finite.*

Proof. Let A have $\operatorname{di}(A) < \infty$. Let $a_0 \in A$. Then

$$a \in A \Rightarrow d(a, a_0) \leqslant \operatorname{di}(A) \Rightarrow a \in N(a_0, \operatorname{di}(A)) \Rightarrow A \subset N(a_0, \operatorname{di}(A)).$$

Conversely let A be bounded. Then there is an $a_0 \in A$ and an $r > 0$ such that $A \subset N(a_0, r)$. Hence

$$a \in A \Rightarrow a \in N(a_0, r) \Rightarrow d(a, a_0) < r \Rightarrow \operatorname{di} A \leqslant r. \blacksquare$$

Theorem 4. *A nonempty set A in a metric space and its closure have the same diameter; that is,* $\operatorname{di}(\bar{A}) = \operatorname{di}(A)$.

Proof. Let $\bar{x}, \bar{y} \in \bar{A}$. Choose $\varepsilon > 0$ arbitrarily. Then there are $x, y \in A$ such that $d(x, \bar{x}) < \varepsilon$ and $d(y, \bar{y}) < \varepsilon$; so that

$$\begin{aligned} d(\bar{x}, \bar{y}) &\leqslant d(\bar{x}, x) + d(x, y) + d(y, \bar{y}) \\ &\leqslant \operatorname{di}(A) + 2\varepsilon. \end{aligned}$$

Hence $\operatorname{di}(\bar{A}) \leqslant \operatorname{di}(\bar{A}) + 2\varepsilon$ and thus $\operatorname{di}(\bar{A}) \leqslant \operatorname{di}(A)$, since $\varepsilon > 0$ is arbitrary. But $A \subset \bar{A}$ so that $\operatorname{di}(A) \leqslant \operatorname{di}(\bar{A})$ and thus $\operatorname{di}(\bar{A}) = \operatorname{di}(A)$. $\blacksquare$

3–2 LIMITS INFERIOR AND SUPERIOR

Definition 1. *Given a real sequence $\{a_n\}$, let*

$$\begin{aligned} \underline{a}_1 &= \inf\{a_1, a_2, a_3, \cdots\}, & \bar{a}_1 &= \sup\{a_1, a_2, a_3, \cdots\}, \\ \underline{a}_2 &= \inf\{a_2, a_3, a_4, \cdots\}, & \bar{a}_2 &= \sup\{a_2, a_3, a_4, \cdots\}, \\ &\cdots & &\cdots \\ \underline{a}_n &= \inf\{a_n, a_{n+1}, \cdots\}, & \bar{a}_n &= \sup\{a_n, a_{n+1}, \cdots\}, \\ &\cdots \end{aligned}$$

Then, by definition,

$$\underline{a} = \sup\{\underline{a}_1, \underline{a}_2, \underline{a}_3, \cdots\} \quad \text{and}$$
$$\bar{a} = \inf\{\bar{a}_1, \bar{a}_2, \bar{a}_3, \cdots\}$$

are called the **limit inferior** *and* **limit superior**, *respectively, of the real sequence* $\{a_n\}$. *The notations*

$$\underline{a} = \underline{\lim}_{n\to\infty} a_n = \liminf_{n\to\infty} a_n \qquad \text{and}$$

$$\bar{a} = \overline{\lim}_{n\to\infty} a_n = \limsup_{n\to\infty} a_n$$

are used.

For convenience, $n \to \infty$ will usually be omitted.

EXAMPLES

1. If $a_n = n$, $n = 1, 2, 3, \cdots$, then

$$\underline{a_n} = n, \qquad \overline{a_n} = \infty,$$

$$\underline{a} = \underline{\lim}\, a_n = \sup\{1, 2, 3, \cdots\} = \infty, \qquad \text{and}$$

$$\bar{a} = \overline{\lim}\, a_n = \inf\{\infty, \infty, \infty, \cdots\} = \infty.$$

2. If $a_n = -n$, then $\underline{\lim}\, a_n = \overline{\lim}\, a_n = -\infty$.
3. $\underline{\lim}\, (-1)^n n = -\infty$, $\overline{\lim}(-1)^n n = \infty$.
4. $\underline{\lim}\, (-1)^n = -1$, $\overline{\lim}\, (-1)^n = 1$.
5. $\underline{\lim}\, \dfrac{(-1)^n}{n} = \overline{\lim}\, \dfrac{(-1)^n}{n} = 0$.
6. If $a_n = (-1)^n(1 + 1/n)$, then $\underline{a} = -1$ and $\bar{a} = 1$.
7. If $r_1, r_2, r_2, \cdots$ is any ordering of the rational numbers between 0 and 1, then

$$\underline{\lim}\, r_n = 0 \qquad \text{and} \qquad \overline{\lim}\, r_n = 1.$$

It should be seen that both $\underline{\lim}\, a_n$ and $\overline{\lim}\, a_n$ exist* for any real sequence $\{a_n\}$. Also, the following theorem and corollaries need little more than being pointed out.

Theorem 1. *In the notation of the definition*

$$\underline{a}_1 \leqslant \underline{a}_2 \leqslant \underline{a}_3 \leqslant \cdots \qquad \textit{and} \qquad \bar{a}_1 \geqslant \bar{a}_2 \geqslant \bar{a}_3 \geqslant \cdots.$$

Corollary 1. *For each positive integer k*

$$\underline{\lim}\, a_n = \sup\{\underline{a}_k, \underline{a}_{k+1}, \cdots\},$$
$$\overline{\lim}\, a_n = \inf\{\bar{a}_k, \bar{a}_{k+1}, \cdots\}.$$

Corollary 2. $\underline{a}_l \leqslant \bar{a}_k$ *for any* $k, l \in J$.

Corollary 3. $\underline{\lim}\, a_n \leqslant \overline{\lim}\, a_n$.

Corollary 4. *If* $b \leqslant a_n \leqslant B$, $n = 1, 2, 3, \cdots$, then

$$b \leqslant \underline{\lim}\, a_n \leqslant \overline{\lim}\, a_n \leqslant B.$$

The signs $<$ and $>$ are used only between elements of E^1 or of E^*, but not E^k for $k \geqslant 2$. Thus $a < b$ needs no additional reminder that $a \in E^*$ and $b \in E^*$. In particular, $-\infty < c < x$ means that $c \in E^1$, but leaves open whether $x \in E^1$ or $x = \infty$. However "$\{a_n\}$ is a real sequence" means that $a_n \in E^1$ for $n \in J$.

Theorem 2. *Given any real sequence $\{a_n\}$, then*

$$\underline{\lim}(-a_n) = -\overline{\lim}\, a_n \qquad \textit{and} \qquad \overline{\lim}(-a_n) = -\underline{\lim}\, a_n. \tag{1}$$

Proof. Let $b_n = -a_n$. Then, from Theorem 3-1.2,

$$\begin{aligned} \underline{b}_n &= \inf\{b_n, b_{n+1}, \cdots\} \\ &= -\sup\{a_n, a_{n+1}, \cdots\} = -\bar{a}_n \qquad \text{and} \end{aligned}$$

$$\begin{aligned} \underline{\lim}\, b_n &= \sup\{\underline{b}_1, \underline{b}_2, \cdots\} = \sup\{-\bar{a}_1, -\bar{a}_2, \cdots\} \\ &= -\inf\{\bar{a}_1, \bar{a}_2, \cdots\} = -\overline{\lim}\, a_n, \end{aligned}$$

which establishes the first equation in (1). Then

$$\underline{\lim}[-(-a_n)] = -\overline{\lim}(-a_n),$$

from which the second equation in (1) follows. ∎

Theorem 3. *A real sequence $\{a_n\}$* has

(i) $\underline{\lim}\, a_n = -\infty$ *iff the sequence is unbounded below.*
(ii) $\underline{\lim}\, a_n = \infty$ *iff for each $G \in E^1$ there is an $N \in J$ such that*

$$n \geqslant N \Rightarrow a_n \geqslant G.$$

(iii) $\underline{\lim}\, a_n = \underline{a}$ *exists* $(-\infty < \underline{a} < \infty)$ *iff for each positive number* ε *both*
 (a) $\{n | a_n < \underline{a} + \varepsilon\}$ *is infinite; and*
 (b) $\exists\, N \in J$ *such that* $n \geqslant N \Rightarrow a_n > \underline{a} - \varepsilon$.

Proof of (i). Let $G \in E^1$ be arbitrary. Then

$$\begin{aligned} \underline{a} = -\infty &\Leftrightarrow \sup\{\underline{a}_1, \underline{a}_2, \cdots\} = -\infty \\ &\Leftrightarrow \underline{a}_n = -\infty \qquad \text{for each } n \in J \\ &\Leftrightarrow \{k | k \geqslant n \text{ and } a_k < G\} \text{ is infinite for each } n \in J \\ &\Leftrightarrow \{n \,|\, a_n < G\} \text{ is infinite} \\ &\Leftrightarrow \{a_n\} \text{ is unbounded below.} \end{aligned}$$

Proof of (ii). Let $G \in E^1$ be arbitrary. Then

$$\begin{aligned} \underline{a} = \infty &\Leftrightarrow \sup\{\underline{a}_1, \underline{a}_2, \cdots\} = \infty \\ &\Leftrightarrow \exists\, N \in J \text{ such that } \underline{a}_N \geqslant G \\ &\Leftrightarrow \exists\, N \in J \text{ such that } \inf\{a_N, a_{N+1}, \cdots\} \geqslant G \\ &\Leftrightarrow \exists\, N \in J \text{ such that } n \geqslant N \Rightarrow a_n \geqslant G. \end{aligned}$$

Proof of (iii). Let ε be an arbitrary positive number. Then, since $-\infty < \underline{a} < \infty$,

$$\sup\{\underline{a}_1, \underline{a}_2, \cdots\} = \underline{a}$$

$$\Leftrightarrow \begin{cases} \underline{a}_n \leqslant \underline{a} & \text{for } n = 1, 2, \cdots \\ \exists\, N \in J \text{ such that } \underline{a}_N \geqslant \underline{a} - \dfrac{\varepsilon}{2} \end{cases}$$

$$\Leftrightarrow \begin{cases} \inf\{a_n, a_{n+1}, \cdots\} \leqslant \underline{a}, & n = 1, 2, 3, \cdots \\ \exists\, N \text{ such that } \inf\{a_N, a_{N+1}, \cdots\} \geqslant \underline{a} - \dfrac{\varepsilon}{2} \end{cases}$$

$$\Leftrightarrow \begin{cases} \{n \mid a_n < \underline{a} + \varepsilon\} \text{ is infinite} \\ \exists\, N \text{ such that } n \geqslant N \Rightarrow a_n \geqslant \underline{a} - \dfrac{\varepsilon}{2} \end{cases}$$

$\Leftrightarrow$ Both (a) and (b) hold. ∎

Corollary $\underline{1}$.

(i) *If $c \in E^1$ and $c < \underline{\lim}\, a_n$, then $\exists\, N \in J$ such that*

$$n \geqslant N \Rightarrow c < a_n.$$

(ii) *If $c \in E^1$ and $\exists\, N \in J$ such that $n \geqslant N \Rightarrow c < a_n$, then $c \leqslant \underline{\lim}\, a_n$.*

Corollary $\underline{2}$.

(i) *If $c \in E^1$ and $\underline{\lim}\, a_n < c$, then $\{n \mid a_n < c\}$ is infinite.*
(ii) *If $c \in E^1$ is such that $\{n \mid a_n < c\}$ is infinite, then $\underline{\lim}\, a_n \leqslant c$.*

The following Theorem $\bar{3}$ follows by applying Theorem $\underline{3}$ to the sequence $\{-a_n\}$ and using Theorem 2.

Theorem $\bar{3}$. *A real sequence $\{a_n\}$ has*

(i) *$\overline{\lim}\, a_n = \infty$ iff the sequence is unbounded above.*
(ii) *$\overline{\lim}\, a_n = -\infty$ iff for each $G \in E^1$ there is an $N \in J$ such that*

$$n \geqslant N \Rightarrow a_n < G.$$

(iii) *$\overline{\lim}\, a_n = \bar{a}$ exists ($-\infty < \bar{a} < \infty$) iff for each positive number ε both*
(a) *$\{n \mid a_n > \bar{a} - \varepsilon\}$ is infinite, and*
(b) *$\exists\, N \in J$ such that $n \geqslant N \Rightarrow a_n < \bar{a} + \varepsilon$.*

Corollary $\bar{1}$.

(i) *If $c \in E^1$ and $c > \overline{\lim}\, a_n$, then $\exists\, N \in J$ such that*

$$n \geqslant N \Rightarrow c > a_n.$$

(ii) *If $c \in E^1$ and $\exists\, N \in J$ such that $n \geqslant N \Rightarrow c > a_n$, then $c \geqslant \overline{\lim}\, a_n$.*

Corollary $\bar{2}$.

(i) *If $c \in E^1$ and $\overline{\lim}\, a_n > c$, then $\{n \mid a_n > c\}$ is infinite.*

(ii) *If $c \in E^1$ is such that $\{n \mid a_n > c\}$ is infinite, then $\overline{\lim}\, a_n \geqslant c$.*

Theorem 4. *Let $\{a_n\}$ and $\{b_n\}$ be real sequences and define $s_n = a_n + b_n$ for* $n = 1, 2, 3, \cdots$. Then

$$\underline{a} + \underline{b} \leqslant \underline{s} \leqslant \begin{Bmatrix} \underline{a} + \bar{b} \\ \bar{a} + \underline{b} \end{Bmatrix} \leqslant \bar{s} \leqslant \bar{a} + \bar{b} \tag{2}$$

whenever the terms involving + have meaning.

Proof of first inequality.

Case 1. $\underline{a} + \underline{b} = -\infty$. Then $-\infty \leqslant \underline{s}$ so the inequality holds.
Case 2. $-\infty < \underline{a}$ and $-\infty < \underline{b}$. Let c_1 and c_2 be such that

$$-\infty < c_1 < \underline{a} \qquad \text{and} \qquad -\infty < c_2 < \underline{b}.$$

Let $N \in J$ be such that [by Theorem $\underline{3}$, (iii)(b)]

$$n \geqslant N \Rightarrow c_1 < a_n \qquad \text{and} \qquad c_2 < b_n.$$

Then

$$\begin{aligned} (n \geqslant N \Rightarrow c_1 + c_2 < s_n) &\Rightarrow (c_1 + c_2 \leqslant \underline{s}) && \text{by Corollary } \underline{1}, \text{(ii)}, \\ &\Rightarrow \underline{a} + \underline{b} \leqslant \underline{s} && \text{by Theorem 3-1.1.} \end{aligned}$$

Proof of second inequality. If $\underline{a} + \bar{b} = \infty$ then, since $\underline{s} \leqslant \infty$, the inequality holds, so consider

$$\underline{a} < \infty \qquad \text{and} \qquad \bar{b} < \infty.$$

Let c_1 and c_2 be such that $\underline{a} < c_1 < \infty$ and $\bar{b} < c_2 < \infty$. Then the set $\{n \mid a_n < c_1\}$ is infinite and there is an $N \in J$ such that $n \geqslant N \Rightarrow b_n < c_2$. Thus $\{n \mid n \geqslant N$ and $a_n + b_n < c_1 + c_2\}$ is infinite and this

$$\begin{aligned} &\Rightarrow \{n \mid s_n < c_1 + c_2\} \text{ is infinite} \\ &\Rightarrow \underline{s} \leqslant c_1 + c_2 \Rightarrow \underline{s} \leqslant \underline{a} + \bar{b} \end{aligned}$$

and the upper part of the second inequality is proved. The lower part follows by interchanging the roles of a and b.

Proof of the remaining inequalities. These inequalities follow by applying those already proved to $\{-a_n\}$, $\{-b_n\}$, $\{-s_n\}$, and then using Theorem 2. ∎

PROBLEM 1. Use the sequences $\{a_n\}$ and $\{b_n\}$ defined by

$$\left.\begin{aligned} a_n &: -1, 0, \quad 1, -1, 0, \quad 1, \ldots \\ b_n &: \quad 0, 1, -1, \quad 0, 1, -1, \ldots \end{aligned}\right\} \text{repeating of period 3}$$

to show that each inequality in (2) may be proper. [Compare with Problem 2-3.4b.]

Corollary. *Whenever the terms have meaning,*

$$\underline{\lim}\, a_n - \overline{\lim}\, b_n \leqslant \underline{\lim}(a_n - b_n) \leqslant \begin{Bmatrix} \underline{\lim}\, a_n - \underline{\lim}\, b_n \\ \overline{\lim}\, a_n - \overline{\lim}\, b_n \end{Bmatrix}$$

$$\leqslant \overline{\lim}(a_n - b_n) \leqslant \overline{\lim}\, a_n - \underline{\lim}\, b_n .$$

EXAMPLE 8. Let $a_n = n + (-1)^n$ and $b_n = -n$. Then

$$\begin{aligned} &\underline{\lim}\, a_n = \infty, \qquad \overline{\lim}\, a_n = \infty, \\ &\underline{\lim}\, b_n = \overline{\lim}\, b_n = -\infty, \\ &\underline{\lim}(a_n + b_n) = \underline{\lim}(-1)^n = -1, \\ &\overline{\lim}(a_n + b_n) = \overline{\lim}(-1)^n = 1. \end{aligned}$$

In this example, none of the terms of (2) involving + have meaning.

EXAMPLE 9. Given $a_n = (-1)^n$ and $b_n = (-1)^{n+1}$, then

$$\underline{\lim}\, a_n = \underline{\lim}\, b_n = -1, \qquad \overline{\lim}\, a_n = \overline{\lim}\, b_n = 1,$$

and

$$\underline{\lim}(a_n + b_n) = \overline{\lim}(a_n + b_n) = 0.$$

The following theorems are stated for use later on, but the proofs are left as exercises.

Theorem 5. *Let $\{a_n\}$ be a real sequence and $c \in E^1$. Then*

(i) $\underline{\lim}(c + a_n) = c + \underline{\lim}\, a_n$ *and* $\overline{\lim}\,(c + a_n) = c + \overline{\lim}\, a_n$.

(ii) *If* $0 < c$, *then*

$$\underline{\lim}(ca_n) = c\, \underline{\lim}\, a_n \qquad \text{and} \qquad \overline{\lim}(ca_n) = c\, \overline{\lim}\, a_n .$$

(iii) *If* $c < 0$, *then*

$$\underline{\lim}(ca_n) = c\, \overline{\lim}\, a_n \qquad \text{and} \qquad \overline{\lim}(ca_n) = c\, \underline{\lim}\, a_n .$$

Theorem 6. *If $\{a_n\}$ and $\{b_n\}$ are real sequences and $a_n \leqslant b_n$, $n = 1, 2, 3, \cdots$, then*

$$\underline{\lim}\, a_n \leqslant \underline{\lim}\, b_n \qquad \text{and} \qquad \overline{\lim}\, a_n \leqslant \overline{\lim}\, b_n .$$

Theorem 7. *If $\{a_n\}$ and $\{b_n\}$ are real nonnegative sequences, then*

$$\underline{\lim}\, a_n\, \underline{\lim}\, b_n \leqslant \underline{\lim}(a_n b_n) \leqslant \overline{\lim}(a_n b_n) \leqslant \overline{\lim}\, a_n\, \overline{\lim}\, b_n$$

whenever the outside terms are defined.

Theorem 8. *If $\{a_n\}$ is a real positive sequence, then*

$$\underline{\lim}\, \frac{1}{a_n} = \frac{1}{\overline{\lim}\, a_n} \qquad \text{and} \qquad \overline{\lim}\, \frac{1}{a_n} = \frac{1}{\underline{\lim}\, a_n} .$$

3–3 LIMIT OF A REAL SEQUENCE

The following definition uses the notation and terminology of the definitions in Section 3–2.

Definition 1. *Given a real sequence* $\{a_n\}$:

(i) *If* $\underline{a} = \bar{a}$, *then* $\lim a_n$ *is said to exist* and have value* $a = \underline{a} = \bar{a}$.
(ii) *If* $\lim a_n = a$ *exists* and* $-\infty < a < \infty$, *the sequence is said to be* **convergent** *and to* **converge** *to* a (or $\lim a_n$ *is said to exist and* $= a$).
(iii) *If the sequence does not converge, it is said to* **diverge** (or be **divergent**).

Theorem 1. *For a real sequence* $\{a_n\}$:

(a) $\lim a_n$ *exists* and* $= \infty$ *iff corresponding to each* $G \in E^1$, $\exists$ *an* $N \in J$ *such that*

$$n \geqslant N \Rightarrow a_n > G.$$

(b) $\lim a_n$ *exists* and* $= -\infty$ *iff corresponding to each* $G \in E^1$, $\exists$ *an* $N \in J$ *such that*

$$n \geqslant N \Rightarrow a_n < G.$$

(c) *The sequence converges and* $\lim a_n = a \in E^1$ *iff corresponding to each number* $\varepsilon > 0$, $\exists$ *an* $N \in J$ *such that*

$$n \geqslant N \Rightarrow |a_n - a| < \varepsilon.$$

Proof (a). Since $\underline{a} \leqslant \bar{a}$ for any real sequence, then $\underline{a} = \bar{a} = \infty$ iff $\underline{a} = \infty$ and hence (by Theorem 3-2.$\underline{3}$) iff corresponding to each $G \in E^1$ there is an $N \in J$ such that

$$n \geqslant N \Rightarrow a_n > G.$$

The proof of (b) is similar.

Proof (c). Given $-\infty < \underline{a} = \bar{a} < \infty$, then corresponding to each $\varepsilon > 0$, $\exists\ N_1 \in J$ and $N_2 \in J$ (by (iii), (b), of Theorems 3-2.$\underline{3}$ and $\bar{3}$, respectively) such that

$$n \geqslant N_1 \Rightarrow a_n > \underline{a} - \varepsilon$$
$$n \geqslant N_2 \Rightarrow a_n < \bar{a} + \varepsilon$$

and hence, with $a = \underline{a} = \bar{a}$ and $N = \max(N_1, N_2)$,

$$n \geqslant N \Rightarrow a - \varepsilon < a_n < a + \varepsilon,$$

that is

$$|a_n - a| < \varepsilon.$$

Conversely, let $a \in E^1$ be such that corresponding to each $\varepsilon > 0$, $\exists\ N \in J$ such that

$$n \geqslant N \Rightarrow a - \varepsilon < a_n < a + \varepsilon.$$

This, from the first corollaries of Theorems 3-2.$\underline{3}$ and $\overline{3}$, implies that for each $\varepsilon > 0$

$$a - \varepsilon \leqslant \underline{a} \leqslant \bar{a} \leqslant a + \varepsilon,$$

which in turn implies $a \leqslant \underline{a} \leqslant \bar{a} \leqslant a$; that is, $a = \underline{a} = \bar{a}$. ∎

Theorem 2. *Let $\{a_n\}$ and $\{b_n\}$ be real sequences such that* $\lim a_n = a$ *and* $\lim b_n = b$ *exist*. If $a + b$ is defined, then*

(1) $$\lim(a_n + b_n) \text{ exists* and } = a + b.$$

If $a - b$ has meaning, then $\lim(a_n - b_n)$ *exists* and* $= a - b$.

Proof. With $s_n = a_n + b_n$, $n = 1, 2, 3, \cdots$, then from Theorem 3-2.4,

$$a + b \leqslant \underline{s} \leqslant \bar{s} \leqslant a + b,$$

since $\underline{a} = \bar{a} = a$, $\underline{b} = \bar{b} = b$, and $a + b$ has meaning. Thus $\underline{s} = \bar{s} = a + b$ so that (1) holds.

In a similar way the remainder of the theorem follows from the corollary of Theorem 3-2.4. ∎

Definition 2. *A real sequence $\{a_n\}$ is*

(i) **Monotonically increasing** *if $a_1 \leqslant a_2 \leqslant a_3 \leqslant \cdots$.*
Monotonically decreasing *if $a_1 \geqslant a_2 \geqslant a_2 \geqslant \cdots$.*
Monotonic *if it is either monotonically increasing or decreasing.*

(ii) **Bounded below** *if $\exists\, b \in E^1$ such that $b \leqslant a_n$ for $n \in J$.*
Bounded above *if $\exists\, B \in E^1$ such that $a_n \leqslant B$ for $n \in J$.*
Bounded *if it is bounded below and above.*

Theorem 3. *The limit of a monotonic sequence exists* and exists iff the sequence is bounded.*

Proof. Let $\{a_n\}$ be monotonically increasing. Then $a_1 \leqslant \underline{\lim}\, a_n \leqslant \overline{\lim}\, a_n$. Choose any $c \in E^1$ such that $c < \overline{\lim}\, a_n$ and then determine $N \in J$ such that $a_N > c$. Hence $n \geqslant N \Rightarrow a_N \leqslant a_n$ so that $\underline{\lim}\ a_n \geqslant a_N > c$. Since c was any number $< \overline{\lim}\, a_n$, then $\underline{\lim}\, a_n \geqslant \overline{\lim}\, a_n$ so that $\underline{\lim}\, a_n = \overline{\lim}\, a_n$ and $\lim a_n$ exists*.

If $\{a_n\}$ is bounded above by $B \in E^1$, then

$$a_1 \leqslant \underline{\lim}\, a_n = \overline{\lim}\, a_n \leqslant B$$

so that $\lim a_n$ exists. Conversely, if $\overline{\lim}\, a_n < \infty$, then $a_1 \leqslant a_n \leqslant \overline{\lim}\, a_k$ for each $n \in J$ and $\{a_n\}$ is bounded. ∎

PROBLEM 1. If $0 \leqslant a_n \leqslant b_n$, $n = 1, 2, 3, \cdots$ and $\overline{\lim}\, b_n = 0$, then $\lim a_n = 0$.

PROBLEM 2. For any real sequence $\{a_n\}$, if $k \in J$, then

$$\underline{\lim}_{n\to\infty} a_{n+k} = \underline{\lim}_{n\to\infty} a_n \quad \text{and} \quad \overline{\lim}_{n\to\infty} a_{n+k} = \overline{\lim}_{n\to\infty} a_n.$$

PROBLEM 3. For any real sequence $\{a_n\}$,

$$\lim a_n = 0 \qquad \text{iff } \lim|a_n| = 0.$$

PROBLEM 4. It is said that a real sequence $\{a_n\}$ is **attempting to increase** if for each $n \in J$ and each $\varepsilon > 0$ there is an $m > n$ such that $a_m > a_n - \varepsilon$.

Given that $\{a_n\}$ is attempting to increase, show that $\lim a_n$ exists* and exists if the sequence is bounded above.

Theorem 4. *For*

$$x \in E^1, \qquad \lim_{n\to\infty} x^n = \begin{cases} 0 & \text{if } |x| < 1 \\ 1 & \text{if } x = 1 \end{cases},$$

but the limit does not exist if $x = -1$ or $|x| > 1$.

Proof. For x such that $|x| < 1$, then

$$1 > |x| \geqslant |x|^2 \geqslant \cdots \geqslant 0 \qquad \text{and} \qquad \lim|x|^n = c \text{ exists.}$$

Thus $c = \lim|x|^{n+1} = |x| \lim|x|^n = |x|c$ so that $c = 0$, since otherwise we obtain the contradiction $|x| = 1$. Hence $\lim x^n = 0$ by Problem 3.

Clearly, $\lim 1^n = 1$.

Since $\underline{\lim}(-1)^n = -1$ and $\overline{\lim}(-1)^n = 1$, the sequence $\{(-1)^n\}$ does not converge.

If $|x| > 1$, then $1 < |x| < |x|^2 < \cdots$ so that $\lim|x|^n$ exists*. Assume the limit exists and $= c \in E^1$. Then

$$1 < c = \lim|x|^{n+1} = |x|c$$

and the contradiction $|x| = 1$ is obtained. Hence, if $x > 1$ then $\lim x^n = \infty$, if $x < -1$ then $\overline{\lim}\, x^n = \infty$, and in either case the limit does not exist. ∎

Theorem 5. $\lim \sqrt[n]{n} = 1$ *and if $p > 0$, then*

$$\lim \sqrt[n]{p} = 1 \qquad \text{and} \qquad \lim \sqrt[n]{np} = 1.$$

Proof. With $n \geqslant 2$, set $a_n = \sqrt[n]{n} - 1$ so that $a_n > 0$,

$$n = (1 + a_n)^n \geqslant \frac{n(n-1)}{2} a_n^{\,2}, \qquad \text{and} \qquad 0 < a_n < \sqrt{\frac{2}{n-1}}.$$

Hence $\lim a_n = 0$, $1 = 1 + \lim a_n = \lim(1 + a_n) = \lim \sqrt[n]{n}$.

Given $p > 0$, then for $n > p$ we have $\sqrt[n]{n} > \sqrt[n]{p} > 0$ so that

$$1 = \lim \sqrt[n]{n} \geqslant \overline{\lim}\, \sqrt[n]{p} \geqslant \underline{\lim}\, \sqrt[n]{p} \geqslant 0.$$

Assume $\underline{\lim}\, \sqrt[n]{p} < 1$ and choose c such that

$$\underline{\lim}\, \sqrt[n]{p} < c < 1.$$

Then $\{n \mid \sqrt[n]{p} < c\} = \{n \mid p < c^n\}$ is infinite. But this second form means that $p < \overline{\lim}\, c^n$ and, from Theorem 4 since $0 < c < 1$, that $p \leqslant \lim c^n = 0$, in contradiction to $p > 0$. Hence $1 \leqslant \underline{\lim}\, \sqrt[n]{p} \leqslant \overline{\lim}\, \sqrt[n]{p} \leqslant 1$ and $\lim \sqrt[n]{p} = 1$ exists.

Then $\lim \sqrt[n]{np} = 1$ follows from Theorems 3-2.7. ∎

Theorem 6. *With $p > 0$ and r a real number, then*

$$\lim_{n\to\infty} \frac{n^r}{(1+p)^n} = 0. \tag{2}$$

Proof. If $r \leqslant 0$, then $0 < n^r \leqslant 1$ and (2) follows from Theorem 3 and Problem 2.

Thus consider $r > 0$. Since $n^r/(1+p)^n > 0$ for each $n \in J$, then

$$\overline{\lim}\, \frac{n^r}{(1+p)^n} \geqslant \underline{\lim}\, \frac{n^r}{(1+p)^n} \geqslant 0. \tag{3}$$

Assume this limit superior is positive and choose c such that

$$\overline{\lim}\, \frac{n^r}{(1+p)^n} > c > 0.$$

Then the following set is infinite:

$$\left\{n \,\middle|\, \frac{n^r}{(1+p)^n} > c\right\} = \{n \mid \sqrt[n]{nc^{-1/r}} > (1+p)^{1/r}\},$$

where the set equality follows since $r > 0$. Thus

$$\overline{\lim}\, \sqrt[n]{nc^{-1/r}} \geqslant (1+p)^{1/r}.$$

But, from Theorem 5, $\overline{\lim}\, \sqrt[n]{nc^{-1/r}} = \lim \sqrt[n]{nc^{-1/r}} = 1$ since $c^{-1/r} > 0$. Thus $1 \geqslant (1+p)^{1/r}$, $1 \geqslant 1 + p$, and $0 \geqslant p$ in contradiction to $0 < p$. Thus both terms of (3) are zero and hence (2) holds. ∎

3–4 SEQUENCES IN METRIC SPACES

In this section, properties of sequences of points in a metric space will be considered. None of the properties of superior or inferior limits is available since no order among the points of an abstract metric space (or E^k with $k > 1$) is defined. Even addition of elements is undefined.

Definition 1. *A sequence $\{a_n\}$ in a metric space S is said to* **converge** *to $a \in S$, or be* **convergent** *in S and we write $a_n \to a$, if corresponding to each $\varepsilon > 0$ there is an $N \in J$ such that*

$$n \geqslant N \Rightarrow d(a_n, a) < \varepsilon; \qquad \text{that is† } n \geqslant N \Rightarrow a_n \in N(a, \varepsilon).$$

If a sequence in S does not converge in S, it is said to **diverge.**

† The $N \in J$ and $N(a, \varepsilon)$ as a neighborhood should not be confused.

For example, the sequence $\{1/n\}$ converges in the metric space E^1, but diverges in (or relative to) the submetric space $\{x \mid x > 0\}$.

Theorem 1. *A sequence converging in S converges to one and only one point of S.*

Proof. Let $\{a_n\}$ be a sequence in S such that $a_n \to a \in S$ and $a_n \to b \in S$. With $\varepsilon > 0$ arbitrary, choose N_1 and N_2 such that

$$n \geqslant N_1 \Rightarrow d(a_n, a) < \varepsilon \qquad \text{and} \qquad n \geqslant N_2 \Rightarrow d(a_n, b) < \varepsilon.$$

Then $N = \max(N_1, N_2)$ is such that

$$n \geqslant N \Rightarrow d(a, b) \leqslant d(a, a_n) + d(a_n, b) < 2\varepsilon.$$

Since $d(a, b) < 2\varepsilon$ for each $\varepsilon > 0$, then $d(a, b) = 0$ and $a = b$. ∎

If $\{a_n\}$ converges in S to a we also write

$$\lim_{n\to\infty} a_n = a \qquad \text{or merely} \qquad \lim a_n = a.$$

Definition 2. *A sequence $\{a_n\}$ in S is said to be* **bounded** *if its range $A = \{a_1, a_2, a_3, \cdots\}$ is bounded.*

Theorem 2. *Every convergent sequence in S is bounded.*

Proof. Let $\{a_n\}$ converge to $a \in S$. *Corresponding to* $\varepsilon = 1$ let $N \in J$ be such that $n \geqslant N \Rightarrow d(a_n, a) < 1$. Then

$$r = \max[2d(a_1, a), 2d(a_2, a), \cdots, 2d(a_N, a), 1]$$

is such that $n \in J \Rightarrow a_n \in N(a, r)$; that is, $A \subset N(a, r)$ so that $\{a_n\}$ is bounded. ∎

PROBLEM 1. Let $\{a_n\}$ converge in S to a and denote the range of $\{a_n\}$ by $A = \{a_1, a_2, a_3, \cdots\}$. Then either

(a) A is finite and there is an $N \in J$ such that $n \geqslant N \Rightarrow a_n = a$, or

(b) A is infinite and $\{a\} = A'$; that is, a is a cluster point, and the only cluster point, of A.

The following theorem is a restatement of Theorem 2-4.6.

Theorem 3. *Given a set $A \subset S$ with $A' \neq \varnothing$, then corresponding to each $a \in A'$ there is at least one sequence $\{a_n\}$ of points of A which converges in S to a.*

Definition 3. *A sequence $\{a_n\}$ in a metric space S is said to be a* **Cauchy sequence** *if for each $\varepsilon > 0$ there is an $N \in J$ such that*

$$(n \geqslant N, m \geqslant N) \Rightarrow d(a_n, a_m) < \varepsilon.$$

Theorem 4. *Every Cauchy sequence in a metric space S is bounded.*

Proof. Let $\{a_n\}$ be a Cauchy sequence in S. Corresponding to $\varepsilon = 1$, let M be such that

$$(n \geqslant M, m \geqslant M) \Rightarrow d(a_n, a_m) < 1.$$

Hence $n \geqslant M \Rightarrow d(a_n, a_M) < 1$. Thus with

$$r = \max[2d(a_1, a_M), 2d(a_2, a_M), \cdots, 2d(a_{M-1}, a_M), 1],$$

then $a_n \in N(a_M, r)$ for each $n \in J$; that is, $\{a_n\}$ is bounded. ∎

The following theorem shows the close relation between convergent sequences and Cauchy sequences.

Theorem 5.

(a) *In a metric space S, convergent sequences are Cauchy sequences.*
(b) *In E^k, Cauchy sequences are convergent sequences.*

[Since E^k is a metric space, then a sequence in E^k converges iff it is a Cauchy sequence. This is called the **Cauchy criterion** for convergence in E^k and is useful in establishing convergence when a candidate for the limit is not known or readily determinable.]

Proof. (a). Let $\{a_n\}$ be a convergent sequence in S with $a_n \to a$. Choose $\varepsilon > 0$ arbitrarily. Corresponding to $\varepsilon/2$, let N be such that $n \geqslant N \Rightarrow d(a_n, a) < \varepsilon/2$. Then

$$(n \geqslant N, m \geqslant N) \Rightarrow d(a_n, a_m) \leqslant d(a_n, a) + d(a, a_m) < \varepsilon.$$

Hence $\{a_n\}$ is a Cauchy sequence.

Proof. (b). Let $\{a_n\}$ be a Cauchy sequence in E^k and set

$$A_n = \{a_n, a_{n+1}, a_{n+2}, \cdots\}, \qquad n = 1, 2, 3, \cdots.$$

Then $A_1 \supset A_n \supset A_{n+1}$ and $\bar{A}_1 \supset \bar{A}_n \supset \bar{A}_{n+1}$, $n = 1, 2, 3, \cdots$. Since A_1 is bounded (Theorem 4), then each $\bar{A}_n$ is a bounded closed set and is therefore compact (Theorem 2-7.5), since the space is E^k. Thus $\bigcap \bar{A}_n$ is nonempty (Corollary 1 of Theorem 2-6.4). Let $a \in E^k$ be such that

$$a \in \bigcap \bar{A}_n.$$

Given an $\varepsilon > 0$, let $N \in J$ be such that (since $\{a_n\}$ is a Cauchy sequence)

$$(n \geqslant N, m \geqslant N) \Rightarrow d(a_n, a_m) < \varepsilon.$$

Since $(n \geqslant N,\ m \geqslant N) \Rightarrow (a_n \in A_N,\ a_m \in A_N)$, then $\operatorname{di}(A_N) \leqslant \varepsilon$. But $\operatorname{di}(\bar{A}_N) = \operatorname{di}(A_N)$, $a \in \bar{A}_N$ and $n \geqslant N \Rightarrow a_n \in A_N$ so that

$$n \geqslant N \Rightarrow d(a_n, a) \leqslant \operatorname{di}(\bar{A}_N) \leqslant \varepsilon.$$

Thus $\{a_n\}$ converges to a, which finishes the proof. ∎

PROBLEM 2. In a metric space S, if $\{C_n\}$ is a monotonically decreasing sequence of nonempty compact sets such that $\lim \operatorname{di}(C_n) = 0$, then $\bigcap C_n$ contains one and only one point.

Definition 4. *A metric space in which every Cauchy sequence converges is said to be a* **complete metric space.**

A complete normed vector space is called a **Banach space**.

A complete inner product space is called a **Hilbert space**.

Thus Theorem 5 (b) may be restated as

E^k is a complete metric space, or
E^k is a Banach space, or
E^k is a Hilbert space.

Theorem 6. *Let S be a complete metric space and let $f: S \to S$ have the property*

(1) $$d(f(x), f(y)) \leqslant \alpha d(x, y) \qquad \text{for all } x, y \in S,$$

where $0 \leqslant \alpha < 1$ (with α independent of x and y). Then $f(x) = x$ has one and only one solution (in S).

[A function having the property (1) is called a **contraction transformation** on S and an $x \in S$ such that $f(x) = x$ is called a **fixed point** for f. Thus the theorem states: *Every contraction transformation on a complete metric space has one and only one fixed point.*]

Proof. Choose any $x_0 \in S$ and define the sequence $\{x_n\}$ inductively by

$$x_1 = f(x_0), \qquad x_2 = f(x_1), \cdots, \qquad x_n = f(x_{n-1}), \cdots.$$

Then for each $n \in J$

$$\begin{aligned} d(x_n, x_{n+1}) &= d(f(x_{n-1}), f(x_n)) \\ &\leqslant \alpha d(x_{n-1}, x_n). \end{aligned}$$

By repeated application of this relation,

$$d(x_n, x_{n+1}) \leqslant \alpha^2 d(x_{n-2}, x_{n-1}) \leqslant \cdots \leqslant \alpha^n d(x_0, x_1).$$

Hence, if $m > n$, then

$$\begin{aligned} d(x_n, x_m) &\leqslant d(x_n, x_{n+1}) + d(x_{n+1}, x_{n+2}) + \cdots + d(x_{m-1}, x_m) \\ &\leqslant (\alpha^n + \alpha^{n+1} + \cdots + \alpha^{m-1})\, d(x_0, x_1) \leqslant \frac{\alpha^n}{1-\alpha} d(x_0, x_1). \end{aligned}$$

Given a number $\varepsilon > 0$, choose N such that $\alpha^N d(x_0, x_1) < \varepsilon$. Then $n \geqslant N$ and $m \geqslant N$ implies

$$d(x_n, x_m) \leqslant \frac{\alpha^{\min(n,m)}}{1-\alpha} d(x_0, x_1) \leqslant \frac{\alpha^N}{1-\alpha} d(x_0, x_1) < \frac{\varepsilon}{1-\alpha}.$$

Since $1 - \alpha > 0$ is a constant, it follows that $\{x_n\}$ is a Cauchy sequence, which is therefore convergent (since S is complete), and we let

$$\lim x_n = \bar{x}.$$

With $\varepsilon > 0$ given, choose N such that

$$n \geqslant N \Rightarrow d(x_n, \bar{x}) < \varepsilon.$$

Then $n \geqslant N + 1$ implies

$$(x_n, f(\bar{x})) = d(f(x_{n-1}), f(\bar{x})) \leqslant \alpha d(x_{n-1}, \bar{x}) \leqslant \alpha\varepsilon < \varepsilon$$

so that also $\lim x_n = f(\bar{x})$. Since the limit of a convergent sequence is unique,

$$f(\bar{x}) = \bar{x}.$$

To establish that $f(x) = x$ has no other solution, let $y \in S$ be such that $f(y) = y$. Then

$$d(\bar{x}, y) = d(f(\bar{x}), f(y)) \leqslant \alpha d(\bar{x}, y)$$

so that $d(\bar{x}, y) = 0$ (since otherwise $1 \leqslant \alpha$) and hence $y = \bar{x}$. ∎

PROBLEM 3. In a complete metric space, let $\{b_n\}$ be a sequence having the property $d(b_{n+1}, b_n) \leqslant c^n$, $n = 1, 2, 3, \cdots$, where $0 \leqslant c < 1$. Then $\{b_n\}$ converges.

PROBLEM 4. Let x_0 be any real number and define the sequence $\{x_n\}$ in E^1 by

$$x_1 = \cos x_0\,, \qquad x_2 = \cos x_1, \cdots, \qquad x_n = \cos x_{n-1}, \cdots.$$

Use an intuitive (that is, geometric) argument to indicate that $\{x_n\}$ is a convergent sequence.

Definition 5. A sequence $\{b_k\}$ is said to be a **subsequence** of $\{a_n\}$ if there is a sequence $\{n_k\}$ of positive integers such that $1 \leqslant n_1 < n_2 < n_3 < \cdots$ and $b_k = a_{n_k}$, $k = 1, 2, 3, \cdots$.

Theorem 7. *If $\{a_n\}$ converges to a, then any subsequence $\{a_{n_k}\}$ also converges to a.*

Proof. A sequence $\{n_k\}$ as in the definition is such that $n_k \geqslant k$ for each $k \in J$. Thus, with $\varepsilon > 0$ given and N such that $n \geqslant N \Rightarrow d(a_n, a) < \varepsilon$, then $k \geqslant N \Rightarrow d(a_{n_k}, a) < \varepsilon$. ∎

Definition 6. *A point $c \in S$ is said to be a* **subsequential limit** *of a sequence $\{a_n\}$ in S iff there is a subsequence converging to c.*

PROBLEM 5. Given that $\{a_1, a_2, a_3, \cdots\}$ is in a compact set.

(a) Then $\{a_n\}$ has at least one subsequential limit.

(b) If all subsequential limits of $\{a_n\}$ are the same point a, then $\lim a_n = a$; that is, if the convergent subsequences of $\{a_n\}$ all converge to the same point, then the sequence itself converges and to the same point.

PROBLEM 6. Given that $\{a_n\}$ is a bounded sequence in E^1, then

(a) $\underline{\lim}\, a_n$ and $\overline{\lim}\, a_n$ are subsequential limits.

(b) $\underline{\lim}\, a_n \leqslant c \leqslant \overline{\lim}\, a_n$ for all subsequential limits c.

3-5 SEQUENCES OF COMPLEX NUMBERS

The complex number notation $z = x + iy$, $x \in E^1$, $y \in E^1$ will be used. Thus products and quotients are available, as are any of the established properties of the Hilbert space E^2. There are places in this section where E^2 could be replaced by E^k for more generality (when products and quotients are not involved) and it would be well to be on the alert for these. Specializations to the sub-metric space E^1 (with products and quotients) should also be observed for new results and reproofs where order properties of E^1 might have been used before.

The definition of a limit of a sequence $\{z_n\}$ of complex numbers needs only adjusted notation.

A sequence $\{z_n\}$ which converges to 0 is called a **null** sequence. Recall that $0 + i0$ is simplified to 0.

Theorem 1. *A sequence $\{z_n\}$ converges to z iff $\{z_n - z\}$ is a null sequence.*

Proof. The definitions of $z_n \to z$ and $z_n - z \to 0$ are the same; that is, corresponding to each number $\varepsilon > 0$ there is an $N \in J$ such that

$$n \geqslant N \Rightarrow |z_n - z| < \varepsilon. \blacksquare$$

Similarly, if any one of $\{z_n\}$, $\{-z_n\}$, $\{|z_n|\}$ is a null sequence, so are the other two. Clearly, if $\{z_n\}$ is a null sequence and $|w_n| \leqslant |z_n|$, then $\{w_n\}$ is a null sequence.

Theorem 2. *If $\{z_n\}$ and $\{w_n\}$ are null sequences, then $\{z_n + w_n\}$ and $\{z_n - w_n\}$ are null sequences.*

Proof. Given $z_n \to 0$ and $w_n \to 0$, then for each $\varepsilon > 0$, there is an $N_1 \in J$ and an $N_2 \in J$ such that

$$n \geqslant N_1 \Rightarrow |z_n| < \varepsilon, \qquad n \geqslant N_2 \Rightarrow |w_n| < \varepsilon, \qquad \text{and}$$
$$n \geqslant \max(N_1, N_2) \Rightarrow |z_n \pm w_n| \leqslant |z_n| + |w_n| < 2\varepsilon.$$

Since 2ε is arbitrary, the conclusion follows. ∎

Since $\{(-1)^n + (-1)^{n+1}\}$ is null, but neither $\{(-1)^n\}$ nor $\{(-1)^{n+1}\}$ is null, it is possible for $\{z_n + w_n\}$ to be a null sequence with both $\{z_n\}$ and $\{w_n\}$ not null sequences.

PROBLEM 1. If both $\{z_n + w_n\}$ and $\{z_n - w_n\}$ are null sequences, then both $\{z_n\}$ and $\{w_n\}$ are null sequences.

Theorem 3. *If $\{z_n\}$ is bounded and $\{w_n\}$ is null, then $\{z_n w_n\}$ is null.* [*This includes the case in which $\{z_n\}$ is a constant sequence.*]

Proof. Let $r > 0$ be such that $|z_n| < r$ for all $n \in J$. Corresponding to $\varepsilon > 0$, let $N \in J$ be such that $n \geqslant N \Rightarrow |w_n| < \varepsilon$. Then

$$n \geqslant N \Rightarrow |z_n w_n| = |z_n|\,|w_n| < r\varepsilon.$$

Since r is independent of n and ε is arbitrary, then $z_n w_n \to 0$. ∎

PROBLEM 2. Given that $\{z_n\}$ is 1, 0, 2, 0, 3, 0, 4, $\cdots$, find an unbounded sequence $\{w_n\}$ such that $\{z_n w_n\}$ is null.

If $\{z_n\}$ has at most a finite number of zero terms (possibly none), then $\{1/z_n\}$ is understood to have domain $\{n \in J \mid z_n \neq 0\}$, which therefore contains all integers greater than some $p \in J$.

Theorem 4. *Let $\{z_n\}$ converge to $z \neq 0$. Then $\{1/z_n\}$ converges to $1/z$.*

Proof. Corresponding to $\frac{1}{2}|z| > 0$, let $N_1 \in J$ be such that

$$n \geqslant N_1 \Rightarrow |z_n - z| < \tfrac{1}{2}|z|.$$

Then $n \geqslant N_1 \Rightarrow |z| \leqslant |z - z_n| + |z_n| < \frac{1}{2}|z| + |z_n|$ so that

$$n \geqslant N_1 \Rightarrow \left[|z_n| > \frac{1}{2}|z|, \frac{1}{|z_n|} < \frac{2}{|z|}\right]$$

and $\{n \in J \mid n \geqslant N_1\}$ is in the domain of $\{1/z_n\}$. Let $\varepsilon > 0$ be arbitrary and choose $N_2 \in J$ such that $n > N_2 \Rightarrow |z_n - z| < \varepsilon$. Then

$$n \geqslant \max(N_1, N_2) \Rightarrow \left|\frac{1}{z_n} - \frac{1}{z}\right| = \frac{|z_n - z|}{|z_n|\,|z|} < \frac{2}{|z|^2}\,\varepsilon.$$

Since $2/|z|^2$ is independent of n, then $1/z_n \to 1/z$. ∎

Theorem 5. *Let $\{z_n\}$ and $\{w_n\}$ be convergent sequences with $z_n \to z$ and $w_n \to w$. Then*

(a) $z_n + w_n \to z + w, \quad z_n - w_n \to z - w.$
(b) $z_n w_n \to zw.$
(c) $c + z_n \to c + z, \quad cz_n \to cz$ *with c a complex constant.*
(d) $w_n/z_n \to w/z$ *if* $z \neq 0.$

Proof (a). Since $z_n + w_n - (z + w) = (z_n - z) + (w_n - w)$, and $\{z_n - z\}$ and $\{w_n - w\}$ are null sequences (by Theorem 1), then

$$\{(z_n + w_n) - (z + w)\}$$

is a null sequence (by Theorem 2). Thus $z_n + w_n \to z + w$ (by Theorem 1).

Proof (b). $z_n w_n - zw = z_n(w_n - w) + w(z_n - z)$. Since $\{z_n\}$ converges, it is bounded (by Theorem 3-4.2, since E^2 is a metric space). Thus $\{z_n(w_n - w)\}$ and $\{w(z_n - z)\}$ are null sequences (Theorem 3) so that $\{z_n w_n - zw\}$ is a null sequence and $z_n w_n \to zw$.

Proof (c). This follows from (a) and (b), with each $w_n = c$.

Proof (d). With $z \neq 0$, then $\{w_n(1/z_n)\}$ converges to $w(1/z)$ by Theorem 4 and part (b); that is, $\{w_n/z_n\}$ converges to w/z. ∎

Theorem 6. *The sequence $\{z_n\} = \{x_n + iy_n\}$ converges to $z = x + iy$ iff both $\{x_n\}$ converges to x and $\{y_n\}$ converges to y.*

Proof. With $z_n = x_n + iy_n$ and $z = x + iy$, then

$$\left.\begin{matrix}|x_n - x| \\ |y_n - y|\end{matrix}\right\} \leqslant |z_n - z| = |(x_n - x) + i(y_n - y)| \leqslant |x_n - x| + |y_n - y|.$$

Thus, if $\{z_n - z\}$ is null, then so are both $\{x_n - x\}$ and $\{y_n - y\}$ by the first inequality. Conversely, if both $\{x_n - x\}$ and $\{y_n - y\}$ are null, then $\{z_n - z\}$ is null by the second inequality. ∎

The vector (in E^2) interpretation of Theorem 6 is

$$\{\mathbf{a}_n\} = \{(a_{n1}, a_{n2})\} \text{ converges to } \mathbf{a} = (a_1, a_2)$$

iff both

$$\{\mathbf{a}_{n1}\} \text{ converges to } a_1 \text{ and } \{a_{n2}\} \text{ converges to } a_2.$$

Theorem 7. *Given that $\{\mathbf{a}_n\}$ and $\{\mathbf{b}_n\}$ are convergent (in E^2) with $\mathbf{a}_n \to \mathbf{a} = (a_1, a_2)$ and $\mathbf{b}_n \to \mathbf{b} = (b_1, b_2)$, then the dot product sequence $\{\mathbf{a}_n \cdot \mathbf{b}_n\}$ converges with $\mathbf{a}_n \cdot \mathbf{b}_n \to \mathbf{a} \cdot \mathbf{b}$, but not conversely.*

Proof. First note that

$$\begin{aligned} \mathbf{a}_n \cdot \mathbf{b}_n - \mathbf{a} \cdot \mathbf{b} &= (a_{n1} b_{n1} + a_{n2} b_{n2}) - (a_1 b_1 + a_2 b_2) \\ &= a_{n1}(b_{n1} - b_1) + b_1(a_{n1} - a_1) + a_{n2}(b_{n2} - b_2) + b_2(a_{n2} - a_2). \end{aligned} \tag{1}$$

Since $\{\mathbf{a}_n\}$ and $\{\mathbf{b}_n\}$ converge, the sequences of their components converge (Theorem 6). Hence each of the four terms in (1) is the nth term of a null sequence. Thus $\{\mathbf{a}_n \cdot \mathbf{b}_n - \mathbf{a} \cdot \mathbf{b}\}$ is a null sequence (in E^1) and the direct conclusion of the theorem follows.

To see that the converse is not so, let

$$\mathbf{a}_n = \left(n, \frac{1}{n}\right) \quad \text{and} \quad \mathbf{b}_n = \left(\frac{1}{n}, n\right).$$

Then neither $\{\mathbf{a}_n\}$ nor $\{\mathbf{b}_n\}$ converges, but $\{\mathbf{a}_n \cdot \mathbf{b}_n\}$ converges to 2. ∎

PROBLEM 3. Display a pair of nonnull sequences $\{\mathbf{a}_n\}$ and $\{\mathbf{b}_n\}$ such that $\{\mathbf{a}_n \cdot \mathbf{b}_n\}$ is null.

PROBLEM 4. If $\{c_n\}$ is a real sequence converging to c and $\{\mathbf{a}_n\}$ converges to $\mathbf{a}$, then $\{c_n \mathbf{a}_n\}$ converges to $c\mathbf{a}$.

3–6 SERIES

Even though the following definition concerns points in E^k, the usual vector notation will not be used.

Definition 1. *Given a sequence $\{a_n\}$ with each $a_n \in E^k$, let*

$$s_n = a_1 + a_2 + \cdots + a_n, \qquad n = 1, 2, 3, \cdots.$$

Then the set $A \subset E^k \times E^k$, *defined by*

(1) $$A = \{(x,y) \mid x = a_n, y = s_n \quad \text{for some } n \in J\},$$

is said to be a **series** (*or infinite series*) *whose nth* **element** *is* (a_n, s_n), *whose nth* **term** *is* a_n, *and whose nth* **partial sum** *is* s_n *for each* $n \in J$. *If* $\{s_n\}$ *converges with* $s_n \to s$, *the series is said to* **converge** *to s* (*or have* **sum** *s*). *If* $\{s_n\}$ *diverges, the series is said to* **diverge**.

Rather than the cumbersome notation (1) for a series determined by $\{a_n\}$, we revert to the customary representations

$$a_1 + a_2 + a_3 + \cdots + a_n + \cdots \qquad \text{and} \qquad \sum_{n=1}^{\infty} a_n$$

even if the series does not converge. Thus we may say that the series $\sum_{n=1}^{\infty} a_n$ converges or diverges as the case may be. If, however, the series converges to s, we follow convention by setting

(2) $$s = \sum_{k=1}^{\infty} a_n,$$

thus using the sigma notation in the double role of representing the series and its sum. Also (2) is written iff the series converges to s; and *when written, no reminder that the series converges need be given.*

Of course, if $\{a_n\}$ has domain $J \cup \{0\}$ or $J - \{1, 2, \cdots, p\}$ for some $p \in J$, then the associated series are denoted by

$$\sum_{n=0}^{\infty} a_n \qquad \text{or} \qquad \sum_{n=p+1}^{\infty} a_n, \qquad \text{respectively.}$$

If $\{a_n\}$ has $\{a_{n_i}\}$ as a subsequence, then

$$\sum_{i=1}^{\infty} a_{n_i} \quad \text{is a } \textbf{subseries} \text{ of } \quad \sum_{n=1}^{\infty} a_n.$$

Hence, with $p \in J$, the series

$$\sum_{n=1}^{\infty} a_n \quad \text{and the subseries} \quad \sum_{n=p+1}^{\infty} a_n$$

either both diverge or both converge and if the first converges to s, the second converges to $s - (a_1 + a_2 + \cdots + a_p)$.

Whenever only convergence or divergence questions are involved, the series will be represented by $\sum a_n$.

Theorem 1. *If* $\sum a_n$ *converges, then* $\lim a_n = 0$ [*actually* $\lim \mathbf{a}_n = \mathbf{0}$].

Proof. Let $\sum a_n$ converge to s. Then $\lim s_n = s$ so that also $\lim s_{n-1} = s$ and hence

$$s - s = \lim s_n - \lim s_{n-1} = \lim(s_n - s_{n-1}) = \lim a_n. \quad \blacksquare$$

Another convention is to write

$$\lim a_n \neq 0$$

either if the limit does not exist or exists but is not 0.

Corollary. *If* $\lim a_n \neq 0$, *then* $\sum a_n$ *diverges.*

By applying the Cauchy criterion (following Theorem 3-4.5) to the sequence of partial sums, it follows that:

> *A series* $\sum a_n$ *with terms in* E^k *converges iff the sequence of partial sums is a Cauchy sequence.*

The following theorem states this result in series-oriented form.

Theorem 2. *A series* $\sum a_n$ *in* E^k *converges iff corresponding to each number* $\varepsilon > 0$ *there is an* $N \in J$ *such that*

$$n \geqslant N \Rightarrow |s_{n+p} - s_n| < \varepsilon \qquad \text{for } p = 1, 2, 3, \cdots,$$

or equivalently

$$n \geqslant N \Rightarrow \left|\sum_{i=n}^{n+p} a_i\right| < \varepsilon \qquad \text{for } p = 1, 2, 3, \cdots.$$

The Cauchy criterion is difficult to apply in some cases. Once the convergence or divergence of a specific series is established, then this series may be used in the **comparison test** of the following theorem.

Theorem 3. *Given a sequence* $\{a_n\}$ *in* E^k *and sequences* $\{c_n\}$ *and* $\{d_n\}$ *of real terms such that for some* $N_0 \in J$,

$$n \geqslant N_0 \Rightarrow 0 \leqslant c_n \leqslant |a_n| \leqslant d_n:$$

(a) If $\sum d_n$ converges, then $\sum a_n$ converges.
(b) If $\sum c_n$ diverges, then $\sum |a_n|$ diverges.

Proof (a). Let $\sum d_n$ converge. The partial sums of $\sum d_n$ therefore form a Cauchy sequence. Let $\varepsilon > 0$ be arbitrary. There is then an $N \in J$ such that

$$n \geqslant N \Rightarrow \left|\sum_{i=n}^{n+p} d_i\right| < \varepsilon \qquad \text{for } p = 1, 2, 3, \cdots.$$

Hence $n \geqslant \max(N, N_0) \Rightarrow d_n \geqslant 0$ and

$$\left|\sum_{i=n}^{n+p} a_i\right| \leqslant \sum_{i=n}^{n+p} |a_i| \leqslant \sum_{i=n}^{n+p} d_i < \varepsilon \qquad \text{for } p = 1, 2, 3, \cdots;$$

that is, the partial sums of $\sum a_n$ form a Cauchy sequence and hence $\sum a_n$ converges.

Proof (b). Let $\sum c_n$ diverge. Then $\sum |a_n|$ diverges, for its convergence and part (a) implies the convergence of $\sum c_n$. ∎

PROBLEM 1. Point out where in the above proof it can be concluded that:

If $\sum |a_n|$ *converges, then* $\sum a_n$ *converges.*

By applying Theorem 3-5.5 to sequences of partial sums, the following theorem for series in E^2 is obtained:

Theorem 4. *If* $\sum a_n$ *and* $\sum b_n$ *are convergent series of complex terms and* c *and* d *are complex numbers, then* $\sum(ca_n + db_n)$ *converges and*

$$\sum_{n=1}^{\infty} (ca_n + db_n) = c \sum_{n=1}^{\infty} a_n + d \sum_{n=1}^{\infty} b_n.$$

3-7 SERIES OF REAL NUMBERS

The convergence portion of the following theorem about geometric series with ratio $r \in E^1$ has been used repeatedly.

Theorem 1

$$\sum_{n=0}^{\infty} r^n \begin{cases} \textit{converges to } \dfrac{1}{1-r} & \text{if } |r| < 1 \\ \textit{diverges} & \textit{if } |r| \geqslant 1. \end{cases}$$

Proof. If $|r| > 1$, then $\lim r^n \neq 0$ and the series diverges. For $|r| < 1$, then $1 - r \neq 0$ and

$$1 + r + r^2 + \cdots + r^n = \frac{1 - r^{n+1}}{1 - r} \to \frac{1}{1 - r}$$

since $|r|^n \to 0$. ∎

Theorem 2. *Given that* $a_n \geqslant 0$, $n \in J$, *then* $\sum a_n$ *converges iff the sequence of partial sums is bounded (above).*

Proof. Since $a_n \geqslant 0$, then $s_n \leqslant s_n + a_{n+1} = s_{n+1}$ so the sequence $\{s_n\}$ is monotonically increasing and hence converges iff it is bounded (Theorem 3-3.3). ∎

Theorem 3. *Given* $a_1 \geqslant a_2 \geqslant a_3 \geqslant \cdots \geqslant 0$, *then*

$$\sum_{n=1}^{\infty} a_n \quad \textit{and} \quad \sum_{n=0}^{\infty} 2^n a_{2^n}$$

either both converge or both diverge.

Proof. For $n = 1, 2, 3, \cdots$ set

$$s_n = a_1 + a_2 + a_3 + \cdots + a_n \quad \text{and}$$
$$t_n = a_1 + 2a_2 + 4a_4 + \cdots + 2^{n-1} a_{2^{n-1}}.$$

We next show that

$$s_n \leqslant t_n \leqslant 2s_{2^n-1}, \qquad n = 1, 2, 3, \cdots. \tag{1}$$

First

$$\begin{aligned} s_n \leqslant s_{2^n-1} &= a_1 + (a_2 + a_3) + (a_4 + a_5 + a_6 + a_7) + \cdots \\ &\quad + (a_{2^{n-1}} + a_{2^{n-1}+1} + \cdots + a_{2^n-1}) \\ &\leqslant a_1 + 2a_2 + 4a_4 + \cdots + 2^{n-1}a_{2^{n-1}} = t_n. \end{aligned}$$

On the other hand,

$$\begin{aligned} s_{2^n} &= a_1 + a_2 + (a_3 + a_4) + \cdots + (a_{2^{n-1}+1} + \cdots + a_{2^n}) \\ &\geqslant \tfrac{1}{2}a_1 + a_2 + 2a_4 + \cdots + 2^{n-1}a_{2^n} = \tfrac{1}{2}t_{n+1} \end{aligned}$$

so that $t_{n+1} \leqslant 2s_{2^n}$ and hence $t_n \leqslant 2s_{2^{n-1}}$.

From (1) it follows that the monotonically increasing sequences $\{s_n\}$ and $\{t_n\}$ are either both bounded or both unbounded; that is, both either converge or diverge and hence the conclusion of the theorem follows. ∎

Theorem 4. *With $p \in E^1$,*

$$\sum_{n=1}^{\infty} \frac{1}{n^p} \quad \textit{converges iff } p > 1.$$

Proof.

Case 1. $0 < p$. Then

$$\frac{1}{n^p} \geqslant \frac{1}{(n+1)^p}, \qquad n = 1, 2, \ldots$$

so that the given series and (from Theorem 3) the series

$$\sum_{n=0}^{\infty} 2^n \frac{1}{(2^n)^p} = \sum_{n=0}^{\infty} \left(\frac{1}{2^{p-1}}\right)^n$$

either both converge or both diverge. But this last series converges if $p > 1$ and diverges if $0 < p < 1$ (Theorem 1).

Case 2. $p \leqslant 0$. Then $1/n^p \geqslant 1$, $\lim(1/n^p) \neq 0$, and the series diverges. ∎

In particular,

$$\sum \frac{1}{n} \text{ diverges}, \qquad \sum \frac{1}{n^2} \text{ converges},$$

and the nth term of each approaches zero.

If there were a most slowly diverging series of positive terms, such a series would be tried first in comparison tests for divergence of other series. The next theorem shows that there is no such series.

Theorem 5. *If $a_n > 0$ and $\sum a_n$ diverges, then*

$$\sum \frac{a_n}{1 + a_n} \text{ diverges.} \tag{2}$$

Proof.
Case 1. $\lim a_n = 0$. Let N be such that $n \geqslant N \Rightarrow a_n < 1$. Then

$$n \geqslant N \Rightarrow \left(1 + a_n < 2, \frac{a_n}{1 + a_n} > \frac{1}{2} a_n > 0\right)$$

and divergence of (2) follows by the comparison test (Theorem 3-6.3).
Case 2. $\overline{\lim}\, a_n = \bar{a} > 0$. Then with $0 < c < \bar{a}$, the following set is infinite:

$$\{n \mid a_n > c\} = \{n \mid a_n(1 + c) > c(1 + a_n)\} = \left\{n \,\middle|\, \frac{a_n}{1 + a_n} > \frac{c}{1 + c}\right\}$$

and hence the terms in (2) do not approach 0. ∎

PROBLEM 1. Given that $a_n > 0$ and $\sum a_n$ diverges, then

$$\sum \frac{a_n}{s_n} \text{ diverges} \qquad \text{and} \qquad \sum \frac{a_n}{s_n^{\,2}} \text{ converges.}$$

[*Hints*: $\displaystyle\sum_{k=n}^{n+p} \frac{a_k}{s_k} \geqslant \frac{1}{s_{n+p}} \sum_{k=n}^{n+p} a_k = 1 - \frac{s_{n-1}}{s_{n+p}}$; $\quad s_k^{\,2} > s_k s_{k-1}, \quad a_k = s_k - s_{k-1}$.]

The following theorem provides a mechanical substitute for the comparison test of Theorem 3-6.3.

Theorem 6. *Given $a_n > 0$ and $b_n > 0$, let*

$$\underline{c} = \underline{\lim} \frac{a_n}{b_n} \qquad \text{and} \qquad \bar{c} = \overline{\lim} \frac{a_n}{b_n}$$

(so that $0 \leqslant \underline{c} \leqslant \bar{c} \leqslant \infty$).

(a) *If $0 < \underline{c}$ and $\sum b_n$ diverges, then $\sum a_n$ diverges. If $\underline{c} = 0$, divergence of $\sum b_n$ does not imply divergence of $\sum a_n$.*
(b) *If $\bar{c} < \infty$ and $\sum b_n$ converges, then $\sum a_n$ converges. If $\bar{c} = \infty$, convergence of $\sum b_n$ does not imply convergence of $\sum a_n$.*

Proof (a). Given that $0 < \underline{c}$ (possibly $\underline{c} = \infty$), let a be such that $0 < a < \underline{c}$. Then there is an N_0 such that

$$n \geqslant N_0 \Rightarrow \left(a < \frac{a_n}{b_n}, 0 < ab_n < a_n\right).$$

Hence, if $\sum b_n$ diverges, then $\sum ab_n$ diverges and $\sum a_n$ diverges by the comparison test.

With $a_n = 1/n^2$ and $b_n = 1/n$, then

$$\underline{c} = \underline{\lim} \frac{a_n}{b_n} = \lim \frac{1}{n} = 0,$$

but $\sum b_n$ diverges and $\sum a_n$ converges.

PROBLEM 2. Prove part (b). ∎

A knowledge of convergence or divergence of some series $\sum b_n$ is necessary before Theorem 6 can be used to test a series $\sum a_n$. The following two theorems involve only the terms of the series being tested.

Theorem 7 (Ratio Test), *Given that each $a_n > 0$, let*

$$\underline{\rho} = \underline{\lim} \frac{a_{n+1}}{a_n} \qquad \textit{and} \qquad \bar{\rho} = \overline{\lim} \frac{a_{n+1}}{a_n}.$$

(a) *If $\bar{\rho} < 1$, then $\sum a_n$ converges.*
(b) *If $\underline{\rho} > 1$, then $\sum a_n$ diverges.*
(c) *If $\underline{\rho} \leqslant 1 \leqslant \bar{\rho}$, there is no conclusion as to convergence or divergence of $\sum a_n$.*

Corollary. *If there is an N_0 such that $a_{n+1}/a_n \geqslant 1$ for $n \geqslant N_0$, then* $\lim a_n \neq 0$ *and $\sum a_n$ diverges.*

Proof (a). With the hypothesis holding, select r such that $\bar{\rho} < r < 1$ and then N such that

$$n \geqslant N \Rightarrow \left[\frac{a_{n+1}}{a_n} < r, \frac{a_{n+1}}{r^{n+1}} < \frac{a_n}{r^n}\right].$$

Hence

$$n \geqslant N + 1 \Rightarrow \frac{a_{n+1}}{r^{n+1}} < \frac{a_n}{r^n} < \frac{a_{n-1}}{r^{n-1}} < \cdots \leqslant \frac{a_N}{r^N}.$$

Thus

$$\overline{\lim} \frac{a_n}{r^n} \leqslant \frac{a_N}{r^N} < \infty,$$

and (since $\sum r^n$ converges), $\sum a_n$ converges by Theorem 6.

Proof (b). Under the hypothesis, select N such that

$$n \geqslant N \Rightarrow \left[\frac{a_{n+1}}{a_n} \geqslant 1, a_{n+1} \geqslant a_n \geqslant a_{n-1} \geqslant \cdots \geqslant a_N\right].$$

Hence $\underline{\lim}\, a_n \geqslant a_N > 0$, $\lim a_n \neq 0$, and the series diverges, thus proving part (b) as well as the corollary.

Proof (c). For $p > 0$ and $a_n = 1/n^p$, then

$$\lim \frac{a_{n+1}}{a_n} = \lim \left(\frac{n}{n+1}\right)^p = 1 \qquad (= \underline{\rho} = \bar{\rho})$$

with $\sum a_n$ converging if $p = 2$ but diverging if $p = 1$. ∎

(Examples with $\underline{\rho} < 1 < \bar{\rho}$ are given after the next theorem.)

Theorem 8 (Root Test). *With $a_n \geqslant 0$,*

(a) *If* $\overline{\lim} \sqrt[n]{a_n} < 1$, *then* $\sum a_n$ *converges.*
(b) *If* $\overline{\lim} \sqrt[n]{a_n} > 1$, *then* $\lim a_n \neq 0$ *and* $\sum a_n$ *diverges.*
(c) $\overline{\lim} \sqrt[n]{a_n} = 1$ *gives no information about convergence or divergence of* $\sum a_n$.

Proof (a). Under the hypothesis, first select r such that $\overline{\lim} \sqrt[n]{a_n} < r < 1$, then N such that

$$n \geqslant N \Rightarrow (\sqrt[n]{a_n} < r,\ a_n < r^n),$$

and obtain convergence of $\sum a_n$ from Theorems 1 and 3-6.3.

Proof (b). Under the hypothesis,

$$\{n \mid \sqrt[n]{a_n} > 1\} = \{n \mid a_n > 1\}$$

is infinite so that $\overline{\lim}\, a_n \geqslant 1$. Hence $\lim a_n \neq 0$ and $\sum a_n$ diverges.

Proof (c). Again $\sum 1/n$ and $\sum 1/n^2$ confirm the statement, since $\lim \sqrt[n]{n} = \lim \sqrt[n]{n^2} = 1$ (Theorems 3-3.5 and 3-5.5). ∎

EXAMPLE 1. For the series

$$\frac{1}{2} + \frac{1}{3} + \frac{1}{2^2} + \frac{1}{3^2} + \frac{1}{2^3} + \frac{1}{3^3} + \cdots$$

in which $a_{2n-1} = 1/2^n$ and $a_{2n} = 1/3^n$ for $n = 1, 2, 3, \cdots$, then

$$\frac{a_{2n}}{a_{2n-1}} = \left(\frac{2}{3}\right)^n \qquad \text{and} \qquad \frac{a_{2n+1}}{a_{2n}} = \frac{1}{2}\left(\frac{3}{2}\right)^n.$$

Thus $\underline{\rho} = 0 < 1 < \infty = \bar{\rho}$ and the ratio test is inconclusive. However,

$$\sqrt[2n-1]{a_{2n-1}} = \frac{1}{2^{n/(2n-1)}} \qquad \text{and} \qquad \sqrt[2n]{a_{2n}} = \frac{1}{\sqrt{3}}$$

so that $\overline{\lim} \sqrt[n]{a_n} = 1/\sqrt{2} < 1$ and the series converges by the root test. [Also $\underline{\lim} \sqrt[n]{a_n} = 1/\sqrt{3}$.]

PROBLEM 3.

$$\frac{1}{2}+\frac{2}{2}+\frac{1}{2^3}+\frac{2}{2^3}+\frac{1}{2^5}+\frac{2}{2^5}+\frac{1}{2^7}+\frac{2}{2^7}+\cdots$$

has $\underline{\lim}(a_{n+1}/a_n) = 1/8$, $\overline{\lim}(a_{n+1}/a_n) = 2$, and $\lim\sqrt[n]{a_n} = 1/2$.

Theorem 9. *Given that $a_n > 0$, then*

$$\underline{\lim}\,\frac{a_{n+1}}{a_n} \leqslant \underline{\lim}\,\sqrt[n]{a_n} \leqslant \overline{\lim}\,\sqrt[n]{a_n} \leqslant \overline{\lim}\,\frac{a_{n+1}}{a_n}. \tag{3}$$

[This theorem shows that the root test is more powerful than the ratio test. Note also that the series in the above example is one for which every $\leqslant$ in (3) is a strict inequality.]

Proof. For the first inequality, if the left member is 0 there is nothing to prove, so consider it to be positive, and take any number c such that

$$0 < c < \underline{\lim}\,\frac{a_{n+1}}{a_n} \tag{4}$$

and then an N such that

$$n > N \Rightarrow \left[\frac{a_{n+1}}{a_n} > c,\ \frac{a_{n+1}}{c^{n+1}} > \frac{a_n}{c^n} > \frac{a_{n-1}}{c^{n-1}} > \cdots > \frac{a_N}{c^N}\right].$$

Thus $n \geqslant N \Rightarrow \sqrt[n]{a_n} \geqslant \sqrt[n]{a_N c^{-N}}\,c$ and hence

$$\underline{\lim}\,\sqrt[n]{a_n} \geqslant \underline{\lim}\,\sqrt[n]{a_N c^{-N}}\,c = c,$$

since $1 = \lim\sqrt[n]{a_N c^{-N}} = \underline{\lim}\sqrt[n]{a_N c^{-N}}$ by Theorem 3-3.5. The fact that $\underline{\lim}\sqrt[n]{a_n} \geqslant c$ holds for any number c satisfying (4) shows the first inequality in (3).

The last inequality follows in much the same way (and the middle one is a property of limits inferior and superior). ∎

Theorem 10. *If $a_1 \geqslant a_2 \geqslant a_3 \geqslant \cdots \geqslant 0$ and $\lim a_n = 0$, then*

$$\sum(-1)^n a_n \quad \textit{converges.}$$

Proof. From

$$\begin{aligned} s_{2n} &= a_1 - a_2 + a_3 - a_4 + \cdots + a_{2n-1} - a_{2n} \\ &= (a_1 - a_2) + (a_3 - a_4) + \cdots + (a_{2n-1} - a_{2n}) \geqslant 0 \end{aligned}$$

and

$$\begin{aligned} s_{2n-2} \leqslant s_{2n-2} + (a_{2n-1} - a_{2n}) &= s_{2n} \\ &= a_1 - (a_2 - a_3) - \cdots - (a_{2n-2} - a_{2n-1}) - a_{2n} \leqslant a_1, \end{aligned}$$

it follows that $\lim s_{2n}$ exists and $0 \leqslant \lim s_{2n} \leqslant a_1$. Then

$$\lim s_{2n} = \lim s_{2n} + \lim a_{2n+1} = \lim(s_{2n} + a_{2n+1}) = \lim s_{2n+1}$$

so that $\lim s_n$ exists ($\geqslant 0$ and $\leqslant a_1$). ∎

As an example, $\sum_{n=1}^{\infty}(-1)^{n+1}(1/n)$ converges to a value between 0 and 1 [which will be shown to be ln 2 in Section 7-9].

PROBLEM 4. With the sequence $\{a_n\}$ as in Theorem 10 and

$$s = \sum_{n=1}^{\infty}(-1)^{n+1}a_n,$$

then $0 \leqslant (-1)^n(s - s_n) \leqslant a_{n+1}$ for $n = 1, 2, 3, \cdots$.

The number e of the following theorem is recognized as the base of natural logarithms.

Theorem 11. *Given the sequences $\{a_n\}$ and $\{s_n\}$ defined by*

$$a_n = \left(1 + \frac{1}{n}\right)^n \quad \textit{and} \quad s_n = \sum_{k=0}^{n}\frac{1}{k!} \quad \textit{for each } n \in J,$$

then the following limits exist and the equality holds

$$\lim a_n = \lim s_n = e \tag{5}$$

where $2 < e \leqslant 3$.

$$\left[\text{Hence } e = \lim_{n\to\infty}\left(1 + \frac{1}{n}\right)^n = 1 + 1 + \frac{1}{2!} + \frac{1}{3!} + \cdots.\right]$$

Proof. For $r > -1$, by induction on $n \in J$,

$$(1 + r)^n > 1 + nr, \qquad n > 1.$$

Then

$$\frac{a_n}{a_{n-1}} = \frac{n}{n-1}\left(1 - \frac{1}{n^2}\right)^n > \frac{n}{n-1}\left(1 - \frac{1}{n}\right) \qquad \text{for } n > 1$$

so that $2 = a_1 < a_2 < a_3 < \cdots$. Clearly, $2 = s_1 < s_2 < s_3 < \cdots$. Also,

$$\begin{aligned} a_n &= 1 + \binom{n}{1}\frac{1}{n} + \binom{n}{2}\frac{1}{n^2} + \cdots + \binom{n}{n}\frac{1}{n^n} \\ &= 1 + 1 + \frac{1}{2}\left(1 - \frac{1}{n}\right) + \frac{1}{3!}\left(1 - \frac{1}{n}\right)\left(1 - \frac{2}{n}\right) + \cdots + \frac{1}{n!}\left(1 - \frac{1}{n}\right)\cdots\left(1 - \frac{n-1}{n}\right) \\ &< 1 + 1 + \frac{1}{2} + \frac{1}{3!} + \cdots + \frac{1}{n!} = s_n \\ &< 1 + 1 + \frac{1}{2} + \frac{1}{2^2} + \cdots + \frac{1}{2^{n-1}} < 1 + 2 = 3. \end{aligned}$$

Hence the limits of both of the bounded monotonic sequences exist, and upon setting $e = \lim a_n$, it follows that

$$2 < e = \lim a_n \leqslant \lim s_n \leqslant 3. \tag{6}$$

With $m \geqslant n$, then

$$a_m \geqslant 1 + 1 + \binom{m}{2}\frac{1}{m^2} + \cdots + \binom{m}{n}\frac{1}{m^n}$$

$$= 2 + \frac{1}{2}\left(1 - \frac{1}{m}\right) + \cdots + \frac{1}{n!}\left(1 - \frac{1}{m}\right)\left(1 - \frac{2}{m}\right)\cdots\left(1 - \frac{n-1}{m}\right).$$

By holding n fixed and letting $m \to \infty$, it follows that

$$e \geqslant 2 + \frac{1}{2} + \frac{1}{3!} + \cdots + \frac{1}{n!} = s_n > s_{n-1}.$$

Thus $s_n < e$ and $\lim s_n \leqslant e$, which with (6) shows that (5) holds. ∎

Corollary. *e is irrational.*

Proof. Suppose e is rational and set $e = m/n$; $m, n \in J$. Then

$$0 < e - s_n = \frac{1}{(n+1)!} + \frac{1}{(n+2)!} + \frac{1}{(n+3)!} + \cdots$$

$$< \frac{1}{(n+1)!}\left[1 + \frac{1}{n+1} + \frac{1}{(n+1)^2} + \cdots\right] < \frac{1}{n!\,n}.$$

Hence

$$0 < \frac{m}{n}n! - \left(1 + \frac{1}{2} + \frac{1}{3!} + \cdots + \frac{1}{n!}\right)n! < \frac{1}{n} \leqslant 1,$$

which is impossible since no positive integer is less than 1. ∎

Theorem 12. *Let $\sum a_n$ be a real series with sequence of partial sums $\{s_n\}$ and define the sequence $\{\sigma_n\}$ by setting*

$$\sigma_n = \frac{1}{n}(s_1 + s_2 + \cdots + s_n), \qquad n = 1, 2, 3, \cdots.$$

Then

(7) $$\underline{\lim}\, s_n \leqslant \underline{\lim}\, \sigma_n \leqslant \overline{\lim}\, \sigma_n \leqslant \overline{\lim}\, s_n.$$

Proof. If $\underline{\lim}\, s_n = -\infty$, the first inequality holds, so consider $-\infty < \underline{\lim}\, s_n$. Take any $c \in E^1$ such that $-\infty < c < \underline{\lim}\, s_n$ and let $N \in J$ be such that $n \geqslant N \Rightarrow c < s_n$. Then for $n > N$,

$$\sigma_n = \frac{1}{n}(s_1 + s_2 + \cdots + s_{N-1} + s_N + s_{N+1} + \cdots + s_n)$$

$$> \frac{1}{n}(s_1 + s_2 + \cdots + s_{N-1}) + \frac{n-N}{n}c.$$

Thus

$$c = \lim_{n\to\infty}\left[\frac{s_1 + s_2 + \cdots + s_{N-1}}{n} + \left(1 - \frac{N}{n}\right)c\right] \leqslant \underline{\lim}\, \sigma_n .$$

Since c was any number less than $\underline{\lim}\, s_n$,

$$\underline{\lim}\, s_n \leqslant \underline{\lim}\, \sigma_n .$$

Then $\underline{\lim}(-s_n) \leqslant \underline{\lim}(-\sigma_n)$ so that $\overline{\lim}\, s_n \geqslant \overline{\lim}\, \sigma_n$, by Theorem 3-2.2. ▮

Corollary.

(a) *If* $\lim s_n$ *exists*, then* $\lim \sigma_n$ *exists* and*

$$\lim s_n = \lim \sigma_n .$$

(b) *If* $\sum a_n$ *converges to s, then* $\lim \sigma_n = s$.

PROBLEM 5. Given the series $\sum_{n=1}^{\infty}(-1)^{n+1}$, then

$$\lim \sigma_n = \tfrac{1}{2}.$$

Definition 1. *In the notation of Theorem* 12, *if* $\lim \sigma_n = \sigma$ *exists, then* $\sum_{n=1}^{\infty} a_n$ *is said to be* $(C, 1)$ *summable to* σ *(or Cesaro summable of order* 1 *to* σ*).*

3–8 REARRANGING AND GROUPING

Recall that the notation $f\colon J \leftrightarrow J$ is used for a function which is 1–1 on J onto J, and hence the inverse function exists with $f^{-1}\colon J \leftrightarrow J$.

Definition 1. *A sequence* $\{b_n\}$ *is said to be a* **rearrangement** *of a sequence* $\{a_n\}$ *if there is a function* $f\colon J \leftrightarrow J$ *such that*

$$b_n = a_{f(n)} \qquad \textit{for each } n \in J$$

(and hence $a_n = b_{f^{-1}(n)}$ for each $n \in J$ so that $\{a_n\}$ is a rearrangement of $\{b_n\}$).

Theorem 1. *Given a sequence* $\{a_n\}$, *then*

(a) *If* $\{a_n\}$ *converges with* $a_n \to a$, *every rearrangement also converges and to a.*
(b) *If* $\{a_n\}$ *diverges so does every rearrangement of it.*

Proof (a) Let $\{a_n\}$ converge to a and corresponding to $\varepsilon > 0$ arbitrary let N be such that

$$n \geqslant N \Rightarrow |a_n - a| < \varepsilon.$$

Take any rearrangement $\{b_n\}$ of $\{a_n\}$ and let $f\colon J \leftrightarrow J$ be the rearranging function. Then

$$\{n \mid f(n) \leqslant N\}$$

is a finite set of positive integers and therefore has a largest integer, which we denote by M. Thus

$$n \geqslant M \Rightarrow f(n) \geqslant N \Rightarrow (|a_{f(n)} - a| < \varepsilon, |b_n - a| < \varepsilon)$$

and hence $\lim b_n = a$.

Proof (b) Let $\{a_n\}$ diverge and take any rearrangement $\{b_n\}$ of it. Then $\{b_n\}$ diverges, since otherwise its rearrangement $\{a_n\}$ would converge by part (a). ∎

Definition 2. *A series $\sum b_n$ is said to be a* **rearrangement** *of a series $\sum a_n$ if $\{b_n\}$ is a rearrangement of $\{a_n\}$.*

EXAMPLE 1. The series $\sum_{n=1}^{\infty} (-1)^{n+1}(1/n)$ converges to a value s (Theorem 3-7.10) and the rearrangement

$$1 - \frac{1}{2} - \frac{1}{4} + \frac{1}{3} - \frac{1}{6} - \frac{1}{8} + \frac{1}{5} - \cdots + \frac{1}{2n-1} - \frac{1}{4n-2} - \frac{1}{4n} + \cdots$$

also converges but to $\frac{1}{2}s$.

To see this, let s_n and t_n be the nth partial sums of the series and its given rearrangement, respectively. Then

$$t_{3n} = \left(1 - \frac{1}{2}\right) - \frac{1}{4} + \left(\frac{1}{3} - \frac{1}{6}\right) - \frac{1}{8} + \cdots + \left(\frac{1}{2n-1} - \frac{1}{4n-2}\right) - \frac{1}{4n}$$

$$= \frac{1}{2} - \frac{1}{4} + \frac{1}{6} - \frac{1}{8} + \cdots + \frac{1}{2(2n-1)} - \frac{1}{4n} = \frac{1}{2} s_{2n}.$$

Since $s_n \to s$, then $s_{2n} \to s$ and hence $t_{3n} \to \frac{1}{2}s$. But also $t_{3n} - t_{3n+1} \to 0$, $t_{3n} - t_{3n+2} \to 0$ so that $t_n \to \frac{1}{2}s$. ∎

[From results of the next section it follows that the series of the above example can be rearranged to have partial sums t_n such that

$$\underline{\lim}\, t_n = \underline{t} \qquad \text{and} \qquad \overline{\lim}\, t_n = \bar{t}$$

with $\underline{t}$ and $\bar{t}$ given arbitrarily (except $\underline{t} \leqslant \bar{t}$) including the possibilities $\underline{t} = -\infty$ and $\bar{t} = \infty$.]

Definition 3. *A rearrangement $\sum b_n$ of $\sum a_n$ is said to be a* **bounded rearrangement** *if there is a $B \in J$ such that the rearranging function $f\colon J \leftrightarrow J$ satisfies*

$$|f(n) - n| \leqslant B \qquad \text{for each } n \in J$$

(and hence $|f^{-1}(n) - n| \leqslant B$ for each $n \in J$).

Theorem 2. *Given a convergent series $\sum_{n=1}^{\infty} a_n = s$, then any bounded rearrangement also converges to s.*

Proof. With f and B as in the definition, note that if $n > B$, then

$$\{1, 2, \cdots, n - B\} \subset \{f(1), f(2), \cdots, f(n)\} \subset \{1, 2, 3, \cdots, n + B\}$$

and

$$\left|\sum_{i=1}^{n} b_i - s\right| = \left|\sum_{i=1}^{n} a_{f(i)} - s\right| \leqslant \left|\sum_{j=1}^{n-B} a_j - s\right| + \sum_{j=n-B+1}^{n+B} |a_j|.$$

Hence, with $\varepsilon > 0$ arbitrary and N such that

$$n \geqslant N \Rightarrow \left[\left|\sum_{j=1}^{n} a_j - s\right| < \varepsilon \text{ and } |a_n| < \varepsilon\right],$$

then

$$n \geqslant N + B \Rightarrow \left|\sum_{i=1}^{n} b_i - s\right| < \varepsilon + 2B\varepsilon = (1 + 2B)\varepsilon$$

and the conclusion follows, since $1 + 2B$ does not depend on n. ∎

Corollary. *Any bounded rearrangement of a series $\sum a_n$ converges (and to the same sum) iff $\sum a_n$ converges.*

Proof. With $\sum b_n$ a bounded rearrangement of $\sum a_n$, then $\sum a_n$ is a bounded rearrangement of $\sum b_n$, and hence convergence of either implies convergence of the other. ∎

Theorem 3. *Any rearrangement of a series $\sum a_n$ of nonnegative terms converges (and to the same sum) iff $\sum a_n$ converges.*

Proof. Let $\sum b_n$ be a rearrangement determined by a function $f\colon J \leftrightarrow J$. For each $m \in J$ there is an $n \in J$ such that

$$n \geqslant \max[f(1), f(2), \cdots, f(m)]$$

and since each $a_i \geqslant 0$,

$$\sum_{i=1}^{m} b_i = \sum_{i=1}^{m} a_{f(i)} \leqslant \sum_{i=1}^{n} a_i \leqslant \sum_{i=1}^{\infty} a_i.$$

Since this holds for each $m \in J$ (and since each $b_i \geqslant 0$), then

$$\sum_{n=1}^{\infty} b_n \leqslant \sum_{n=1}^{\infty} a_n.$$

But $\sum a_n$ is a rearrangement of $\sum b_n$ and hence the argument need not be repeated to know that

$$\sum_{n=1}^{\infty} a_n \leqslant \sum_{n=1}^{\infty} b_n.$$

Hence both sums exist*, are equal elements of E^*, and if either is finite so is the other. ∎

The series

$$(1) \qquad 1 - 1 + 1 - 1 + 1 - 1 + \cdots$$

diverges, but the series obtained by inserting parentheses

$$(2) \qquad (1 - 1) + (1 - 1) + (1 - 1) + \cdots$$

converges.

Definition 4. *A series* $\sum b_n$ *is said to be obtained from* $\sum a_n$ *by* **inserting parentheses** *if there is a function* $g\colon J \cup \{0\} \to J \cup \{0\}$ *such that* $g(0) = 0 < g(1) < g(2) < g(3) < \cdots$ *and*

$$b_n = a_{g(n-1)+1} + a_{g(n-1)+2} + \cdots + a_{g(n)}.$$

[We think of the nth parenthesis as containing $g(n) - g(n-1)$ terms, $n = 1, 2, 3, \cdots$. For example, $g(n) = 2n$, $n = 0, 1, 2, \cdots$ defines the function relating (1) and (2).]

Theorem 4. *Given a series* $\sum_{n=1}^{\infty} a_n$, *insert parentheses to obtain* $\sum_{n=1}^{\infty} b_n$, *according to a function g as defined above.*

(a) *If* $\sum_{n=1}^{\infty} a_n = s$, *then also* $\sum_{n=1}^{\infty} b_n = s$.

(b) *Given in addition that* $\lim a_n = 0$, *that there is a* $B \in J$ *for which* $g(n) - g(n-1) \leqslant B$, $n = 1, 2, 3, \cdots$ *and that* $\sum_{n=1}^{\infty} b_n = t$. *Then also* $\sum_{n=1}^{\infty} a_n = t$.

Proof (a). Let $\{s_n\}$ and $\{t_n\}$ be the sequences of partial sums of $\sum a_n$ and $\sum b_n$, respectively. Then

$$g(n) \geqslant n \qquad \text{and} \qquad t_n = s_{g(n)} \qquad \text{for each } n \in J.$$

Thus $\{t_n\}$ is a subsequence of $\{s_n\}$. Hence if $\{s_n\}$ converges, then so does $\{t_n\}$ and to the same value.

Proof (b). For convenience set $s_{g(0)} = 0$. Given any $m \in J$ there is an $n \in J \cup \{0\}$ such that

$$g(n) < m \leqslant g(n+1) \leqslant g(n) + B$$

and hence

$$\begin{aligned} |s_m - t| &= |s_{g(n)} + a_{g(n)+1} + a_{g(n)+2} + \cdots + a_m - t| \\ &\leqslant |t_n - t| + \sum_{i=1}^{B} |a_{g(n)+i}|. \end{aligned}$$

Let $\varepsilon > 0$ be arbitrary and then take N such that

$$n \geqslant N \Rightarrow (|t_n - t| < \varepsilon \text{ and } |a_n| < \varepsilon).$$

Then for any $m > g(N)$ the n for which $g(n) < m \leqslant g(n+1)$ is such that $n > N$. Thus

$$m > g(N) \Rightarrow |s_m - t| < \varepsilon + B\varepsilon;$$

that is, $\lim t_m = t$, as we wished to prove. ∎

Corollary. *Under condition* (b), $\sum_{n=1}^{\infty} a_n$ *and* $\sum_{n=1}^{\infty} b_n$ *either both converge, and to the same value, or else both diverge.*

3-9 ABSOLUTE AND CONDITIONAL CONVERGENCE

Definition 1. *A series* $\sum a_n$, *with each* $a_n \in E^k$, *is said to be*

(i) **Absolutely convergent** *if* $\sum |a_n|$ *converges.*
(ii) **Conditionally convergent** *if* $\sum a_n$ *converges and* $\sum |a_n|$ *diverges.*

For example, the convergent series $\sum (-1)^{n+1}(1/n)$ converges conditionally, since its absolute value series $\sum (1/n)$ diverges.

The definition (i) does not say specifically that an absolutely convergent series converges, but the next theorem does.

Theorem 1. *If* $\sum |a_n|$ *converges, then* $\sum a_n$ *converges.*

Proof. For each $n, p \in J$,

$$\left|\sum_{i=n}^{n+p} a_i\right| \leqslant \sum_{i=n}^{n+p} |a_i|.$$

Thus if $\sum |a_n|$ converges, then its partial sums form a Cauchy sequence, which implies that the partial sums of $\sum a_n$ form a Cauchy sequence and hence that $\sum a_n$ converges. ∎

It follows from Theorem 1 that tests for convergence of real series may be applied to $\sum |a_n|$ as tests for absolute convergence of $\sum a_n$.

Given that $\{a_n\}$ is a real sequence, we define a_n^+ and a_n^- for each $n \in J$ by

(1) $$a_n^+ = \frac{|a_n| + a_n}{2} = \begin{cases} a_n & \text{if } a_n \geqslant 0 \\ 0 & \text{if } a_n < 0, \end{cases}$$

(2) $$a_n^- = \frac{|a_n| - a_n}{2} = \begin{cases} 0 & \text{if } a_n \geqslant 0 \\ -a_n & \text{if } a_n < 0. \end{cases}$$

Hence $a_n^+ \geqslant 0$, $a_n^- \geqslant 0$,

(3) $$a_n = a_n^+ - a_n^-, \quad \text{and} \quad |a_n| = a_n^+ + a_n^-.$$

Theorem 2. *Given that* $\{a_n\}$ *is a real sequence, then*

(a) $\sum a_n$ *converges absolutely, iff both* $\sum a_n^+$ *and* $\sum a_n^-$ *converge, and if they converge then*

(4) $$\sum a_n = \sum a_n^+ - \sum a_n^- \quad \text{and} \quad \sum |a_n| = \sum a_n^+ + \sum a_n^-.$$

(b) *If* $\sum a_n$ *converges conditionally, then* $\lim a_n = 0$ *but* $\sum a_n^+$ *diverges to* ∞ *and* $\sum a_n^-$ *diverges to* ∞.

Proof (a). Let $\sum a_n$ converge absolutely. Then both $\sum a_n$ and $\sum |a_n|$ converge so that, from (1),

$$\tfrac{1}{2} \sum (a_n + |a_n|) = \sum a_n^+$$

converges and, similarly, $\sum a_n^-$ converges from (2).

Conversely, let $\sum a_n^+$ and $\sum a_n^-$ converge. Then not only does convergence of $\sum a_n$ and $\sum |a_n|$ follow from (3), but the equalities (4) as well.

Proof (b). Let $\sum a_n$ converge conditionally. Then $\sum a_n$ converges (so $\lim a_n = 0$) and $\sum |a_n|$ diverges. Thus (1), written as $|a_n| = 2a_n^+ - a_n$, shows that $\sum a_n^+$ diverges (since if it converged so would $\sum |a_n|$). But $\sum a_n^+$ has no negative terms, and hence its monotonically increasing sequence of partial sums is unbounded above. Similarly for $\sum a_n^-$. ∎

Theorem 3. *Given that $\sum a_n$ is an absolutely convergent series in E^k and $\sum b_n$ is a rearrangement of it. Then $\sum b_n$ also converges absolutely with*

$$\sum_{n=1}^{\infty} b_n = \sum_{n=1}^{\infty} a_n \quad \textit{and} \quad \sum_{n=1}^{\infty} |b_n| = \sum_{n=1}^{\infty} |a_n|. \tag{5}$$

Proof. First consider $\sum a_n$ to be a real series. Then $\sum b_n^+$ is a rearrangement of $\sum a_n^+$ so that

$$\sum_{n=1}^{\infty} b_n^+ = \sum_{n=1}^{\infty} a_n^+$$

by Theorem 3-8.3. Similarly, $\sum (b_n^-) = \sum (a_n^-)$. Hence, from Theorem 2(a),

$$\sum b_n = \sum b_n^+ - \sum b_n^- = \sum a_n^+ - \sum a_n^- = \sum a_n$$

and in the same way $\sum |b_n| = \sum |a_n|$.

In case $a_n = (a_{n1}, a_{n2}, \cdots, a_{nk}) \in E^k$, then $a_{ni} \in E^1; n \in J, i = 1, 2, \cdots, k$. Hence $\sum b_{ni}$ and $\sum a_{ni}$ are real, are rearrangements of one another and, by the first part of the proof,

$$\sum b_{ni} = \sum a_{ni}; \qquad i = 1, 2, \cdots, k.$$

Thus the first equation of (5) follows by summing components and using the extension of Theorem 3-5.6 from E^2 to E^k. The second equation in (5) follows, since the series are already real and rearrangements of one another. ∎

Theorem 4. *Given a real conditionally convergent series $\sum a_n$ and $\underline{t}, \bar{t} \in E^*$ with $\underline{t} \leq \bar{t}$, then there is a rearrangement $\sum b_n$ with sequence of partial sums $\{t_n\}$ such that*

$$\underline{t} = \underline{\lim}\, t_n \quad \textit{and} \quad \bar{t} = \overline{\lim}\, t_n. \tag{6}$$

Proof. The details are messy, but the idea is simple so we sketch the method when $-\infty < \underline{t} < \bar{t} < \infty$.

(1) Take enough of the nonnegative terms of $\sum a_n$ as they come so that the sum of those selected just exceeds $\bar{t}$; that is, if the last positive term were left off the sum would be $\leq \bar{t}$ (possible, since $\sum a_n^+$ diverges to $+\infty$).

(2) Take enough of the negative terms of $\sum a_n$ as they occur so that their sum added to the sum in (1) is just less than $\underline{t}$.

(3) Start with the first nonnegative term not selected in (1), add nonnegative terms to the sum in (2) until a sum is obtained just exceeding $\bar{t}$.

(4) Add more negative terms, etc.

It should be seen that this rearranging of $\sum a_n$ is possible (since $\sum a_n^+$ diverges to ∞ and $-\sum a_n^-$ diverges to $-\infty$) and that the sequence of partial sums of the rearrangement satisfies (6), since $\lim a_n = 0$. ∎

PROBLEM I. Think through a similar process in case $\underline{t} = -\infty$ and $\bar{t} = \infty$ (and see that $\lim a_n = 0$ need not be used).

3–10 THE SPACE l^2

Theorem 1. *Let $\{a_n\}$ and $\{b_n\}$ be real sequences such that both $\sum a_k^2$ and $\sum b_k^2$ converge. Then $\sum (a_k b_k)$ and $\sum (a_k + b_k)^2$ both converge,*

$$(1) \qquad \left[\sum_{k=1}^{\infty} (a_k b_k)\right]^2 \leqslant \left[\sum_{k=1}^{\infty} a_k^2\right]\left[\sum_{k=1}^{\infty} b_k^2\right], \qquad \text{and}$$

$$(2) \qquad \left[\sum_{k=1}^{\infty} (a_k + b_k)^2\right]^{1/2} \leqslant \left[\sum_{k=1}^{\infty} a_k^2\right]^{1/2} + \left[\sum_{k=1}^{\infty} b_k^2\right]^{1/2}$$

Proof. By applying Schwarz inequality to the absolute values of the first n terms of $\sum a_k$ and $\sum b_k$, the first inequality in

$$\sum_{k=1}^{n} |a_k b_k| \leqslant \left\{\left[\sum_{k=1}^{n} a_k^2\right]\left[\sum_{k=1}^{n} b_k^2\right]\right\}^{1/2}$$
$$\leqslant \left\{\left[\sum_{k=1}^{\infty} a_k^2\right]\left[\sum_{k=1}^{\infty} b_k^2\right]\right\}^{1/2}$$

is obtained. Hence $\sum |a_k b_k|$ converges since its sequence of partial sums is monotonically increasing and bounded above. Then $\sum (a_k b_k)$ converges with

$$\pm \sum_{k=1}^{\infty} a_k b_k \leqslant \sum_{k=1}^{\infty} |a_k b_k|.$$

Hence $|\sum a_k b_k| \leqslant \sum |a_k b_k|$ and (1) is seen to hold.

In a similar way (2) follows from the finite Minkowski inequality. ∎

The above results assure the convergence of all the series necessary to validate the concepts in the following definition.

Definition 1. *With a single Greek letter denoting a real sequence, then $\alpha = \{a_k\}$ is said to be* **square summable** *if $\sum a_k^2$ converges. The space of all square summable sequences is denoted by l^2. With $\alpha \in l^2$, $\beta \in l^2$, and $c \in E^1$, then by definition,*

$$\alpha + \beta = \{a_k + b_k\} \text{ and } c\alpha = \{ca_k\},$$

the **dot product** $\alpha \cdot \beta$ *is defined by*

$$\alpha \cdot \beta = \sum_{k=1}^{\infty} a_k b_k,$$

the l^2**-norm** *of* α *is denoted and defined by*

$$\|\alpha\|_2 = \sqrt{(\alpha \cdot \alpha)} = \sqrt{\sum_{k=1}^{\infty} a_k{}^2},$$

and the l^2**-distance** *between* α *and* β *is defined to be*

$$\|\alpha - \beta\|_2 = \sqrt{(\alpha - \beta) \cdot (\alpha - \beta)} = \sqrt{\sum_{k=1}^{\infty} (a_k - b_k)^2}.$$

The sequence all of whose terms are 0 *is denoted by* θ *and is called the* **zero point** (*or* **origin**) *of* l^2.

PROBLEM I. Check that l^2, with the above definitions, is a normed vector space.

It is thus natural to ask whether l^2 is complete. The following discussion shows that some properties of E^k have analogies in l^2, while others do not, and then a proof of completeness of l^2 is given.

The following definition is given to avoid any misunderstanding.

Definition 2. *A sequence* $\alpha_1, \alpha_2, \alpha_3, \cdots$ *with each* $\alpha_n \in l^2$ *is said to* **converge in** l^2 *if there is an* $\alpha \in l^2$ *such that corresponding to each* $\varepsilon > 0$ *an* $N \in J$ *exists for which*

$$n \geqslant N \Rightarrow \|\alpha_n - \alpha\|_2 = 0.$$

If $\{\alpha_n\}$ *converges in* l^2 to α, *we write either* $\alpha_n \underset{l^2}{\rightarrow} \alpha$ *or* $\lim_{l^2} \alpha_n = \alpha$ [and may drop the l^2 when it is clear that convergence in l^2 is being considered.]

Note that if $\alpha = (a_1, a_2, a_3, \cdots) \in l^2$ and $\beta = (b_1, b_2, b_3, \cdots) \in l^2$, then for $k = 1, 2, 3, \cdots$,

$$|a_k - b_k| = \sqrt{(a_k - b_k)^2} \leqslant \sqrt{\sum_{i=1}^{\infty} (a_i - b_i)^2} = \|\alpha - \beta\|_2. \tag{3}$$

Theorem 2. *Let* $\alpha_1, \alpha_2, \alpha_3, \cdots$ *converge in* l^2 *to* α. *Then with the designations*

$$\alpha_n = (a_{n1}, a_{n2}, \cdots, a_{nk}, \cdots), \qquad n = 1, 2, 3, \cdots \qquad \text{and}$$

$$\alpha = (a_1, a_2, \cdots, a_k, \cdots),$$

it follows that for each k *the real sequence* (*of* kth *components*)

$$(a_{1k}, a_{2k}, a_{3k}, \cdots, a_{nk}, \cdots)$$

converges to a_k; *that is*,

$$\lim_{n \to \infty} a_{nk} = a_k \qquad \textit{for } k = 1, 2, 3, \cdots. \tag{4}$$

Proof. The convergence in l^2 of $\{\alpha_n\}$ to α carries the specific requirement that each $\alpha_n \in l^2$ and $\alpha \in l^2$. Hence $\alpha_n - \alpha \in l^2$ and $\|\alpha_n - \alpha\|_2$ is defined. Corresponding to $\varepsilon > 0$ arbitrary, let N be such that $n \geqslant N \Rightarrow \|\alpha_n - \alpha\|_2 < \varepsilon$. Then, from (3),

$$n \geqslant N \Rightarrow |a_{nk} - a_k| \leqslant \|\alpha_n - \alpha\|_2 < \varepsilon, \qquad k = 1, 2, 3, \cdots.$$

This states that (4) holds. ∎

However, the converse of Theorem 2 does not hold.

COUNTER EXAMPLE. Consider the sequence

$$\alpha_1 = (1, 0, 0, 0, \cdots),$$
$$\alpha_2 = (1, 1, 0, 0, \cdots),$$
$$\alpha_3 = (1, 1, 1, 0, \cdots),$$
$$\cdots\cdots\cdots\cdots\cdots\cdots,$$

where for each $n \in J$, $\alpha_n = (a_{n1}, a_{n2}, \cdots, a_{nk}, \cdots)$ with

$$a_{nk} = \begin{Bmatrix} 1 & \text{if } k \leq n \\ 0 & \text{if } k > n \end{Bmatrix}.$$

Then $\sum_{k=1}^{\infty} a_{nk}{}^2 = \sum_{k=1}^{n} (1) = n$ and $\alpha_n \in l^2$ for each $n \in J$. The kth components form a sequence (going down the kth column of the above display) with the first $k - 1$ elements 0 and all other elements 1. Thus the sequence of limits of components is

$$\alpha = (1, 1, 1, \cdots),$$

which is not an element of l^2. Hence $\|\alpha_n - \alpha\|_2$ does not exist and convergence in l^2 of α_n to α is not possible.

Theorem 3. *The space l^2 is a Hilbert space.*

[From Problem 1, we need prove only that every Cauchy sequence in l^2 converges in l^2. Actually, complete vector spaces with norm defined in terms of an inner product are called Hilbert spaces because of the pioneering work of D. Hilbert on l^2.]

Proof. Let $\alpha_1, \alpha_2, \alpha_3, \cdots$ be a Cauchy sequence in l^2 and accordingly, with $\varepsilon > 0$ given arbitrarily, take an $N \in J$ such that

$$n \geqslant N, m \geqslant N \Rightarrow \|\alpha_n - \alpha_m\|_2 < \varepsilon. \tag{5}$$

Set $\alpha_n = (a_{n1}, a_{n2}, \cdots, a_{nk}, \cdots)$ for each $n \in J$. Then, for each $k \in J$, the real sequence $a_{1k}, a_{2k}, \cdots, a_{nk}, \cdots$, of kth components, is such that

$$n, m \geqslant N \Rightarrow |a_{nk} - a_{mk}| \leqslant \|\alpha_n - \alpha_m\|_2$$

and is hence a Cauchy sequence in E^1 which therefore converges to an $a_k \in E^1$; that is,

$$\lim_{n \to \infty} a_{nk} = a_k \qquad \text{for } k = 1, 2, 3, \cdots. \tag{6}$$

We thus have a real sequence $\alpha = (a_1, a_2, \cdots, a_k, \cdots)$ as a candidate for the limit in l^2 of $\{\alpha_n\}$.

Two things must be proved: that $\alpha \in l^2$, and that $\lim_{n\to\infty} \|\alpha_n - \alpha\|_2 = 0$.

Any Cauchy sequence in any metric space is bounded (Theorem 3-4.4) and we let $B \in E^1$ be a bound for our sequence $\{\alpha_n\}$ so that

$$\|\alpha_n\|_2 \leqslant B; \qquad \text{that is, } \sum_{k=1}^{\infty} a_{nk}^2 \leqslant B^2 \qquad \text{for } n = 1, 2, 3, \cdots.$$

Hence, with $p, n \in J$,

$$\sum_{k=1}^{p} a_{nk}^2 \leqslant B^2.$$

But the limit of a finite sum is the sum of the limits when they exist separately. Thus, by using (6),

$$B^2 \geqslant \lim_{n\to\infty} \sum_{k=1}^{p} a_{nk}^2 = \sum_{k=1}^{p} a_k{}^2 \qquad \text{for each } p \in J.$$

Thus the partial sums of $\sum a_k{}^2$ are bounded above so that

$$\sum_{k=1}^{\infty} a_k{}^2 \text{ converges;} \qquad \text{that is, } \alpha \in l^2.$$

Since $\alpha \in l^2$, we know that $\alpha_n - \alpha \in l^2$ and $\|\alpha_n - \alpha\|_2$ exists.

With $\varepsilon > 0$ given anew, we proceed as in obtaining (5) and have an $N \in J$ such that

$$n, m \geqslant N \Rightarrow \left(\|\alpha_n - \alpha_m\|_2 < \varepsilon, \sum_{k=1}^{\infty} (a_{nk} - a_{mk})^2 < \varepsilon^2 \right).$$

Hence, with $p \in J$, $n \geqslant N$, and $m \geqslant N$,

$$\sum_{k=1}^{p} (a_{nk} - a_{mk})^2 < \varepsilon^2,$$

a finite sum in which we let $m \to \infty$ to obtain

$$n \geqslant N \Rightarrow \left(\sum_{k=1}^{p} (a_{nk} - a_k)^2 \leqslant \varepsilon^2, \sum_{k=1}^{\infty} (a_{nk} - a_k)^2 \leqslant \varepsilon^2 \right).$$

This fact in l^2-norm notation is

$$n \geqslant N \Rightarrow \|\alpha_n - \alpha\|_2 \leqslant \varepsilon,$$

which means that $\lim_{l^2} \alpha_n = \alpha$. Hence l^2 is a Hilbert space. ∎

PROBLEM 2. (a). If $\alpha = \{a_k\}$ is such that $\sum |a_k|$ converges, then $\alpha \in l^2$.

(b) Find an $\alpha = \{a_k\} \in l^2$ such that $\sum |a_k|$ diverges.

PROBLEM 3. For each $n \in J$ let ε_n be the sequence with all terms 0 except the nth which is 1; that is,

$$\varepsilon_n = (\delta_{n1}, \delta_{n2}, \delta_{n3}, \cdots) \qquad \text{where} \qquad \delta_{nk} = \begin{cases} 0 & \text{if } k \neq n \\ 1 & \text{if } k = n \end{cases}.$$

Then:

(a) $\|\varepsilon_n\|_2 = 1$.

(b) $\|\varepsilon_n - \varepsilon_m\|_2 = \sqrt{2}$ if $m \neq n$.

(c) $\{\varepsilon_1, \varepsilon_2, \varepsilon_3, \cdots\}$ is an orthonormal set; that is,

$$(\varepsilon_n \cdot \varepsilon_m) = \begin{Bmatrix} 0 & \text{if } m \neq n \\ 1 & \text{if } m = n \end{Bmatrix} = \delta_{nm}$$

(d) If $\alpha = \{a_k\} \in l^2$, then

$$\sum_{n=1}^{\infty} (\varepsilon_n \cdot \alpha)\varepsilon_n \text{ converges in } l^2 \text{ to } \alpha.$$

(e) Even if $\alpha \notin l^2$, is there a formal sense in which

$$\alpha = \sum_{n=1}^{\infty} (\varepsilon_n \cdot \alpha)\varepsilon_n?$$

(f) For each integer n, then (with each a_k real)

$$a_1\varepsilon_1 + a_2\varepsilon_2 + \cdots + a_n\varepsilon_n = \theta$$

iff $a_1 = a_2 = \cdots = a_n = 0$.

The following theorem gives another property that l^2 has in common with any Euclidean space.

Theorem 4. *The space l^2 is separable.*

Proof. Let R be the subset of l^2 defined by

$$\rho = \{r_1, r_2, r_3, \cdots\} \in R$$

iff each r_k is rational and $\sum r_k^2$ converges. Then R is a countable subset of l^2. To see that R is dense in l^2, select any $\alpha = \{a_i\} \in l^2$ and let $\varepsilon > 0$ be arbitrary. For each k let r_k be a rational number such that

$$|a_k - r_k| < \frac{\varepsilon}{2^k}.$$

Then $|r_k| < |a_k| + \varepsilon/2^k$ and

$$r_k^2 < a_k^2 + \frac{\varepsilon}{2^{k-1}}c + \frac{\varepsilon^2}{4^k},$$

where c is any constant such that $|a_k| < c$, $k = 1, 2, 3, \cdots$. Thus $\sum r_k^2$ converges, $\rho = \{r_k\} \in R \subset l^2$, and

$$\|\alpha - \rho\|_2^2 = \sum_{k=1}^{\infty} (a_k - r_k)^2 < \tfrac{1}{3}\varepsilon^2.$$

Since $\varepsilon > 0$ is arbitrary, R is everywhere dense in l^2. ∎

3–11 DOUBLE SEQUENCES AND SERIES

A double sequence is a function whose domain is $J \times J$ and may be displayed in matrix form as

$$\begin{array}{cccccc} a_{11} & a_{12} & a_{13} & \cdots & a_{1j} & \cdots \\ a_{21} & a_{22} & a_{23} & \cdots & a_{2j} & \cdots \\ \cdots & \cdots & \cdots & \cdots & \cdots & \cdots \\ a_{i1} & a_{i2} & a_{i3} & \cdots & a_{ij} & \cdots \\ \cdots & \cdots & \cdots & \cdots & \cdots & \cdots \end{array}$$

We shall use matrix terminology and say that a double sequence $\{a_{ij}\}$ has ***i*th row**

$$a_{i1} \quad a_{i2} \quad a_{i3} \quad \cdots \quad a_{ij} \quad \cdots,$$

***j*th column** (but written in transpose form as a row)

$$a_{1j} \quad a_{2j} \quad a_{3j} \quad \cdots \quad a_{ij} \quad \cdots,$$

and **principal diagonal** $a_{11}, a_{22}, a_{33}, \cdots, a_{ii}, \cdots$.

Thus a single sequence $\alpha_1, \alpha_2, \alpha_3, \cdots$ in the space l^2 of the preceding section is closely related to a double sequence in which the ith row is a real square summable single sequence.

Definition 1. *A double sequence $\{a_{ij}\}$ each of whose elements is a point of a metric space S, is said to converge in S if there is an $\alpha \in S$ such that corresponding to each number $\varepsilon > 0$ an $N \in J$ exists such that*

$$i, j \geqslant N \Rightarrow d(a_{ij}, a) < \varepsilon.$$

If $\{a_{ij}\}$ converges in S to a, we write

$$a_{ij} \to a \qquad \text{or} \qquad \lim_{(i,j)\to(\infty,\infty)} a_{ij} = a,$$

but usually omit the (∞, ∞) and sometimes all $(i, j) \to (\infty, \infty)$. If the ith row converges to r_i and the jth column to c_j, we write

$$\lim_j a_{ij} = r_i \qquad \text{and} \qquad \lim_i a_{ij} = c_j, \qquad \text{respectively.}$$

EXAMPLE 1. A double sequence $\{a_{ij}\}$ has main diagonal defined by

$$a_{ii} = \begin{cases} 1 & \text{if } i \text{ is odd} \\ 0 & \text{if } i \text{ is even} \end{cases},$$

and all other terms are zero. For this sequence,

$$\lim_j a_{ij} = 0, \qquad \lim_i a_{ij} = 0, \qquad \text{and} \qquad \lim_j (\lim_i a_{ij}) = 0, \qquad \text{but}$$

$$\lim_{(i,j)} a_{ij} \text{ does not exist.}$$

EXAMPLE 2. Let the first row be 1, 2, 3, $\cdots$ and all other terms be 0. Then

$$\lim_{(i,j)} a_{ij} = 0.$$

Hence this double sequence converges, but it is not bounded.

Theorem 1. *Given a real double sequence $\{a_{ij}\}$ converging in E^1 to $a \in E^1$ and set*

$$\underline{r}_i = \varliminf_j a_{ij} \qquad \text{and} \qquad \bar{r}_i = \varlimsup_j a_{ij} \qquad \text{for } i = 1, 2, 3, \cdots$$

[which are therefore in E^* and some of them may be ∞ or $-\infty$]. Then the two sequences $\{\underline{r}_i\}$ and $\{\bar{r}_i\}$ converge and converge to a.

Proof. With $\varepsilon > 0$ given, let N be such that

$$i, j \geqslant N \Rightarrow |a_{ij} - a| < \varepsilon.$$

Hence $i \geqslant N \Rightarrow a - \varepsilon < a_{ij} < a + \varepsilon$ for $j = N, N+1, N+2, \cdots$. Thus

$$i \geqslant N \Rightarrow a - \varepsilon \leqslant \varliminf_j a_{ij} \leqslant \varlimsup_j a_{ij} \leqslant a + \varepsilon;$$

that is,

$$i \geqslant N \Rightarrow (|\underline{r}_i - a| < \varepsilon, |\bar{r}_i - a| < \varepsilon),$$

which says that $\lim \underline{r}_i = a$ and $\lim \bar{r}_i = a$. ∎

The above result may be stated succinctly as:
If $\{a_{ij}\}$ is a real convergent double sequence, then

$$\lim_i \left(\varliminf_j a_{ij}\right) = \lim_i \left(\varlimsup_j a_{ij}\right) = \lim_{(i,j)} a_{ij},$$

and then in a similar way (in terms of columns first),

$$\lim_j \left(\varliminf_i a_{ij}\right) = \lim_j \left(\varlimsup_i a_{ij}\right) = \lim_{(i,j)} a_{ij}.$$

Corollary 1. *If $\{a_{ij}\}$ is a real convergent double sequence, then*

$$\lim_i \left(\varliminf_j a_{ij}\right) = \lim_j \left(\varliminf_i a_{ij}\right) = \lim_{(i,j)} a_{ij}$$

with other qualities having $\varlimsup$ in either or both parentheses.

Corollary 2. *If $\{a_{ij}\}$ is a real convergent double sequence and if all row and column single sequences converge, then the double limit may be obtained by using either iterated limit; that is,*

$$\lim_i \left(\lim_j a_{ij}\right) = \lim_j \left(\lim_i a_{ij}\right) = \lim_{(i,j)} a_{ij}.$$

EXAMPLE 3. The accompanying double sequence shows that convergence (in this case to 0), row-convergence, and convergence of row-limits is not enough to ensure column-convergence.

$$\begin{matrix} 1 & \frac{1}{2} & \frac{1}{3} & \cdots \\ -1 & -\frac{1}{2} & -\frac{1}{3} & \cdots \\ 1 & \frac{1}{2} & \frac{1}{3} & \cdots \\ -1 & -\frac{1}{2} & -\frac{1}{3} & \cdots \\ \cdots & \cdots & \cdots & \cdots \end{matrix}$$

The usual complex number notation is used in the following problem.

PROBLEM 1. (a) A double sequence $\{z_{mn}\}$ of complex numbers converges to $z = x + iy$ iff the real sequences $\{x_{mn}\}$, $\{y_{mn}\}$ converge to x and y, respectively.

(b) If $\{z_{mn}\}$ converges and if each of its rows and columns converges, then

$$\lim_m \left(\lim_n z_{mn}\right) = \lim_n \left(\lim_m z_{mn}\right) = \lim_{(m,n)} z_{mn}.$$

It should be seen that the following result may be obtained by the method used in solving Problem 1.

Theorem 2.

(a) *A double sequence* $\{\mathbf{a}_{ij}\}$ *in* E^k, *with*

$$\mathbf{a}_{ij} = (a_{ij}(1), a_{ij}(2), \cdots, a_{ij}(k))$$

converges iff all of the double sequences

$$\{a_{ij}(l)\};\ l = 1, 2, \cdots, k \textit{ converge.}$$

(b) *If* $\{\mathbf{a}_{ij}\}$ *converges, if*

$\mathbf{a}_{i1}, \mathbf{a}_{i2}, \mathbf{a}_{i3}, \cdots$ *converges for each* i *and if*
$\mathbf{a}_{j1}, \mathbf{a}_{j2}, \mathbf{a}_{j3}, \cdots$ *converges for each* j,

then

$$\lim_i \left(\lim_j \mathbf{a}_{ij}\right) = \lim_j \left(\lim_i \mathbf{a}_{ij}\right) = \lim_{(i,j)} \mathbf{a}_{ij}.$$

Definition 2. *Given the double sequence* $\{a_{ij}\}$, *with elements in a normed vector space, form the double sequences* $\{s_{mn}\}$ *and* $\{S_{mn}\}$, *where for each* $m, n \in J$,

$$s_{mn} = \sum_{i=1}^{m} \sum_{j=1}^{n} a_{ij} \quad \text{and} \quad S_{mn} = \sum_{i=1}^{m} \sum_{j=1}^{n} \|a_{ij}\|.$$

The **double series** *with* (i, j) *term* a_{ij}, *is denoted by*

$$\sum_{(i,j)=(1,1)}^{(\infty,\infty)} a_{ij} \quad \textit{or simply} \sum_{(i,j)} a_{ij}, \quad \textit{or even} \sum a_{ij},$$

is said to **converge** *to* s *if* $\lim_{(m,n)} s_{mn} = s$ *exists, and is said to* **converge absolutely** *if* $\{S_{mn}\}$ *converges.*

Theorem 3. *Let* $\sum a_{ij}$, *with each term in* E^k, *converge absolutely. Then*

(a) $\sum a_{ij}$ *converges; that is,*

$$\sum_{(i,j)} a_{ij} = s \text{ exists.}$$

(b) *The single series in each row, and in each column, converges absolutely; that is, there are sequences* $\{s_i\}$ *and* $\{t_j\}$ *such that*

$$\sum_j a_{ij} = s_i \qquad \text{and} \qquad \sum_i a_{ij} = t_j$$

with convergence absolute.

(c) *The series* $\sum s_i$ *and* $\sum t_j$ *both converge to* s; *that is,*

$$\sum_{i=1}^{\infty}\left[\sum_{j=1}^{\infty} a_{ij}\right] = \sum_{j=1}^{\infty}\left[\sum_{i=1}^{\infty} a_{ij}\right] = \sum_{(i,j)=(1,1)}^{(\infty,\infty)} a_{ij},$$

and also $\sum s_i$ *and* $\sum t_j$ *converge absolutely.*

Proof (a). To avoid subscripts on subscripts we replace s_{mn} by $s(m, n)$ and S_{mn} by $S(m, n)$. Also, we denote the limit of the convergent sequence $\{S(m, n)\}$ by S and note that

$$0 \leqslant S(m, n) \leqslant S \qquad \text{for all } m, n \in J.$$

With $m_1 < m_2$ and $n_1 < n_2$ all in J, then

$$\text{(1)} \qquad \|s(m_2, n_2) - s(m_1, n_1)\| \leqslant \sum_{i=1}^{m_1} \sum_{j=n_1+1}^{n_2} \|a_{ij}\| + \sum_{i=m_1+1}^{m_2} \sum_{j=1}^{n_2} \|a_{ij}\|$$
$$= S(m_2, n_2) - S(m_1, n_1)$$

as illustrated in diagrammatic figure.

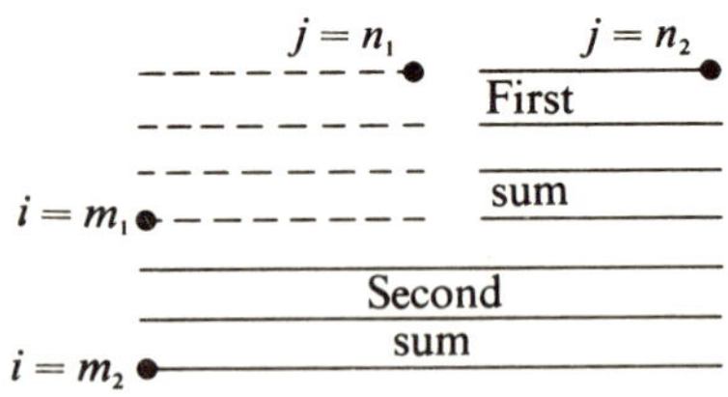

Since $S(1, 1) \leqslant S(2, 2) \leqslant S(3, 3) \leqslant \cdots \leqslant S$, the single sequence $\{S(n, n)\}$ converges and hence is a Cauchy sequence. Corresponding to a given $\varepsilon > 0$, let $N \in J$ be such that

$$n \geqslant N \Rightarrow S(n + p, n + p) - S(n, n) < \varepsilon, \qquad p = 1, 2, 3, \cdots.$$

Then from (1),

$$n \geqslant N \Rightarrow \|s(n + p, n + p) - s(n, n)\| < \varepsilon, \qquad p = 1, 2, 3, \cdots.$$

Thus $\{s(n, n)\}$ is a Cauchy sequence and hence converges (since the space is E^k).

Let

$$\lim_{n \to \infty} s(n, n) = s.$$

Again, with $\varepsilon > 0$ given, let $N \in J$ be such that both

$$n \geqslant N \Rightarrow \|s(n, n) - s\| < \varepsilon \qquad \text{and}$$
$$m, n \geqslant N \Rightarrow [S - \varepsilon < S(m, n) \leqslant S,\ \|s(m, m) - s(n, n)\| < \varepsilon].$$

Thus for any $m, n \geqslant N$,

$$\begin{aligned}\|s(m, n) - s\| &\leqslant \|s(m, n) - s(N, N)\| + \|s(N, N) - s\| \\ &\leqslant S(m, n) - S(N, N) + \varepsilon, \qquad \text{by using (1)}, \\ &\leqslant \|S(m, n) - S\| + \|S - S(N, N)\| + \varepsilon < 3\varepsilon.\end{aligned}$$

Thus $\lim_{(m,n)} s(m, n) = s$, which establishes part (a).

Proof (b). For $i \in J$, then (as soon as $n \geqslant i$)

$$\sum_{j=1}^{n} \|a_{ij}\| \leqslant \sum_{i=1}^{n} \sum_{j=1}^{n} \|a_{ij}\| \leqslant S(n, n) \leqslant S.$$

Hence $\sum_j a_{ij}$ converges absolutely and in the same way $\sum_i a_{ij}$ converges absolutely, thus establishing part (b).

Proof (c). Given a number $\varepsilon > 0$, let $N \in J$ be such that

$$m, n \geqslant N \Rightarrow \left(\|s(m, n) - s\| < \varepsilon, \left\| \sum_{i=1}^{m} \sum_{j=1}^{n} a_{ij} - s \right\| < \varepsilon \right).$$

With m fixed ($m \geqslant N$), let $n \to \infty$. Since we have a finite sum $\sum_{i=1}^{m}$ of terms, each having a limit as $n \to \infty$, it follows that

$$m \geqslant N \Rightarrow \left(\left\| \sum_{i=1}^{m} \left[\lim_{n \to \infty} \sum_{j=1}^{n} a_{ij} \right] - s \right\| < \varepsilon, \left\| \sum_{i=1}^{m} s_i - s \right\| \leqslant \varepsilon \right).$$

Thus $\sum s_i$ converges to s and in the same way $\sum t_j$ converges to s. The following problem then finishes the proof of part (c).

PROBLEM 2. Both $\sum s_i$ and $\sum t_j$ converge absolutely. ∎

EXAMPLE 4. With M and Z, the 2×2 matrices

$$M = \begin{pmatrix} 1 & -1 \\ -1 & 1 \end{pmatrix} \qquad \text{and} \qquad Z = \begin{pmatrix} 0 & 0 \\ 0 & 0 \end{pmatrix}$$

form a double sequence $\{a_{ij}\}$ according to the pattern

$$\begin{matrix} M & Z & Z & Z & \cdots \\ Z & M & Z & Z & \cdots \\ Z & Z & M & Z & \cdots \\ \cdots & \cdots & \cdots & \cdots & \cdots \end{matrix}$$

Then the series of elements in each row converges to 0, and the same is true of columns, so that

$$\sum_i \left(\sum_j a_{ij}\right) = \sum_j \left(\sum_i a_{ij}\right).$$

Even though these iterated series converge to the same value, the double series $\sum_{(i,j)} a_{ij}$ does not converge, since $s(2n, 2n) = 0$ and $s(2n+1, 2n+1) = 1$.

The following theorem gives a criterion under which the double series converges absolutely, and thus may be evaluated by using either iterated series according to Theorem 3.

Theorem 4. *Let $\{a_{ij}\}$, with each $a_{ij} \in E^k$, be such that $\sum_j a_{ij}$ converges absolutely with $\sum_j |a_{ij}| = R_i$ for each $i \in J$ and $\sum_i R_i$ converges. Then $\sum_{(i,j)} a_{ij}$ converges absolutely.*

Proof. Set $\sum_i R_i = S$. Then for each $m, n \in J$,

$$S \geqslant \sum_{i=1}^{m} R_i, \qquad R_i \geqslant \sum_{j=1}^{n} \|a_{ij}\|, \qquad \text{and} \qquad S \geqslant S(m, n).$$

Given an $\varepsilon > 0$, first choose $N \in J$ such that

$$\sum_{i=1}^{N} R_i > S - \varepsilon$$

and then M_i for $i = 1, 2, \cdots, N$ such that

$$\sum_{j=1}^{M_i} \|a_{ij}\| > R_i - \frac{\varepsilon}{N}.$$

Hence for $M = \max(M_1, M_2, \cdots, M_N)$ it follows that

$$m, n \geqslant \max(M, N) \Rightarrow S(m, n) \geqslant S(M, N) > S - 2\varepsilon.$$

Thus the double sequence $\{S(m, n)\}$ converges and, by definition, the double series $\sum_{(i,j)} a_{ij}$ converges absolutely. ∎

PROBLEM 3. If $\sum a_{ij}$, with each $a_{ij} \in E^1$, converges absolutely, then

$$\sum_{(1,1)}^{(\infty,\infty)} a_{ij} = \sum_{n=1}^{\infty} \sum_{k=1}^{n} a_{n+1-k,k}.$$

(*Hint*: Sum up the 45° diagonals of the matrix on page 159.)

3–12 POWER SERIES

In this section we consider a special type of series of functions, namely, the generalization of polynomials to power series,

$$\sum_{n=0}^{\infty} a_n z^n,$$

in which each coefficient a_n is a complex constant and z is a complex variable.

Any power series converges at $z = 0$, and there are power series which converge only if $z = 0$. For example $\sum n!\, z^n$ diverges if $z \neq 0$, since then $\lim n!\,|z|^n \neq 0$ by the corollary of Theorem 3-7.7. On the other hand, the ratio test applied to the absolute value series of $\sum z^n/n!$ gives

$$\overline{\lim_{n\to\infty}} \left| \frac{n!\, z^{n+1}}{(n+1)!\, z^n} \right| = |z| \lim_{n\to\infty} \frac{1}{n+1} = 0 < 1 \qquad \text{if } z \neq 0,$$

thus showing that $\sum z^n/n!$ converges (absolutely) for all complex values of z.

The result in the following theorem is as definitive as could be expected of a single test.

Theorem 1. *Given a power series $\sum a_n z^n$, let*

$$\alpha = \overline{\lim} \sqrt[n]{|a_n|} \qquad \text{and} \qquad R = \begin{cases} 0 & \text{if } \alpha = \infty \\ 1/\alpha & \text{if } 0 < \alpha < \infty. \\ \infty & \text{if } \alpha = 0 \end{cases}$$

Then $\sum a_n z^n$

(a) *Converges (absolutely) if $R > 0$ and $|z| < R$.*
(b) *Diverges if $R < \infty$ and $|z| > R$.*
(c) *The conditions $0 < R < \infty$ and $|z| = R$ give no information about convergence or divergence.*

Proof. This follows from the root test (for nonnegative series), since the limit superior there applied to $\sum |a_n z^n|$ takes the form

$$\overline{\lim} \sqrt[n]{|a_n|\,|z^n|} = |z|\, \overline{\lim} \sqrt[n]{|a_n|} = |z|\alpha.$$

Under (a), convergence of $\sum a_n z^n$ follows from convergence of $\sum |a_n z^n|$. Under (b), however, divergence of $\sum a_n z^n$ does not follow from divergence of $\sum |a_n z^n|$, but does follow from $\lim |a_n z^n| \neq 0$.

The statement in (c) is seen since $\sum z^n/n$ and $\sum z^n/n^2$ both have $R = 1$. ∎

Definition 1. *The $R \in E^*$ defined in Theorem* 1 *is called the* **radius of convergence** *of $\sum a_n z^n$. The* **circle of convergence** *of $\sum a_n z^n$ is defined to be the origin if $R = 0$, E^2 if $R = \infty$, and $\{z \mid |z| = R\}$ if $0 < R < \infty$.*

Hence, any power series converges (absolutely) at all points interior to its circle of convergence and diverges at all points of E^2 exterior to its circle of convergence. Whether a power series converges or diverges at a point on its circle of convergence requires further investigation.

For example, both

$$\sum z^n \qquad \text{and} \qquad \sum \frac{1}{n^2} z^n$$

have $R = 1$, but the first diverges whereas the second converges (absolutely) on the common circle of convergence (since if $|z| = 1$, then $|\lim z|^n \neq 0$ for the first and $|z^n|/n^2 \leqslant 1/n^2$ for the second).

EXAMPLE $\sum_{n=1}^{\infty} (1/n)\, z^n$ has $R = 1$, diverges if $z = 1$, and converges if $|z| = 1$, but $z \neq 1$.

Proof (for $|z| = 1$, but $z \neq 1$). Let $t_n = z + z^2 + \cdots + z^n$ and note that

$$|t_n| = \frac{|z|\,|1 - z^n|}{|1 - z|} \leqslant \frac{2}{|1 - z|} \qquad \text{for } n = 1, 2, 3, \cdots.$$

Also for $p, N \in J$,

$$\begin{aligned}\sum_{n=N}^{N+p} \frac{1}{n} z^n &= \sum_{n=N}^{N+p} \frac{1}{n}(t_n - t_{n-1}) = \sum_{n=N}^{N+p} \frac{1}{n} t_n - \sum_{n=N-1}^{N+p-1} \frac{1}{n+1} t_n \\ &= \sum_{n=N}^{N+p-1} \left(\frac{1}{n} - \frac{1}{n+1}\right) t_n + \frac{1}{N+p} t_{N+p} - \frac{1}{N} t_{N-1},\end{aligned}$$

and thus

$$\begin{aligned}\left|\sum_{n=N}^{N+p} \frac{1}{n} z^n\right| &\leqslant \frac{2}{|1 - z|} \left[\sum_{n=N}^{N+p-1} \left(\frac{1}{n} - \frac{1}{n+1}\right) + \frac{1}{N+p} + \frac{1}{N}\right] \\ &= \frac{1}{N} \frac{4}{|1 - z|}.\end{aligned}$$

Given a number $\varepsilon > 0$ choose N such that

$$\frac{1}{N} \frac{4}{|1 - z|} < \varepsilon$$

and thus see that the partial sums of $\sum (1/n) z^n$ form a Cauchy sequence, provided $z = 1$ but $z \neq 1$. Thus, under the stated conditions, $\sum (1/n) z^n$ converges (since E^2 is a complete metric space). ∎

PROBLEM 1. If $\{a_n\}$ is monotonically decreasing (hence real) with $\lim a_n = 0$ and if a series of complex terms $\sum b_n$ has bounded partial sums, then $\sum a_n b_n$ converges.

The following theorem shows what an influence powers of z with $|z| < 1$ have on a sequence.

Theorem 2. *For any bounded sequence $\{a_n\}$ in E^2 and any $z \in E^2$ with $|z| < 1$, the power series*

$$\sum a_n z^n \text{ converges absolutely.}$$

Proof. Select a $B \in E^1$ such that $|a_n| < B$ for $n = 0, 1, 2, \cdots$. Then

$$|a_n z^n| \leqslant B |z|^n$$

and hence $\sum a_n z^n$ converges (for $|z| < 1$) by the comparison test. ∎

Corollary. *Let* $\sum a_n$ *be a convergent series in* E^2 *with partial sums* $\{s_n\}$. *Then for* $|z| < 1$,

$$\sum a_n z^n, \qquad \sum s_n z^n, \qquad \text{and} \qquad \sum s_{n-1} z^n$$

converge absolutely and also

$$\sum_{n=0}^{\infty} a_n z^n = (1 - z) \sum_{n=0}^{\infty} s_n z^n \qquad (|z| < 1). \tag{1}$$

Proof. Convergence follows easily from the theorem, and then

$$\begin{aligned}\sum_{n=0}^{\infty} a_n z^n &= a_0 + \sum_{n=1}^{\infty} (s_n - s_{n-1}) z^n \\ &= a_0 + \sum_{n=1}^{\infty} s_n z^n - \sum_{n=1}^{\infty} s_{n-1} z^n \\ &= \sum_{n=0}^{\infty} s_n z^n - z \sum_{n=0}^{\infty} s_n z^n, \qquad \text{for } |z| < 1,\end{aligned}$$

and hence (1) follows. ∎

Recall from elementary calculus that, for

$$f : (a, b) \to E^1 \qquad \text{and} \qquad L \in E^1,$$

by definition $\lim_{x \to b-} f(x) = L$ means

Corresponding to each number $\varepsilon > 0$ there is a $\delta > 0$ such that

$$a < b - \delta < x < b \Rightarrow |f(x) - L| < \varepsilon.$$

Upon replacing E^1, in the two places it appears, by E^2 (but keeping the open interval (a, b)), the definition of $\lim_{x \to b-} f(x) = L$ is exactly the same as before. We use this definition in the following theorem and later consider limits of $f: (a, b) \to E^2$ in more detail.

ABEL

Theorem 3. (Abel) *Given that* $\sum a_n$ *converges in* E^2, *then (with* $x \in E^1$*)*

$$\lim_{x \to 1-} \sum_{n=0}^{\infty} a_n x^n = \sum_{n=0}^{\infty} a_n. \tag{2}$$

[This is an example of the limit of an infinite sum being the infinite sum of the limits. Since $\sum |a_n x^n|$ converges for $-1 < x < 1$, the limit (2) shows "how nearly" a convergent series comes to being absolutely convergent.]

Proof. Since we are eventually going to let $x \to 1-$, we take $0 < x < 1$. With z replaced by x in (1), then

$$\sum_{n=0}^{\infty} a_n x^n = (1 - x) \sum_{n=0}^{\infty} s_n x^n, \qquad 0 < x < 1.$$

Since $1 = (1 - x) \sum_{n=0}^{\infty} x^n$, $0 < x < 1$, and $s = \sum_{n=0}^{\infty} a_n$,

$$s = s(1 - x) \sum_{n=0}^{\infty} x^n; \qquad \text{that is,} \qquad \sum_{n=0}^{\infty} a_n = (1 - x) \sum_{n=0}^{\infty} s x^n.$$

From the convergence of the series involved, we thus have

$$\sum_{n=0}^{\infty} a_n x^n - \sum_{n=0}^{\infty} a_n = (1-x) \sum_{n=0}^{\infty} (s_n - s)x^n.$$

From the boundedness of $\{s_n - s\}$, select $B \in E^1$ such that

$$|s_n - s| < B \qquad \text{for } n = 0, 1, 2, 3, \cdots.$$

Let $\varepsilon > 0$ be arbitrary. Since $\{s_n - s\}$ converges to 0, choose $N \in J$ such that $n \geqslant N \Rightarrow |s_n - s| < \varepsilon/2$. Hence

$$\begin{aligned}\left|\sum_{n=0}^{\infty} a_n x^n - \sum_{n=0}^{\infty} a_n\right| &\leqslant (1-x) \sum_{n=0}^{N-1} |s_n - s| x^n + (1-x) \sum_{n=N}^{\infty} |s_n - s| x^n \\ &\leqslant (1-x) B \sum_{n=0}^{N-1} x^n + \frac{\varepsilon}{2} x^N \\ &< (1-x) BN + \frac{\varepsilon}{2}, \qquad 0 < x < 1.\end{aligned}$$

Note that this holds uniformly for all x such that $0 < x < 1$, since neither N nor B depends in any way on x. Now choose $\delta > 0$ such that $1 - \delta < x < 1 \Rightarrow (1 - x) < \varepsilon/(2BN)$. Consequently

$$1 - \delta < x < 1 \Rightarrow \left|\sum_{n=0}^{\infty} a_n x^n - \sum_{n=0}^{\infty} a_n\right| < \varepsilon,$$

which establishes the existence of the limit and the equality in (2). ∎

As motivation for the next definition, we formally multiply

$$\begin{aligned} &a_0 + a_1 z + a_2 z^2 + a_3 z^3 + \cdots \qquad \text{by} \\ &b_0 + b_1 z + b_2 z^2 + b_3 z^3 + \cdots \end{aligned}$$

and collect coefficients of powers of z as

$$a_0 b_0 + (a_0 b_1 + a_1 b_0) z + (a_0 b_2 + a_1 b_1 + a_2 b_0) z^2 + \cdots.$$

Definition 2. *Two series* $\sum a_n$ *and* $\sum b_n$ *are said to have* $\sum c_n$ *as their* **Cauchy product** *series where*

$$c_n = \sum_{k=0}^{n} a_k b_{n-k}, \qquad n = 0, 1, 2, \cdots.$$

Convergence of $\sum a_n$ to s and $\sum b_n$ to t might be expected to yield convergence of $\sum c_n$ to st, but it does not. For example,

$$a_n = b_n = \frac{(-1)^n}{\sqrt{n+1}}, \qquad n = 0, 1, 2, \cdots$$

are such that $\sum a_n$ and $\sum b_n$ converge, but $\sum c_n$ diverges, since

$$|c_n| = \left| \sum_{k=0}^{n} \frac{(-1)^k}{\sqrt{k+1}} \frac{(-1)^{n-k}}{\sqrt{n-k+1}} \right| = \sum_{k=0}^{n} \frac{1}{\sqrt{k+1}\sqrt{n-k+1}}$$

$$\geqslant (n+1)\frac{2}{n+2} \qquad (\text{from } 2\sqrt{ab} \leqslant a+b \text{ for } a>0,\ b>0).$$

The following discussion leads to a sufficient condition for convergence of $\sum c_n$, and to the "right value":

Given that $\sum b_n$ converges, we set

$$t_n = \sum_{k=0}^{n} b_k \qquad \text{and} \qquad t = \sum_{n=0}^{\infty} b_n .$$

With $\sum a_n$ a series and $\sum c_n$ the Cauchy product series, then

$$\begin{aligned}
\sum_{k=0}^{n} c_k &= \sum_{k=0}^{n} \sum_{i=0}^{k} a_i b_{k-i} \\
&= a_0 b_0 \\
&\quad + a_0 b_1 + a_1 b_0 \\
&\quad + a_0 b_2 + a_1 b_1 \quad + a_2 b_0 \\
&\quad + \cdots \\
&\quad + a_0 b_n + a_1 b_{n-1} + a_2 b_{n-2} + \cdots + a_n b_0 \\
&= a_0 \sum_{i=0}^{n} b_i + a_1 \sum_{i=0}^{n-1} b_i + a_2 \sum_{i=0}^{n-2} b_i + \cdots + a_n \sum_{i=0}^{0} b_i \\
&= \sum_{k=0}^{n} \left[a_k \sum_{i=0}^{n-k} b_i \right] = \sum_{k=0}^{n} a_k t_{n-k} = \sum_{k=0}^{n} a_k [t - (t - t_{n-k})]
\end{aligned}$$

$$= t \sum_{k=0}^{n} a_k - \sum_{k=0}^{n} a_k (t - t_{n-k}). \tag{3}$$

Thus, if $\sum a_n$ converges to s, the first term in (3) approaches st and $\sum c_n$ will converge to st, **provided** the second term in (3) approaches 0 as $n \to \infty$ (as it does under the conditions of the next theorem).

Theorem 4. *Given that $\sum a_n$ converges absolutely and $\sum b_n$ converges, then the Cauchy product series converges and*

$$\sum_{n=0}^{\infty} c_n = \left(\sum_{n=0}^{\infty} a_n \right) \left(\sum_{n=0}^{\infty} b_n \right). \tag{4}$$

Proof. Let A and B be such that

$$A = \sum_{n=0}^{\infty} |a_n| \qquad \text{and} \qquad |t - t_n| \leqslant B \qquad \text{for } n = 0, 1, 2, 3, \cdots .$$

Given a number $\varepsilon > 0$, take $N \in J$ such that

$$m \geqslant N \Rightarrow \left(|t - t_m| < \varepsilon \text{ and } \sum_{k=m}^{\infty} |a_k| < \varepsilon \right).$$

Then $(n \geqslant 2N$ and $0 \leqslant k \leqslant N) \Rightarrow n - k \geqslant 2N - N = N$, and hence $n \geqslant 2N$ implies

$$\begin{aligned} \left| \sum_{k=0}^{n} a_k(t - t_{n-k}) \right| &\leqslant \sum_{k=0}^{N} |a_k|\,|t - t_{n-k}| + \sum_{k=N+1}^{n} |a_k|\,|t - t_{n-k}| \\ &\leqslant \varepsilon \sum_{k=0}^{N} |a_k| + B \sum_{k=N+1}^{n} |a_k| \\ &\leqslant \varepsilon A + B\varepsilon = (A + B)\varepsilon. \end{aligned}$$

Since A and B are independent of n, the second term in (3) approaches 0 as $n \to \infty$, and thus (4) holds under the given conditions. ∎

Corollary. *Given that $\sum a_n z^n$ and $\sum b_n z^n$ have radii of convergence $R_1 > 0$ and $R_2 > 0$, respectively, let $r = \min(R_1, R_2)$. Then*

$$\sum_{n=0}^{\infty} c_n z^n = \left(\sum_{n=0}^{\infty} a_n z^n \right) \left(\sum_{n=0}^{\infty} b_n z^n \right) \qquad \text{for } |z| < r.$$

Given that $f\colon (a, b) \to E^2$ that $g\colon (a, b) \to E^2$, and that $\lim_{x \to b-} f(x)$ and $\lim_{x \to b-} g(x)$ both exist, then

$$\lim_{x \to b-} [f(x)g(x)] = \lim_{x \to b-} f(x) \lim_{x \to b-} g(x).$$

The existence of this limit and the equality follow by the same methods used in elementary calculus for the corresponding result about functions from (a, b) to E^1. Also, a proof will be given later, but we use the result in the next theorem, which trades the absolute convergence of $\sum a_n$ or $\sum b_n$ in Theorem 4 for convergence of $\sum c_n$.

Theorem 5. *Given that all three of $\sum a_n$, $\sum b_n$, and the Cauchy product series $\sum c_n$ converge, then*

$$\sum_{n=0}^{\infty} c_n = \sum_{n=0}^{\infty} a_n \sum_{n=0}^{\infty} b_n. \tag{5}$$

Proof. From the above corollary,

$$\sum_{n=0}^{\infty} c_n x^n = \sum_{n=0}^{\infty} a_n x^n \sum_{n=0}^{\infty} b_n x^n \qquad \text{for } 0 < x < 1.$$

Let $x \to 1-$, use Abel's theorem and the above-stated property of the limit of a product, and see that (5) follows. ∎

PROBLEMS

1. $\sum_{n=1}^{\infty} \frac{n-1}{n!} = 1.$

2. $\sum_{n=1}^{\infty} \frac{1}{n^2} = 4 \sum_{n=1}^{\infty} \frac{1}{(2n)^2} = \frac{4}{3} \sum_{n=1}^{\infty} \frac{1}{(2n-1)^2}.$

3. (a). $\frac{4k-1}{4k^2-2k} > \frac{1}{k}$, $k = 1, 2, 3, \cdots$.

(b). With $s_n = 1 + \frac{1}{2} + \frac{1}{3} + \cdots + 1/n$, then

$$s_{2n} \geqslant \tfrac{1}{2} + s_n, \qquad n = 1, 2, 3, \cdots.$$

(c). $\lim s_n = \infty$.

4. If $\lim a_n = 0$, then Σa_n and $\Sigma(a_n + 2a_{n+1})$ are both convergent or both divergent.

5. If each $a_n \geqslant 0$ and Σa_n converges, then $\Sigma(\sqrt{a_n}/n)$ converges.

6. Given that $a_n > 0$ and Σa_n converges, let $r_n = \Sigma_{k=n}^{\infty} a_k$. Then

(a). $\sum \frac{a_n}{r_n}$ diverges. (b) $\sum \frac{a_n}{\sqrt{r_n}}$ converges.

7. Given that $0 < a_1 < a_2 < \cdots$ and $\lim a_n = \infty$, then

$$\sum_{n=1}^{\infty} \frac{a_{n+1} - a_n}{a_{n+1}} \text{ diverges.}$$

[*Hint:* $\sum_{k=n}^{m-1} \frac{a_{k+1} - a_k}{a_{k+1}} \geqslant \sum_{k=n}^{m-1} \frac{a_{k+1} - a_k}{a_m} = \frac{a_m - a_n}{a_m} \to 1 \quad \text{as } m \to \infty.$]

[For other such properties see M. E. Taylor, "Some Divergence Properties o Series," *Am. Math. Monthly*, Vol. 73 (1966), p. 872.]

8. $\lim_{m\to\infty} \left(\lim_{n\to\infty} \cos^{2n}(m!\pi x) \right) = \begin{cases} 1 & \text{if } x \text{ is rational} \\ 0 & \text{if } x \text{ is irrational.} \end{cases}$

9. Any sequence $\{a_n\}$ has a monotonic subsequence.
[*Hint: Am. Math. Monthly*, Vol. 71, No. 6 (June–July 1964), p. 665.]

10. (a). If Σa_n is a convergent series of positive terms and

$$\sigma_n = \frac{1}{n}(a_1 + a_2 + \cdots + a_n), \qquad n = 1, 2, 3, \cdots,$$

then $\Sigma \sigma_n$ diverges.
(b). If Σa_n is a convergent series of real terms and

$$\tau_n = \frac{a_1 + 2a_2 + \cdots + na_n}{n(n+1)}, \qquad n = 1, 2, 3, \cdots,$$

then

$$\sum_{n=1}^{\infty} \tau_n \qquad \text{converges with sum} = \sum_{n=1}^{\infty} a_n.$$

11. If $\{c_n\}$ is such that $|nc_n| \leqslant M$, $n = 1, 2, 3, \cdots$, then

$$s_n = c_0 + c_1 + \cdots + c_n \qquad \text{and} \qquad \sigma_n = \frac{1}{n+1}(s_0 + s_1 + \cdots + s_n)$$

are such that $|s_n - \sigma_n| \leqslant M$.

12. (a). $\displaystyle\sum_{k=n+1}^{\infty} \frac{1}{k(k-1)} = \frac{1}{n}.$

(b). $\displaystyle\sum_{k=n}^{\infty} \frac{1}{k^2} < \frac{2}{n}.$

(c). If $h > 0$, then

$$\sum_{k=1}^{\infty} \frac{\sin^2 kh}{k^2} < 3h.$$

[*Hint:* Take n such that $n - 1 < 1/h \leqslant n$ and use (b).]

(d). If $\lim_{n\to\infty} a_n = 0$, then

$$\lim_{h\to 0} \sum_{k=1}^{\infty} a_k \frac{\sin^2 kh}{k^2 h} = 0.$$

[*Hint:* $\exists\, n$ such that $k \geqslant n \Rightarrow |a_k| < \varepsilon$.

$$\left|\sum_{n}^{\infty}\right| \leqslant \frac{\varepsilon}{|h|} 3|h|$$

$$\overline{\lim_{h\to 0}} \left|\sum_{1}^{\infty}\right| \leqslant \overline{\lim_{h\to 0}} \left|\sum_{1}^{n-1}\right| + 3\varepsilon.]$$

13. Let Σa_n be a series of positive terms. Prove:

(a). (Kummer's Test). Given a sequence $\{p_n\}$ of positive terms, let

$$P_n = p_n \frac{a_n}{a_{n+1}} - p_{n+1}.$$

If $\underline{\lim}\, P_n > 0$, then Σa_n converges.
If $\Sigma(1/p_n)$ diverges and $\overline{\lim}\, P_n < 0$, then Σa_n diverges.

(b). (Raabe's Test). Let

$$R_n = n\left(\frac{a_n}{a_{n+1}} - 1\right).$$

If $\underline{\lim}\, R_n > 1$, then Σa_n converges.
If $\overline{\lim}\, R_n < 1$, then Σa_n diverges.

(c). Test

$$\Sigma \frac{1}{n}, \quad \Sigma \frac{1}{n^2}, \quad \Sigma \frac{3!\,n!}{(n+3)!}$$

for convergence or divergence.

14. With $a_0 = 1$, $a_1 = 2$, and $a_{n+1} = a_{n-1} + a_n$, $n = 1, 2, 3, \cdots$, then

$$a_n = \frac{5 + 3\sqrt{5}}{10}\left(\frac{1+\sqrt{5}}{2}\right)^n + \frac{5 - 3\sqrt{5}}{10}\left(\frac{1-\sqrt{5}}{2}\right)^n.$$

15. The following statement is false: A sequence $\{x_n\}$ whose range S is infinite has a limit, if, and only if, S has exactly one cluster point, in which case the cluster point is also the limit of the sequence.

16. (a). If $b_1 \geqslant b_2 \geqslant \cdots > 0$ and $\Sigma b_n = s < \infty$, then, corresponding to each number z, $0 \leqslant z \leqslant s$, there is a sequence $\{\varepsilon_n\}$ of 0's and 1's such that $z = \Sigma \varepsilon_n b_n$.

(b). Given $a_1 \geqslant a_2 \geqslant \cdots > 0$ such that $\Sigma a_n = 1$, let S be the set of all numbers representable as $\Sigma\, \varepsilon_n a_n$ for some sequence $\{\varepsilon_n\}$ of 0's and 1's. Then, corresponding to each number z, $0 \leqslant z \leqslant 2$ there is a pair $x, y \in S$ such that $x + y = z$ iff

$$a_n \leqslant 2r_n \qquad \text{where } r_n = \sum_{k=n+1}^{\infty} a_k, \qquad n = 1, 2, 3, \cdots.$$

[*Hint:* "Some Properties of Sets of the Cantor Type," *J. London Math. Soc.*, Vol. 16 (1941), pp. 38–42.]

17. If $\{x_n\}$ and $\{y_n\}$ are Cauchy sequences in a metric space, then $\{d(x_n, y_n)\}$ converges.

18. (a). Given a number $0 < \alpha < 1$, then

$$t^\alpha \leqslant \alpha t + 1 - \alpha \qquad \text{for all } t \geqslant 0.$$

[*Hint:* Use elementary calculus to find the maximum value of $f(t) = t^\alpha - \alpha t$, $t \geqslant 0$.]

(b). Set $\beta = 1 - \alpha$ so that $0 < \beta < 1$. If $u > 0$ and $v > 0$, then

$$u^\alpha v^\beta \leqslant \alpha u + \beta v.$$

[*Hint:* $t = u/v$.]

(c). Given $r > 1$, set $s = r/(r-1)$. Then $s > 1$ and $1/r + 1/s = 1$. If $A \geqslant 0$ and $B \geqslant 0$, then

$$AB \leqslant \frac{1}{r} A^r + \frac{1}{s} B^s.$$

(d). If $a_1, a_2, \cdots, a_n$ and $b_1, b_2, \cdots, b_n$ are nonnegative numbers then **Hölder's inequality**,

$$\sum_{i=1}^{n} a_i b_i \leqslant \left\{\sum_{i=1}^{n} a_i^r\right\}^{1/r} \left\{\sum_{i=1}^{n} b_i^s\right\}^{1/s},$$

holds. [*Hint:* $A_i = a_i / \{\Sigma_{j=1}^n a_j^r\}^{1/r}$ in case the sum $\neq 0$.]

(e). Minkowski's inequality,

$$\left\{\sum_{i=1}^{n} (a_i + b_i)^r\right\}^{1/r} \leqslant \left\{\sum_{i=1}^{n} a_i^r\right\}^{1/r} + \left\{\sum_{i=1}^{n} b_i^r\right\}^{1/r},$$

holds. [*Hint:* $(a+b)^r = a(a+b)^{r/s} + b(a+b)^{r/s}$.]

(f). $\displaystyle \frac{1}{n}\sum_{i=1}^{n} a_i \leqslant \left\{\frac{1}{n}\sum_{i=1}^{n} a_i^r\right\}^{1/r}$

19. In this problem, $0 < \alpha < 1$ and $\beta = 1 - \alpha$.

(a). Use elementary calculus to find the maximum and minimum of

$$f(t) = \frac{1 + t^\alpha}{(1+t)^\alpha}, \qquad t \geqslant 0,$$

and obtain $(1+t)^\alpha \leqslant 1 + t^\alpha \leqslant 2^\beta (1+t)^\alpha$.

(b). If $a_k \geqslant 0$ and $b_k \geqslant 0$, $k = 1, 2, \cdots, n$, then

$$\left\{\sum_{k=1}^{n} (a_k + b_k)^\alpha\right\}^{1/\alpha} \leqslant \left\{\sum_{k=1}^{n} (a_k^\alpha + b_k^\alpha)\right\}^{1/\alpha} \leqslant 2^{\beta/\alpha}\left\{\left(\sum_{k=1}^{n} a_k^\alpha\right)^{1/\alpha} + \left(\sum_{k=1}^{n} b_k^\beta\right)^{1/\alpha}\right\}.$$

[*Hint:* Replace t by b_k/a_k and again by $(\Sigma b_k^\alpha / \Sigma a_k^\alpha)^{1/\alpha}$.]

20. Let Σa_n be a convergent series of real terms.

(a). If Σd_n is a divergent series of positive terms, then

$$\varliminf \frac{a_n}{d_n} \leqslant 0 \leqslant \varlimsup \frac{a_n}{d_n}.$$

In particular, $\varliminf n a_n \leqslant 0 \leqslant \varlimsup n a_n$.

(b). If there is a constant c such that

$$na_n \leqslant (n+c)a_{n-1}, \qquad n = 2, 3, \cdots,$$

then $\lim na_n = 0$.

[*Hint:* $na_n \leqslant (n-1)a_{n-1} + (1+c)a_{n-1}$,

$$(n+k)a_{n+k} \leqslant (n-1)a_{n-1} + (1+c)\sum_{i=0}^{k} a_{n+i-1}.$$

Take $\overline{\lim}_{k\to\infty}$ and then $\underline{\lim}_{n\to\infty}$.]

(c). If $na_n \geqslant (n+c)a_{n-1}$, $n = 2, 3, \cdots$, then $\lim na_n = 0$.

21. (a). If $\{a_n\}$ is any sequence of positive terms, then

$$\overline{\lim_{n\to\infty}}\, n\left(\frac{1+a_{n+1}}{a_n} - 1\right) \geqslant 1.$$

[*Hint:* Otherwise there is an N such that

$$n \geqslant N \Rightarrow \frac{1}{n+1} < \frac{a_n}{n} - \frac{a_{n+1}}{n+1}$$

and summation from $n = N$ to $N+k$ leads to convergence of $\Sigma\, 1/n$.]

(b). Given any $\varepsilon > 0$, there are sequences $\{a_n\}$ of positive terms such that

$$\overline{\lim}\, n\left(\frac{1+a_{n+1}}{a_n} - 1\right) < 1 + \varepsilon.$$

[*Hint:* Try $a_n = cn$.]

(c). Using elementary properties of logarithms, show that the limit in part (a) exists and equals 1 for $a_n = n \ln n$.

22. If $\{x_n\}$ is such that all three subsequences $\{x_{2n+1}\}$, $\{x_{2n}\}$, and $\{x_{3n}\}$ converge, then $\{x_n\}$ converges.

23. For $a_n = (-1)^{n+1}n$, $n = 1, 2, 3, \cdots$,

$$\underline{\lim}\, \sigma_n = -\tfrac{1}{2} \qquad \text{and} \qquad \overline{\lim}\, \sigma_n = \tfrac{1}{2}.$$

24. The series

$$\Sigma_n \left[\frac{(-1)^n}{\sqrt{n}} + \frac{i}{n^2}\right]$$

is conditionally convergent; $i = \sqrt{-1}$.

25. If $\{a_n\}$ is a real sequence, $a_n \neq 0$, $a_n \to 0$,

$$-1 \leqslant \frac{a_{n+1}}{a_n}, \qquad \text{and} \qquad \overline{\lim_{n\to\infty}}\, \frac{a_{n+1}}{a_n} < 1,$$

then $\Sigma\, a_n$ converges. [Note that this includes the ratio test and the test for series of terms with alternating sign.]

[*Hint:* There is an r, $0 < r < 1$, and an N such that if $n \geqslant N$, then (consider the ratio negative or positive separately)

$$\left(1 + \frac{a_{n+1}}{a_n}\right)(1-r) \leqslant \left(1 - \left|\frac{a_{n+1}}{a_n}\right|\right)(1+r),$$

and hence, with $\lambda = (1+r)/(1-r)$, (consider $a_n < 0$ or >0)

$$|a_n + a_{n+1}| \leqslant \lambda(|a_n| - |a_{n+1}|).$$

With $N \leqslant n < m$ and $R_{nm} = a_n + a_{n+1} + \cdots + a_m$, then

$$2|R_{nm}| \leqslant |a_n| + |a_n + a_{n+1}| + \cdots + |a_{m-1} + a_m| + |a_m| \leqslant \cdots \leqslant (1+\lambda)|a_n|.]$$

26. If $a_1 \geqslant a_2 \geqslant \cdots \geqslant a_n \geqslant 0$ and $b_1, b_2, \cdots, b_n$ are such that

$$\sum_{j=1}^{k} a_j \leqslant \sum_{j=1}^{k} b_j \qquad \text{for } k = 1, 2, \cdots, n,$$

then

$$\sum_{j=1}^{n} a_j^2 \leqslant \sum_{j=1}^{n} b_j^2.$$

[*Hint:* Set $a_{n+1} = 0$. Sum

$$(a_k - a_{k+1}) \sum_{j=1}^{k} a_j \leqslant (a_k - a_{k+1}) \sum_{j=1}^{n} b_j$$

from $k = 1$ to n, interchange order of summation, obtain

$$\sum_{j=1}^{n} a_j^2 \leqslant \sum_{j=1}^{n} a_j b_j,$$

and apply Schwarz inequality.]

27. (a).
$$\sum_{k=1}^{\infty} \frac{n}{k(n+k)} = \sum_{k=1}^{n} \frac{1}{k}, \qquad \text{for } n = 1, 2, 3, \cdots.$$

(b).
$$\sum_{n=1}^{\infty} \sum_{k=1}^{\infty} \frac{1}{k(n+k)^2} = \sum_{n=1}^{\infty} \frac{1}{(n+1)^2} \sum_{k=1}^{n} \frac{1}{k}.$$

[*Hint:* Write the terms of the left iterated sum in matrix form with the top row for $n = 1$, the second row for $n = 2$, etc., and then sum up the 45° diagonals.]

(c). Let $s_0 = 0$ and $\sum_1^n 1/k = s_n$, $n = 1, 2, 3, \cdots$. From $s_n = (1/n) + s_{n-1}$ obtain

$$\sum_{n=1}^{\infty} \frac{s_n}{n^2} = \sum_{n=1}^{\infty} \frac{1}{n^3} + \sum_{n=1}^{\infty} \frac{s_n}{(n+1)^2}.$$

(d). $\displaystyle\sum_{n=1}^{\infty} \frac{s_n}{n^2} = 2 \sum_{n=1}^{\infty} \frac{1}{n^3}.$

[*Hint:*
$$\sum_{n=1}^{\infty} \frac{s_n}{n^2} = \sum_{n=1}^{\infty} \frac{1}{n^2} \sum_{k=1}^{\infty} \frac{n}{k(n+k)}$$
$$= \sum_{n=1}^{\infty} \sum_{k=1}^{\infty} \frac{n+k}{nk(n+k)^2}$$
$$= \sum_{n=1}^{\infty} \sum_{k=1}^{\infty} \frac{1}{k(n+k)^2} + \sum_{n=1}^{\infty} \sum_{k=1}^{\infty} \frac{1}{n(n+k)^2}$$
$$= 2 \sum_{n=1}^{\infty} \sum_{k=1}^{\infty} \frac{1}{k(n+k)^2}$$

by symmetry, and then use (b) and (c).]

28. If $\{a_{ij}\}$ converges to a, all in a metric space, then

$$\lim_{i} \overline{\lim_{j}}\, d(a_{ij}, a) = \lim_{j} \overline{\lim_{i}}\, d(a_{ij}, a) = 0.$$

29. Given that $\{a_n\}$ and $\{b_n\}$ are real sequences, then $\Sigma\,(a_n b_n)$ converges if:
(a). Σa_n converges and $\Sigma(b_n - b_{n+1})$ converges absolutely, or
(b). Σa_n has bounded partial sums, $b_n \to 0$, and $\Sigma(b_n - b_{n+1})$ converges absolutely.

30. For any sequence $\{A_n\}$ of sets, upon using $\overline{A} = \overline{\lim}\, A_n$ and $\underline{A} = \underline{\lim}\, A_n$

$$\chi_{\overline{A}}(x) = \overline{\lim}\, \chi_{A_n}(x) \text{ and } \chi_{\underline{A}} = \underline{\lim}\, \chi_{A_n}(x).$$

31. If $\{G_n\}$ is a sequence of open sets such that each G_n contains all rational points on $(0, 1)$, then the intersection $G = \bigcap G_n$ contains at least one irrational number. [*Note:* A little more work will show that G is noncountable.]

32. Let $\{a_n\}$ be a real Cauchy sequence and set

$$\underline{L} = \underline{\lim}\, a_n, \ \overline{L} = \overline{\lim}\, a_n.$$

Given an $\varepsilon > 0$ let N be such that

$$n \geqslant N \Rightarrow |a_{n+p} - a_n| < \varepsilon \qquad \text{for } p = 1, 2, 3, \cdots.$$

Justify each step in:

(1). $n \geqslant N \Rightarrow a_n - \varepsilon \leqslant \underline{L} \leqslant \overline{L} \leqslant a_n + \varepsilon.$

(2). $\overline{L} - 2\varepsilon \leqslant \underline{L} \leqslant \overline{L} \leqslant \underline{L} + 2\varepsilon.$

(3). $\underline{L} = \overline{L}.$

33. If $\{a_n\}$, $\{b_n\}$ are sequences of positive numbers with $\{a_n\}$ monotonically decreasing, then

$$\sum_{n=1}^{\infty} a_n b_n = \sum_{n=1}^{\infty} \left[(a_n - a_{n+1}) \sum_{k=1}^{n} b_k\right].$$

[*Hint:* Both represent the sum (finite or infinite) of the areas of the rectangles in the figure.]

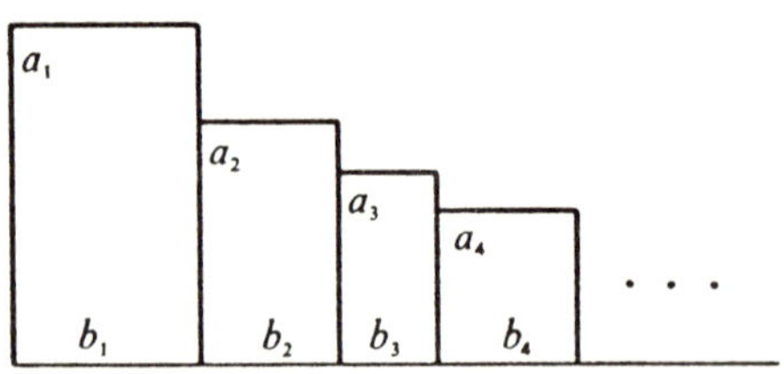

34. With $\{a_n\}$ a sequence of positive numbers, the **geometric-arithmetic** mean inequality is

$$(a_1 a_2 \cdots a_n)^{1/n} \leqslant \frac{1}{n}(a_1 + a_2 + \cdots + a_n), \quad n = 1, 2, 3, \cdots.$$

[*Hint 1.* Fix a positive integer n. Let g be the left side and σ the right side. Check the following steps for a proof that $g \leqslant \sigma$.

$$a_1 a_2 \leqslant \left(\frac{a_1 + a_2}{2}\right)^2$$

$$a_1 a_2 a_3 a_4 \leqslant \left(\frac{a_1 + a_2}{2}\right)^2 \left(\frac{a_3 + a_4}{2}\right)^2 \leqslant \left(\frac{a_1 + a_2 + a_3 + a_4}{4}\right)^4$$

and by induction

$$a_1 a_2 \cdots a_{2^m} \leqslant \left(\frac{a_1 + a_2 + \cdots + a_{2^m}}{2^m}\right)^{2m}, \quad m = 1, 2, \cdots.$$

Take $2^m > n$ and redefine $a_k = \sigma$ for $k = n + 1, n + 2, \cdots, 2^m$. Then

$$a_1 a_2 \cdots a_n \sigma^{2m-n} \leqslant \left(\frac{a_1 + a_2 + \cdots + a_n + (2^m - n)\sigma}{2^m}\right)^{2^m} = \sigma^{2m},$$

etc.

Hint 2. In Theorem 1-3.1, set $y = x - 1$ and write the result as

$$n - 1 \geqslant x(n - x^{n-1}), \quad x > 0.$$

With $n > 1$, $a > 0$, and $A \geqslant 0$, set

$$x = \left(\frac{na}{a + A}\right)^{1/(n-1}$$

Then

$$(n-1)\left(\frac{a+A}{na}\right)^{1/(n-1)} \geqslant n - \frac{na}{a+A} = \frac{na}{a+A}\frac{A}{a},$$

$$(n-1)\left(\frac{a+A}{na}\right)^{n/(n-1)} \geqslant \frac{A}{a}, \quad \left(\frac{a+A}{n}\right)^n \geqslant a\left(\frac{A}{n-1}\right)^{n-1}.$$

Hence

$$\left(\frac{a_1 + a_2 + \cdots + a_n}{n}\right)^n \geqslant a_1\left(\frac{a_2 + a_3 + \cdots + a_n}{n-1}\right)^{n-1}$$

$$\geqslant a_1 a_2\left(\frac{a_3 + \cdots + a_n}{n-2}\right)^{n-2} \geqslant \cdots.]$$

Problems on Infinite Products

Formal multiplication of the first of the expressions

$$\frac{2 \cdot 3 \cdot 4 \cdot 5 \cdots}{1 \cdot 2 \cdot 3 \cdot 4 \cdots} \quad \text{and} \quad \frac{1 \cdot 2 \cdot 3 \cdot 4 \cdots}{1 \cdot 2 \cdot 3 \cdot 4 \cdots}$$

by 1 yields the second. If, in analogy with infinite series, we interpret these to mean

$$\lim_{n\to\infty} \frac{2 \cdot 3 \cdot 4 \cdots (n+1)}{1 \cdot 2 \cdot 3 \cdots n} \quad \text{and} \quad \lim_{n\to\infty} \frac{1 \cdot 2 \cdot 3 \cdots n}{1 \cdot 2 \cdot 3 \cdots n},$$

respectively, we do not obtain the same result. According to the following definition, the first two expressions as they stand are not infinite products.

Definition 1. *Given a sequence $\{u_n\}$ of real numbers such that $u_n \neq 0, n = 1, 2, 3, \cdots$, we call*

$$u_1 u_2 u_3 \cdots = \prod_{n=1}^{\infty} u_n$$

an **infinite product** *and*

$$p_n = u_1 u_2 \cdots u_n = \prod_{k=1}^{n} u_k$$

its nth **partial product.** *The infinite product is said to*

(i). **converge** *to $p \neq 0$ if* $\lim_{n\to\infty} p_n = p \neq 0$;
(ii). **diverge** *to* 0 if $\lim_{n\to\infty} p_n = 0$; *and*
(iii). **diverge** *if $\{p_n\}$ does not have a limit.*

[Some books do not have the restriction $u_n \neq 0$, but add to the above definition that $\Pi\, u_n$ also diverges to 0 if $u_n = 0$ for infinitely many n, and that $\Pi\, u_n$ converges to 0 if there is an N such that $u_N = 0$, $u_n \neq 0$ for $n > N$, and

$$\lim_{n\to\infty} \prod_{k=N+1}^{N+n} u_k \qquad \text{exists and is not zero.}$$

It is to avoid such nonessential complications that, for this discussion, we make the restriction

$$u_n \neq 0, \qquad n = 1, 2, 3, \cdots.]$$

PROBLEM 1. If $\Pi\, u_n$ converges, then

(a). $\lim_{n\to\infty} u_n = 1$.
(b). There is an N such that $n \geqslant N \Rightarrow u_n > 0$ and

$$\lim_{n\to\infty} p_n = (u_1 u_2 \cdots u_N) \lim_{n\to\infty} \prod_{k=N+1}^{N+n} u_k \neq 0.$$

Note that if $u_n = -0.9$, then $\Pi\, u_n$ diverges to zero even though u_n does not $\to 0$ as $n \to \infty$.

PROBLEM 2. For each positive integer N the infinite product $\Pi\, u_n$

(a). Converges iff $\lim_{n\to\infty} p_{N+n}$ exists and is not zero.
(b). Diverges to 0 iff $\lim_{n\to\infty} (|u_1|\,|u_2| \cdots |u_n|) = 0$.

Hence, without losing any generality, we make the further restriction

$$u_n > 0, \qquad n = 1, 2, 3, \cdots.$$

It is now convenient to change notation and set

$$u_n = 1 + a_n,\ a_n > -1, \qquad n = 1, 2, 3, \cdots.$$

Thus, from Problem 1, if $\Pi(1 + a_n)$ converges, then $\lim_{n\to\infty} a_n = 0$.

In this discussion we also use the properties of natural logarithms,

$$\lim_{x\to 0} \frac{\ln(1+x)}{x} = 1 \qquad \text{and} \qquad \lim_{x\to 0} \frac{\ln(1+x) - x}{-\frac{1}{2}x^2} = 1,$$

obtained in elementary calculus from l'Hospital's rules.

PROBLEM 3. Whenever $a_n \geqslant 0$ (or else $a_n \leqslant 0$), $n = 1, 2, 3, \cdots$, then

$$\Pi(1 + a_n) \text{ converges iff } \Sigma a_n \text{ converges.}$$

[*Hint:* $\ln p_n = \Sigma_{k=1}^{n} \ln(1 + a_n)$ and, with $a_n \geqslant 0$, there is an N such that

$$n \geqslant N \Rightarrow \tfrac{1}{2} a_n \leqslant \ln(1 + a_n) \leqslant \tfrac{3}{2} a_n .]$$

Definition 2. An infinite product $\Pi(1 + a_n)$ is said to **converge absolutely** provided $\Pi(1 + |a_n|)$ converges.

PROBLEM 4. If $\Pi(1 + a_n)$ converges absolutely, then

(a). $\Pi(1 + a_n)$ converges.
(b). Every rearrangement converges and to the same value.

PROBLEM 5. (a) If any two of $\Pi(1 + a_n)$, Σa_n, and $\Sigma a_n{}^2$ converge, then the third also converges.

(b). If Σa_n converges absolutely, then both $\Pi(1 + a_n)$ and $\Sigma a_n{}^2$ converge.
(c). If Σa_n converges conditionally, then $\Pi(1 + a_n)$ converges or diverges to 0 according as $\Sigma a_n{}^2$ converges or diverges.

[*Hint:* Under any of the hypotheses there is an N such that

$$n \geqslant N \Rightarrow a_n - \tfrac{3}{4} a_n{}^2 \leqslant \ln(1 + a_n) \leqslant a_n - \tfrac{1}{4} a_n{}^2,$$

which is obtainable from the second logarithmic limit.]

An application in elementary calculus is the infinite product representation

$$\tfrac{\pi}{2} = \tfrac{2}{1} \cdot \tfrac{2}{3} \cdot \tfrac{4}{3} \cdot \tfrac{4}{5} \cdot \tfrac{6}{5} \cdot \tfrac{6}{7} \cdots. \tag{1}$$

This is derived by using integration by parts and mathematical induction to obtain

$$I_{2n} = \int_0^{\pi/2} \sin^{2n} x \, dx = \frac{\pi}{2} \frac{1 \cdot 3 \cdots (2n-1)}{2 \cdot 4 \cdots 2n} \quad \text{and}$$

$$I_{2n+1} = \int_0^{\pi/2} \sin^{2n+1} x \, dx = \frac{2 \cdot 4 \cdots 2n}{3 \cdot 5 \cdots (2n+1)}.$$

Since $0 \leqslant \sin x \leqslant 1$ for $0 \leqslant x \leqslant \pi/2$, then $I_{2n} \geqslant I_{2n+1} \geqslant I_{2n+2}$.
Therefore

$$1 \geqslant \frac{I_{2n+1}}{I_{2n}} \geqslant \frac{I_{2n+2}}{I_{2n}} = \frac{2n+1}{2n+2} \quad \text{and}$$

$$\frac{\pi}{2} = \frac{\pi}{2} \lim_{n\to\infty} \frac{I_{2n+1}}{I_{2n}} = \lim_{n\to\infty} \left[\frac{2}{1} \cdot \frac{2}{3} \cdot \frac{4}{3} \cdots \frac{2n}{2n-1} \cdot \frac{2n}{2n+1}\right].$$

Hence the even partial products in (1) approach $\pi/2$ and thus they all do, since the nth factor approaches 1.

PROBLEM 6. If the 1 in the denominator of the preceding equation had been ignored and the limit of "partial products" of

$$\frac{2 \cdot 2 \cdot 4 \cdot 4 \cdot 6 \cdot 6 \cdots}{3 \cdot 3 \cdot 5 \cdot 5 \cdot 7 \cdot 7 \cdots}$$

taken, what would be the result?

Chapter 4

MEASURE AND INTEGRATION

4-1 OUTER LEBESGUE MEASURE IN E^1

Whether an interval I is of the form

$$(a, b),\ -\infty \leqslant a < b \leqslant \infty,$$
$$(a, b],\ -\infty \leqslant a < b < \infty,$$
$$[a, b),\ -\infty < a < b \leqslant \infty, \quad \text{or}$$
$$[a, b],\ -\infty < a \leqslant b < \infty$$

will be left indefinite unless specifically qualified, but for any of these the length of I will be denoted by $|I|$, where

$$|I| = b - a.$$

For each $a \in E^1$, the set $\{a\}$ is an interval of zero length. We also admit the empty set as an interval of zero length and usually will not warn of its possible presence in a union or its length in a sum.

The following theorem and corollary are examples of obvious results which are somewhat bothersome to prove. We shall make the proof depend upon the fact that if a, b, a_i, and b_i are numbers such that

$$(1) \qquad a_1 < a < b < b_n \qquad \text{and} \qquad a_i < a_{i+1} < b_i < b_{i+1}$$

for $i = 1, 2, \cdots, n-1$, then

$$(2) \qquad \begin{aligned} b - a &= (a_2 - a) + (a_3 - a_2) + \cdots + (a_n - a_{n-1}) + (b - a_n) \\ &< (b_1 - a_1) + (b_2 - a_2) + \cdots + (b_{n-1} - a_{n-1}) + (b_n - a_n). \end{aligned}$$

Theorem 1. *If $I = [a, b]$, $-\infty < a < b < \infty$ is covered by a countable family $\{I_i\}$ of intervals (that is, $I \subset \bigcup I_i$), then*

$$|I| \leqslant \sum |I_i|.$$

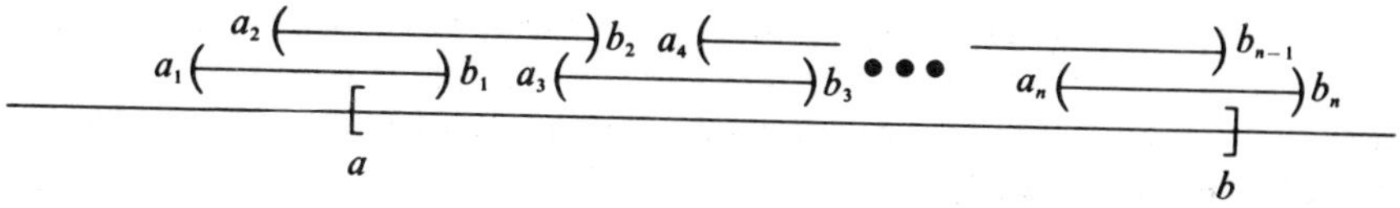

FIG. 1

Proof. Assume there is a finite closed interval $I = [a, b]$ with a countable covering $\{I_i\}$ of intervals such that

$$|I| > \sum |I_i|.$$

Then $\sum |I_i|$ is a finite sum or a convergent series of nonnegative terms, and we set

$$\lambda = |I| - \sum |I_i| > 0.$$

Just in case some of the intervals of $\{I_i\}$ are not open, let J_i be an open interval containing I_i such that $|J_i| < |I_i| + \lambda/2^i$. Then

$$\sum |J_i| < \sum |I_i| + \lambda = |I|.$$

Also, $\{J_i\}$ is an open covering of the bounded closed (compact) set I. Accordingly, select a finite subfamily which also covers I (by Theorem 2–7.4 or more directly by Problem 2–5.3). Discard any member of this finite subfamily that is a subset of another member. What remains is a finite number of open intervals whose union covers $[a, b]$, and they may be named $J_1, J_2 \cdots J_n$ with end points satisfying (1), since none of these is a subset of another. Hence, from (2), the contradiction

$$|I| \leqslant \sum_{j=1}^{n} |J_j| \leqslant \sum |J_i| < |I|$$

is obtained. ∎

Corollary. *The conclusion of the theorem also holds if I is a semiopen or an open interval.*

Proof. Consider $I = (a, b]$, $-\infty < a < b < \infty$ and take any countable family $\{I_i\}$ of intervals covering I. Augment this family by the interval $I_0 = \{a\}$. The augmented family covers $\bar{I} = [a, b]$, and hence by the theorem,

$$|I| = |\bar{I}| = (b - a) \leqslant \sum |I_i| + |I_0| = \sum |I_i|,$$

since $|I_0| = 0$. A similar proof may be made if $I = [a, b)$ or $I = (a, b)$, $-\infty < a < b < \infty$. ∎

PROBLEM 1. If I is an infinite interval (and hence $|I| = \infty$) and $\{I_i\}$ is a countable family of intervals covering I, then $\sum |I_i| = \infty$.

Let $A \subset E^1$ be the union of a finite number of pairwise disjoint intervals. In a sense, the "length" of this set A is the sum of the lengths of its separate intervals. Even if A is the countably infinite union of pairwise disjoint intervals $I_1, I_2, \cdots$, then the series

$$\sum_{i=1}^{\infty} |I_i|$$

either converges or diverges to ∞; in either case the result is intuitively "the length of the set." Since an open set $G \subset E^1$ is either an open interval or a countable union of pairwise disjoint open intervals (Theorem 2–5.5), then G seems to have, theoretically at least, a length. But is there a reasonable (and useful) way of assigning a "length" to a set which is not necessarily a countable union of intervals?

In 1904 Henri Lebesgue introduced a generalization of the notion of length, which is both intuitive and has many applications, extensions, and abstractions. The following definition is the first step in a generalization, equivalent to Lebesgue's, of length.

Definition 1. *The outer Lebesgue measure*† *of a set $A \subset E^1$ is denoted by m^*A and defined by*

$$m^*A = \begin{cases} 0 \text{ if } A = \varnothing \\ \inf\{y \mid y = \sum |I_n| \quad \text{where } A \subset \bigcup I_n\} \quad \text{if } A \neq \varnothing. \end{cases}$$

Hence, for each set $A \subset E^1$, $m^*A \geqslant 0$ is a unique element of E^*, and for $A \neq \varnothing$:

(i) If $\{I_i\}$ is any countable family of intervals such that

$$A \subset \bigcup I_n, \qquad \text{then } m^*A \leqslant \sum |I_n|;$$

and

(ii) Corresponding to each number $\varepsilon > 0$, there is a countable family $\{I_n\}$ of intervals such that

$$A \subset \bigcup I_n \qquad \text{and} \qquad \sum |I_n| \leqslant m^*A + \varepsilon.$$

[In this last line the symbol $\leqslant$ is used instead of $<$ to take care of the cases in which $m^*A = \infty$.]

Theorem 2. *Any interval $I \subset E^1$ has $m^*I = |I|$.*

Proof. With $|I|$ in place of m^*A, (i) holds by Theorem 1, its corollary, or Problem 1; and (ii) holds for the covering family whose single element is the interval I itself. ∎

Corollary 1. *If $a \in E^1$, then $m^*\{a\} = 0$.*

Theorem 3. *If $A \subset B \subset E^1$, then $m^*A \leqslant m^*B$.*

Proof. This holds if $A = \varnothing$ and also if $A \neq \varnothing$ since then

$$\begin{aligned} m^*A &= \inf\{y \mid y = \sum |I_n| \quad \text{with } A \subset \bigcup I_n\} \\ &\leqslant \inf\{y \mid y = \sum |I_n| \quad \text{with } A \subset B \subset \bigcup I_n\} = m^*B. \end{aligned}$$ ∎

† Sometimes called the exterior Lebesgue measure.

Theorem 4. *For any countable family $A_1, A_2, A_3, \cdots$ of sets in E^1,*

$$m^*(\bigcup A_n) \leqslant \sum m^*A_n.$$

Proof. For convenience, set $A = \bigcup A_n$. In case $\sum m^*A_n$ diverges (to ∞) there is nothing to prove. Hence consider $\sum m^*A_n$ converging and choose $\varepsilon > 0$ arbitrarily. Each A_n then has $m^*A_n < \infty$, and for each nonempty A_n we choose a countable family $I_{n1}, I_{n2}, I_{n3}, \cdots$ of intervals such that

$$A_n \subset \bigcup_k I_{nk} \qquad \text{and} \qquad m^*A_n > \sum_k |I_{nk}| - \frac{\varepsilon}{2^n}.$$

Then the following series converge and

$$\sum m^*A_n > \sum_n \sum_k |I_{nk}| - \varepsilon.$$

Hence (Theorem 3–11.3] the double series $\sum_{(n,k)}|I_{nk}|$ converges absolutely to $\sum_n \sum_k |I_{nk}|$ and the terms may be added in any order. Hence, upon renaming the sets in the double sequences $\{I_{nk}\}$ as a single sequence $I_1, I_2, \cdots, I_j, \cdots$,

$$A \subset \bigcup I_j \qquad \text{and} \qquad m^*A \leqslant \sum_j |I_j| = \sum_{(n,k)} |I_{nk}| < \sum_n m^*A_n + \varepsilon.$$

Since $\varepsilon > 0$ is arbitrary, $m^*A \leqslant \sum_n m^*A_n$. ∎

Corollary 1. If $A \subset E^1$ *is countable, then* $m^*A = 0$.

Proof. A may be written as $A = \bigcup A_n$ with each A_n a set of one element. ∎

Corollary 2. *The set of rational numbers has outer Lebesgue measure zero.*

It follows from Theorem 4 that if A and B are any sets in E^1, then

$$m^*(A \cup B) \leqslant m^*A + m^*B. \tag{3}$$

There are conditions on A and B under which the equality holds. In fact a considerable portion of measure theory is concerned directly or indirectly with conditions under which an outer measure of a union is the sum of the outer measures. In (3) if $A = B$ and $0 < m^*A < \infty$, then $A \cup B = A$ and the left side has value half the right side. Thus overlapping of A and B may result in a strict inequality in (3), but even if $A \cap B = \varnothing$, it is possible (as shown later) for the strict inequality to hold.

4–2 OUTER MEASURES AND MEASURABILITY

In the preceding section the outer Lebesgue measure function m^*, on the subsets of E^1 into $\{y \mid 0 \leqslant y \leqslant \infty\}$, was defined and some of its properties obtained. Further properties of m^* will be developed in the present section without, however, specific use of the definition of m^*A. The following definition places this further development in a more general setting.

Definition 1. *With S a space, a function μ^* on the subsets of S into $\{y \mid 0 \leqslant y \leqslant \infty\}$ is called an* **outer measure** *on S, provided the following properties hold:*

(i) *$\mu^*\varnothing = 0$ and $0 < \mu^*A < \infty$ for at least one set $A \subset S$.*
(ii) *If $A \subset B \subset S$, then $\mu^*A \leqslant \mu^*B$.*
(iii) *If $A_1, A_2, A_3, \cdots$ is a countable family of sets in S, then*

$$\mu^*(\bigcup A_n) \leqslant \sum \mu^*A_n.$$

From the preceding section it is seen that outer Lebesgue measure on E^1 is an outer measure on $S = E^1$. There are, however, outer measures other than m^*, as the following examples and problems show.

Recall that if A and C are sets, then we may write AC or $A \cap C$, whichever seems convenient. Thus one of the distributive laws may appear as

$$(A \cup B)C = AC \cup BC.$$

EXAMPLE 1. Let $C \subset E^1$ have $0 < m^*C < \infty$ and define μ^*A for each $A \subset E^1$ by

$$\mu^*A = m^*(AC).$$

Then μ^* is a metric outer measure on E^1.

First $A \subset E^1 \Rightarrow 0 \leqslant m^*(AC) < \infty \Rightarrow 0 \leqslant \mu^*A \leqslant \infty$. We next check (i)–(iii) in turn. For (i),

$$\mu^*\varnothing = m^*(\varnothing \cap C) = m^*\varnothing = 0 \qquad \text{and}$$
$$\mu^*C = m^*(C \cap C) = m^*C \Rightarrow 0 < \mu^*C < \infty.$$

For (ii), $A \subset B \Rightarrow AC \subset BC \Rightarrow m^*(AC) \leqslant m^*(BC)$, and hence

$$\mu^*A \leqslant \mu^*B.$$

For (iii), $(\bigcup A_n)C = \bigcup(A_n C)$ so that

$$\begin{aligned}\mu^*(\bigcup A_n) &= m^*[(\bigcup A_n)C] = m^*[\bigcup (A_n C)] \\ &\leqslant \sum m^*(A_n C) = \sum \mu^*A_n.\end{aligned}$$

PROBLEM 1. Let S be a space of at least two points and $s_0 \in S$. For each $A \subset S$, define

$$\mu^*A = \begin{cases} 0 & \text{if } s_0 \notin A \\ 1 & \text{if } s_0 \in A. \end{cases}$$

Then μ^* is an outer measure on S.

PROBLEM 2. Let S be a space of at least two elements and take $s_0 \in S$. Define $\nu^*\varnothing = 0$ and for $A \subset S$, but $A \neq \varnothing$, by

$$\nu^*A = \begin{cases} 1 & \text{if } s_0 \notin A \\ 2 & \text{if } s_0 \in A. \end{cases}$$

Then ν^* is an outer measure on S.

PROBLEM 3. Let $S = E^2$. The outer Lebesgue plane measure m_2^*A of a set $A \subset E^2$ is defined by

$$m_2^*A = \begin{cases} 0 & \text{if } A = \varnothing \\ \inf\{y \mid y = \sum |R_n| \quad \text{with } A \subset (\bigcup R_n)\} & \text{if } A \neq \varnothing \end{cases}$$

where each R_n is a rectangle in E^2 and $|R_n|$ is its area. Show that outer Lebesgue plane measure is an outer measure on E^2.

Hereafter we shall not bother to say "let μ^* be an outer measure on a space S." Hence this phrase (including the "a space S") is to be understood whenever μ^* is used. Also, "every set W" means "every set $W \subset S$," and $A^c = S - A$.

Given a set A, then for every set W we have that

$$W = WA \cup WA^c,$$

$\mu^*W = \mu^*(WA \cup WA^c)$ and hence, from (iii),

$$\mu^*W \leqslant \mu^*(WA) + \mu^*(WA^c).$$

Definition 2. *A set is said to be μ-***measurable** *if for every set W the equality*

(1) $$\mu^*W = \mu^*(WA) + \mu^*(WA^c)$$

holds. Any μ-measurable set A is assigned the μ-measure μA by setting $\mu A = \mu^*A$.

[*Note:* In Lebesgue's original work† he confined all sets to an arbitrary, but fixed, interval I of finite length; that is, he considered I essentially as a universal space. Then, with I_1 an interval interior to I, the set $I - I_1$ is the union of two intervals, the sum of whose length is $|I| - |I_1|$ so that

$$|I| = m^*I_1 + m^*(I - I_1).$$

Also, if $A = I_1 \cup I_2 \cup \cdots \cup I_n$, with no two of these intervals having more than end points in common, then

$$|I| = m^*A + m^*(I - A),$$

and the equation was proved to hold if A is the union of countably many non-overlapping intervals. This additive property was then used as the definition of measurability of any set $A \subset I$.

The restriction to bounded sets was too confining and the problem arose of defining measurability of unbounded sets; i.e., with I replaced by the whole Euclidean line E^1. Clearly,

$$|E^1| = m^*A + m^*(E^1 - A)$$

could not be used, since both sides are infinite for all sets $A \subset E^1$. If A and W are any sets, then

$$m^*W \leqslant m^*(WA) + m^*(W - A)$$

† H. Lebesgue, "Integrale," Longueur, Aire. *These. Ann. di Matem.* (3), 7, 1902.

as noted before the definition. If A fails to have the desirable additive property, then there is at least one set W_1 for which

$$m^*W_1 < m^*(W_1A) + m^*(W_1 - A).$$

It was therefore quite natural to consider a set A to be measurable in case the equality holds for all sets W; hence the above definition.

The first use of this definition of measurability was by Carathéodory† in considering an outer measure of sets in the plane as a generalization of arc length of curves. See problems at the end of Chapter 5. Carathéodory developed his notion of linear measure of plane sets (including the now prominent definition of measurability) in searching for a fresh topic to present to the undergraduate mathematics club at the University of Munich, where the members were already familiar with standard Lebesgue theory.]

As noted above, if W and A are any sets whatever, then $W = (WA) \cup (WA^c)$ and hence

(2) $$\mu^*W = \mu^*[(WA) \cup (WA^c)] \leqslant \mu^*(WA) + \mu^*(WA^c)$$

by (iii). There are two simple consequences of this fact which are used so often that they are stated as theorems for easy reference.

Theorem 1. *A set is μ-measurable iff for every set W*

(3) $$\mu^*W \geqslant \mu^*(WA) + \mu^*(WA^c).$$

Proof. Since (2) always holds, then (1) holds iff (3) holds. ∎

Theorem 2. *A set A is μ-measurable iff* (3) *holds for every set W having $\mu^*W < \infty$.*

Proof. If $\mu^*W = \infty$ and A is any set whatever, then $\mu^*(WA) + \mu^*(WA^c) = \infty$ from (2). Hence if $\mu^*W = \infty$, then (1) holds as $\infty = \infty$ whether A is μ-measurable or not. Thus a set A is μ-measurable iff (3) holds for all sets W with $\mu^*W < \infty$. ∎

PROBLEM 4. If A is a μ-nonmeasurable set, then there is a subset of A with positive finite μ^*-measure.

[*Hint:* If A is μ-nonmeasurable, there is a set W with $\mu^*W < \infty$ and $\mu^*W < \mu^*WA + \mu^*WA^c$. Assume $\mu^*WA = 0$ and get a contradiction, and then assume $\mu^*WA = \infty$.]

Theorem 3. *A set A is μ-measurable iff A^c is μ-measurable.*

Proof. Equation (1) is symmetric in A and A^c ; that is, (1) may be written

$$\begin{aligned} \mu^*W &= \mu^*(WA^c) + \mu^*(WA) \\ &= \mu^*(WA^c) + \mu^*[W(A^c)^c], \end{aligned}$$

which is (1) with A replaced by A^c.

† C. Carathéodory, "Über das lineare Mass von Punktmengen, eine Verallgemeinerung des Längenbegriffs," *Gött. Nach.*, 1914, p. 404.

PROBLEM 5. Both the empty set $\varnothing$ and the whole space S are μ-measurable.

PROBLEM 6. If $\mu^*A = 0$, then A is μ-measurable.

Theorem 4. *Two μ-measurable sets A and B have μ-measurable intersection AB and μ-measurable union $A \cup B$.*

Proof. Let W be any set. From the μ-measurability of A and B we wish to obtain (1), with AB in place of A and $(AB)^c$ in place of A^c. We may use

$$\mu^*U = \mu^*(UB) + \mu^*(UB^c) \qquad \text{for any set } U \qquad \text{and}$$
$$\mu^*V = \mu^*(VA) + \mu^*(VA^c) \qquad \text{for any set } V$$

as well as

$$\mu^*W = \mu^*(WA) + \mu^*(WA^c). \tag{4}$$

With $U = WA$ and $V = W(AB)^c$, then

$$\mu^*(WA) = \mu^*(WAB) + \mu^*(WAB^c) \qquad \text{and} \tag{5}$$

$$\begin{aligned} \mu^*[W(AB)^c] &= \mu^*[W(AB)^cA] + \mu^*[W(AB)^cA^c] \\ &= \mu^*[W(A^c \cup B^c)A] + \mu^*[W(A^c \cup B^c)A^c] \\ &= \mu^*(WAB^c) + \mu^*(WA^c). \end{aligned} \tag{6}$$

By substituting $\mu^*(WA)$ from (5) into (4) we have

$$\mu^*W = \mu^*(WAB) + \mu^*(WAB^c) + \mu^*(WA^c)$$

and this with (6) gives

$$\mu^*W = \mu^*(WAB) + \mu^*[W(AB)^c].$$

Thus AB is μ-measurable.

Since A^c and B^c are μ-measurable (Theorem 3), then A^cB^c are μ-measurable, by the first part of the present theorem, and hence

$$(A^cB^c)^c = A \cup B$$

is μ-measurable. ∎

Corollary. *If $A_1, A_2, \cdots, A_n$ with $n \in J$ are each μ-measurable, then*

$$\bigcup_{k=1}^{n} A_k \qquad \text{and} \qquad \bigcap_{k=1}^{n} A_k$$

are μ-measurable.

Theorem 5. *Let A be a μ-measurable subset of a set B. If W is any set with $\mu^*W < \infty$, then*

$$\mu^*(WBA^c) = \mu^*(WB) - \mu^*(WA).$$

Proof From the μ-measurability of A we have

$$\mu^*U = \mu^*(UA) + \mu^*(UA^c)$$

for any set U. Hence, with $U = WB$,

$$\begin{aligned}\mu^*(WB) &= \mu^*(WBA) + \mu^*(WBA^c)\\ &= \mu^*(WA) + \mu^*(WBA^c), \qquad \text{from } A \subset B,\end{aligned}$$

and the desired equation follows since $\mu^*(WA) \leqslant \mu^*W < \infty$.

PROBLEM 7. Let A be a μ-measureable set with $\mu A < \infty$. If B is any set such that $B \supset A$, then

$$\mu^*(BA^c) = \mu^*B - \mu A; \qquad \text{that is,}$$
$$\mu^*(B - A) = \mu^*B - \mu A.$$

Lemma 1. *A monotonically decreasing sequence*

$$B_1 \supset B_2 \supset B_3 \supset \cdots$$

of μ-measurable sets has μ-measurable intersection

$$D = \bigcap_{n=1}^{\infty} B_n.$$

Proof. Let W be any set with $\mu^*W < \infty$. The numerical sequence $\{\mu^*(WB_n)\}$ has finite-valued terms, is convergent (since it is monotonically decreasing), and

$$\mu^*(WD) \leqslant \lim_{n\to\infty} \mu^*(WB_n), \tag{7}$$

since $D \subset B_n$. Also $D^c = B_1{}^c \cup (B_1B_2{}^c) \cup (B_2B_3{}^c) \cup \cdots$ and

$$\begin{aligned}\mu^*(WD^c) &\leqslant \mu^*(WB_1{}^c) + \sum_{n=1}^{\infty} \mu^*(WB_nB_{n+1}^c) \qquad \text{by (iii)}\\ &= \mu^*(WB_1{}^c) + \lim_{n\to\infty} \sum_{k=1}^{n} \mu^*(WB_kB_{k+1}^c)\\ &= \mu^*(WB_1{}^c) + \lim_{n\to\infty} \sum_{k=1}^{n} [\mu^*(WB_k) - \mu^*(WB_{k+1})] \qquad \text{by Theorem 5}\\ &= \mu^*(WB_1{}^c) + \mu^*(WB_1) - \lim_{n\to\infty} \mu^*(WB_{n+1})\\ &\leqslant \mu^*W - \mu^*(WD)\end{aligned}$$

from the μ-measurability of B_1 and from (7). Thus

$$\mu^*W \geqslant \mu^*(WD) + \mu^*(WD^c)$$

for every set W having $\mu^*W < \infty$. Hence D is μ-measurable by Theorem 2. ∎

Theorem 6. *A sequence $\{A_n\}$ of μ-measurable sets has μ-measurable intersection, union, limit superior, and limit inferior.*

Proof. By the corollary of Theorem 4, each of the sets

$$B_n = \bigcap_{k=1}^{n} A_k, \qquad n = 1, 2, 3, \cdots$$

is μ-measurable. Since $B_1 \supset B_2 \supset B_3 \supset \cdots$ and

$$\bigcap B_n = \bigcap A_n,$$

then $\bigcap A_n$ is μ-measurable by Lemma 1. Hence

$$\bigcup A_n = (\bigcap A_n{}^c)^c$$

is μ-measurable.

With B_n redefined as $B_n = \bigcup_{k=n}^{\infty} A_k$, then B_n is μ-measuable and

$$\bigcap_{n=1}^{\infty} B_n = \overline{\lim}\, A_n$$

is μ-measurable. Hence $\underline{\lim}\, A_n = (\overline{\lim}\, A_n{}^c)^c$ is also μ-measurable. ▮

Theorem 7. *Let A be a μ-measurable set. If B is any set, then*

$$\mu^*(A \cup B) + \mu^*(AB) = \mu A + \mu^* B, \tag{8}$$

and if in addition $A \cap B = \varnothing$, then

$$\mu^*(A \cup B) = \mu A + \mu^* B. \tag{9}$$

Proof. If $\mu^* B = \infty$, then also $\mu^*(A \cup B) = \infty$ and (8) holds.

Hence consider $\mu^* B < \infty$.

Since $\mu^* W = \mu^* WA + \mu^* WA^c$ for any set W, we first use $W = A \cup B$ and then $W = B$ to obtain

$$\begin{aligned} \mu^*(A \cup B) &= \mu^*[(A \cup B)A] + \mu^*[(A \cup B)A^c] \\ &= \mu A + \mu^*(BA^c) \qquad \text{and} \\ \mu^* B &= \mu^*(BA) + \mu^*(BA^c), \end{aligned}$$

respectively. These equations yield (8), since both $\mu^*(AB)$ and $\mu^*(BA^c)$ are finite under the condition $\mu^* B < \infty$ being considered.

If $A \cap B = \varnothing$, then $\mu^*(A \cap B) = 0$ and (9) follows from (8). ▮

Theorem 8. *A sequence $\{A_n\}$ of pairwise disjoint μ-measurable sets has*

$$\mu\left(\bigcup_{n=1}^{\infty} A_n\right) = \sum_{n=1}^{\infty} \mu A_n.$$

Proof. From $\bigcup A_n \supset \bigcup_{k=1}^n A_k$ and from Theorem 7 by induction,

$$\mu\left(\bigcup A_n\right) \geqslant \mu\left(\bigcup_{k=1}^{n} A_k\right) = \sum_{k=1}^{n} \mu A_k, \qquad n = 1, 2, 3, \cdots.$$

By letting $n \to \infty$, we have $\mu(\bigcup A_n) \geqslant \sum \mu A_n$. But the reverse inequality holds by (iii) and thus the equality holds. ∎

Theorem 9. *A monotonically increasing sequence*

$$A_1 \subset A_2 \subset A_3 \subset \cdots$$

of μ-measurable sets has

$$\mu\left(\bigcup_{n=1}^{\infty} A_n\right) = \lim_{n\to\infty} \mu A_n.$$

Proof. If $\mu A_N = \infty$ for some $N \in J$, then $\mu(\bigcup A_n) \geqslant \mu(A_N) = \infty$, $\mu(A_n) = \infty$ for each $n \geqslant N$, and the equality holds.

Hence consider $\mu(A_n) < \infty$, $n = 1, 2, 3, \cdots$. Since

$$\bigcup A_n = A_1 \cup (A_2 A_1^c) \cup (A_3 A_2^c) \cup \cdots$$

is the union of pairwise disjoint μ-measurable sets, then by Theorem 8,

$$\begin{aligned}\mu\left(\bigcup A_n\right) &= \mu A_1 + \mu(A_2 A_1^c) + \mu(A_3 A_2^c) + \cdots \\ &= \mu A_1 + \lim_{n\to\infty} \sum_{k=1}^{n} [\mu(A_{k+1}) - \mu(A_k)] \qquad \text{by Problem 7} \\ &= \lim_{n\to\infty} \mu(A_{n+1}).\end{aligned}$$ ∎

Theorem 10. *A monotonically decreasing sequence*

$$A_1 \supset A_2 \supset A_3 \supset \cdots$$

of μ-measurable sets, with $\mu A_n < \infty$ for at least one n, has

$$\mu\left(\bigcap_{n=1}^{\infty} A_n\right) = \lim_{n\to\infty} \mu A_n.$$

[*Note:* In E^1 let $A_n = \{x \mid x \geqslant n\}$, $n = 1, 2, 3, \cdots$. Then

$$A_1 \supset A_2 \supset A_3 \supset \cdots, \; m(A_n) = \infty \qquad \text{and} \qquad \lim_{n\to\infty} m(A_n) = \infty.$$

However, $\bigcap A_n = \varnothing$ and thus $m(\bigcap A_n) = 0$. This shows that "with $\mu A_n < \infty$ for at least one $n \in J$" is a necessary condition.]

Proof. Let $N \in J$ be such that $\mu A_N < \infty$ and note that $n \geqslant N \Rightarrow \mu A_N \geqslant \mu A_n \geqslant \mu A_{n+1} \geqslant \cdots \geqslant 0$. Therefore the existence of the limit and the inequalities

$$\lim_{n\to\infty} \mu A_n \leqslant \mu A_N < \infty$$

follow. Let $D = \bigcap_{n=1}^{\infty} A_n$. Then

$$A_N = D \cup (A_N A_{N+1}^c) \cup (A_{N+1} A_{N+2}^c) \cup \cdots$$

expresses A_N as the union of pairwise disjoint μ-measurable sets. Hence, from Theorem 8,

$$\begin{aligned}\mu A_N &= \mu D + \sum_{k=N}^{\infty} \mu(A_k A_{k+1}^c) \\ &= \mu D + \lim_{n\to\infty} \sum_{k=N}^{N+n} [\mu A_k - \mu A_{k+1}] \qquad \text{by Problem 7} \\ &= \mu D + \mu A_N - \lim_{n\to\infty} \mu(A_{N+n+1}).\end{aligned}$$

Since all terms are finite, the desired result follows. ∎

Theorem 11. *A sequence $\{A_n\}$ of μ-measurable sets has*

$$\mu(\underline{\lim}\, A_n) \leqslant \underline{\lim}\, \mu A_n.$$

If in addition $\mu(A_n \cup A_{n+1} \cup A_{n+2} \cup \cdots) < \infty$ for at least one $n \in J$, then

$$\mu(\overline{\lim}\, A_n) \geqslant \overline{\lim}\, \mu A_n.$$

Proof. for $\underline{\lim}$. Let

$$B_n = \bigcap_{k=n}^{\infty} A_k.$$

By definition, $\underline{\lim}\, A_n = \bigcup B_n$. Also, $\mu(\bigcup B_n) = \lim \mu B_n$ by Theorem 9. Thus

$$\mu(\underline{\lim}\, A_n) = \lim \mu B_n.$$

For each $n \in J$, $B_n \subset A_{n+k}$ and $\mu B_n \leqslant \mu A_{n+k}$ for $k = 1, 2, 3, \cdots$. Thus

$$\mu B_n \leqslant \underline{\lim}_{k\to\infty}\, \mu(A_{n+k}) = \underline{\lim}_{k\to\infty}\, \mu A_k$$

for each $n \in J$. Consequently

$$\lim_{n\to\infty} \mu B_n \leqslant \underline{\lim}_{n\to\infty}\, \mu A_n$$

and the desired inequality follows.

PROBLEM 8. Prove the second part of the theorem and show that the additional condition is necessary. ∎

Corollary. *If $\{A_n\}$ is a convergent sequence of μ-measurable sets with each $A_n \subset B$ where $\mu^* B < \infty$, then $\lim \mu A_n$ exists and*

$$\mu(\lim A_n) = \lim \mu A_n.$$

PROBLEM 9. Let S be a nonempty set of points. For each $A \subset S$, define

$$\mu^*A = \text{the number of points of } A,$$

interpreted to mean that $\mu^*\varnothing = 0$ and $\mu^*A = \infty$ if A has infinitely many points. Then μ^* is an outer measure on the space S and every subset of S is μ-measurable.

At this point in the development, the only sets in E^1 which have been proved to be Lebesgue measurable are the whole space E^1 and the sets A having $m^*A = 0$. In Section 5–1 we shall prove that all open sets in E^1 are Lebesgue measurable. This fact is not necessary for the further development of the present chapter, but does indicate that the further theory is concerned with more than a rather trivial family of sets. In fact, as soon as open sets are shown to be Lebesgue measurable, we shall know immediately that all closed sets are Lebesgue measurable (Theorem 3). Then the family of measurable sets includes all F_σ and G_δ sets (Theorem 6), and hence all countable unions, intersections, limits superior and inferior of these, and these, and these, $\cdots$ countably many times. Now, with the family of Lebesgue measurable sets so very extensive, one may wonder if perhaps all sets of E^1 are Lebesgue measurable. Lebesgue nonmeasurable sets are discussed in Section 5–2.

4–3 MEASURABLE FUNCTIONS

The definitions and theorems of this section do not depend upon the specific Lebesgue outer measures m^* (or m_2^* of Problem 4–2.3). We thus consider an abstract space S and outer measure μ^*. Examples and illustrations will, however, be given with $S = E^1$ and $\mu^* = m^*$.

In addition to the set function μ^*, we shall consider point functions having subsets of S as domains but ranges in the extended real number system E^*. Hence $-\infty$ or ∞ may be a value of a function f for some $x \in S$.

A common phraseology is to say that a function f is "defined μ-almost everywhere in a set A" if

$$\mu\{x \in A \mid f(x) \text{ is not defined}\} = 0.$$

Also, functions f and g are "equal μ-almost everywhere in A" if both are defined μ-almost everywhere in A and $f(x) = g(x)$ except for x in a subset (possibly empty) of μ-measure zero.

Definition 1. *A function f is said to be μ-**measurable** on a μ-measurable set A if f is defined μ-almost everywhere in A and*

$$\{x \in A \mid f(x) > a\}$$

is a μ-measurable set for each $a \in E^1$. In case $A = S$, we merely say that f is μ-measurable.

For example, with $S = E^1$ and $\mu = m$, the function defined by $f(x) = x^2$ is such that

$$\{x \mid f(x) > a\} = \begin{cases} (-\infty, -\sqrt{a}) \cup (\sqrt{a}, \infty) & \text{if } a \geqslant 0 \\ E^1 & \text{if } a < 0 \end{cases}.$$

Hence this function is m-measurable (considering that we shall prove in Sections 5–1 that all open sets in E^1 are m-measurable).

It follows readily that if f and g are equal μ-almost everywhere, then f is μ-measurable iff g is μ-measurable.

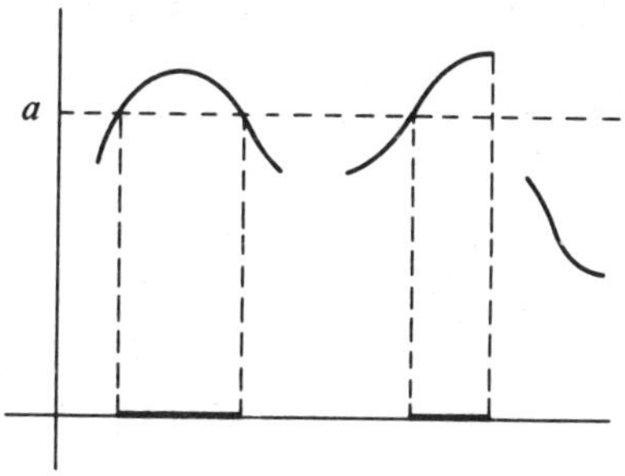

FIG. 1

Theorem 1. *For a function f, each of the conditions implies the others:*

(1) *f is μ-measurable.*

(2) *$\{x \mid f(x) \geqslant a\}$ is μ-measurable for each $a \in E^1$.*

(3) *$\{x \mid f(x) < a\}$ is μ-measurable for each $a \in E^1$.*

(4) *$\{x \mid f(x) \leqslant a\}$ is μ-measurable for each $a \in E^1$.*

(5) *$\{x \mid f(x) > r\}$ is μ-measurable for each rational r.*

Also, $>$ in (5) may be replaced by $\geqslant$, $<$, or $\leqslant$.

Proof. Let f be μ-measurable and select any $a \in E^1$. Then for each $n \in J$, the set $\{x \mid f(x) > a - (1/n)\}$ is μ-measurable. Thus

$$\{x \mid f(x) \geqslant a\} = \bigcap_{n=1}^{\infty} \left\{x \,\middle|\, f(x) > a - \frac{1}{n}\right\}$$

is a μ-measurable set (by Theorem 4–2.6). Hence (1) $\Rightarrow$(2).

Since $\{x \mid f(x) < a\} = S - \{x \mid f(x) \geqslant a\}$, then (2) $\Rightarrow$ (3).

The implications (3) $\Rightarrow$ (4) and (4) $\Rightarrow$ (1) are left as problems.

Clearly (1) $\Rightarrow$ (5). Select any $a \in E^1$ and order all rationals less than a as $r_1, r_2, r_3, \cdots$. Then

$$\{x \mid f(x) \geqslant a\} = \bigcap_{n=1}^{\infty} \{x \mid f(x) > r_n\}.$$

Thus (5) $\Rightarrow$ (2) and hence (5) implies all of the others. ∎

Theorem 2. *If f is a μ-measurable function, then*

$$\{x \mid f(x) = a\}$$

is a μ-measurable set for each $a \in E^*$, but not conversely.

Proof. $\{x \mid f(x) = a\} = \{x \mid f(x) \leqslant a\} \cap \{x \mid f(x) \geqslant a\}$ for each $a \in E^1$,

$$\{x \mid f(x) = -\infty\} = \bigcap_{n=1}^{\infty} \{x \mid f(x) < -n\},$$

and

$$\{x \mid f(x) = \infty\} = \bigcap_{n=1}^{\infty} \{x \mid f(x) > n\}.$$

To see that the converse does not hold, let $A \subset E^1$ be an m-nonmeasurable set (as given in Section 5–2) and define f by

$$f(x) = \begin{cases} x^2 & \text{if} \quad x \in A \\ -x^2 & \text{if} \quad x \in A^c. \end{cases}$$

Then for each $a \in E^1$, the set $\{x \mid f(x) = a\}$ consists of at most two points and is thus an m-measurable set. However, $\{x \mid f(x) > 0\} = A - \{0\}$ and hence f is not an m-measurable function. ∎

Theorem 3. *If f is a μ-measurable function, then $|f|$ is a μ-measurable function, but not conversely.*

Proof. Let f be a μ-measurable function. Then

$$\{x \mid |f(x)| < 0\} = \varnothing, \{x \mid |f(x)| = 0\} = \{x \mid f(x) = 0\}, \qquad \text{and}$$
$$\{x \mid |f(x)| < a\} = \{x \mid f(x) < a\} \cap \{x \mid f(x) > -a\}, \qquad a > 0.$$

Thus $|f|$ is a μ-measurable function.

PROBLEM 1. Show that the converse is not true. ∎

PROBLEM 2. If f is a μ-measurable function, then

(a) f^2 is a μ-measurable function,

(b) cf is a μ-measurable function for each $c \in E^1$, and

(c) $\{x \mid a \leqslant f(x) < b\}$ is a μ-measurable set.

Theorem 4. *If f and g are μ-measurable functions, then $f + g$ and fg are μ-measurable functions.*

Proof. It is understood that $f + g$ is defined at an $x \in S$ iff $f(x)$ and $g(x)$ are not only defined but are not opposite infinities.

Select any $a \in E^1$, let $x \in S$ be such that $f(x) + g(x)$ is defined, and note that:

(a) If there is a rational number r such that $f(x) > r$ and $g(x) > a - r$, then $f(x) + g(x) > a$.

(b) If $f(x) + g(x) > a$, then neither $f(x)$ nor $g(x)$ is $-\infty$, there is a rational number r such that $f(x) > r > a - g(x)$, and for this r both $f(x) > r$ and $g(x) > a - r$.

Thus for any $a \in E^1$,

$$\{x \mid f(x) + g(x) > a\} = \bigcup_{r \in R} [\{x \mid f(x) > r\} \cap \{x \mid g(x) > a - r\}],$$

where R is the set of all rational numbers. Hence, if f and g are μ-measurable functions, then $f + g$ is a μ-measurable function. This result, the equation

$$4f(x)g(x) = [f(x) + g(x)]^2 - [f(x) - g(x)]^2$$

and previous results (including Problem 2) show that if f and g are μ-measurable, then so is fg. ∎

Theorem 5. *If f and g are μ-measurable functions, then*

(a) $\{x \mid f(x) > g(x)\}$, $\{x \mid f(x) < g(x)\}$, and $\{x \mid f(x) = g(x)\}$ *are measurable sets.*
(b) *$\max(f, g)$ and $\min(f, g)$ are measurable functions.*

Proof. The function $f - g$ is μ-measurable by Theorem 4 and Problem 2(b). Hence

$$\{x \mid f(x) - g(x) > 0\} = \{x \mid f(x) > g(x)\}$$

is a μ-measurable set, and then μ-measurability of the other sets in (a) follows easily.

The first part of (b) follows from

$$\max(f, g) = \tfrac{1}{2}[|f - g| + f + g]$$

and the second part from

$$\min(f, g) = -\max(-f, -g). \ ∎$$

Theorem 6. *Let $\{f_n\}$ be a sequence of μ-measurable functions. Then*

(a) *$\inf\{f_n\}$, $\sup\{f_n\}$, $\underline{\lim}\, f_n$ and $\overline{\lim}\, f_n$ are μ-measurable functions.*
(b) *$L = \{x \mid \lim f_n(x)$ exists*$\}$ is a μ-measurable set.*
(c) *The function f defined by $f(x) = \lim f_n(x)$ iff $x \in L$ is μ-measurable on L.*

Proof. With $\underline{g} = \inf\{f_n\}$, then

$$\{x \mid \underline{g}(x) \geqslant a\} = \bigcap_{n=1}^{\infty} \{x \mid f_n(x) \geqslant a\}$$

and hence $\underline{g}$ is a μ-measurable function. In a similar way, $\bar{g} = \sup\{f_n\}$ is a μ-measurable function. Hence

$$\underline{g}_m = \inf_{n \geqslant m}\{f_n\} \qquad \text{and} \qquad \sup\{\underline{g}_m\} = \underline{\lim}\, f_n$$

are μ-measurable functions. Then $\overline{\lim}\, f_n = -\underline{\lim}\,(-f_n)$ is a μ-measurable function.

$$\mathrm{L} = \{x \mid \underline{\lim} f_n(x) = \overline{\lim} f_n(x)\}$$

is a μ-measurable set by Theorem 5. From

$$\{x \mid f(x) > a\} = \mathrm{L} \cap \{x \mid \overline{\lim} f_n(x) > a\}$$

it follows that f is a μ-measurable function on L. ∎

Definition 2.

(i) *Associated with each set A there is a function* χ_{A} *defined by*

$$\chi_{\mathrm{A}}(x) = \begin{cases} 1 & \text{if } x \in A \\ 0 & \text{if } x \in A^c \end{cases}$$

and is called the **characteristic function** *of A.*

(ii) *A function s having domain S and range a finite number of values in* E^1 *is called a* **step function**.

Thus s is a step function iff for some $n \in J$ there are numbers $c_1, c_2, \cdots, c_n$ and sets $A_1, A_2, \cdots, A_n$ such that†

$$s(x) = \sum_{i=1}^{n} c_i \chi_{A_i}(x) \qquad \text{with } c_i \neq c_j,\ A_i \cap A_j = \varnothing \quad \text{for } 1 \leqslant i < j \leqslant n.$$

It follows readily that a step function s is μ-measurable iff each set A_i in this expression is a μ-measurable set.

Also, if s_1 and s_2 are step functions, then $s_1 + s_2$ and $s_1 s_2$ have a finite number of values and are thus step functions. Hence, if $s_1, s_2, \cdots, s_n$ are step functions, then $\sum_{i=1}^{n} s_i$ is a step function.

The following theorem will be used in the next section on Lebesgue integration.

Theorem 7. *Let f be a nonnegative μ-measurable function. Then there is a monotonically increasing sequence* $\{s_n\}$ *of nonnegative μ-measurable step functions such that* $\lim s_n(x)$ *exists* and*

$$\lim_{n\to\infty} s_n(x) = f(x)$$

for each x in the domain of f.

Proof. With $n \in J$ let $A_n = \{x \mid f(x) \geqslant n\}$,

$$A_{n,i} = \left\{x \,\middle|\, \frac{i-1}{2^n} \leqslant f(x) < \frac{i}{2^n}\right\} \qquad \text{for } i = 1, 2, \cdots, n2^n,$$

and define s_n by

$$s_n = n\chi_{A_n} + \sum_{i=1}^{n2^n} \frac{i-1}{2^n} \chi_{A_{n,i}}.$$

Each A_n and $A_{n,i}$ is a measurable set and thus s_n is a measurable function.

† In some books, "step function" is used only if $A_1, A_2, \cdots, A_n$ are intervals, and "simple functions" or "finitely valued functions" is used for what are called step functions here.

If x is such that $f(x) = \infty$, then

$$s_n(x) = n < n + 1 = s_{n+1}(x) \qquad \text{and} \qquad s_n(x) \to \infty.$$

If x is such that $f(x) < \infty$, then for $n > f(x)$,

$$f(x) - 2^{-n} \leqslant s_n(x) \leqslant s_{n+1}(x) < f(x)$$

so that $\{s_n(x)\}$ is monotonically increasing and $s_n(x) \to f(x)$. ∎

In case $S = E^1$ and $f(x) = x^{-2}$, the sequence $\{s_n\}$ of step functions associated with f in the above proof is such that $\lim s_n$ is not a step function.

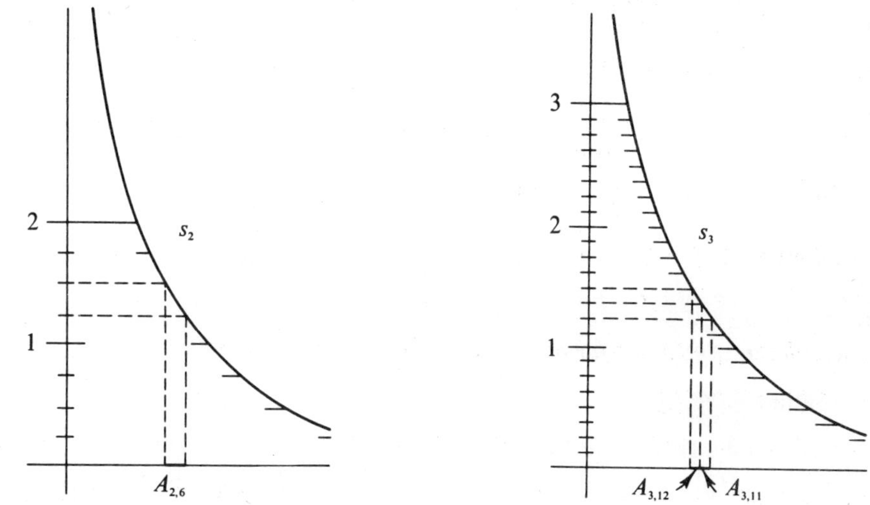

FIG. 2

4–4 RINGS AND ADDITIVITY

In Section 4–2 we obtained some properties of the family $\mathfrak{M}$ of μ-measurable sets. A modicum of these properties is abstracted in the following definitions. These definitions focus attention on properties sufficient for, and which facilitate, proofs of some theorems on Lebesgue integrals as defined in the next section.

Definition 1. *A nonempty family $\mathscr{R}$ of sets is said to be*

(i) *A* **ring** *if whenever $A \in \mathscr{R}$ and $B \in \mathscr{R}$, then also*

$$A \cup B \in \mathscr{R} \qquad \text{and} \qquad A - B \in \mathscr{R}.$$

(ii) *A* **σ-ring** *if, in addition, whenever $A_n \in \mathscr{R}$, $n = 1, 2, 3, \cdots$, then also*

$$\bigcup_{n=1}^{\infty} A_n \in \mathscr{R}.$$

Even if $\mathscr{R}$ is a σ-ring of sets in a space S, the set S itself might not be a member of $\mathscr{R}$.

The following observations are simple theorems; those without accompanying reasons are considered as problems.

1. $\varnothing \in \mathscr{R}$ for any ring $\mathscr{R}$, since $A \in \mathscr{R} \Rightarrow (A - A) \in \mathscr{R}$.
2. If $\mathscr{R}$ is a ring, then

$$(A \in \mathscr{R}, B \in \mathscr{R}) \Rightarrow A \cap B \in \mathscr{R},$$

since $A \cap B = A - (A - B)$.

3. If $\mathscr{R}$ is a ring and $A_k \in \mathscr{R}$ for $k = 1, 2, \cdots, n$, then

$$\bigcup_{k=1}^{n} A_k \in \mathscr{R} \qquad \text{and} \qquad \bigcap_{k=1}^{n} A_k \in \mathscr{R}.$$

4. If $\mathscr{R}$ is a σ-ring and $A_n \in \mathscr{R}$ for $n = 1, 2, 3, \cdots$, then

(a) $$\bigcap_{n=1}^{\infty} A_n = A_1 - \bigcup_{n=1}^{\infty} (A_1 - A_n) \in \mathscr{R}.$$

(b) $$\underline{\lim}\, A_n \in \mathscr{R} \qquad \text{and} \qquad \overline{\lim}\, A_n \in \mathscr{R}.$$

Definition 2. *Let $\mathscr{R}$ be a ring and ν a function such that either*

$$\nu: \mathscr{R} \to (-\infty, \infty] \qquad \textit{or} \qquad \nu: \mathscr{R} \to [-\infty, \infty)$$

and such that $|\nu(A)| < \infty$ for at least one $A \in \mathscr{R}$. Then ν is said to be

(i) **Additive** *on $\mathscr{R}$ if*

$$(A \in \mathscr{R}, B \in \mathscr{R}, A \cap B = \varnothing) \Rightarrow \nu(A \cup B) = \nu(A) + \nu(B).$$

(ii) **Countably additive** *on $\mathscr{R}$ if whenever $\{A_n\}$ is a pairwise disjoint sequence of members of $\mathscr{R}$, and $\bigcup A_n \in \mathscr{R}$ then*

$$\nu\left(\bigcup_{n=1}^{\infty} A_n\right) = \sum_{n=1}^{\infty} \nu(A_n).$$

[*Note:* In (ii) the ring $\mathscr{R}$ need not be a σ-ring and in case it is not a σ-ring the equality is meaningless unless the condition "$\bigcup A_n \in \mathscr{R}$" is satisfied.]

Observations

5. If ν is additive on a ring $\mathscr{R}$, then
   (a) $\nu(\varnothing) = 0$.
   (b) $(A \in \mathscr{R}, B \in \mathscr{R}, A \subset B, |\nu(A)| < \infty) \Rightarrow \nu(B - A) = \nu(B) - \nu(A)$.
   (c) If $B_k \in \mathscr{R}$ for $k = 1, 2, \cdots, n$ and $B_i \cap B_j = \varnothing$ for $1 \leqslant i < j \leqslant n$, then

$$\nu\left(\bigcup_{k=1}^{n} B_k\right) = \sum_{k=1}^{n} \nu(B_k).$$

6. If ν is additive on a ring $\mathscr{R}$ and $\nu(A) \geqslant 0$ for each $A \in \mathscr{R}$, then

$$(A \in \mathscr{R}, B \in \mathscr{R}, A \subset B) \Rightarrow \nu(A) \leqslant \nu(B).$$

7. If the conditions of (ii) are satisfied and if the series there
   (a) Converges, then it converges absolutely.
   (b) Diverges, then it diverges
   to ∞ if $v: \mathscr{R} \to (-\infty, \infty]$;
   to $-\infty$ if $v: \mathscr{R} \to [-\infty, \infty)$.

Theorem 1. *Let v be countably additive on a ring $\mathscr{R}$ and let*

$$A_1 \subset A_2 \subset \cdots \subset A_n \subset \cdots \qquad \text{with each } A_n \in \mathscr{R}$$

be such that also $\bigcup A_n \in \mathscr{R}$. Then

$$v\left(\bigcup_{n=1}^{\infty} A_n\right) = \lim_{n\to\infty} v(A_n).$$

Proof. Define $B_1 = A_1$ and $B_n = A_n - A_{n-1}$ for $n = 2, 3, \cdots$. Then $\{B_n\}$ is a pairwise disjoint sequence of members of $\mathscr{R}$ such that $\bigcup_{n=1}^{\infty} B_n = \bigcup_{n=1}^{\infty} A_n$. Hence

$$\begin{aligned} v\left(\bigcup_{n=1}^{\infty} A_n\right) &= \sum_{n=1}^{\infty} v(B_n) \qquad \text{by (ii)} \\ &= \lim_{n\to\infty} \sum_{k=1}^{n} v(B_k) \\ &= \lim_{n\to\infty} v\left(\bigcup_{k=1}^{n} B_k\right) \qquad \text{from 5(c)} \\ &= \lim_{n\to\infty} v(A_n). \ \blacksquare \end{aligned}$$

4–5 LEBESGUE INTEGRATION

Henri Lebesgue defined outer measure m^* for sets in E^1 and then measurability of sets and functions, in order to introduce a concept of integration applicable to a wider class of functions than in Riemann integration.† Any outer measure may, however, be used instead of the specific m^* in this process of integration. Hence the integrals defined in this section are called "Lebesgue integrals" even though an abstract outer measure μ^* on an abstract space S is used.

It is customary to use "μ-almost everywhere" to mean "except possibly for a set of μ-measure zero."

Definition 1. *Let A be a μ-measurable set.*

(i) *A nonnegative μ-measurable step function*

$$s = \sum_{i=1}^{n} c_i \chi_{B_i}, \ B_i \ \mu\text{-measurable}, \ B_i B_j = \varnothing, \ 1 \leqslant i < j \leqslant n, \ c_i \geqslant 0$$

† Riemann integration is given in Section 8–3. We have thus not followed the historical development of integration.

is assigned the unique element $\mathscr{I}(s, A)$ *of* E^*, *called the* **Lebesgue sum** *of* s *over* A, *by setting*

$$\mathscr{I}(s, A) = \sum_{i=1}^{n} c_i \mu(AB_i),$$

where 0 *is used for the* ith *term if* $c_i = 0$ *even if* $\mu(AB_i) = \infty$. (See the definition of $0 \cdot \infty$ on p. 119.)

(ii) *Let* f *be a nonnegative* μ-*measurable function defined* μ-*almost everywhere in* A. *The* **Lebesgue integral** $\int_A f\, d\mu$ *of* f *over* A *with respect to* μ *is the unique element of* E^* *defined by*

$$\int_A f\, d\mu = \sup \mathscr{I}(s, A),$$

where the supremum is taken over all μ-*measurable step functions* s *such that* $0 \leqslant s \leqslant f$ *on* A.

(iii) *If* f *is a nonpositive* μ-*measurable function defined* μ-*almost everywhere in* A, *then*

$$\int_A f\, d\mu = -\int_A (-f)\, d\mu.$$

PROBLEM 1. Let f be the function defined by

$$f(x) = \begin{cases} 0 & \text{if } x \text{ is rational} \\ 1 & \text{if } x \text{ is irrational} \end{cases}$$

and let $A \subset E^1$ be m-measurable. Show that

$$\int_A f\, dm = mA.$$

PROBLEM 2. If $f \geqslant 0$ and $=0$ at m-almost all points of an m-measurable set A, then

$$\int_A f\, dm = 0.$$

PROBLEM 3. If $f \geqslant 0$, if f is m-measurable on an m-measurable A, and if $\int_A f\, dm = 0$, then $f = 0$ at m-almost all points of A.

A function f is said to have **positive part** f^+ and **negative part** f^-, where for each x in the domain of f,

$$f^+(x) = \begin{cases} f(x) & \text{if } f(x) \geqslant 0 \\ 0 & \text{if } f(x) < 0, \end{cases}$$

$$f^-(x) = \begin{cases} 0 & \text{if } f(x) > 0 \\ -f(x) & \text{if } f(x) \leqslant 0. \end{cases}$$

Hence $f^+ \geqslant 0, f^- \geqslant 0, |f| = f^+ + f^-$, and $f(x) = f^+(x) - f^-(x)$ for all x in the domain of f for which at least one of $f^+(x)$ or $f^-(x)$ is not $+\infty$. Also it follows that f is μ-measurable iff f^+ and f^- are both μ-measurable.

Definition 2. *Let f be a μ-measurable function, let A be a μ-measurable set on which f is defined μ-almost everywhere, and consider the integrals*

$$(1) \qquad \int_A f^+ \, d\mu \quad \text{and} \quad \int_A f^- \, d\mu,$$

(which exist* by Definition 1).

(i) *If at least one of these integrals is finite, then f is said to be* **Lebesgue integrable** *with* **Lebesgue integral** *defined by*

$$\int_A f \, d\mu = \int_A f^+ \, d\mu - \int_A f^- \, d\mu.$$

(ii) *If both integrals in* (1) *are finite, then $\int_A f \, d\mu$ is finite and f is said to be* **summable** *over A with respect to μ.*

(iii) *The set of all functions summable over A with respect to μ is denoted by*

$$\mathscr{L}(A, \mu).$$

[*Note:* If f is summable over A, then both $\int_A f^+ \, d\mu$ and $\int_A f^- \, d\mu$ are nonnegative finite numbers, and presently we shall see that

$$\int_A |f| \, d\mu = \int_A (f^+ + f^-) \, d\mu = \int_A f^+ \, d\mu + \int_A f^- \, d\mu < \infty.$$

Thus, in conformity with series terminology, it would be better to say that f is "absolutely summable."]

From the above definition it is seen that sets of μ-measure zero may be neglected in all considerations of Lebesgue integrals. Also, if f is μ-measurable and

$$A_0 = \{x \mid f(x) = 0\}, \qquad \text{then} \int_{A_0} f \, d\mu = 0.$$

We shall therefore avoid much repetition of nonessential conditions by stating once and for all that henceforth:

The domains of all functions are extended to all of S by redefining (*if necessary*) *any function to be zero at all points of the complement of its original domain.*

Now that the definitions are given, we shall shorten "μ-measurable" to "measurable," $\mathscr{L}(A, \mu)$ to $\mathscr{L}(A)$, and

$$\int_A f \, d\mu \quad \text{to} \quad \int_A f.$$

With s and t nonnegative measurable step functions, f and g nonnegative measurable functions, and A and B measurable sets, then each of the following statements is a simple consequence of Definition 1:

1. $\int_A s = \mathscr{I}(s, A)$.
2. $s + t$ is a nonnegative measurable step function and
$$\int_A (s+t) = \int_A s + \int_A t.$$
3. If $g \leqslant f$, then $0 \leqslant \int_A g \leqslant \int_A f$. (Recall that $0 \leqslant g$.)
4. If $B \subset A$, then $0 \leqslant \int_B f \leqslant \int_A f$.
5. If $0 < c < \infty$, then $\int_A cf = c \int_A f$.

The family of measurable sets will be denoted by $\mathfrak{M}$. Hence (see Section 4–2) $\mathfrak{M}$ is a σ-ring which, moreover, has the whole space S as one of its elements. The next four theorems use only Definition 1.

Theorem 1. *Let f be a nonnegative measurable function. For each set $M \in \mathfrak{M}$, define*

$$v_f(M) = \int_M f.$$

Then the set function v_f is countably additive on $\mathfrak{M}$.

Proof. Take any pairwise disjoint sequence $\{A_k\}$ of measurable sets and let $A = \bigcup A_k$. We then wish to prove that $\lim_{n\to\infty} \sum_{k=1}^{n} v_f(A_k)$ exists* and is equal to $v_f(A)$; that is, we wish to establish the equality

$$\lim_{n\to\infty} \sum_{k=1}^{n} \int_{A_k} f = \int_A f. \tag{2}$$

First consider the case in which f is a nonnegative step function s and note that

$$v_s(M) = \int_M s = \mathscr{I}(s, M), \qquad M \in \mathfrak{M}.$$

With s expressed as in (i) of Definition 1,

$$\begin{aligned} v_s(A) &= \mathscr{I}\left(s, \bigcup_{k=1}^{\infty} A_k\right) \\ &= \sum_{i=1}^{n} c_i \mu\left[B_i\left(\bigcup_{k=1}^{\infty} A_k\right)\right] \\ &= \sum_{i=1}^{n} c_i \mu\left[\bigcup_{k=1}^{\infty} (A_k B_i)\right] \\ &= \sum_{i=1}^{n} c_i \left[\sum_{k=1}^{\infty} \mu(A_k B_i)\right], \end{aligned}$$

since μ is countably additive on the σ-ring $\mathfrak{M}$. The summations may be interchanged, since all terms are nonnegative:

$$v_s(A) = \sum_{k=1}^{\infty} \left[\sum_{i=1}^{n} c_i \mu(A_k B_i) \right]$$

$$= \sum_{k=1}^{\infty} \mathscr{I}(s, A_k) = \sum_{i=1}^{\infty} v_s(A_k).$$

This shows that v_s is countably additive on $\mathfrak{M}$ for each nonnegative measurable step function s.

Hence, for every measurable step function s such that $0 \leqslant s \leqslant f$, we have

$$\mathscr{I}(s, A) = \sum_{k=1}^{\infty} \mathscr{I}(s, A_k) \leqslant \sum_{k=1}^{\infty} \int_{A_k} f.$$

Therefore, by taking the supremum for all such s,

(3) $$\int_A f \leqslant \sum_{k=1}^{\infty} \int_{A_k} f; \qquad \text{that is, } v_f(A) \leqslant \sum_{k=1}^{\infty} v_f(A_k).$$

If $v_f(A_n) = \infty$ for some n, then the equality holds. since

$$v_f(A) = \int_A f \geqslant \int_{A_n} f = v_f(A_n) = \infty = \sum_{k=1}^{\infty} v_s(A_k).$$

Hence consider $v_f(A_k) < \infty$, $k = 1, 2, 3, \cdots$. Let $\varepsilon > 0$ be arbitrary and select measurable step functions s_1 and s_2 such that

$$0 \leqslant s_1 \leqslant f \text{ on } A_1, \qquad 0 \leqslant s_2 \leqslant f \text{ on } A_2,$$

$$\mathscr{I}(s_1, A_1) \geqslant \int_{A_1} f - \varepsilon, \qquad \text{and} \qquad \mathscr{I}(s_2, A_2) \geqslant \int_{A} f - \varepsilon.$$

Since $A_1 A_2 = \varnothing$, then $s = \chi_{A_1} s_1 + \chi_{A_2} s_2$ is a measurable step function such that $0 \leqslant s \leqslant f$ on $A_1 \cup A_2$. Hence

$$\mathscr{I}(s_1, A_1) + \mathscr{I}(s_2, A_2) = \mathscr{I}(s, A_1) + \mathscr{I}(s, A_2)$$

$$= \mathscr{I}(s, A_1 \cup A_2) \leqslant \int_{A_1 \cup A_2} f.$$

Thus, in terms of v_f,

$$v_f(A_1 \cup A_2) \geqslant v_f(A_1) + v_f(A_2) - 2\varepsilon.$$

Hence $v_f(A_1 \cup A_2) \geqslant v_f(A_1) + v_f(A_2)$ and then, for any integer n,

$$v_f\left(\bigcup_{k=1}^{n} A_k\right) \geqslant \sum_{k=1}^{n} v_f(A_k).$$

From statement 4, written in terms of v_f, we have

$$v_f(A) \geqslant v_f\left(\bigcup_{k=1}^{n} A_k\right) \geqslant \sum_{k=1}^{n} v_f(A_k).$$

Consequently

$$\nu_f(A) \geqslant \sum_{k=1}^{\infty} \nu_f(A_k),$$

which with (3) shows that ν_f is countably additive on $\mathfrak{M}$. ∎

Corollary. *If f is a nonnegative measurable function, if*

$$A_1 \subset A_2 \subset A_3 \subset \cdots \qquad \text{with each } A_n \in \mathfrak{M}$$

and we set $A = \bigcup A_n$, then

$$\int_A f = \lim_{n\to\infty} \int_{A_n} f.$$

Proof. Since ν_f is countably additive on $\mathfrak{M}$, this corollary is a special case of Theorem 4–4.1. ∎

It is frequently desirable to interchange limits and integrals. Hence the next result, known as "Lebesgue's monotone convergence theorem," is used in many practical and theoretical situations.

Theorem 2. *Let A be a measurable set and let $\{f_n\}$ be a sequence of measurable functions such that*

$$0 \leqslant f_1(x) \leqslant f_2(x) \leqslant f_3(x) \leqslant \cdots \qquad \text{for each } x \in A. \tag{4}$$

Then

$$\lim_{n\to\infty} \int_A f_n = \int_A \left(\lim_{n\to\infty} f_n\right).$$

Proof. We shall consider that (4) holds for all $x \in S$ (for if it did not, we would replace f_n by $g_n = \chi_A f_n$). Thus $\lim f_n$ exists* and we define

$$f(x) = \lim_{n\to\infty} f_n(x).$$

Hence f is a measurable function (by Theorem 4–3.6(c)), and since it is nonnegative, $\int_A f$ is defined. Also $\lim \int_A f_n$ exists*, since $\int_A f_n \leqslant \int_A f_{n+1} \leqslant \int_A f$ from (4) and Statement 3. We set

$$\alpha = \lim_{n\to\infty} \int_A f_n \leqslant \int_A f. \tag{5}$$

Thus we need only to prove that $\alpha \geqslant \int_A f$.

Take any measurable step function s satisfying $0 \leqslant s \leqslant f$. Choose any number λ such that $0 < \lambda < 1$ and let

$$A_n = A \cap \{x \mid f_n(x) \geqslant \lambda s(x)\}, \qquad n = 1, 2, 3, \cdots.$$

Each A_n is a measurable set (by Theorem 4–3.5),

$$A_1 \subset A_2 \subset A_3 \subset \cdots, \qquad \text{and} \qquad A = \bigcup_{n=1}^{\infty} A_n.$$

Hence, for each n,

$$\int_A f_n \geqslant \int_{A_n} f_n \geqslant \lambda \int_{A_n} s.$$

As $n \to \infty$, the last term approaches $\lambda \int_A s$, by the corollary of Theorem 1, and the first term approaches α. Hence

$$\alpha \geqslant \lambda \int_A s$$

for each λ such that $0 < \lambda < 1$, and thus

$$\alpha \geqslant \int_A s = \mathscr{I}(s, A).$$

By taking the supremum over all measurable s with $0 \leqslant s \leqslant f$, we have

$$\alpha \geqslant \int_A f$$

and therefore, as noted under (5), the proof is complete. ∎

Corollary. *If $f \geqslant 0$ and $g \geqslant 0$ are measurable functions, then*

$$\int_A (f+g) = \int_A f + \int_A g.$$

Proof. By Theorem 4–3.7, let $\{s_n\}$ and $\{t_n\}$ be monotonically increasing sequences of nonnegative step functions such that $s_n \to f$ and $t_n \to g$. Then $\{s_n + t_n\}$ is the same type of sequence, $(s_n + t_n) \to (f + g)$, and the result follows from the theorem since

$$\int_A (s_n + t_n) = \int_A s_n + \int_A t_n, \qquad \text{by 2, p. 202.}$$ ∎

Theorem 3. (Fatou's Lemma). *Let A be a measurable set and let $\{f_n\}$ be a sequence of nonnegative measurable functions. Then*

$$\int_A \left(\varliminf_{n\to\infty} f_n \right) \leqslant \varliminf_{n\to\infty} \int_A f_n.$$

Proof. The functions g_m defined by

$$g_m = \inf_{n \geqslant m} f_n,\ m = 1, 2, 3, \cdots, \quad \text{and} \quad \varliminf_{n\to\infty} f_n = \sup g_m = \lim_{m\to\infty} g_m$$

are measurable (by Theorem 4–3.6), and since each is nonnegative,

$$\int_A g_m \quad \text{and} \quad \int_A \varliminf_{n\to\infty} f_n = \int_A \lim_{m\to\infty} g_m$$

are defined. Also, $\{g_m\}$ is a nonnegative monotonically increasing sequence and

$$\lim_{m\to\infty} \int_A g_m = \int_A \left(\lim_{m\to\infty} g_m \right)$$

by Theorem 2. But $\int_A g_n \leqslant \int_A f_n$, since $0 \leqslant g_n \leqslant f_n$, and thus

$$\lim_{n\to\infty} \int_A g_n = \varliminf_{n\to\infty} \int_A g_n \leqslant \varliminf_{n\to\infty} \int_A f_n.$$

Hence the conclusion of the theorem is seen to follow. ∎

PROBLEM 3. If $f_n = \begin{Bmatrix} \chi_{(0,\,2)} & \text{for } n \text{ odd} \\ \chi_{(1,\,3)} & \text{for } n \text{ even} \end{Bmatrix}$, then $\int (\varliminf f_n) < \varliminf \int f_n$.

Theorem 4. *Given a sequence $\{f_n\}$ of nonnegative measurable functions and a measurable set A, then*

$$\int_A \sum_{n=1}^{\infty} f_n = \sum_{n=1}^{\infty} \int_A f_n.$$

Proof. Apply Theorem 2 and its corollary to the monotonically increasing sequence of partial sums of $\sum f_n$. ∎

The next theorem is the first one in which Definition 2 is used.

Theorem 5. *Let f be a measurable function and A a measurable set.*

(a) *Then $\int_A |f|$ exists* and $\int_A |f| = \int_A f^+ + \int_A f^-$.*

(b) *If $\int_A f$ exists* [and in particular if $f \in \mathscr{L}(A)$], then*

$$\left|\int_A f\right| \leqslant \int_A |f|.$$

Proof. Only the measurability of f implies that $\int_A f^+ \geqslant 0$ and $\int_A f^- \geqslant 0$ are defined and thus that

$$\int_A |f| = \int_A (f^+ + f^-) = \int_A f^+ + \int_A f^-.$$

Under the additional condition that $\int_A f$ exists*, then (according to Definition 2) both $\int_A f^+ \pm \int_A f^-$ have meaning, and (b) follows from

$$\int_A f = \int_A f^+ - \int_A f^- \leqslant \int_A f^+ + \int_A f^- = \int_A |f| \qquad \text{and}$$

$$-\int_A f = -\int_A f^+ + \int_A f^- \leqslant \int_A f^+ + \int_A f^- = \int_A |f|. \quad ∎$$

The following theorem is analogous to the comparison test for (absolute) convergence of series.

Theorem 6. (Dominated Summability) *Let $A \in \mathfrak{M}$ and let f be a measurable function on A. If there is a function $g \in \mathscr{L}(A)$ such that*

$$|f(x)| \leqslant g(x) \qquad \textit{for almost all } x \in A,$$

then $f \in \mathscr{L}(A)$.

Proof. For such a function g,

$$\left|\int_A f\right| \leqslant \int_A |f| \leqslant \int_A g < \infty. \quad \blacksquare$$

Corresponding to each measurable function f we denote by $\mathfrak{M}_f$ the family of measurable sets A over which f is summable, that is, for which $f \in \mathscr{L}(A)$. $\mathfrak{M}_f$ is not an empty family, since any set of measure zero belongs to it.

Theorem 7. *Given a measurable function f, then*

(a) *$\mathfrak{M}_f$ is a ring.*
(b) *The set function ν_f defined for each $M \in \mathfrak{M}_f$ by*

$$\nu_f(M) = \int_M f$$

is σ-additive on $\mathfrak{M}_f$; that is, if $\{A_n\}$ is a pairwise disjoint sequence of members of $\mathfrak{M}_f$ such that also $A = \bigcup A_n \in \mathfrak{M}_f$, then

$$\int_A f = \sum_{n=1}^{\infty} \int_{A_n} f,$$

with the series converging absolutely.

Proof. Since $f^+ \geqslant 0$ and $f^- \geqslant 0$, the set functions ν_{f^+} and ν_{f^-} defined in Theorem 1 are countably additive on the σ-ring $\mathfrak{M}$ and hence on the subfamily $\mathfrak{M}_f$ of $\mathfrak{M}$. Take any $A \in \mathfrak{M}_f$ and $B \in \mathfrak{M}_f$. Then

$$0 \leqslant \nu_{f^+}(A) < \infty, \qquad 0 \leqslant \nu_{f^+}(B) < \infty,$$
$$0 \leqslant \nu_{f^+}(A - B) \leqslant \nu_{f^+}(A) < \infty,$$

and

$$\begin{aligned} 0 \leqslant \nu_{f^+}(A \cup B) &= \nu_{f^+}[B \cup (A - B)] \\ &\leqslant \nu_{f^+}(B) + \nu_{f^+}(A - B) \\ &\leqslant \nu_{f^+}(B) + \nu_{f^+}(A) < \infty \end{aligned}$$

by using observation 7 of Section 4–4. In the same way,

$$0 \leqslant \nu_{f^-}(A - B) < \infty \qquad \text{and} \qquad 0 \leqslant \nu_{f^-}(A \cup B) < \infty.$$

Thus $A - B \in \mathfrak{M}_f$ and $A \cup B \in \mathfrak{M}_f$.

Hence $\mathfrak{M}_f$ is a ring. [But $\mathfrak{M}_f$ need not be a σ-ring. For example, if $f(x) = 1$ for all $x \in E^1$ and $\mu = m$, then each interval $I_n = (-n, n) \in \mathfrak{M}_f$, but $\bigcup I_n = E^1 \notin \mathfrak{M}_f$.]

In order to establish (b), take a sequence $\{A_k\}$ as described in the theorem. Then

$$\nu_{f^+}(A) = \sum_{n=1}^{\infty} \nu_{f^+}(A_n),$$

from Theorem 1, but we know that the series converges to the finite value $\nu_{f^+}(A)$, since $A \in \mathfrak{M}_f$. The series, moreover, converges absolutely since any rearrangement of $\{A_n\}$ gives the same result. Since the same holds with f^+ replaced by f^-, we have

$$\begin{aligned}
\nu_f(A) &= \int_A f = \int_A f^+ - \int_A f^- = \nu_{f^+}(A) - \nu_{f^-}(A) \\
&= \sum_{n=1}^{\infty} \nu_{f^+}(A_n) - \sum_{n=1}^{\infty} \nu_{f^-}(A_n) = \sum_{n=1}^{\infty} [\nu_{f^+}(A_n) - \nu_{f^-}(A_n)] \\
&= \sum_{n=1}^{\infty} \left[\int_{A_n} f^+ - \int_{A_n} f^-\right] = \sum_{n=1}^{\infty} \int_{A_n} f = \sum_{n=1}^{\infty} \nu_f(A_n),
\end{aligned}$$

and moreover, by Theorem 3–9.2, the series converges absolutely. ∎

Theorem 8. *If $f \in \mathscr{L}(A)$ and $g \in \mathscr{L}(A)$, then $f + g \in \mathscr{L}(A)$ and*

$$\int_A (f + g) = \int_A f + \int_A g. \tag{6}$$

Proof. We first divide A into four sets:

A^{++} on which $f \geqslant 0$, $g \geqslant 0$; A^{+-} on which $f \geqslant 0$, $g < 0$;
A^{-+} on which $f < 0$, $g \geqslant 0$; A^{--} on which $f < 0$, $g < 0$.

All four sets are known to be measurable. With B any one of the sets,

$$\int_B f^+, \quad \int_B f^-, \quad \int_B g^+, \quad \text{and} \quad \int_B g^-$$

are all finite, since each is $\geqslant 0$ and $\leqslant$ the integral over A of the corresponding function. Hence $f + g \in \mathscr{L}(A^{++})$, since

$$(f + g)^+ = f^+ + g^+ = f + g \qquad \text{on } A^{++}.$$

Also (6), with A^{++} in place of A, holds by the corollary on p. 205, and similarly for A^{--}. Next consider A^{+-}. Let

$$B_1 = A^{+-} \cap \{x \mid f(x) \geqslant -g(x)\} \qquad \text{and} \qquad B_2 = A^{+-} - B_1.$$

Then f, $f + g$, and $-g$ are all nonnegative on B_1 and hence the Lebesgue integral of each exists* on B_1. Thus,

$$\begin{aligned}
\int_{B_1} f &= \int_{B_1} [(f + g) + (-g)] = \int_{B_1} (f + g) + \int_{B_1} (-g) \\
&= \int_{B_1} (f + g) - \int_{B_1} g \qquad \text{by Definition 4-5.1 (iii).}
\end{aligned}$$

But $\int_{B_1} g$ and $\int_{B_1} f$ are finite so that (6) holds with A replaced by B_1. Similarly for B_2, and thus $f+g \in \mathscr{L}(A^{+-})$, with the desired equality of integrals over A^{+-} Now interchange f and g to obtain the same result for A^{-+}. Hence, by Theorem 7, $f+g \in \mathscr{L}(A)$ and the equality (6) holds. ∎

The next theorem is known as Lebesgue's dominated convergence principle.

Theorem 9. *Let $A \in \mathfrak{M}$, let $\{f_n\}$ be a sequence of measurable functions converging on A, and set*

$$\lim_{n\to\infty} f_n(x) = f(x), \qquad x \in A.$$

If there is a function $g \in \mathscr{L}(A)$ such that

$$(7) \qquad |f_n(x)| \leqslant g(x), \qquad x \in A,\ n = 1, 2, 3, \cdots,$$

then each $f_n \in \mathscr{L}(A), f \in \mathscr{L}(A)$, $\left\{\int_A f_n\right\}$ converges, and

$$(8) \qquad \lim_{n\to\infty} \int_A f_n = \int_A f.$$

[*Note:* Since a countable union of sets of measure zero has measure zero, it would be sufficient to require that each f_n be defined almost everywhere in A and that $\{f_n\}$ converge almost everywhere in A.]

Proof. Since (7) holds, then each $f_n \in \mathscr{L}(A)$, by Theorem 6. Also, $f \in \mathscr{L}(A)$, since

$$0 \leqslant \begin{Bmatrix} f^-(x) \\ f^+(x) \end{Bmatrix} \leqslant |f(x)| = \lim_{n\to\infty} |f_n(x)| \leqslant g(x).$$

Hence $f+g \in \mathscr{L}(A)$, and by using Theorem 8,

$$\int_A f + \int_A g = \int_A (f+g) = \int_A \lim(f_n + g) = \int_A \underline{\lim}(f_n + g).$$

Theorem 3 (Fatou's lemma) may be applied to the sequence $\{f_n + g\}$, since $f_n + g \geqslant 0$ on A, and yields

$$\int_A \underline{\lim}(f_n + g) \leqslant \underline{\lim} \int_A (f_n + g) = \underline{\lim} \int_A f_n + \int_A g.$$

Therefore, since $0 \leqslant \int_A g < \infty$,

$$\int_A f \leqslant \underline{\lim} \int_A f_n.$$

Also, $g - f \in \mathscr{L}(A)$ and Fatou's lemma may be applied to the nonnegative sequence $\{g - f_n\}$ to yield

$$-\int_A f \leqslant \underline{\lim}\left[-\int_A f_n\right]$$

from which, by Theorem 3–2.2,

$$\int_A f \geqslant \overline{\lim} \int_A f_n .$$

Thus the limit in (8) exists and the equality holds. ∎

Theorem 10. *Let* $A \in \mathfrak{M}$ *and* $f \in \mathscr{L}(A)$. *Then corresponding to each given* $\varepsilon > 0$ *there is a* $\delta > 0$ *such that*

$$(B \in \mathfrak{M},\ B \subset A,\ \mu B < \delta) \Rightarrow \int_B |f| < \varepsilon.$$

Proof. Set $g = |f|$ and for each $x \in A$ define

$$g_n(x) = \begin{Bmatrix} g(x) & \text{if } g(x) \leqslant n \\ n & \text{if } g(x) > n \end{Bmatrix}, \qquad n = 1, 2, 3, \cdots.$$

Then g and each g_n is summable over A, with

$$0 \leq g_1(x) \leqslant g_2(x) \leqslant g_3(x) \leqslant \cdots \qquad \text{and} \qquad \lim_{n\to\infty} g_n(x) = g(x)$$

for each $x \in A$. Hence, from Theorem 2,

$$\lim_{n\to\infty} \int_A g_n = \int_A g \qquad \left(= \int_A |f| < \infty \right).$$

Let N be a positive integer such that

$$\int_A g - \int_A g_N < \frac{\varepsilon}{2} \qquad \text{and hence} \qquad 0 \leqslant \int_A (g - g_N) < \frac{\varepsilon}{2}.$$

Choose $\delta = \varepsilon/2N$. Hence, for any measurable set $B \subset A$ with $\mu B < \delta$, we have

$$\int_B g_N \leqslant \int_B N = N\mu(B) < \frac{\varepsilon}{2}$$

and therefore

$$\int_B g = \int_B (g - g_N) + \int_B g_N \leqslant \int_A (g - g_N) + \frac{\varepsilon}{2} < \varepsilon. \ ∎$$

In the following problem, "almost everywhere" is abbreviated to a.e.

PROBLEM 4. (a) If $A \subset E^1$ is a bounded m-measurable set and f is a measurable function such that

$$c \leqslant f(x) \leqslant d \text{ a.e., then } c\, mA \leqslant \int_A f\, dm \leqslant d\, mA.$$

(b) If $f(x) \geqslant 0$ a.e. and $\int_{E^1} f\, dm = 0, \qquad$ then $f(x) = 0$ a.e.

[*Hint:* $\{x \mid f(x) > 0\} = \bigcup \{x \mid f(x) \geqslant 1/n\}$.]

(c) If $A \subset E^1$ is a measurable set and $f: A \to E^1$ is a measurable function such that

$$-\infty < c \leqslant \frac{1}{mB} \int_B f\, dm \leqslant d < \infty$$

for all measurable sets $B \subset A$ having $0 < mB < \infty$, then $c \leqslant f(x) \leqslant d$ for almost all $x \in A$.

[*Hint:* Suppose $m\{x \in A \mid f(x) < c\} > 0$. Then there is a positive integer n and a measurable set $B_n \subset A$ such that $0 < mB_n < \infty$ and $f(x) < c - 1/n$ for all $x \in B_n$, etc.]

4–6 REAL $\mathscr{L}^2$ SPACES

Definition 1. *Given an m_k-measurable set $A \subset E^k$, then $\mathscr{L}^2(A, m_k)$ denotes the family of all functions $f: A \to E^*$ satisfying*

(i) *f is defined m_k-almost everywhere in A and is m_k-measurable on A.*
(ii) *$f^2 \in \mathscr{L}(A, m_k)$; that is $\int_A |f(x)|^2\, dm_k < \infty$.*

Throughout the remainder of this section we take a fixed m_k-measurable set $A \subset E^k$ and shorten

$$\int_A f(x)\, dm_k \quad \text{to} \quad \int f,$$

$$\mathscr{L}^2(A, m_k) \quad \text{to} \quad \mathscr{L}^2,$$

m_k-almost everywhere in A to a.e., and

$f \in \mathscr{L}^2$ and $g \in \mathscr{L}^2$ to $f, g \in \mathscr{L}^2$.

Theorem 1. *Given $f, g \in \mathscr{L}^2$ and $a, b \in E^1$, then*

$$af + bg \in \mathscr{L}^2.$$

Proof. Since $\int |f|^2 < \infty$, then $|f(x)| < \infty$ a.e. and the same is true of $|g(x)|$. Thus $af(x) + bg(x)$ is defined a.e. and the m_k-measurability of $af + bg$ follows from Theorem 4–3.4 and Problem 4–3.2. Let $c = \max(|a|, |b|)$. Then

$$|af(x) + bg(x)|^2 \leqslant 2c^2[|f(x)|^2 + |g(x)|^2] \text{ a.e.}$$

Since $|f|^2, |g|^2 \in \mathscr{L}$, then

$$\int 2c^2[|f|^2 + |g|^2] = 2c^2\left[\int |f|^2 + \int |g|^2\right] < \infty,$$

and we have that $|af + bg|^2 \in \mathscr{L}$ by Theorem 4–5.6; that is,

$$af + bg \in \mathscr{L}^2. \blacksquare$$

PROBLEM 1. If $f, g \in \mathscr{L}^2$, then the product $fg \in \mathscr{L}^2$.

This problem establishes the existence of the integrals in the following definition.

Definition 2. *For $f, g \in \mathscr{L}^2$, then*

(i) *The* **inner product** *of f and g is defined by*

$$f \cdot g = \int (fg).$$

(ii) *The $\mathscr{L}^2$* **norm** *of f is defined by*

$$\|f\| = \sqrt{(f \cdot f)} = \left\{\int f^2\right\}^{1/2}.$$

(iii) *The $\mathscr{L}^2$* **distance** *between f and g is*

$$\|f - g\|.$$

Theorem 2. *If $f, g \in \mathscr{L}^2$, then*

$$|f \cdot g| \leqslant \|f\| \, \|g\| \qquad \textit{(Schwarz inequality)} \qquad \textit{and}$$
$$\|f + g\| \leqslant \|f\| + \|g\| \qquad \textit{(Minkowski's inequality)}.$$

Proof. The proof of Theorem 1–9.2 was written in such a way that it may now be reread with integrals in place of sums. Also see Theorem 2–12.1. ∎

A function $f \in \mathscr{L}^2$ will now be considered as a point in the space $\mathscr{L}^2$, with $\mathscr{L}^2$ distance to a point $g \in \mathscr{L}^2$ defined by $\|f - g\|$. Thus, if also $h \in \mathscr{L}^2$, then

$$\|f - g\| \leqslant \|(f - h) + (h - g)\| \leqslant \|f - h\| + \|h - g\|$$

and a check will show that $\| \ \|$ satisfies Definition 2–4.1 (i), (ii), and (iii). Part (a) of the following theorem has thus been established.

Theorem 3. *The space $\mathscr{L}^2$ is*

(a) *A vector, inner product, normed space.*
(b) *A complete space.*

[*Note:* The whole theorem may be stated as: $\mathscr{L}^2$ is a Hilbert space.]

Proof. Since only (b) needs additional proof we take a Cauchy (in the $\mathscr{L}^2$ norm) sequence

$$f_1, f_2, f_3, \cdots, \qquad \text{with each } f_n \in \mathscr{L}^2.$$

Hence, given a number $\varepsilon > 0$, there is an N such that

$$i, j \geqslant N \Rightarrow \|f_i - f_j\| < \varepsilon.$$

The objective is to prove that $\{f_n\}$ converges in the $\mathscr{L}^2$ norm to an $f \in \mathscr{L}^2$; that is, we wish to find an $f \in \mathscr{L}^2$ for which we can prove that $\lim \|f_n - f\| = 0$.

Toward this end let n_1 be such that

$$n \geqslant n_1 \Rightarrow \|f_{n_1} - f_n\| < \tfrac{1}{2},$$

and with $n_1 < n_2 < \cdots < n_{i-1}$ determined, let $n_i > n_{i-1}$ be such that

$$n > n_i \Rightarrow \|f_{n_i} - f_n\| < \frac{1}{2^i}.$$

Hence $n_i \geqslant i$. Set

$$g_i = f_{n_i}, \qquad i = 1, 2, 3, \cdots.$$

Just in case $m_k A$ is not finite, let $A_n = A \cap I_n{}^k$, where $I_n = [-n, n]$, $n = 1, 2, 3, \cdots$, and let h be the characteristic function of A_n; that is, $h = \chi_{A_n}$. Then $h \in \mathscr{L}^2$; in fact

$$\int h^2 = \int_{A_n} h^2 = m_k A_n < \infty.$$

Upon applying Schwarz inequality to h and $|g_i - g_{i+1}|$ over A_n, we have

$$\begin{aligned}\int_{A_n} |g_i - g_{i+1}| &= \int_{A_n} h\,|g_i - g_{i+1}| \\ &\leqslant (m_k A_n)\,\|g_i - g_{i+1}\| \leqslant \frac{1}{2^i} m_k A_n.\end{aligned}$$

Thus

$$\sum_{i=1}^{\infty} \int_{A_n} |g_i - g_{i+1}| \leqslant m_k A_n.$$

Upon interchanging summation and integration, which is permissible for non-negative functions by Theorem 4–5.4, we have

$$\int_{A_n} \sum_{i=1}^{\infty} |g_i - g_{i+1}| \leqslant m_k A_n < \infty.$$

Thus $\sum_{i=1}^{\infty} |g_i(x) - g_{i+1}(x)| < \infty$ a.e. in A_n, $n = 1, 2, 3, \cdots$, and hence a.e. The absolute convergence of this series a.e. implies convergence of $\sum [g_i(x) - g_{i+1}(x)]$ a.e. If this series diverges at $x \in A$, we set $f(x) = 0$, but for $x \in A$ where there is convergence, we set

$$\begin{aligned}f(x) &= g_1(x) - \sum_{i=1}^{\infty} [g_i(x) - g_{i+1}(x)] \\ &= \lim_{n\to\infty} g_{n+1}(x).\end{aligned}$$

This shows that $\{g_n(x)\}$ converges a.e. and hence that

$$\lim_{j\to\infty} |g_i(x) - g_j(x)|^2 = |g_i(x) - f(x)|^2 \text{ a.e.}, \qquad i = 1, 2, 3, \cdots.$$

Corresponding to $\varepsilon > 0$ arbitrary, let N be such that

$$i, j \geqslant N \Rightarrow \|f_i - f_j\| < \varepsilon.$$

Hence, by using Fatou's lemma (Theorem 4–5.3) to obtain the first inequality below, we have that

$$\begin{aligned} i \geqslant N \Rightarrow \|g_i - f\|^2 &= \int |g_i(x) - f(x)|^2 \\ &= \int \lim_{j\to\infty} |g_i(x) - g_j(x)|^2 \\ &\leqslant \varliminf_{j\to\infty} \int |g_i(x) - g_j(x)|^2 = \varliminf_{j\to\infty} \|g_i - g_j\|^2 \leqslant \varepsilon^2, \end{aligned}$$

since $g_i = f_{n_i}$ and $n_i \geqslant i$. This shows that $f \in \mathscr{L}^2$, since $f = g_i + (f - g_i)$ and $g_i \in \mathscr{L}^2$ for all i, while $f - g_i \in \mathscr{L}^2$ if $i \geqslant N$. Also

$$i \geqslant N \Rightarrow \|f_i - f\| \leqslant \|f_i - g_i\| + \|g_i - f\| < 2\varepsilon.$$

Thus, as we wished to show,

$$f \in \mathscr{L}^2 \qquad \text{and} \qquad \lim_{n\to\infty} \|f_n - f\| = 0. \blacksquare$$

If $f \in \mathscr{L}^2(A, m_k)$, $g \in \mathscr{L}^2(A, m_k)$, and $\|f - g\| = 0$, it may be that $f(x) \neq g(x)$ for some $x \in A$, but $m_k\{x \in A \mid f(x) \neq g(x)\} = 0$. Hence $\mathscr{L}^2(A, m_k)$ is an example of a normed vector space (which is also complete) for which the property (iv) on p. 70 does not hold. It is therefore customary to interpret a point of $\mathscr{L}^2(A, m_k)$ as a whole family of functions, any two of which differ on A at most in a set of m_k-measure zero.

4–7 COMPLEX $\mathscr{L}^2$ SPACES

We take an interlude to give an intuitive orientation of the previous work on real inner product spaces and to lead into complex $\mathscr{L}^2$ spaces.

The dot (inner) product of ordinary plane vectors **u** and **v** is

$$\mathbf{u} \cdot \mathbf{v} = |\mathbf{u}|\ |\mathbf{v}| \cos \theta,$$

where θ is the angle (undirected and lying in a straight angle) between **u** and **v**. This itself grew out of the concept of

$$\text{work} = (\text{force}) \cdot (\text{distance})$$

where the force is constant and in line of motion. If a force **v**, constant except for point application, is not in line of motion but at the angle θ to it, then the effective force is the signed component $|\mathbf{v}| \cos \theta$ in the line of motion along **u** and the work through the distance $|\mathbf{u}|$ is $|\mathbf{u}|\ |\mathbf{v}| \cos \theta$.

In particular, $\mathbf{u} \cdot \mathbf{u} = |\mathbf{u}|\ |\mathbf{u}| \cos 0°$ so that $\mathbf{u} \cdot \mathbf{u} \geqslant 0$ and

(a) $$|\mathbf{u}| = \sqrt{(\mathbf{u} \cdot \mathbf{u})}.$$

Since θ is the angle between the vectors, then

(b) $$\mathbf{u} \cdot \mathbf{v} = \mathbf{v} \cdot \mathbf{u}.$$

If **u** and **v** are perpendicular, then $\mathbf{u} \cdot \mathbf{v} = 0$.

If $a \neq 0$ is a real number (scalar), then the angle between $a\mathbf{u}$ and $\mathbf{v}$ is θ if $a > 0$, but is $180° - \theta$ if $a < 0$. Thus, since $|a\mathbf{u}| = |a|\,|\mathbf{u}|$, which is $a|\mathbf{u}|$ if $a > 0$, but $-a|\mathbf{u}|$ if $a < 0$, then

$$(a\mathbf{u}) \cdot \mathbf{v} = \begin{cases} a|\mathbf{u}|\,|\mathbf{v}| \cos\theta & \text{if } a > 0 \\ -a|\mathbf{u}|\,|\mathbf{v}| \cos(180 - \theta) & \text{if } a < 0. \end{cases}$$

Consequently $(a\mathbf{u}) \cdot \mathbf{v} = a|\mathbf{u}|\,|\mathbf{v}| \cos\theta$ in either case. Hence

(c) $$(a\mathbf{u}) \cdot \mathbf{v} = a(\mathbf{u} \cdot \mathbf{v}) \quad \text{and} \quad \mathbf{u} \cdot (a\mathbf{v}) = a(\mathbf{u} \cdot \mathbf{v}).$$

By geometric considerations, the distributive law

(d) $$(\mathbf{u}_1 + \mathbf{u}_2) \cdot \mathbf{v} = (\mathbf{u}_1 \cdot \mathbf{v}) + (\mathbf{u}_2 \cdot \mathbf{v})$$

is seen to hold.

Upon introducing a rectangular coordinate system in the plane with unit vectors $\mathbf{i}$ and $\mathbf{j}$ on the axes, the component form

$$\mathbf{u} = x\mathbf{i} + y\mathbf{j}$$

of a vector appears. Thus the distributive law together with $\mathbf{i} \cdot \mathbf{i} = \mathbf{j} \cdot \mathbf{j} = 1$ and $\mathbf{i} \cdot \mathbf{j} = \mathbf{j} \cdot \mathbf{i} = 0$ lead to

(1) $$\begin{aligned} \mathbf{u} \cdot \mathbf{v} &= (a_1\mathbf{i} + a_2\mathbf{j}) \cdot (b_1\mathbf{i} + b_2\mathbf{j}) \\ &= a_1 b_1 + a_2 b_2 . \end{aligned}$$

This formula is seized upon in generalizing the dot product to vectors

$$\mathbf{u} = (a_1, a_2, \cdots, a_k) \quad \text{and} \quad \mathbf{v} = (b_1, b_2, \cdots, b_k)$$

in E^k by defining

(1_1) $$\mathbf{u} \cdot \mathbf{v} = \sum_{i=1}^{k} a_i b_i .$$

Since k may be as large as desired, a next step is to let $k \to \infty$. Then vectors

$$\mathbf{u} = (a_1, a_2, a_3, \cdots)$$

with countably many real components are obtained, but the resulting series corresponding to (1_1) need not converge. Cognizant of Schwarz inequality,

$$\sum_{i=1}^{k} |a_i b_i| \leqslant \sqrt{\sum_{i=1}^{k} a_i^2 \sum_{i=1}^{k} b_i^2}$$

Hilbert restricted consideration to vectors $\mathbf{u} = (a_1, a_2, a_3, \cdots)$ for which $\sum a_i^2$ converges, and for such sequences could let $k \to \infty$ first on the right and then on the left. Hence, with $\sum b_i^2$ also converging, $\sum (a_i b_i)$ converges absolutely and thus converges, permitting the definition

(1_2) $$\mathbf{u} \cdot \mathbf{v} = \sum_{i=1}^{\infty} a_i b_i .$$

In the space of such vectors, the length (norm) of **u** is then defined by

$$\|\mathbf{u}\| = \sqrt{\sum_{i=1}^{\infty} a_i^2} = \sqrt{(\mathbf{u} \cdot \mathbf{u})}$$

which gives formula (a) above. Also (b), (c), and (d) are easily checked.

The replacement of series square summable by functions square integrable (i.e., functions $f \in \mathscr{L}^2$) was carried out in the preceding section.

In some situations, vectors with complex components are used to advantage. Thus another generalization is to have vectors

$$\mathbf{u} = (z_1, z_2, \cdots, z_k), \qquad \mathbf{v} = (w_1, w_2, \cdots w_k)$$

with components z_j and w_j complex numbers.

Before proceeding we recall that $z = x + iy$, x and y real, has $|z| = \sqrt{x^2 + y^2}$ and complex conjugate $\bar{z} = x - iy$. Thus

$$z\bar{z} = (x + iy)(x - iy) = x^2 + y^2 = |z|^2.$$

Also, if z_1, z_2, w_1, and w_2 are complex numbers, then

$$\overline{(z_1 \bar{w}_1 + \bar{z}_2 w_2)} = \bar{z}_1 w_1 + z_2 \bar{w}_2 .$$

The dot product of the above vectors **u** and **v** might be defined by setting

$$\mathbf{u} \cdot \mathbf{v} \quad \text{equal to} \quad \sum_{j=1}^{k} z_j w_j .$$

If this were done, then $\mathbf{u} \cdot \mathbf{u}$ would be equal to $\sum_{j=1}^{k} (z_j)^2$ which need not be real and would not be a natural value for the square of the "length" of **u**.

It is found that the dot product defined by

$$(1_3) \qquad \mathbf{u} \cdot \mathbf{v} = \sum_{j=1}^{h} z_j \bar{w}_j$$

has advantages, and this is taken as the definition. Then

$$\mathbf{u} \cdot \mathbf{u} = \sum_{j=1}^{k} z_j \bar{z}_j = \sum_{j=1}^{k} |z_j|^2.$$

At least this is a real nonnegative number and the definition

$$(\bar{\text{a}}) \qquad \|\mathbf{u}\| = \sqrt{(\mathbf{u} \cdot \mathbf{u})}$$

of the norm $\|\mathbf{u}\|$ of **u** is made. However, by reversing order,

$$\mathbf{v} \cdot \mathbf{u} = \sum_{j=1}^{k} w_j \bar{z}_j$$
$$= \overline{\sum_{j=1}^{k} z_j \bar{w}_j} = \overline{(\mathbf{u} \cdot \mathbf{v})}$$

and therefore the commutativity of (b) is replaced by

$$(\bar{\text{b}}) \qquad \mathbf{u} \cdot \mathbf{v} = \overline{(\mathbf{v} \cdot \mathbf{u})}.$$

From $a\mathbf{u} = (az_1, az_2, \cdots, az_k)$, with a complex, then a check using the definition (1_3) shows that

$$(\bar{\text{c}}) \qquad (a\mathbf{u}) \cdot \mathbf{v} = a(\mathbf{u} \cdot \mathbf{v}) \qquad \text{but}^\dagger \qquad \mathbf{u} \cdot (a\mathbf{v}) = \bar{a}(\mathbf{u} \cdot \mathbf{v}).$$

Clearly the distributive law

$$(\bar{\text{d}}) \qquad (\mathbf{u} + \mathbf{v}) \cdot \mathbf{w} = (\mathbf{u} \cdot \mathbf{w}) + (\mathbf{v} \cdot \mathbf{w})$$

holds.

Note that $(\bar{\text{a}})$–$(\bar{\text{d}})$ reduce to (a)–(d) when imaginary parts are all zero. Schwarz inequality also has a complex form.

Theorem 1. *Given* $\mathbf{u} = (a_1, a_2, \cdots, a_k)$ *and* $\mathbf{v} = (b_1, b_2, \cdots, b_k)$, *each* a_j *and* b_j *complex, then*

$$(2) \qquad \left| \sum_{j=1}^{k} a_j \bar{b}_j \right|^2 \leqslant \sum_{j=1}^{k} |a_j|^2 \sum_{j=1}^{k} |b_j|^2; \text{ that is,}$$

$$|\mathbf{u} \cdot \mathbf{v}| \leqslant \|\mathbf{u}\| \, \|\mathbf{v}\|.$$

Proof. The first form of (2) is in Theorem 1–10.1 and the second in Theorem 2–12.1. ∎

We now represent a complex valued function f as $f = u + iv$; specifically, if f has domain A, then

$$u: A \to E^* \qquad \text{and} \qquad v: A \to E^*.$$

Also we shall consider $A \subset E^k$ and use Lebesgue outer measure m_k^* (although any space X with an outer measure μ^* could be used for all except the very last result about completeness).

Definition. *Given a measurable set* A *on which* $f = u + iv$ *is defined* a.e., *then we say that*

(i) *f is measurable on A if both u and v are measurable on A.*

(ii) $\int_A f$ *exists* if* $\int_A u$ *and* $\int_A v$ *both exist* and when they do*

$$\int_A f = \int_A u + i \int_A v.$$

(iii) $f \in \mathscr{L}(A)$ *if* $u \in \mathscr{L}(A)$ *and* $v \in \mathscr{L}(A)$.

(iv) $f \in \mathscr{L}^2(A)$ *if f is measurable and*

$$\int_A |f|^2 < \infty.$$

† Some physics books have $\mathbf{u} \cdot \mathbf{v} = \sum_{j=1}^{k} (\bar{z}_j w_j)$ instead of (1_3). Then

$$(a\mathbf{u}) \cdot \mathbf{v} = \bar{a}(\mathbf{u} \cdot \mathbf{v}) \qquad \text{and} \qquad \mathbf{u} \cdot (a\mathbf{v}) = a(\mathbf{u} \cdot \mathbf{v})$$

appear in place of $(\bar{\text{c}})$, but $(\bar{\text{a}})$, $(\bar{\text{b}})$, and $(\bar{\text{d}})$ remain the same.

Since $|f| \leqslant |u| + |v|$, if $f \in \mathscr{L}(A)$, then $|f|$ is real, nonnegative, and dominated by the real $|u| + |v| \in \mathscr{L}(A)$ so that

$$\int_A |f| \leqslant \int_A |u| + \int |v| < \infty,$$

and hence $|f| \in \mathscr{L}(A)$. Conversely, if f is measurable and $\int_A |f| < \infty$, then

$$\int_A |u| < \infty \qquad \text{and} \qquad \int_A |v| < \infty$$

from $|u| \leqslant |f|$ and $|v| \leqslant |f|$. Thus part (a) of the following theorem has been proved.

Theorem 2. *Given that $f = u + iv$ is measurable on a measurable set A, then*

(a) $f \in \mathscr{L}(A)$ *iff* $|f| \in \mathscr{L}(A)$.
(b) *If* $f \in \mathscr{L}(A)$, *then*

$$\left|\int_A f\right| \leqslant \int_A |f|.$$

Proof (b). With $f \in \mathscr{L}(A)$, then a and b defined by

$$a = \int u \qquad \text{and} \qquad b = \int v \qquad \text{(with integrals over } A\text{)}$$

are real, $\int f = a + ib$ and $|\int f| = \sqrt{a^2 + b^2}$. From Schwarz inequality for real numbers,

$$(au + bv)^2 \leqslant (a^2 + b^2)(u^2 + v^2) = (a^2 + b^2)|f|^2$$

and

$$au + bv \leqslant |au + bv| \leqslant \sqrt{a^2 + b^2}\,|f|.$$

Upon integrating both sides over A, everything being real,

$$a \int u + b \int v \leqslant \sqrt{a^2 + b^2} \int |f|,$$

$$a^2 + b^2 \leqslant \sqrt{a^2 + b^2} \int |f|,$$

$$\sqrt{a^2 + b^2} \leqslant \int |f|; \qquad \text{that is,}$$

$$\left|\int f\right| \leqslant \int |f|. \quad \blacksquare$$

By applying previous results about real functions to the real functions u and v, many of these results carry over to complex-valued functions. For example, if

$$f = u_1 + iv_1 \in \mathscr{L}(A) \qquad \text{and} \qquad g = u_2 + iv_2 \in \mathscr{L}(A),$$

then $u_1 + u_2$ and $v_1 + v_2$ are measurable, both belong to $\mathscr{L}(A)$, in each case the integral of the sum is the sum of the integrals, and hence

$$\int_A f + \int_A g = \int_A u_1 + i\int_A v_1 + \int_A u_2 + i\int_A v_2$$

$$= \int_A (u_1 + u_2) + i\int_A (v_1 + v_2)$$

$$= \int_A (f + g).$$

Similarly, if c is a complex constant, then

$$c\int_A f = \int_A cf.$$

We shall now make a quick rundown of properties of complex $\mathscr{L}^2$ by considering, without further reminder, that f and g are measurable complex-valued functions.

Given $f, g \in \mathscr{L}^2$, then $f\bar{g} \in \mathscr{L}$, since

$$|f\bar{g}| \leqslant \tfrac{1}{2}[|f|^2 + |\bar{g}|^2] = \tfrac{1}{2}[|f|^2 + |g|^2],$$

so the definition of inner product and norm by

$$f \cdot g = \int f\bar{g} \qquad \text{and} \qquad \|f\| = \sqrt{(f \cdot f)}$$

are justified. The first inequality below is in Theorem 2, the second follows by the same method used in the real case

$$\left|\int fg\right| \leq \int |fg| \leq \|f\|\ \|g\|,$$

and any f may be replaced by $\bar{f}$ and/or any g by $\bar{g}$. Hence

$$0 \leq \|f+g\|^2 = \int (f+g)(\bar{f}+\bar{g})$$

$$= \int f\bar{f} + \int f\bar{g} + \int \bar{f}g + \int g\bar{g}$$

$$\leq \|f\|^2 + \left|\int f\bar{g}\right| + \left|\int \bar{f}g\right| + \|g\|^2$$

$$\leq \|f\|^2 + 2\|f\|\ \|g\| + \|g\|^2$$

$$= (\|f\| + \|g\|)^2$$

and

$$\|f+g\| \leq \|f\| + \|g\|.$$

Thus complex $\mathscr{L}^2$ is a vector, inner product, normed space, and we shall see that it is also complete, provided (as we are considering) the domain space of the function is E^k. With measurable $f_n = u_n + iv_n$, then $\{f_n\}$ is a Cauchy sequence in the (real) norm of complex $\mathscr{L}^2$ iff $\{u_n\}$ and $\{v_n\}$ are both Cauchy sequences in the norm of real $\mathscr{L}^2$, since

$$\left.\begin{matrix}\|u_n - u_j\| \\ \|v_n - v_j\|\end{matrix}\right\} \leq \| f_n - f_j\| \leq \|u_n - u_j\| + \|v_n - v_j\|$$

for $n, j = 1, 2, 3, \cdots$. Hence, given a Cauchy sequence $\{f_n\}$, we let (by Theorem 4–6.3) $u_0, v_0 \in \mathscr{L}^2$ be such that

$$\lim_{n\to\infty} \|u_n - u_0\| = \lim_{n\to\infty} \|v_n - v_0\| = 0.$$

Hence $u_0 + iv_0 = f_0 \in \mathscr{L}^2$ and (put $j = 0$ in the above inequalities)

$$\lim_{n\to\infty} \| f_n - f_0\| = 0.$$

PROBLEMS

1. Let A and B be μ-measurable sets. Theorem 4–2.4 showed first that $A \cap B$ is μ-measurable and used this to obtain that $A \cup B$ is μ-measurable. Check that the following holds for each set W and shows directly that $A \cup B$ is μ-measurable.

$$\begin{aligned}\mu^* W &= \mu^* WA + \mu^* WA^c \\ &= \mu^* WA + \mu^* WA^c B + \mu^* WA^c B^c \\ &\geqslant \mu^* W(A \cup A^c B) + \mu^* WA^c B^c \\ &= \mu^* W(A \cup B) + \mu^* W(A \cup B)^c.\end{aligned}$$

2. If A, B, and C are μ-measurable sets, then

$$\mu(A) + \mu(B) + \mu(C) + \mu(ABC) = \mu(A \cup B \cup C) + \mu(AB) + \mu(AC) + \mu(BC).$$

3. For this problem we say that a number x, $0 < x < 1$, is 7-free if it has a decimal expansion not using the digit 7. Show that the set of 7-free numbers has Lebesgue measure zero.

4. A set $A \subset E^1$ has measure zero iff A can be covered by a sequence of open intervals, the sum of whose lengths is finite in such a way that each point of A is an interior point of infinitely many of these intervals.

5. (a) Let J be the space of all positive integers and define $\varphi^*(A)$ for each set $A \subset J$ by

$$\varphi^*(A) = \begin{cases} 0 & \text{if } A = \varnothing \\ \sum\limits_{n \in A} 1/n^2 & \text{if } A \neq \varnothing. \end{cases}$$

Then φ^* is an outer measure and all subsets of J are φ-measurable.

(b) Does the same conclusion hold if $1/n^2$ is replaced by $1/n$?

(c) What if $1/n^2$ is replaced by n?

(d) The function ψ^* defined by $\psi(\varnothing) = 0$ and $\psi^*(A) = 1 + \varphi^*(A)$ if $A \neq \varnothing$ is an outer measure, but $\varnothing$ and J are the only ψ-measurable sets.

6. If f is nonnegative, $f \in \mathscr{L}$, and $A_n = \{x \mid n \leqslant f(x) < n + 1\}$, $n = 0, 1, 2, \cdots$, then $\sum_{n=0}^{\infty} n(mA_n)$ converges.

7. Given a number $\eta > 0$ and a sequence $\{A_n\}$ of measurable sets such that

$$A_n \subset [a, b] \qquad \text{and} \qquad mA_n \geqslant \eta, \qquad n = 1, 2, 3, \cdots,$$

then there is at least one $x \in [a, b]$ such that $x \in A_n$ for infinitely many n.

8. Definition. A nonempty family $\mathscr{S}$ of sets is said to be a **semiring**, provided
(i) $(A \in \mathscr{S}, B \in \mathscr{S}) \Rightarrow AB \in \mathscr{S}$; and
(ii) If $A \in \mathscr{S}$, $A_1 \in \mathscr{S}$, and $A_1 \subset A$, then there are elements $A_2, A_3, \cdots, A_n$ of $\mathscr{S}$ such that $A_1, A_2, \cdots, A_n$ are pairwise disjoint and

$$A = A_1 \cup A_2 \cup \cdots \cup A_n.$$

Theorem 1. *In E^1 the family of open-closed intervals and $\varnothing$ is a semiring.*

Theorem 2. *Let $\mathscr{S}$ be a semiring. If $A \in \mathscr{S}$ and $A_1, \cdots, A_k$ are disjoint elements of $\mathscr{S}$ and subsets of A, then there are elements $B_{k+1}, \cdots, B_n$ of $\mathscr{S}$ such that $A_1, \cdots, B_n$ are disjoint and*

$$A = A_1 \cup \cdots \cup A_k \cup B_{k+1} \cup \cdots \cup B_n.$$

[*Hint:* Mathematical induction on k.]

9. Let $\mathscr{T}$ be a nonempty family of sets having the properties
(i) $(A \in \mathscr{T}$ and $B \in \mathscr{T}) \Rightarrow AB \in \mathscr{T}$; and
(ii) If $A \in \mathscr{T}$, $B \in \mathscr{T}$, and $A \subset B$, there is a finite number of sets $C_0, C_1, \cdots, C_n$ such that

$$A = C_0 \subset C_1 \subset \cdots \subset C_n = B$$

and $(C_i - C_{i-1}) \in \mathscr{T}$, $\quad i = 1, 2, \cdots, n$. Then $\mathscr{T}$ is a semiring (Problem 8), and conversely.

10. (a) Given a set $A \subset E^1$, let h be a positive number and define the function f_h by

$$f_h(x) = m^*A[x - h, x + h], \qquad x \in E^1.$$

Then the set $A_a = \{x \mid f_h(x) \geqslant a\}$ is closed for each number a.

(b) For each $x \in E^1$,

$$\varliminf_{h \to 0^+} \frac{f_h(x)}{2h} = \varliminf_{n \to \infty} \frac{f_{1/n}(x)}{2/n}$$

and the same is true for limit superiors.

11. Given a measurable set A with $\mu A < \infty$, let $\mathscr{A}$ be the space of measurable functions each having domain almost all of A and finite a.e. in A. Define the distance between $f \in \mathscr{A}$ and $g \in \mathscr{A}$ by

$$d(f, g) = \int_A \frac{|f - g|}{1 + |f - g|}\, d\mu.$$

Then $\mathscr{A}$ is a pseudometric space.
[*Hint:* Problem 26, page 45.]

12. Given a space S on which an outer measure μ^* is defined, let $\mathscr{S}$ be the space of all subsets of S. With $A \in \mathscr{S}$ and $B \in \mathscr{S}$, define the μ^*-distance between A and B by

$$d(A, B) = m^*(A \triangle B)$$

where $A \triangle B$ is the symmetric difference of $A \subset S$ and $B \subset S$ as given on page 48. Then $\mathscr{S}$, with this distance function, is a pseudometric space.

13. Let $\{f_n\}$ be a sequence of functions, each measurable and finite a.e. on a measurable set A, and let C be the subset of A on which the sequence converges. Then

$$C = \bigcap_k \bigcup_n \bigcap_m \left\{x \in A \,\middle|\, |f_n(x) - f_m(x)| < \frac{1}{k}\right\},$$

and thus C is measurable.

14. Let $\{\mu_n^*\}$ be a sequence of outer measures on a space X. If $\{a_n\}$ is any sequence of positive numbers, then the function μ^* defined for each $A \subset X$ by

$$\mu^*(A) = \sum_1^\infty a_n \mu_n^*(A)$$

is an outer measure on X provided there is at least one $A \subset X$ for which the series converges to a positive number.

15. (Alternative Proof of Theorem 4–6.3(b)).

(a) Instead of the sequence $\{n_i\}$ chosen in the proof, take $n_i < n_2 < \cdots$ such that

$$n \geqslant n_i \Rightarrow \|f_{n_i} - f_n\| < \frac{1}{4^i}.$$

(b) Set $g_i = f_{n_i}$ and define the sets

$$A_i = \left\{x \in A \,\middle|\, |g_i(x) - g_{i+1}(x)| \geqslant \frac{1}{2^i}\right\},$$

and

$$M = A_1 \cap A_2 \cap A_3 \cap \cdots.$$

(c) Show that $m_k A_i < 1/4^i$ by proving that

$$\int_{A_i} |g_i(x) - g_{i+1}(x)|^2 \geqslant \frac{1}{4^i} m_k A_i.$$

(d) Obtain an upper bound of $m_k A_i$ and then show that

$$m_k M = \lim_{i \to \infty} m_k A_i = 0.$$

(e) Show that $i \leqslant j < l$ and $x \in A - A_i$ imply

$$|g_j(x) - g_l(x)| < \frac{1}{2^{i-1}}.$$

(f) Justify setting $f(x) = \lim_{n \to \infty} g_n(x)$ for $x \in A - A_i$, $i = 1, 2, \cdots$ and then that

$$f(x) = \lim_{n \to \infty} g_n(x) \qquad \text{for } x \in \bigcup_{i=1}^\infty (A - A_i) = A - M.$$

(g) Proceed with this f as in the proof of the theorem.

Problems on $\mathscr{L}^p$ Spaces

In this group of problems the domain space is the interval [0, 1], Lebesgue m-measure is used, all functions are real and defined a.e. in [0, 1], and all integrals $\int$ mean $\int_0^1$.

Definition 1. With p a positive number, $\mathscr{L}^p$ is the set of all measurable functions f such that $|f|^p \in \mathscr{L}$. The $\mathscr{L}^p$ **norm** of $f \in \mathscr{L}^p$ is defined by

$$\|f\|_p = \left\{\int |f|^p\right\}^{1/p}.$$

PROBLEM 1. For this problem, $p > 1$ and $q = p/(p-1)$ so that $q > 1$ and

$$\frac{1}{p} + \frac{1}{q} = 1.$$

The spaces $\mathscr{L}^p$ and $\mathscr{L}^q$ are called **conjugates** of one another.

(a) If $f \in \mathscr{L}^p$ and $g \in \mathscr{L}^q$, then $(fg) \in \mathscr{L}$ and

$$\int |fg| \leqslant \|f\|_p \, \|g\|_q \qquad \text{(Hölder's inequality)}$$

(b) If f and g are both in $\mathscr{L}^p$, then $(f+g) \in \mathscr{L}^p$, the triangular inequality

$$\|f+g\|_p \leqslant \|f\|_p + \|g\|_p \qquad \text{(Minkowski's inequality)}$$

holds, and thus $\mathscr{L}^p$ (for $p \geqslant 1$) is a pseudometric space.

[*Hint:* Problem 18, Chap. 3.]

PROBLEM 2. Let $f(x) = 2$, $g(x) = 0$ for $0 \leqslant x \leqslant \frac{1}{2}$, and $f(x) = 0$, $g(x) = 3$ for $\frac{1}{2} < x \leqslant 1$. Then (with $p = \frac{1}{2}$),

$$\|f+g\|_{0.5} > \|f\|_{0.5} + \|g\|_{0.5}.$$

PROBLEM 3. If $0 < p < 1$ and $f, g \in \mathscr{L}^p$, then

$$\begin{aligned}\|f+g\|_p &\leqslant \left\{\int |f|^p + \int |g|^p\right\}^{1/p} \\ &\leqslant 2^{(1-p)/p}\{\|f\|_p + \|g\|_p\}.\end{aligned}$$

[*Hint:* Problem 19, Chap. 3.]

Definition 2. For $p > 0$, a sequence $\{f_n\}$ in $\mathscr{L}^p$ is said

(i) To **converge in** $\mathscr{L}^p$ to $f \in \mathscr{L}^p$ (or converge in mean of order p to $f \in \mathscr{L}^p$) if

$$\lim_{n\to\infty} \|f_n - f\|_p = 0.$$

(ii) To be a **Cauchy sequence in** $\mathscr{L}^p$ (or a Cauchy sequence in mean of order p) if

$$\|f_m - f_n\|_p \to 0 \qquad \text{as} \qquad (m, n) \to (\infty, \infty).$$

PROBLEM 4. For $p > 0$, if $\{f_n\}$ is a Cauchy sequence in $\mathscr{L}^p$, then there is an $f \in \mathscr{L}^p$ such that $\{f_n\}$ converges in $\mathscr{L}^p$ to f.

[*Hint:* Proof of Theorem 4–6.3 (b) modified appropriately.]

Hence $\mathscr{L}^p$, $p > 0$, is a Banach space and in particular is a Hilbert space if $p = 2$ (by Theorem 4–6.3).

PROBLEM 5. Given $0 < \alpha < \beta < \infty$.

(a) Then $\mathscr{L}^\beta \subset \mathscr{L}^\alpha$ and if $f \in \mathscr{L}^\beta$, then $\|f\|_\beta \leqslant \|f\|_\alpha$.

(b) If $A \subset [0, 1]$ is measurable and $f \in \mathscr{L}^\alpha$, then

$$\left[\int_A |f|^\beta\right]^{1/\beta} \leqslant \left[\int_A |f|^\alpha\right]^{1/\alpha} [mA]^{(\beta-\alpha)/\alpha}.$$

[*Hint:* Set $r = \beta/\alpha$ and show that $|f|^\alpha \in \mathscr{L}^r$.]

Definition 3. (i) A measurable function f is said to be **essentially bounded** if there is a number b such that $\{x \mid |f(x)| > b\}$ has measure zero and such a b is called an **essential upper bound** of f.
(ii) The set of all essentially bounded functions is denoted by $\mathscr{L}^\infty$.
(iii) If $f \in \mathscr{L}^\infty$, then $\|f\|_\infty$ is defined as the infimum of all essential upper bounds of f.
Note that $\|f\|_\infty$ is not defined in terms of integrals.

PROBLEM 6. (a) If $f \in \mathscr{L}^\infty$, then $f \in \mathscr{L}^p$ for all $0 < p < \infty$.
(b) If $f \in \mathscr{L}^\infty$, then

$$\|f\|_\infty = \lim_{n\to\infty} \|f\|_n .$$

[*Hint:* Let b be an essential upper bound of f. Then $\int |f|^n \leqslant b^n$. Take $\overline{\lim}$ of the nth root. For any $0 < \lambda < \|f\|_\infty$, there is a measurable set $A \subset [0, 1]$ such that $\int |f|^n \geqslant \lambda^n \, mA > 0$. Take $\underline{\lim}$ of nth root.]

If $(1/p) + (1/q) = 1$ and $p \to 1$, then $q \to \infty$. For this reason $\mathscr{L}$ and $\mathscr{L}^\infty$ are said to be **conjugates** of one another.

PROBLEM 7. (a) $\mathscr{L}^\infty$ is a pseudometric space.
(b) $\mathscr{L}^\infty$ is a Banach space.
(c) For $f \in \mathscr{L}^\infty$ and $\{f_n\}$ is a sequence in $\mathscr{L}^\infty$, then

$$\lim_{n\to\infty} \|f_n - f\|_\infty = 0$$

iff there is a set A of measure zero such that corresponding to each number $\varepsilon > 0$ there is a positive integer N such that

$$n \geqslant N \Rightarrow |f_n(x) - f(x)| < \varepsilon \qquad \text{for all } x \in [0, 1] - A.$$

Problems on Convergence to Measure

An abstract space X and a measure μ are given.

Definition 1. Let $f: X \to E^*$ and $f_n : X \to E^*$, $n = 1, 2, 3, \cdots$ be μ-measurable functions, each of which is finite a.e.
(i) The sequence $\{f_n\}$ is said to **converge to f in measure** if for each number $\eta > 0$,

$$\lim_{n\to\infty} \mu\{x \mid |f_n(x) - f(x)| \geqslant \eta\} = 0,$$

and we write $\underset{n\to\infty}{li\mu}\ f_n = f$.
(ii) $\{f_n\}$ is said to be a μ-**Cauchy sequence** (or a Cauchy sequence in μ-measure) if corresponding to each pair of numbers $\eta > 0$ and $\varepsilon > 0$ there is an N such that

$$N \leqslant n < m \Rightarrow \mu\{x \mid |f_n(x) - f_m(x)| \geqslant \eta\} < \varepsilon.$$

PROBLEM 1. (a) The functions in the following sequence are called the Rademacher functions. Take the characteristic function of $[0, \frac{1}{2}]$ for f_1; of $[\frac{1}{2}, 1]$ for f_2; of $[(k-1)/2^2, k/2^2]$ for f_{2+k}, $k = 1, 2, 3, 4$; of $[(k-1)/2^n, k/2^n]$ for f with subscript $2^n - 2 + k$, $k = 1, 2, \cdots, 2^n$; etc.; for $n = 1, 2, 3, \cdots$.
The sequence of Rademacher functions is divergent on [0,1] but converges in Lebesgue measure,

(b) The sequence $\{\chi_{[n,n+1]}\}$ converges on E^1, but does not converge in Lebesgue measure.

PROBLEM 2. (a) If $li\mu\, f_n = f$ and $f = g$ a.e., then $li\mu\, f_n = g$.
(b) If $li\mu\, f_n = f$ and $li\mu\, f_n = g$, then $f = g$ a.e.
(c) *If* $li\mu\, f_n = f$ and $li\mu\, g_n = g$, then

$$li\mu\,(f_n + g_n) = f + g.$$

(d) If a sequence converges in μ-measure, then it is a μ-Cauchy sequence.

The following theorem is the converse of Problem 2(d).

Theorem. *If $\{f_n\}$ is a μ-Cauchy sequence, then there is a function f which is finite everywhere and*

$$\underset{n\to\infty}{li\mu}\; f_n = f.$$

Proof. Set $X_0 = \bigcup_n \{x \mid |f_n(x)|\} = \infty$ so that $\mu(X_0) = 0$.

Let $\Sigma\, \eta_i$ be a convergent series of positive numbers. For $i = 1, 2, 3, \cdots$ determine N_i such that

$$N_i \leqslant n < m \Rightarrow \mu\{x \mid |f_n(x) - f_m(x)| \geqslant \eta_i\} < \eta_i.$$

Select $n_1 < n_2 < n_3 < \cdots$ such that $n_i > N_i$. Set $g_i = f_{n_i}$ and

$$F_i = \{x \mid |g_i(x) - g_{i+1}(x)| \geqslant \eta_i\}.$$

Therefore $\mu(F_i) < \eta_i$ and

$$x \notin (F_i \cup X_0) \Rightarrow |g_i(x) - g_{i+1}(x)| < \eta_i.$$

Set $G_k = \bigcup_{i\geqslant k} F_i$. Hence, for any integer k,

$$[k \leqslant p < q, x \notin (G_k \cup X_0)] \Rightarrow$$

$$|g_p(x) - g_q(x)| = \left|\sum_{i=p}^{q-1} [g_i(x) - g_{i+1}(x)]\right| \leqslant \sum_{i\geqslant p} \eta_i.$$

Since $\sum_{i\geqslant p} \eta_i \to 0$ as $p \to \infty$, we have that

$$\{g_i\} \text{ is a uniform Cauchy sequence on } (G_k \cup X_0)^c, \qquad k = 1, 2, 3, \cdots.$$

We next set

$$G = \cap\, G_k$$

and have that $x \notin G \cup X_0 \Rightarrow \{g_i(x)\}$ converges to a finite limit. Hence the function g defined by

$$g(x) = \lim_{i\to\infty} g_i(x), \qquad x \notin G \cup X_0$$

has all its values finite and is defined a.e., since

$$\mu(G \cup X_0) \leqslant \mu G_k + \mu X_0 \leqslant \sum_{i\geqslant k} \mu F_i < \sum_{i\geqslant k} \mu_i \to 0 \qquad \text{as } k \to \infty.$$

We shall show that

$$\underset{i\to\infty}{li\mu}\; g_i = g. \tag{1}$$

To do so let $\eta > 0$ and $\varepsilon > 0$ be given. Select an integer k such that

$$\mu G_k < \sum_{i\geqslant k} \eta_i < \varepsilon \qquad \text{and} \qquad \eta_k < \eta.$$

Since $\{g_i\}$ is a uniform Cauchy sequence on $(G_k \cup X_0)^c$, it converges uniformly to g on $(G_k \cup X_0)^c$ and we select an N such that

$$i \geqslant N \Rightarrow |g_i(x) - g(x)| < \eta \qquad \text{for all } x \notin G_k \cup X_0 .$$

Hence, if $i \geqslant N$, if $g_i(x) - g(x)$ is defined, and if $|g_i(x) - g(x)| \geqslant \eta$, then $x \in G_k$. Thus

$$i \geqslant N \Rightarrow \mu\{x \mid |g_i(x) - g(x) \geqslant \eta\} \leqslant \mu G_k < \varepsilon$$

and (1) is established.

Upon defining the function f by

$$f(x) = \begin{cases} 0 & \text{if } x \in G \cup X_0 \\ g(x) & \text{if } x \notin G \cup X_0 \end{cases}$$

the f is finite everywhere, is equal to g a.e., and

(2) $$\operatorname*{li\mu}_{i \to \infty} g_i = f; \qquad \text{that is, } \operatorname*{li\mu}_{i \to \infty} f_{n_i} = f$$

by Problem 2(a).

Again let $\eta > 0$ and $\varepsilon > 0$ be given. If n and x are such that

$$\eta \leqslant |f_n(x) - f(x)|,$$

then for any i either $|f_n(x) - f_{n_i}(x)| \geqslant \eta/2$ or $|f_{n_i}(x) - f(x)| \geqslant \eta/2$, since the assumption that both are $<\eta/2$ gives a contradiction. Hence the set $\{x \mid |f_n(x) - f(x)| \geqslant \eta\}$ is included in the union

$$\left\{x \,\middle|\, |f_n(x) - f_{n_i}(x)| \geqslant \frac{\eta}{2}\right\} \cup \left\{x \,\middle|\, |f_{n_i}(x) - f(x)| \geqslant \frac{\eta}{2}\right\}, \qquad i = 1, 2, 3, \cdots.$$

Upon determining N such that

$$m, n \geqslant N \Rightarrow \mu\left\{x \,\middle|\, |f_n(x) - f_m(x)| \geqslant \frac{\eta}{2}\right\} < \frac{\varepsilon}{2},$$

we have that

$$n \geqslant N \Rightarrow \mu\{x \mid |f_n(x) - f(x)| \geqslant \eta\} < \varepsilon,$$

since from (2) there are i's satisfying both

$$n_i \geqslant N \qquad \text{and} \qquad \mu\left\{x \,\middle|\, |f_{n_i}(x) - f(x)| \geqslant \frac{\eta}{2}\right\} < \frac{\varepsilon}{2}.$$

Consequently $\operatorname{li\mu} f_n = f$. ∎

Definition 2. *If* $f \in \mathscr{L}(X, \mu)$ *and* $f_n \in \mathscr{L}(X, \mu)$, $n = 1, 2, 3, \cdots$, *then* $\{f_n\}$ *is said to* **converge in mean** *to* f *if*

$$\lim_{n \to \infty} \int_X |f_n - f| = 0,$$

and we write l.i.m. $f_n = f$. *A Cauchy sequence in mean is then defined as in the problems on* $\mathscr{L}^p$ *spaces.*

PROBLEM 3. (a) If l.i.m. $f_n = f$, then $\operatorname{li\mu} f_n = f$.

(b) If $\{f_n\}$ is a Cauchy sequence in mean, then it is also a μ-Cauchy sequence.

(c) The converses of (a) and (b) are not true.

Chapter 5

MEASURE THEORY

There are topics in this chapter that are not preliminary to anything else in this book, and some of the material required for further work will not be referred to until late in the book. Thus a possible procedure (even recommended) is that all except the first two sections of this chapter be skipped for the time being and returned to piecemeal as topics are referred to. One reason for including the various topics at this place in the book is to emphasize that very little machinery is required for any of them even though some of them have been considered too esoteric for an elementary course. A beginning graduate student would, however, do well to have all the topics and methods of this chapter in his kit of mathematical tools.

5–1 METRIC OUTER MEASURE

If S is any space and μ^* is any outer measure on S, then μ^* has the additivity property expressed by

$$\mu^*(A \cup B) = \mu^*A + \mu^*B,$$

provided $A \cap B = \varnothing$ *and at least one of* A *or* B *is* μ*-measurable*. (Theorem 4–2.7). The next definition specializes S to a metric space and imposes an additivity condition on the outer measure. Recall that $d(A, B)$ denotes the distance between sets A and B of a metric space (Section 2–11).

Definition 1. *Let* S *be a metric space and* μ^* *an outer measure on* S (so that (i), (ii), and (iii) of Definition 4–2.1 hold). *Then* μ^* *is said to be a* **metric outer measure** *if for* $A \subset S$ *and* $B \subset S$,

(iv) $\qquad \mu^*(A \cup B) = \mu^*A + \mu^*B \qquad$ *provided* $\qquad d(A, B) > 0.$

We know that E^1 is a metric space. Theorem 3 below states that m^* is a metric outer measure, but the intermediate results will also be used later.

Theorem 1. *Let $\overline{m}(A)$ be defined for each set $A \subset E^1$ by*

$$\overline{m}(A) = \begin{cases} 0 & \text{if } A = \varnothing \\ \inf\{y \mid y = \sum |I_n|, A \subset \bigcup I_n, \text{ each } I_n \text{ closed}\} & \text{if } A \neq \varnothing. \end{cases}$$

Then $\overline{m}(A) = m^*A$ for each $A \subset E^1$.

Proof. Because the condition "I_n closed" is in the definition of $\overline{m}$ but not in the definition of m^*, it follows that

$$m^*A \leqslant \overline{m}A \qquad \text{for each } A \subset E^1.$$

Hence, if $m^*A = \infty$, then also $\overline{m}A = \infty$, and thus consider any set A having $m^*A < \infty$. Given an $\varepsilon > 0$, take a sequence $\{I_n\}$ of intervals such that

$$A \subset \bigcup I_n \qquad \text{and} \qquad \sum |I_n| < m^*A + \varepsilon.$$

Then each $|I_n| < \infty$ and the closure $\bar{I}_n$ of I_n has $|\bar{I}_n| = |I_n|$. Also,

$$A \subset \bigcap \bar{I}_n \qquad \text{so that} \qquad \overline{m}A \leqslant \sum |\bar{I}_n| = \sum |I_n| < m^*A + \varepsilon.$$

Hence $\overline{m}A \leqslant m^*A$ and thus $\overline{m}A = m^*A$. ∎

PROBLEM 1. Show that Theorem 1 holds with "closed" replaced by "open."

Theorem 2. *Given a positive number λ, let m_λ^* be defined by replacing "each I_n closed" in Theorem 1 by "$|I_n| < \lambda$." Then $m_\lambda^*A = m^*A$ for each $A \subset E^1$.*

Proof. Again $m_\lambda^* A \geqslant m^*A$. Consider $m^*A < \infty$ and choose any $\varepsilon > 0$. This time we take a sequence $\{I_n\}$ of closed intervals covering A and such that

$$\sum |I_n| < m^*A + \varepsilon.$$

Each I_n here has $|I_n| < \infty$ and because it is closed, we divide I_n into a finite number of closed segments, each of length less than λ. By summing the lengths of all such segments for all the I_n, we see that $m_\lambda^*A < m^*A + \varepsilon$ and that $m_\lambda^*A = m^*A$. ∎

PROBLEM 2. Let $\mathscr{I}$ be the family of all intervals having both end points rational. Then for $A \subset E^1$ and $A \neq \varnothing$,

$$m^*A = \inf\{y \mid y = \sum |I_n|, \qquad \text{where } A \subset \bigcup I_n \text{ and each } I_n \in \mathscr{I}\}.$$

Theorem 3. *Lebesgue outer measure m^* is a metric outer measure.*

Proof. Take two nonempty sets A and B of E^1 such that $d(A, B) > 0$. Let

$$\lambda = \tfrac{1}{3}d(A, B)$$

and choose any number $\varepsilon > 0$. Cover $A \cup B$ by a countable family $\{I_n\}$ of intervals each of length less than λ and such that

$$m^*(A \cup B) + \varepsilon \geqslant \sum |I_n|.$$

Rename the members of $\{I_n\}$ that contain at least one point of A as $I_1', I_2', I_3', \cdots$ and the remaining as $I_1'', I_2'', I_3'', \cdots$. Then $A \subset \bigcup I_n'$, but $A(\bigcup I_n'') = \varnothing$ because $|I_n| < \frac{1}{3}d(A, B)$. Hence $B \subset \bigcup I_n''$ and

$$m^*(A \cup B) + \varepsilon \geqslant \sum |I_n| = \sum |I_n'| + \sum |I_n''| \geqslant m^*A + m^*B,$$

even if some or all three of the series diverge. Hence

$$m^*(A \cup B) \geqslant m^*A + m^*B.$$

But we know that $m^*(A \cup B) \leqslant m^*A + m^*B$, and thus equality holds. ∎

PROBLEM 3. If $A \subset I$ and $B \subset J$ with I and J intervals such that $I \cap J = \varnothing$, then $m^*(A \cup B) = m^*A + m^*B$.

We return to the more general situation in which the space S is any metric space and μ^* is a metric outer measure.

Lemma. *Let G be a nonempty open set such that also $G^c \neq \varnothing$ and let*

$$B_n = \left\{x \,\middle|\, d(x, G^c) > \frac{1}{n}\right\}, \qquad n = 1, 2, 3, \cdots.$$

If $A \subset G$ and $\mu^*A < \infty$, then

$$\mu^*A = \lim_{n\to\infty} \mu^*AB_n.$$

Proof. Set $R_1 = B_1$ and $R_n = B_{n+1} - B_n, \qquad n = 1, 2, 3, \cdots$. Then

$$\mu^*A(R_n \cup R_{n+2}) = \mu^*AR_n + \mu^*AR_{n+2},$$

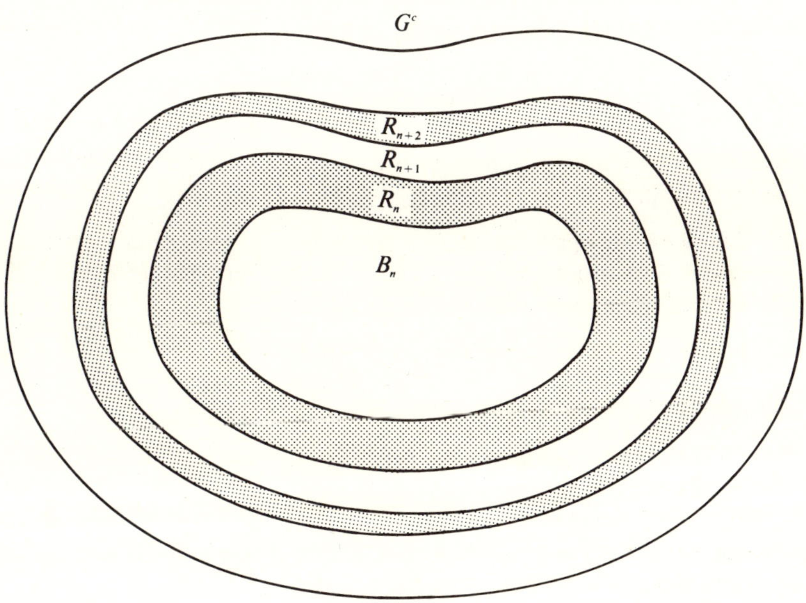

FIG. 1

since this holds if either AR_n or AR_{n+2} is empty and if both are nonempty, then

$$d(AR_n, AR_{n+2}) \geqslant d(R_n, R_{n+2}) \geqslant \frac{1}{n+1} - \frac{1}{n+2} > 0.$$

Hence, for each positive integer n,

$$\mu^*A \geqslant \mu^*\left(A\bigcup_{k=n}^{n+m} R_{2k-1}\right) = \sum_{k=n}^{n+m} \mu^*AR_{2k-1}, \qquad \text{and}$$

$$\mu^*A \geqslant \mu^*\left(A\bigcup_{k=n}^{n+m} R_{2k}\right) = \sum_{k=n}^{n+m} \mu^*AR_{2k}.$$

Upon letting $m \to \infty$ in the first of these, we obtain convergence of the series and the equality in

$$\mu^*\left[A\bigcup_{k=n}^{\infty} R_{2k-1}\right] = \sum_{k=n}^{\infty} \mu^*AR_{2k-1}.$$

Since $d(B_{2n-1}, R_{2n+j}) > 0$, $j = 1, 2, 3, \cdots$, if $B_{2n-1} \neq \varnothing$, we now have

$$\mu^*A \geqslant \mu^*\left[A\left(B_{2n-1} \cup \bigcup_{k=n}^{\infty} R_{2k+1}\right)\right] = \mu^*AB_{2n-1} + \mu^*\bigcup_{k=n}^{\infty} AR_{2k+1}$$

$$= \mu^*AB_{2n-1} + \sum_{k=n}^{\infty} \mu^*AR_{2k+1} \qquad \text{and}$$

$$\mu^*A \geqslant \lim_{n\to\infty} \mu^*AB_{2n-1}.$$

In a similar way, $\mu^*A \geqslant \lim \mu^*AB_{2n}$, and therefore

$$\mu^*A \geqslant \lim_{n\to\infty} \mu^*AB_n.$$

But $AB_n \subset AB_{n+1} \subset A$, and hence $\lim \mu^*AB_n \leqslant \mu^*A$. ∎

Theorem 4. *If μ^* is a metric outer measure on a metric space S, then every open set is μ-measurable.*

Proof. Let G be an open set. If $G = \varnothing$ or $G = S$, then G is μ-measurable. Hence consider $G \neq \varnothing$ and $G^c \neq \varnothing$. Let W be any set with $\mu^*W < \infty$. In the notation of the lemma, and with $WG = A$, we have

$$\mu^*WG = \lim_{n\to\infty} \mu^*WGB_n.$$

Since $W \supset WB_n \cup WG^c$ and $d(B_n, G^c) > 0$ if $B_n \neq \varnothing$, we have

$$\mu^*W \geqslant \mu^*WGB_n + \mu^*WG^c.$$

Consequently $\mu^*W \geqslant \mu^*WG + \mu^*WG^c$ and hence G is μ-measurable. ∎

Euclidean k-space E^k was discussed in Chapter 2 and a k-cell was defined as any set of the form

$$R = I_1 \times I_2 \times \cdots \times I_k$$

with I_n an interval of E^1 for $n = 1, 2, \cdots, k$. The k-volume $|R|$ of R is now defined by

$$|R| = |I_1|\,|I_2| \cdots |I_k|.$$

It is thus only a change in notation and terminology to assign the Lebesgue outer k-measure $m_k^* A$ to each $A \subset E^k$ as the infimum of all sums

$$\sum |R_n| \qquad \text{for which } A \subset \bigcup R_n .$$

The same techniques used for m^* show that m_k^* is a metric outer measure and thus that all theorems about μ-measurability carry over to m_k-measurability.

Most of the results about the specific outer measure m^* hold for m_k^*. In particular, every open set in E^k is m_k-measurable and hence so is every closed set in E^k.

All the following material is given for sets in E^1, m^*, and m-measurability, although all results hold for E^k, m_k^*, and m_k-measurability. Familiarity with E^1 may, however, make visualization easier than if E^k were used. In fact, good practice in generalization will be gained if some thought is given to how much of the discussion after Theorem 5 could be freed to a metric space and a metric outer measure for which (2) holds.

Theorem 5. *Let A be any set in E^1. Then corresponding to each number $\varepsilon > 0$ there is an open set G such that*

$$(1) \qquad A \subset G \qquad \text{and} \qquad mG \leqslant m^*A + \varepsilon.$$

Also there is an m-measurable set $\hat{A}$ (actually a G_δ set) such that†

$$(2) \qquad A \subset \hat{A} \qquad \text{and} \qquad m\hat{A} = m^*A.$$

Proof. If $m^*A = \infty$, then $G = \hat{A} = E^1$ is open, m-measurable,

$$A \subset G = \hat{A} \qquad \text{and} \qquad m^*A = mG - \varepsilon = m\hat{A} = \infty$$

for every finite number $\varepsilon > 0$.

Hence consider $m^*A < \infty$ and $A \neq \varnothing$. With $n \in J$, let $I_1, I_2, \cdots, I_k, \cdots$ be open intervals such that

$$A \subset \bigcup I_k \qquad \text{and} \qquad \sum |I_k| < m^*A + \frac{1}{n},$$

which is possible by Problem 1. Then each I_k is m-measurable and $mI_k = |I_k|$. The set G_n defined by $G_n = \bigcup I_k$ is open, since each I_k is open,

$$A \subset G_n, \qquad \text{and} \qquad mG_n \leqslant \sum mI_k < m^*A + \frac{1}{n}.$$

† A set $\hat{A}$ having these properties is sometimes called an **equimeasurable superset** of A.

Set $\hat{A} = \bigcap G_n$. Then $\hat{A}$ is m-measurable (Theorem 4–2.6),

$$A \subset \hat{A} \subset G_n, \qquad \text{and} \qquad m^*A \leqslant m\hat{A} \leqslant m^*G_n < m^*A + \frac{1}{n}, \qquad n = 1, 2, 3, \cdots.$$

Thus $m\hat{A} = m^*A$ and (2) is established. Also with $\varepsilon > 0$ arbitrary and $n \in J$ such that $(1/n) < \varepsilon$, the open set G_n satisfies (1). ∎

Theorem 6. *If $A_1 \subset A_2 \subset A_3 \subset \cdots$, then*

$$m^*\left(\bigcup A_n\right) = \lim_{n\to\infty} m^*A_n.$$

Proof. Let $\hat{A}, \hat{A}_1, \hat{A}_2, \cdots$ be m-measurable sets such that

$$\bigcup A_n \subset \hat{A}, \qquad m^*\left(\bigcup A_n\right) = m\hat{A},$$
$$A_n \subset \hat{A}_n, \qquad \text{and} \qquad m^*A_n = m\hat{A}_n, \qquad n = 1, 2, \cdots.$$

Set

$$B_n = \bigcap_{i=n}^{\infty} (\hat{A}_i \hat{A}).$$

Then B_n and $\bigcup B_n$ are m-measurable, $B_1 \subset B_2 \subset B_3 \subset \cdots$,

$$A_n \subset B_n \subset \hat{A}_n, \qquad \text{and} \qquad \bigcup A_n \subset \bigcup B_n \subset \hat{A}$$

so that $m^*A_n = mB_n$ and $m^*(\bigcup A_n) = m(\bigcup B_n)$. Since

$$m\left(\bigcup B_n\right) = \lim_{n\to\infty} mB_n,$$

by Theorem 4-2.9, the desired result is seen to hold. ∎

Theorem 7. *If A has m^*A finite and there is an m-measurable subset*† $\underset{\smile}{A} \subset A$ *with $m\underset{\smile}{A} = m^*A$, then A is m-measurable.*

Proof. Since $\underset{\smile}{A}$ is an m-measurable subset of A, then (Problem 4–2.7)

$$m^*(A - \underset{\smile}{A}) = m^*A - m\underset{\smile}{A} = m^*A - m^*A = 0,$$

since $m^*A < \infty$. Since $m^*(A - \underset{\smile}{A}) = 0$, then $A - \underset{\smile}{A}$ is m-measurable and hence $A = \underset{\smile}{A} \cup (A - \underset{\smile}{A})$ is m-measurable. ∎

The m-measurability of a set A requires that the additivity property $m^*W = m^*WA + m^*WA^c$ hold for all sets W. The following theorem gives a test for m-measurability of a set A in terms of a single set M, provided this test set M is known to be m-measurable and $A \subset M$.

† Called an **equimeasurable kernel** of A.

Theorem 8. *Given an m-measurable set M with $mM < \infty$, then a set $A \subset M$ is m-measurable iff*

$$mM = m^*A + m^*(M - A).$$

[If A were bounded and M were an interval I, this would be Lebesgue's original definition of measurability of A.]

Proof. If A is m-measurable, then in particular (with $W = M$),

$$mM = m^*MA + m^*MA^c = m^*A + m^*(M - A).$$

Conversely, consider that the equation holds. Take $\hat{B}$, an m-measurable set such that $M - A \subset \hat{B}$ and $m^*(M - A) = m\hat{B}$, which is possible by Theorem 5. Then

$$M - M\hat{B} \text{ is an } m\text{-measurable subset of } A,$$

$$mM = mM\hat{B} + mM\hat{B}^c = mM\hat{B} + m(M - M\hat{B}).$$

But $M - A \subset M\hat{B} \subset \hat{B}$, $m^*(M - A) \leqslant mM\hat{B} \leqslant m\hat{B} = m^*(M - A)$ so that $mM\hat{B} = m^*(M - A)$. Consequently,

$$\begin{aligned} mM &= m^*A + m^*(M - A) \\ &= m^*A + mM\hat{B} \\ &= m^*A + mM - m(M - M\hat{B}). \end{aligned}$$

Thus $m^*A = m(M - M\hat{B})$ and hence A is m-measurable, by Theorem 7. ∎

Theorem 9. *Let $A \subset E^1$ be an m-measurable set. Then corresponding to each number $\varepsilon > 0$ there is a closed set F and an open set G such that*

$$F \subset A \subset G, \qquad m(A - F) < \varepsilon, \qquad \textit{and} \qquad m(G - A) < \varepsilon.$$

Proof. Label the closed intervals

$$[0, 1], [-1, 0], [1, 2], [-2, -1], \cdots, [k - 1, k], [-k, -k + 1], \cdots$$

in some order as $I_1, I_2, I_3, \cdots$. With $n \in J$, set

$$A_n = A \cap I_n \qquad \text{and} \qquad B_n = I_n - A_n.$$

By Theorem 5, let G_n and H_n be open sets such that

$$A_n \subset G_n, \qquad mG_n < mA_n + \frac{\varepsilon}{2^n},$$

$$B_n \subset H_n, \qquad \text{and} \qquad mH_n < mB_n + \frac{\varepsilon}{2^n}.$$

From Problem 4–2.7, the measurability of A_n and B_n, and $mA_n < \infty$ and $mB_n < \infty$, it follows that

$$m(G_n - A_n) = mG_n - mA_n < \frac{\varepsilon}{2^n} \qquad \text{and}$$

(3) $$0 \leqslant mI_nH_n - mB_n \leqslant mH_n - mB_n < \frac{\varepsilon}{2^n}.$$

Set $G = \bigcup G_n$. Then G is an open set, $G \supset A$, and

$$\begin{aligned} x \in G - A &= \bigcup G_n - \bigcup A_n \\ \Rightarrow x \in G_n &\quad \text{for some } n \text{ and } x \notin A_n \text{ for any } n \\ \Rightarrow x \in \bigcup &(G_n - A_n). \end{aligned}$$

Thus $G - A \subset \bigcup (G_n - A_n)$, and

$$m(G - A) \leqslant \sum m(G_n - A_n) < \sum \frac{\varepsilon}{2^n} = \varepsilon.$$

Hence the open set G has the properties stated in the theorem.

Next let $F_n = I_n - I_n H_n = I_n \cap H_n^c$. Then F_n is closed, since I_n and H_n^c are both closed. Also $I_n H_n \supset B_n$, and hence $F_n \subset I_n - B_n = A_n$; that is, F_n is a closed subset of A_n. Moreover,

$$\begin{aligned} mF_n &= mI_n - m(I_n H_n) && \text{by Problem 4–2.7} \\ &> mI_n - mB_n - \frac{\varepsilon}{2^n} && \text{by (3)} \\ &= m(I_n - B_n) - \frac{\varepsilon}{2^n} = mA_n - \frac{\varepsilon}{2^n}. \end{aligned}$$

Hence, again by Problem 4–2.7,

$$m(A_n - F_n) = mA_n - mF_n < \frac{\varepsilon}{2^n}.$$

Set $F = \bigcup F_n$. A countable union of closed sets need not be closed, but F is closed. For if x is a cluster point of F, then either x is an interior point or an end point of one of the closed intervals $I_1, I_2, I_3, \cdots$, each of length 1; hence x is a cluster point of at least one of the closed sets in $\{F_n\}$, and thus is a point of at least one of these sets so that $x \in \bigcup F_n = F$. Also, $F \subset \bigcup A_n = A$,

$$A - F = \bigcup A_n - \bigcup F_n \subset \bigcup (A_n - F_n), \qquad \text{and}$$
$$m(A - F) \leqslant \sum m(A_n - F_n) < \varepsilon. \blacksquare$$

Theorem 10. *A set $A \subset E^1$ is m-measurable iff corresponding to each number $\varepsilon > 0$ there is a closed set $F \subset A$ for which $m^*(A - F) < \varepsilon$.*

Proof. If A is m-measurable, then such closed subsets F exist by Theorem 9 and thus we need to prove only the converse.

For each integer n, let F_n be closed $F_n \subset A, m^*(A - F_n) < 1/n$, and set $\underline{A} = \bigcup F_n$. Then $\underline{A}$ is a measurable

$$A - \underline{A} \subset A - F_n \qquad \text{and} \qquad m^*(A - \underline{A}) \leqslant m^*(A - F_n) < \frac{1}{n}.$$

Thus $m^*(A - \underline{A}) = 0$ so that $A - \underline{A}$ is m-measurable. Consequently

$$A = A \cup (A - \underline{A}) \text{ is } m\text{-measurable. } \blacksquare$$

5–2 PROPERTIES OF LEBESGUE MEASURE

The order properties of E^1 do not enter in an essential way before the last theorem of this section. Thus the material prior to the last theorem might as well be for E^k except for some complications in terminology and notation.

Definition 1. *A set $A \subset E^1$ is said to be covered in the Vitali sense by a family $\mathscr{F}$ of intervals if each interval of $\mathscr{F}$ is closed with positive length and each point of A is in an interval of $\mathscr{F}$ with arbitrarily small length; that is, given any $a \in A$ and any $\delta > 0$, there is an interval I such that*

$$a \in I \in \mathscr{F} \quad \text{and} \quad 0 < |I| < \delta.$$

Theorem 1. (Vitali) *Let A be a set with $m^*A < \infty$ and $\mathscr{F}$ a family of closed intervals covering A in the Vitali sense.*

(i) *Then there is a countable pairwise disjoint subfamily of $\mathscr{F}$ which covers almost all of A.*

(ii) *Given a number $\varepsilon > 0$, there is a finite number of intervals $I_1, I_2, \cdots, I_n$ in $\mathscr{F}$, no two having a point in common, such that*

$$m^*\left(A - \bigcup_{k=1}^{n} I_k\right) < \varepsilon.$$

Proof. Determine (by Theorem 5–1.5) an open set G such that

$$A \subset G \quad \text{and} \quad mG < m^*A + 1.$$

Discard from $\mathscr{F}$ all intervals which intersect G^c and let $\mathscr{F}_1$ be the remaining subfamily. Since G^c is closed, $\mathscr{F}_1$ also covers A in the Vitali sense. Let

$$b_1 = \sup\{|I| \mid I \in \mathscr{F}_1\}$$

and note that $0 < b_1 \leqslant mG < \infty$, since $I \in \mathscr{F}_1 \Rightarrow I \subset G$. Select an interval $I_1 \in \mathscr{F}_1$ with $|I_1| > \frac{1}{2}b_1$.

If $m^*(A - I_1) = 0$ we are already through. If $m^*(A - I_1) > 0$, then discard from $\mathscr{F}_1$ all intervals intersecting I_1 and let $\mathscr{F}_2$ be the remaining subfamily of $\mathscr{F}_1$. Then, since I_1 is closed, $\mathscr{F}_2$ covers $A - I_1$ in the Vitali sense. Let $b_2 = \sup\{|I| \mid I \in \mathscr{F}_2\}$ and select an interval $I_2 \in \mathscr{F}_2$ such that $|I_2| > \frac{1}{2}b_2$.

With $I_1, I_2, \cdots, I_n$ pairwise disjoint intervals selected from the successively decreasing subfamilies $\mathscr{F}_1, \mathscr{F}_2, \cdots, \mathscr{F}_n$, we make the inductive step as follows:

(a) If $m^*[A - (I_1 \cup I_2 \cup \cdots \cup I_n)] = 0$, then stop the process.

(b) If $m^*[A - \bigcup_{k=1}^{n} I_k] > 0$, then discard from $\mathscr{F}_n$ all intervals which intersect I_n and let $\mathscr{F}_{n+1}$ be the remaining subfamily. Since $\bigcup_{k=1}^{n} I_k$ is closed, $\mathscr{F}_{n+1}$ covers $A - \bigcup_{k=1}^{n} I_k$ in the Vitali sense. Let

$$b_{n+1} = \sup\{|I| \mid I \in \mathscr{F}_{n+1}\}$$

and select $I_{n+1} \in \mathscr{F}_{n+1}$ such that

$$|I_{n+1}| > \tfrac{1}{2}b_{n+1}.$$

If the process stops at some stage n, then by (a) the finite pairwise disjoint subfamily $\{I_1, I_2, \cdots, I_n\}$ of $\mathscr{F}$ covers almost all A.

Hence, consider that the process does not stop. We now have a sequence $\{I_n\}$ of pairwise disjoint intervals selected from $\mathscr{F}_1$ and thus $I_n \subset G$, $n = 1, 2, 3, \cdots$. Therefore,

$$\infty > m(G) \geqslant m\left(\bigcup_{n=1}^{\infty} I_n\right) = \sum_{n=1}^{\infty} m(I_n) \qquad \text{by Theorem 4–2.8.}$$

This series therefore converges, from which we also note that $\lim |I_n| = 0$ and hence

$$\lim_{n \to \infty} b_n = 0.$$

Let J_n be the interval (not necessarily in $\mathscr{F}$) concentric with I_n but with $|J_n| = 5|I_n|$, $n = 1, 2, 3, \cdots$. We next show that

$$(1) \qquad A - \bigcup_{k=1}^{N-1} I_k \subset \bigcup_{k=N}^{\infty} J_k$$

for each positive integer $N \geqslant 2$. The set on the left, having positive outer measure, is nonempty and we select any point x_0 of it. Also, this set is covered in the Vitali sense by $\mathscr{F}_N$ and we select an interval $I_0 \in \mathscr{F}_N$ such that $x_0 \in I_0$. Hence

$$0 < |I_0| \leqslant b_N \qquad \text{and} \qquad I_0 \cap \left(\bigcup_{k=1}^{N-1} I_k\right) = \varnothing.$$

But I_0 intersects at least one of the intervals $I_N, I_{N+1}, I_{N+2}, \cdots$, for otherwise I_0 would be an element of $\mathscr{F}_N, \mathscr{F}_{N+1}, \mathscr{F}_{N+2}, \cdots$, and therefore $0 < |I_0| \leqslant b_{N+k}$, $k = 0, 1, 2, \cdots$, contrary to $\lim b_n = 0$. Let $k_0 \geqslant 0$ be such that

$$I_0 \cap I_{N+k} = \varnothing \qquad \text{for } k = -1, 0, 1, \cdots, k_0 - 1, \qquad \text{but}$$

$$I_0 \cap I_{N+k_0} \neq \varnothing.$$

Thus both I_0 and I_{N+k_0} are intervals in $\mathscr{F}_{N+k_0}$, with I_{N+k_0} the interval selected at the stage $N + k_0$ so that

$$|I_0| \leqslant b_{N+k_0} \qquad \text{but} \qquad |I_{N+k_0}| > \tfrac{1}{2} b_{N+k_0}.$$

Hence $|I_{N+k_0}| > \frac{1}{2}|I_0|$, and since $I_0 \cap I_{N+k_0} \neq \varnothing$,

$$x_0 \in I_0 \subset J_{N+k_0}.$$

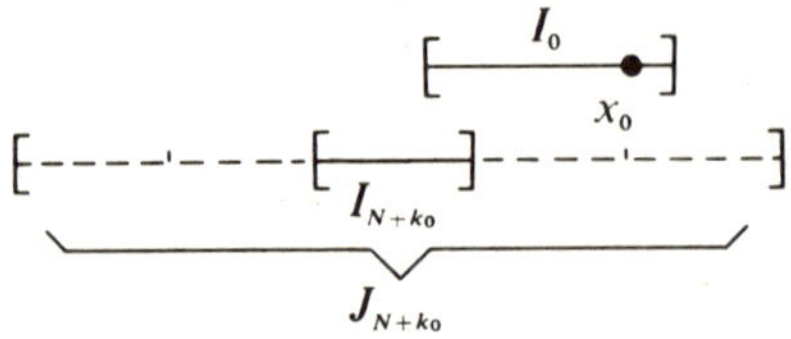

Thus (1) is established, since x_0 was any point of $A - \bigcup_{k=1}^{N-1} I_k$, and

$$m^*\left(A - \bigcup_{n=1}^{\infty} I_n\right) \leqslant m^*\left(A - \bigcup_{k=1}^{N-1} I_k\right) \leqslant m\left(\bigcup_{k=N}^{\infty} J_k\right) \leqslant \sum_{k=N}^{\infty} m(J_k) = 5 \sum_{k=N}^{\infty} |I_k|.$$

The last sum approaches zero as $N \to \infty$, since $\sum |I_n|$ converges, and thus

$$m^*\left(A - \bigcup_{n=1}^{\infty} I_n\right) = 0.$$

Hence $\{I_1, I_2, I_3, \cdots\}$ is a countable pairwise disjoint subfamily of $\mathscr{F}$ covering almost all of A.

This finishes the proof of part (i) of the theorem. Part (ii) follows, since

$$\lim_{n\to\infty} m^*\left(A - \bigcup_{k=1}^{n} I_k\right) = m^*\left(A - \bigcup_{n=1}^{\infty} I_n\right) = 0. \blacksquare$$

Notice how easy it would have been to replace E^1 by E^k and each interval by a closed k-ball or a closed k-cube in the whole Vitali discussion.

Definition 2. *Given a set $A \subset E^1$, then for each $x \in E^1$, the limits inferior and superior as $h \to 0+$ of*

$$\frac{m^*(A \cap [x - h, x + h])}{2h}$$

are denoted by $\underline{D}(A, x)$ and $\overline{D}(A, x)$ and are called the **lower** *and* **upper densities** *of A at x. If these are equal, their common value is denoted by $D(A, x)$ and called the* **density** *of A at x.*

Theorem 2

(i) *For any set $A \subset E^1$, the density exists and $D(A, x) = 1$ at almost all points of A.*

(ii) *If A is measurable, then $D(A, x)$ exists for almost all $x \in E^1$ and*

$$D(A, x) = \begin{cases} 1 & \text{for almost all } x \in A \\ 0 & \text{for almost all } x \in A^c. \end{cases}$$

Proof. It is clear that all these densities are $\leqslant 1$ at all points of E^1 (and if $mA = 0$, there is nothing to prove). Let $B_0 = \{x \in A \mid \underline{D}(A, x) < 1\}$ and

$$B_n = \left\{x \in A \,\middle|\, \underline{D}(A, x) < 1 - \frac{1}{n}\right\}$$

for $n = 2, 3, 4, \cdots$. Then $B_0 = \bigcup B_n$ and hence

$$m^* B_0 \leqslant \sum_{n=2}^{\infty} m^* B_n. \tag{2}$$

First consider A bounded. Since $B_n \subset A$, then $m^* B_n \leqslant m^* A < \infty$. Take a fixed integer $n \geqslant 2$, let $\varepsilon > 0$ be given, and take an open set G such that

$$B_n \subset G \qquad \text{and} \qquad mG < m^* B_n + \varepsilon.$$

With $x \in B_n$, let $\mathscr{F}(x)$ be the family of all closed intervals $I = [x - h, x + h]$ such that

$$\frac{m^*(AI)}{|I|} < 1 - \frac{1}{n} \qquad \text{and} \qquad I \subset G.$$

There are such intervals with arbitrarily small $h > 0$. Thus the family

$$\mathscr{F} = \bigcup_{x \in B_n} \mathscr{F}(x)$$

covers B_n in the sense of Vitali. By Vitali's theorem, let $\{I_i\}$ be a sequence of intervals in $\mathscr{F}$ such that

$$m\left(B_n - \bigcap I_i\right) = 0, \qquad I_i \cap I_j = \varnothing \qquad \text{or } i \neq j.$$

Hence $m^*B_n = m^*\left(B\left[\bigcup I_i\right]\right)$, since $\bigcup I_i$ is m-measurable, and

$$m^*B_n \leqslant \sum_i m^*(B_n I_i) \leqslant \sum_i m^*(AI_i) < \left(1 - \frac{1}{n}\right) \sum_i |I_i|$$

$$= \left(1 - \frac{1}{n}\right) m\left(\bigcup I_i\right) \leqslant \left(1 - \frac{1}{n}\right) m(G) < \left(1 - \frac{1}{n}\right)(m^*B_n + \varepsilon).$$

Therefore $m^*B_n < (n - 1)\varepsilon$.

Consequently, $mB_n = 0$, $n = 2, 3, 4, \cdots$. Hence $mB_0 = 0$ from (2). This says that $D(A, x) = 1$ at almost all points of A if A is bounded.

In case A is not bounded, apply the above result to $A \cap [-k, k]$, $k = 1, 2, 3, \cdots$. Since a countable union of sets of measure zero has measure zero, part (i) of the theorem is proved.

Hence $D(A^c, x) = 1$ for almost all $x \in A^c$. In case A is m-measurable, we have

$$m(AI) + m(A^cI) = |I|$$

for any interval I and (ii) follows easily from this. ∎

PROBLEM 1. (a) Given a set $A \subset E^1$, let $\underline{D}_+(A, x)$ and $\overline{D}_+(A, x)$ be the limits inferior and superior as $h \to 0+$ of

$$\frac{m^*(A \cap [x, x + h])}{h}, \qquad x \in E^1.$$

Then $\underline{D}_+(A, x) = \overline{D}_+(A, x) = 1$ for almost all $x \in A$.

(b) For almost all $x \in A$

$$\underline{\lim}_{|I| \to 0} \frac{m^*AI}{|I|} = \overline{\lim}_{|I| \to 0} \frac{m^*AI}{|I|} = 1$$

where the intervals I involved have $x \in I$ and $|I| > 0$.

Given a set A and a number $x \in E^1$, we use the notation $x + A$ (or $A + x$) for the **translate** of A by x:

$$\begin{aligned} x + A &= \{y \mid y = x + a \quad \text{for some } a \in A\} \\ &= \{y \mid y - x \in A\}. \end{aligned}$$

Thus an interval $I = (a, b)$ and its translate $x + I = (x + a, x + b)$ have the same length $b - a$. Also

$$\begin{aligned} x + A^c &= \{y \mid y - x \in A^c\} = \{y \mid y - x \notin A\} \\ &= \{y \mid y \notin x + A\} = (x + A)^c. \end{aligned} \tag{3}$$

In addition, for any set W,

$$\begin{aligned} x + A(-x + W) &= \{y \mid y - x \in A(-x + W)\} \\ &= \{y \mid y - x \in A \text{ and } y - x \in -x + W\} \\ &= \{y \mid y - x \in A \text{ and } y \in W\} \\ &= \{y \mid y - x \in A\} \cap \{y \mid y \in W\} \\ &= (x + A)W. \end{aligned} \tag{4}$$

Theorem 3. *Any set A and any translate of A have the same outer m^*-measure. Also A is m-measurable iff each of its translates is m-measurable.*

Proof. With $\varepsilon > 0$ and $I_1, I_2, I_3, \cdots$ such that

$$A \subset \bigcup I_n \qquad \text{and} \qquad \sum |I_n| \leqslant m^*A + \varepsilon,$$

then $x + A \subset \bigcup (x + I_n)$ and

$$m^*(x + A) \leqslant \sum |x + I_n| = \sum |I_n| \leqslant m^*A + \varepsilon.$$

Hence $m^*(x + A) \leqslant m^*A$. But A is the translate by $-x$ of $x + A$ and thus $m^*A \leqslant m^*(x + A)$. Therefore

$$m^*(x + A) = m^*(A). \tag{5}$$

Let A be m-measurable and select any set W. Then

$$\begin{aligned} m^*W(x + A) &+ m^*W(x + A)^c \\ &= m^*W(x + A) + m^*W(x + A^c) \qquad \text{by (3)} \\ &= m^*[x + A(-x + W)] + m^*[x + A^c(-x + W)] \qquad \text{by (4)} \\ &= m^*A(-x + W) + m^*A^c(-x + W) \qquad \text{by (5)} \\ &= m^*(-x + W) \qquad \text{since } A \text{ is measurable} \\ &= m^*W \qquad \text{by (5),} \end{aligned}$$

and thus $x + A$ is m-measurable.

Conversely, if $x + A$ is m-measurable, then its translate $-x + (x + A) = A$ is m-measurable. ∎

Lemma. *Let A be an m-measurable set having the origin as a point of density (that is, $D(A, 0) = 1$), and let n be a positive integer. Then there is a number $\eta > 0$ such that*

$$(|h_i| < \eta, \qquad i = 1, 2, \cdots, n) \Rightarrow m[(A + h_1)(A + h_2)\cdots(A + h_n)] > 0.$$

Proof. Select a number $\varepsilon > 0$. [For the moment consider that $0 < \varepsilon < 1$, but we shall obtain a smaller upper bound presently.] Choose a number $b > 0$ such that, for the interval $I = (-b, b)$, we have

$$\frac{m(A \cap I)}{2b} > (1 - \varepsilon).$$

Set $A \cap I = B$ and therefore have that $B \subset I$ and

$$m(IB^c) = mI - mB < 2b\varepsilon.$$

Given a number h, we note that any point of the translate $B + h$ falling outside the interval I will be in the interval $[b, b + h]$ if $h > 0$, but in $[-b + h, -b]$ if $h < 0$ so that in either case,

$$m[I^c(B + h)] \leqslant |h|.$$

By expressing $I \cup (B + h)$ in two ways, we see that

$$\begin{aligned} m[I \cup (B + h)] &= mI + m[I^c(B + h)] \\ &= m(B + h) + m[I(B + h)^c]. \end{aligned}$$

Since $m(B + h) = mB$, we thus have

$$\begin{aligned} m[I(B + h)^c] &= mI - mB + m[I^c(B + h)] \\ &< 2b\varepsilon + |h|. \end{aligned}$$

Now let $h_1, h_2, \cdots, h_n$ be given numbers and note that

$$\begin{aligned} m\left[I\bigcap_{i=1}^{n}(B + h_i)\right] &= mI - m\left\{I\left[\bigcap_{i=1}^{n}(B + h_i)\right]^c\right\} \\ &= 2b - m\left[\bigcup_{i=1}^{n} I(B + h_i)^c\right] \\ &\geqslant 2b - \sum_{i=1}^{n} m[I(B + h_i)^c] \\ &> 2b - \sum_{i=1}^{n} (2b\varepsilon + |h_i|) \\ &= 2b - 2nb\varepsilon - \sum_{i=1}^{n} |h_i|. \end{aligned}$$

This final number would be $\geqslant b > 0$ had we chosen ε and η such that

$$0 < \varepsilon < \frac{1}{4n}, \qquad 0 < \eta < \frac{b}{2n}, \qquad \text{and required} \qquad |h_i| < \eta, \qquad i = 1, 2, \cdots, n.$$

Since $A \supset B$, and hence $A + h_i \supset I(B + h_i)$, the stated result is seen to hold. ∎

Theorem 4. *Let $A \subset E^1$ be a measurable set with $mA > 0$ and let n be a positive integer. Then there is a number $\eta > 0$ such that*

$$(|h_i| < \eta,\ i = 1, 2, \cdots, n) \Rightarrow m[(A + h_1)(A + h_2) \cdots (A + h_n)] > 0.$$

[Note that all the above work on translation could be stated in E^k by using k-component vectors $\mathbf{x}$ as translators. An interpretation of Theorem 4 in E^2 is that any plane measurable set A having $m_2 A > 0$ contains three points, which are the vertices of a triangle similar to any given triangle; or, in more general terms, given any random scattering of a finite number of points in the plane, there is a similar scatter of points, all belonging† to A.]

Proof. Take any point a which is a density point of A. Then the origin is a density point of the translate $A - a$, and we apply the lemma to this set. ∎

The following very special corollary will be used in Example 6–9.1.

Corollary. *If $A \subset E^1$ is measurable and $mA > 0$, then there is an $\eta > 0$ such that corresponding to each number h, where $|h| < \eta$, there is an x satisfying all three of*

$$x \in A, \qquad x + h \in A, \qquad \textit{and} \qquad x - h \in A.$$

Proof. Merely take $n = 3$, $h_1 = 0$, $h_2 = h$, and $h_3 = -h$ with $|h| < \eta$. There is an $x \in A(A + h)(A - h)$, since the set has positive measure. Hence there is a $y_1 \in A$ and $y_2 \in A$ such that $y_1 - h = x$ and $y_2 + h = x$. Thus $x \in A$, $y_1 = x + h \in A$, and $y_2 = x - h \in A$. ∎

PROBLEM 2.‡ Let $A_1, A_2, \cdots, A_n$ be measurable sets, each having positive measure, and let a_i be a point of density of A_i, $i = 1, 2, \cdots, n$. Then there exists a number $\eta > 0$ such that

$$(|x_i - a_i| < \eta,\ i = 1, 2, \cdots, n) \Rightarrow m[(A_1 - x_1)(A_2 - x_2) \cdots (A_n - x_n)] > 0.$$

Associated with each set A is the **difference set**

$$\Delta(A) = \{a - b \mid a \in A \text{ and } b \in A\}.$$

With $h_1 = 0$ and $n = 2$ in Theorem 4, we may state the following special result, which will then be used in establishing the nonmeasurability of a certain set.

Theorem 5. *If $A \subset E^1$ is measurable with $mA > 0$, then $\Delta(A)$ contains an interval [in fact, there is a number $\eta > 0$ such that*

$$(-\eta, \eta) \subset \Delta(A)].$$

[*Note:* Recall that in Section 2–9 it was shown that the Cantor set Γ has $\Delta(\Gamma) = [-1, 1]$. This result was obtained after Theorem 5 was known (by a

† For a proof of this using a different method see Theodore A. Bick, "Configurations in Plane Measurable Sets," *Am. Math. Monthly*, to appear soon.

‡ For this result, other similar results, and applications, see J. H. B. Kemperman, "A General Functional Equation," *Trans. Am. Math. Soc.* **86**, No. 1 (September 1957), pp. 28–56.

method different from that given here). Knowing Theorem 5, Steinhaus† wondered if there is any set of measure zero whose difference set fills an interval. The Cantor set was a natural candidate to try.]

Theorem 6. *There is an m-nonmeasurable set.*

Proof. Corresponding to each number $x \in E^1$, let

$$B_x = \{t \mid t - x \text{ is rational}\}.$$

Hence, with x and y any numbers, there is a set B_x and a set B_y. We first show (and it is crucial in the proof to follow) that:

(6) If $x - y = r_0$ is rational, then $B_x = B_y$, but

(7) If $x - y$ is irrational, then $B_x \cap B_y = \varnothing$.

For, under the conditions in (6), let $b \in B_x$. Then there is a rational number r_1 such that

$$b = x + r_1 = (y + r_0) + r_1 = y + (r_0 + r_1).$$

Since $r_0 + r_1$ is rational, $b \in B_y$. Thus $B_x \subset B_y$, and in the same way, $B_y \subset B_x$.

Under the conditions in (7), suppose there is an element $d \in B_x \cap B_y$. Then there are rational numbers r_x and r_y such that $b = x + r_x$ and $b = y + r_y$. Hence $x - y = r_y - r_x$, which contradicts $x - y$ being irrational.

A family $\mathscr{F}$ of sets is therefore defined such that if $B' \in \mathscr{F}$ and $B'' \in \mathscr{F}$, then either‡

$$B' = B'' \qquad \text{or else} \qquad B' \cap B'' = \varnothing.$$

Each set $B \in \mathscr{F}$ is nonempty, in fact is countably infinite. Select one and only one point from each B and let A be the set of points selected.§

From the definition of A, we see that $a \in A$ iff there is a $B \in \mathscr{F}$ such that

$$A \cap B = \{a\}.$$

Thus, with $a' \in A$, $a'' \in A$, and $a' \neq a''$, the sets $B' \in \mathscr{F}$ and $B'' \in \mathscr{F}$, for which

$$A \cap B' = \{a'\} \qquad \text{and} \qquad A \cap B'' = \{a''\},$$

are such that $B' \cap B'' = \varnothing$ and hence, from (6) and (7), $a' - a''$ is irrational. The difference set $\Delta(A)$, therefore, contains no nonzero rational, and thus

$$\Delta(A) \text{ does not contain an interval.}$$

† H. Steinhaus, *Nowa vlasnośe mnogości G. Cantor*, Wektor, Poland, 1917, p. 105.

‡ Those familiar with group theory will recognize the elements of $\mathscr{F}$ as the cosets of the additive group E^1 over the subgroup of the rationals.

§ Whether this innocuous-sounding instruction can be followed depends upon the so-called "choice axiom," which most mathematicians accept but others reject. For a thorough discussion, see A. P. Morse, *Set Theory*, McGraw-Hill, New York, 1966. Actually we have already used the choice axiom, in hidden or mild form, several times.

This means, by using Theorem 5, that

A cannot be m-measurable with positive measure; that is,

if A is m-measurable, then it has $mA = m^*A = 0$.

We next show that $m^*A > 0$ (and as soon as this is done we shall have that A is m-nonmeasurable).

Let $r_0 = 0$ and let $r_1, r_2, r_3, \cdots$ be an ordering, without duplication, of the nonzero rational numbers. Let A_n denote the translate of A by r_n; that is, for $n = 0, 1, 2, 3, \cdots$,

$$A_n = r_n + A = \{x \mid x - r_n \in A\}.$$

Then $m^*A_n = m^*A$, $n = 0, 1, 2, 3, \cdots$, and

$$\sum_{k=0}^{n} m^*A_k = (n+1)m^*A.$$

Also $\bigcup A_n = E^1$. For take any $x \in E^1$. Then $B_x \in \mathscr{F}$ and we let a_x be the unique number such that

$$B_x \cap A = \{a_x\}.$$

Since x and a_x are both elements of B_x, there is a nonnegative integer n such that

$$x - a_x = r_n.$$

Therefore $x \in A_n$, since $x = r_n + a_x$ and $a_x \in A$, and hence $E^1 = \bigcup A_n$. Consequently

$$\infty = mE^1 \leqslant \sum_{n=0}^{\infty} m^*A_n = \lim_{n\to\infty} (n+1)m^*A$$

and it follows that $m^*A > 0$. Thus, as parenthetically noted above, A is m-nonmeasurable. ∎

Corollary. *Given any set M having $m^*M > 0$, then there is a nonmeasurable set $N \subset M$.*

Proof. Among the things already known about the sets $A_0, A_1, A_2, \cdots$ are

(8) $$\bigcup_{n=0}^{\infty} A_n = E^1.$$

(9) For each n, if $a \in A_n$, $b \in A_n$, and $a \neq b$, then $b - a$ is irrational.

First $\bigcup_{n=0}^{\infty} (MA_n) = ME^1 = M$ from (8), and then

$$0 < m^*M \leqslant \sum_{n=0}^{\infty} m^*(MA_n).$$

All terms in the sum cannot be zero and we let k be such that

$$m^*(MA_k) > 0.$$

Assume MA_k is measurable. Then its difference set $\Delta(MA_k)$ contains an interval (Theorem 5). Select r rational in this interval and then $a \in MA_k$ and $b \in MA_k$ such that $b - a = r$. This violates (9). Thus MA_k is nonmeasurable and we have a nonmeasurable subset of M. ∎†

PROBLEM 3. The sets $A_0, A_1, A_2, \cdots$ defined in the preceding proof are such that

$$A_n \cap A_k = \varnothing \qquad \text{for } n \neq k;$$

that is, $E^1 = \bigcup_{n=0}^{\infty} A_n$ expresses E^1 as a countable union of pairwise disjoint m-nonmeasurable sets.

PROBLEM 4. With A the set defined in the proof of Theorem 6, if B is m-measurable and $B \subset A$, then $mB = 0$.

Theorem 7. *Given any m-nonmeasurable set A and any m-measurable set* $M \supset A$, *then* $m^*(M - A) > 0$.

[*Note:* Even if $m^*A < \infty$ and $\hat{A} \supset A$ is m-measurable with $m\hat{A} = m^*A$ (as shown to exist in Theorem 5–1.5), then $m^*(\hat{A} - A) > 0$ and in particular $m^*(\hat{A} - A) \neq m\hat{A} - m^*A$.

In contrast, if A is m-measurable with $mA < \infty$ and B is any set such that $B \supset A$, then $m^*(B - A) = m^*B - mA$ as Problem 4–2.7 states.]

Proof. Suppose there is an m-measurable set M such that

$$M \supset A \qquad \text{and} \qquad m^*(M - A) = 0.$$

Then $M - A$ is m-measurable and $A = M - (M - A)$ expresses A as the difference of two m-measurable sets. Thus A is m-measurable, contradicting that A is m-nonmeasurable. ∎

5–3 σ-ALGEBRAS

In a space S on which an outer measure μ^* is defined, the family $\mathfrak{M}$ of all μ-measurable sets is a prototype of σ-algebras.

Definition 1. *In a space S, a family* $\mathscr{A}$ *of sets is said to be a σ-algebra, provided*:
(i) $S \in \mathscr{A}$;
(ii) *If* $A \in \mathscr{A}$, *then also* $A^c \in \mathscr{A}$;
(iii) *If* $A_n \in \mathscr{A}$, $n = 1, 2, 3, \cdots$, *then* $\bigcup A_n \in \mathscr{A}$ [and hence also $\bigcap A_n = (\bigcup A_n{}^c)^c \in \mathscr{A}$].

Note that any σ-ring is a σ-algebra if the whole space is an element of the ring.

† For a construction of nonmeasurable sets without using density properties, see Problem 2 at the end of this chapter.

Any σ-algebra has S as one of its elements and therefore, from (ii), $\varnothing$ is also an element. Moreover, the family consisting only of $\varnothing$ and S is a σ-algebra (for which (iii) is trivially satisfied). Thus the σ-algebra $\{\varnothing, S\}$ is contained in all σ-algebras of subsets of S. On the other hand, the family of all subsets of S is a σ-algebra which is maximal in the sense that it contains all σ-algebras of S.

Theorem 1. *Let $\mathscr{H}$ be a nonempty family of subsets of S.* [$\mathscr{H}$ need not be a σ-algebra.] *Then there is smallest σ-algebra $\mathscr{H}^*$ containing all elements (sets) of $\mathscr{H}$.*

[This means that $\mathscr{H}^*$ is a σ-algebra, that $\mathscr{H} \subset \mathscr{H}^*$, and if $\mathscr{A}$ is any σ-algebra such that $\mathscr{H} \subset \mathscr{A}$, then $\mathscr{H}^* \subset \mathscr{A}$. $\mathscr{H}^*$ is said to be the σ-algebra **generated** by $\mathscr{H}$.]

Proof. There is at least one σ-algebra containing $\mathscr{H}$, namely, the family of all subsets of S. Let $\mathscr{H}^*$ be the intersection of all σ-algebras containing $\mathscr{H}$; that is, a set H is in $\mathscr{H}^*$ iff each and every σ-algebra containing $\mathscr{H}$ has H as one of its elements. Hence, if $\mathscr{A}$ is a σ-algebra and $\mathscr{H} \subset \mathscr{A}$, then $\mathscr{H}^* \subset \mathscr{A}$, since $\mathscr{A}$ is among the σ-algebras whose intersection is defined to be $\mathscr{H}^*$. Also $A \in \mathscr{H} \Rightarrow A \in \mathscr{H}^*$, since A is an element of every σ-algebra containing $\mathscr{H}$. Thus, to complete the proof, we must show that $\mathscr{H}^*$ is a σ-algebra.

Since $\{\varnothing, S\}$ is in every σ-algebra, we have that S is in the intersection of all σ-algebras containing $\mathscr{H}$ so that

$$S \in \mathscr{H}^*.$$

If $H \in \mathscr{H}^*$, then H is in every σ-algebra containing $\mathscr{H}$ so that H^c is in every σ-algebra containing $\mathscr{H}$ and hence also

$$H^c \in \mathscr{H}^*.$$

Finally, take $H_n \in \mathscr{H}^*$, $n = 1, 2, 3, \cdots$, and let $\mathscr{A}$ be any σ-algebra containing $\mathscr{H}$. Then $H_n \in \mathscr{A}$, $n = 1, 2, \cdots$, so that $\bigcup H_n \in \mathscr{A}$ by (iii). Since this is true of every σ-algebra $\mathscr{A}$ containing $\mathscr{H}$, then $\bigcup H_n$ is in the intersection of all such σ-algebras; that is,

$$\bigcup H_n \in \mathscr{H}^*. \quad \blacksquare$$

Definition 2. *Let $\mathscr{G}$ be the family of all open sets in a metric space S and set $\mathscr{B} = \mathscr{G}^*$, where $\mathscr{G}^*$ is the σ-algebra generated by $\mathscr{G}$. The sets of the σ-algebra $\mathscr{B}$ are said to be the* **Borel sets.**

Hence every open set is a Borel set, and then every closed set is a Borel set, since the complement of every set in $\mathscr{G}^*$ is also in $\mathscr{G}^*$. Then, by (iii), all F_σ sets and all G_δ sets are Borel sets.

PROBLEM 1. Let $\mathscr{C}$ be the family of all compact sets in a metric space S. Then $\mathscr{C}^* = \mathscr{B}$.

PROBLEM 2. Confining attention to E^1, let $\mathscr{I}$ be the family of all open intervals. Then $\mathscr{I}^* = \mathscr{B}$.

Throughout the remainder of this section we shall consider that:

The space is E^1 and measurable means Lebesgue measurable.

Since the family $\mathfrak{M}$ of all measurable sets is a σ-algebra containing all open sets and the family $\mathscr{B}$ of Borel sets is the smallest σ-algebra containing all open sets, we have that

$$\mathscr{B} \subset \mathscr{M}.$$

Among other things, this inclusion means that

All Borel sets are Lebesgue measurable.

Borel (the professor) contended that Lebesgue (his student) had really not done anything new and thus believed (in present σ-algebra terminology) that $\mathfrak{M} = \mathscr{B}$. Lebesgue disagreed and the controversy became rather bitter, but neither could settle it. Somewhat later Souslin† constructed what he called analytic sets, some of which are Lebesgue measurable, but not Borel sets. Hence Souslin gave a constructive proof that $\mathscr{B}$ is a proper subalgebra of $\mathfrak{M}$, thus upholding Lebesgue.

The description of analytic sets is extremely detailed. Fortunately there are easy existence proofs, using transfinite arguments, that

$\mathscr{B}$ is a proper subalgebra of $\mathfrak{M}$.

One such proof goes as follows:

1. The Cantor set Γ has measure zero, but Γ has the cardinal $\mathfrak{c}$ of the continuum (the same as E^1 by Theorem 2–2.10).
2. The family of all subsets of Γ has cardinal $2^{\mathfrak{c}}$, by the definition (in Problem 2–9.1, page 96) of the cardinal of the set of subsets of any set.
3. Each subset of Γ is measurable (since it has measure zero).
4. The family $\mathfrak{M}$ of all measurable subsets of E^1 has cardinal $\geqslant 2^{\mathfrak{c}}$, since $\mathfrak{M}$ includes all subsets of Γ. (See Definition 2–9.1 for the use of $\leqslant$ and $<$ between cardinals.)
5. $\mathfrak{M}$ has cardinal $\leqslant 2^{\mathfrak{c}}$ (since $\mathfrak{M}$ is a subfamily of the family of all subsets of E^1 and this family has cardinal $2^{\mathfrak{c}}$).
6. The cardinal of $\mathfrak{M}$ is $2^{\mathfrak{c}}$.
7. The cardinal of $\mathscr{B}$ is $\mathfrak{c}$. [This is the only step not already proved; the following lemmas and theorem prove it.]
8. $\mathfrak{c} < 2^{\mathfrak{c}}$, by Problem 2–9.1.
9. Thus there are sets in $\mathfrak{M}$ which are not in $\mathscr{B}$; that is, there are measurable sets which are not Borel sets.

† M. Souslin "Sur une définition des ensembles B sans nombres transfinis," *Comptes Rendus des Séances de l'Academie des Sciences*, **164** (1917) p. 88.

Lemma 1. *With $\mathscr{I}$ the family of all open intervals in E^1, the cardinal of $\mathscr{I}$ is $\mathfrak{c}$.*

Proof. Each point of E^1 is the lower end point of at least one interval of $\mathscr{I}$ and hence the set of lower end points has cardinal $\mathfrak{c}$. Once a lower end point is selected, there are $\mathfrak{c}$ possibilities for the upper end point of an interval of $\mathscr{I}$. Thus the cardinal of $\mathscr{I}$ is $\mathfrak{c}\cdot\mathfrak{c}$. But $\mathfrak{c}\cdot\mathfrak{c} = \mathfrak{c}$, by (1) page 92. ∎

Let $\mathscr{H}$ be a nonempty family of subsets of E^1. For the following lemmas and theorem we temporarily use $\mathscr{H}'$ and $\mathscr{H}''$ for the families defined by

$$\mathscr{H}' = \{H \mid H^c \in \mathscr{H}\}$$

and say $H \in \mathscr{H}''$ iff there is a countable (finite or infinite) sequence $\{H_n\}$ with each $H_n \in \mathscr{H} \cup \mathscr{H}'$ such that

$$H = \bigcup H_n.$$

Note in particular that $\mathscr{H} \subset \mathscr{H}''$.

Starting with the family $\mathscr{I}$ of open intervals of E^1, we define families

$$\mathscr{F}_1 = \mathscr{I}, \qquad \mathscr{F}_{n+1} = \mathscr{F}_n'', \qquad n = 1, 2, 3, \cdots.$$

Lemma 2. *The cardinal of $\mathscr{F}_n$ is $\mathfrak{c}$.*

Proof. $\mathscr{F}_1$ has cardinal $\mathfrak{c}$, by Lemma 1. Let $\mathscr{F}_n$, $n \geqslant 1$ have cardinal $\mathfrak{c}$. Then $\mathscr{F}_n'$ has cardinal $\mathfrak{c}$ and thus so does $\mathscr{F}_n \cup \mathscr{F}_n'$. Since each member of $\mathscr{F}_n''$ is the union of at most $\aleph_0$ members of $\mathscr{F}_n \cup \mathscr{F}_n'$, the cardinal of $\mathscr{F}_n''$ is $\aleph_0 \cdot \mathfrak{c}$. But $\aleph_0 \cdot \mathfrak{c} = \mathfrak{c}$ by (2) page 93. Hence $\mathscr{F}_{n+1} = \mathscr{F}_n''$ also has cardinal $\mathfrak{c}$. ∎

Lemma 3. *$\bigcup_{n=1}^{\infty} \mathscr{F}_n$ has cardinal $\mathfrak{c}$.*

Proof. The cardinal is $\aleph_0 \cdot \mathfrak{c} = \mathfrak{c}$. ∎

Lemma 4. *$\bigcup \mathscr{F}_n$ is a σ-algebra and hence $\mathscr{I}^* \subset \bigcup \mathscr{F}_n$.*

Proof. $E^1 \in \mathscr{F}_2 \subset \bigcup \mathscr{F}_n$.

Select any $F \in \bigcup \mathscr{F}_n$. Then there is an n such that $F \in \mathscr{F}_n$. Thus $F^c \in \mathscr{F}_n' \subset \mathscr{F}_n \cup \mathscr{F}_n' \subset \mathscr{F}_{n+1}$. Hence $F^c \in \bigcup \mathscr{F}_n$.

Take any sequence $\{F_k\}$ of sets in $\bigcup \mathscr{F}_n$. Then there is a sequence $\{n_k\}$ such that $F_k \in \mathscr{F}_{n_k}$ and hence

$$\bigcup_k F_k \in \bigcup_k \mathscr{F}_{n_k} \subset \bigcup_n \mathscr{F}_n. \quad ∎$$

PROBLEM 3. $\bigcup \mathscr{F}_n \subset \mathscr{I}^*$.

Theorem 2. *$\mathscr{B}$ has cardinal $\mathfrak{c}$.*

Proof. $\mathscr{B} = \mathscr{I}^*$, by Problem 2, $\mathscr{I}^* = \bigcup \mathscr{F}_n$ by Lemma 4 and Problem 3, and the cardinal of $\bigcup \mathscr{F}_n$ is $\mathfrak{c}$ by Lemma 3. ∎

5-4 LEBESGUE OUTER k-MEASURE

A point of E^{k+1} may be denoted by $(\mathbf{x}, z)$, where

$$\mathbf{x} = (x_1, x_2, \cdots, x_k) \in E^k \qquad \text{and} \qquad z \in E^1.$$

Also, if R is a k-cell and I an interval, then $R \times I$ is a $(k + 1)$-cell with $(k + 1)$-volume

$$|R \times I| = |R|\,|I|,$$

whether I is open, closed, or semiopen. Also $R \times I$ is m_{k+1}-measurable and

$$|R \times I| = m_{k+1}(R \times I) = \int_R |I|\, dm_k$$

in the sense that the step function s defined by

$$s(\mathbf{x}) = \begin{cases} |I| & \text{if } \mathbf{x} \in R \\ 0 & \text{if } \mathbf{x} \in E^k - R \end{cases}$$

is m_k-measurable with

$$\int_R s(\mathbf{x})\, dm_k = \mathscr{I}(s, R) = |I|\,|R| = m_{k+1}(R \times I).$$

Theorem 1. *If $A \subset E^k$ and $a \in E^1$, then the set*

$$B = \{(\mathbf{x}, a) \,|\, \mathbf{x} \in A\} \textit{ has } m_{k+1}B = 0.$$

[Intuitively, any k-dimensional set considered as embedded in $(k + 1)$-dimensional space has m_{k+1} measure zero.]

Proof. Take a k-cube $R_n = \{\mathbf{x} \,|\, -n \leqslant x_i \leqslant n,\ i = 1, 2, \cdots, n\}$ and let $B_n = \{(\mathbf{x}, a) \,|\, \mathbf{x} \in AR_n\}$. Choose $\varepsilon > 0$ arbitrarily. Then the $(k + 1)$-cell

$$S_n = \{(\mathbf{x}, z) \,|\, \mathbf{x} \in R_n,\ a \leqslant z \leqslant a + \varepsilon\} = R_n \times [a, a + \varepsilon]$$

is such that $B_n \subset S_n$ and $m^*_{k+1}B_n \leqslant |S_n| = (2n)^k\varepsilon$. Thus $m^*_{k+1}B_n = 0$, $n = 1, 2, 3, \cdots$. Hence $m_{k+1}\, B = 0$, since B is the countable union $\bigcup B_n$ of sets of m_{k+1}-measure zero. ∎

PROBLEM I. If $A \subset E^k$ and I is an interval of E^1, then

$$m^*_{k+1}(A \times I) \leqslant |I|\, m^*_k A.$$

This problem will be used in the following two lemmas, in each of which

$A \subset E^k$ and I is an interval of E^1.

Lemma 1. *If A is m_k-measurable, then $A \times I$ is m_{k+1}-measurable and*

$$m_{k+1}(A \times I) = |I|\, m_k A. \tag{1}$$

Proof. Let $a \in E^1$, $b \in E^1$, and $a < b$. Then the four sets

$$A \times (a, b), \qquad A \times [a, b), \qquad A \times (a, b], \qquad \text{and} \qquad A \times [a, b]$$

differ by sets of m_{k+1}-measure zero (Theorem 1). Hence, if any of these sets is m_{k+1}-measurable and (1) holds for the interval, then the same is true of the others. We shall prove the result with A an m_k-measurable set and $I = [a, b]$.

Given a number $\varepsilon > 0$, take a closed set $F \subset A$ such that $m_k(A - F) < \varepsilon$, which is possible by Theorem 5–1.9. Then $F \times I$ is a closed subset of $A \times I$ and, from Problem 1,

$$m_{k+1}^*(A \times I - F \times I) = m_{k+1}^*[(A - F) \times I] \leqslant |I|\, m_k(A - F) < |I|\, \varepsilon.$$

Hence $A \times I$ is an m_{k+1}-measurable set, by Theorem 5–1.10.

Let $I_n = [-n, n]$ and $R_n = I_n{}^k$. Then $A_n = AR_n$ is m_k-measurable and $m_k(A_n) \leqslant (2n)^k < \infty$. Thus

$$\begin{aligned} m_{k+1}(R_n \times I) - m_{k+1}(A_n \times I) & \\ &= m_{k+1}[(R_n - A_n) \times I] \qquad \text{by Problem 4–2.7} \\ &\leqslant |I|\, m_k(R_n - A_n) = |I|\,(m_k R_n - m_k A_n). \end{aligned}$$

Subtract the finite value $m_k(R_n \times I) = |I|\, m_k R_n$ from both sides and obtain

$$m_{k+1}(A_n \times I) \geqslant |I|\, m_k A_n.$$

Hence, by Problem 1, the equality holds. Consequently, by Theorem 4–2.9,

$$m_{k+1}(A \times I) = \lim_{n\to\infty} m_{k+1}(A_n \times I) = |I| \lim_{n\to\infty} m_k A_n = |I|\, m_k A. \ \blacksquare$$

Lemma 2. *Whether A is m_k-measurable or not,*

$$m_{k+1}^*(A \times I) = |I|\, m_k^* A.$$

Proof. As remarked in the beginning of the proof of Lemma 1, we may consider $I = [a, b]$. Corresponding to $\varepsilon > 0$ arbitrary, take an open set $H \subset E^{k+1}$ such that

$$A \times I \subset H \qquad \text{and} \qquad mH \leqslant m_{k+1}^*(A \times I) + \varepsilon.$$

For each $\mathbf{x} \in A$ let $T_{\mathbf{x}} = \{\mathbf{x}\} \times I$. Then $T_{\mathbf{x}} \subset E^{k+1}$ is a closed subset of H, and thus has a positive distance $r_{\mathbf{x}}$ to the closed set H^c. Let $R_{\mathbf{x}}$ be the open k-cell centered at $\mathbf{x}$ and having diagonal length $\frac{1}{2}r_{\mathbf{x}}$. Note that $R_{\mathbf{x}} \times I \subset H$. The set B defined by

$$B = \bigcup_{\mathbf{x} \in A} R_{\mathbf{x}}$$

is an m_k-measurable subset of E^k, since it is open. Thus

$$m_{k+1}(B \times I) = |I|\, m_k B$$

by Lemma 1. Also $A \subset B$ and $B \times I \subset H$ so that

$$|I|\, m_k^* A \leqslant |I|\, m_k B = m_{k+1}(B \times I) \leqslant m_{k+1} H \leqslant m_{k+1}^*(A \times I) + \varepsilon.$$

Thus $|I|\, m_k^* A \leqslant m_{k+1}(A \times I)$ which, with Problem 1, establishes the desired result. ∎

The next theorem includes Lemmas 1 and 3 and the converse of Lemma 1.

Theorem 2. *Given $A \subset E^k$ and $I \subset E^1$, then $A \times I$ is m_{k+1}-measurable iff A is m_k-measurable. Also*

$$m_{k+1}^*(A \times I) = |I|\, m_k^* A.$$

Proof. First consider A bounded and take a bounded k-cell $R \supset A$. Then

$$m_{k+1}^*(A \times I) + m_{k+1}^*[(R - A) \times I] = |I|\, [m_k^* A + m_k^*(R - A)]$$

by Lemma 2. Also, $m_{k+1}(R \times I) = |I|\, m_k R$. The desired result then follows by applying Theorem 5–1.8.

In case A is not bounded, let

$$R_n = I_n{}^k \qquad \text{where } I_n = [-n, n], \quad n = 1, 2, 3, \cdots.$$

Then $A \cap R_n$ is bounded and hence is m_k-measurable iff $(A \cap R_n) \times I$ is m_{k+1}-measurable. Since

$$A = \bigcup (AR_n) \qquad \text{and} \qquad A \times I = \bigcup\, [(AR_n) \times I],$$

the theorem is seen to hold. ∎

5–5 FUBINI'S THEOREM

Given a set $A \subset E^2$, the set $A(x)$ defined by

$$A(x) = \{y \mid (x, y) \in A\}, \qquad x \in E^1$$

is called the **vertical x-section** of A. Then $A(x) \subset E^1$, but is represented in Fig. 1 as the intersection of A and a line perpendicular to the x-axis. Even if A is m_2-measurable, it may be that $A(x)$ is m-nonmeasurable for some x. For example, with $N \subset [0, 1]$ nonmeasurable, the set

$$A = [0, 1) \times [0, 1] \cup \{1\} \times N$$

is m_2-measurable, but $A(1)$ is m-nonmeasurable.

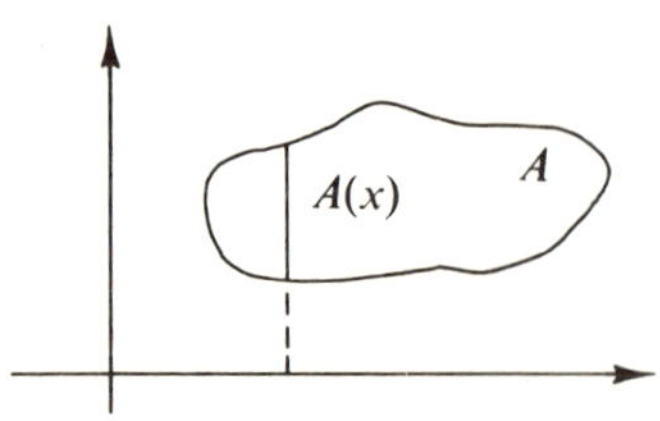

FIG. 1

Theorem 1. *If $A \subset E^2$ is m_2-measurable, then $A(x)$ is m-measurable for m-almost all $x \in E^1$ and*

$$\int_{E^1} mA(x)\,dx = m_2 A.$$

By symmetry, the same is true for horizontal sections.

Proof. The lemma holds if A is a 2-cell (by Section 5–4) and then if A is the union of a finite number of 2-cells.

Let A be an open set. Take a sequence $\{C_n\}$ of sets, each the union of a finite number of 2-cells such that $C_n \subset C_{n+1}$ and $A = \bigcup C_n$. Then

$$A(x) = \lim_{n\to\infty} C_n(x), \qquad mA(x) = \lim_{n\to\infty} mC_n(x),$$

and

$$m_2 A = \lim_{n\to\infty} m_2 C_n = \lim_{n\to\infty} \int_{E^1} mC_n(x)\,dx = \int_{E^1} mA(x)\,dx,$$

where the last equality follows from Theorem 4–5.2. Thus the theorem holds for open sets.

Let A be closed. Set $R_n = \{(x, y) \mid |x| < n,\ |y| < n\}$ and $A_n = A \cap R_n$. The lemma holds for the open set $R_n - A_n$. Hence

$$\int_{E^1} m[R_n(x) - A_n(x)]\,dx = m_2[R_n - A_n], \qquad \int_{E^1} mA_n(x) = m_2 A_n,$$

and we let $n \to \infty$ to obtain the theorem for closed sets.

Let A be m_2-measurable with $m_2 A < \infty$. Take a decreasing sequence $\{G_n\}$ of open sets containing A and an increasing sequence $\{F_n\}$ of closed sets contained in A such that

$$\lim m_2 G_n = \lim m_2 F_n = m_2 A$$

by Theorem 5–1.9 (see comment preceding Theorem 5–1.5). Then

$$\begin{aligned}\lim_{n\to\infty} \int_{E^1} mG_n(x)\,dx &= \lim_{n\to\infty} m_2 G_n = m_2 A \\ &= \lim_{n\to\infty} m_2 F_n = \lim_{n\to\infty} \int_{E^1} mF_n(x)\,dx.\end{aligned}$$

Hence, all terms being finite,

$$0 = \lim_{n\to\infty} \int_{E^1} [mG_n(x) - mF_n(x)]\,dx = \int_{E^1} \lim_{n\to\infty} m[G_n(x) - F_n(x)]\,dx,$$

and since the integrand is a nonnegative nonincreasing function of n, we have

$$\lim_{n\to\infty} [mG_n(x) - mF_n(x)] = 0 \qquad \text{almost everywhere,}$$

by Theorem 4–5.2 and Problem 4–5.4(b). Thus $A(x)$ is m-measurable a.e. Therefore the integrand in $\int_{E^1} mA(x)\,dx$ is defined a.e., so the integral has meaning and is equal to $m_2 A$ because it lies between two sequences of integrals, each tending to $m_2 A$.

If A is m_2-measurable, but $m_2 A = \infty$, we apply the above result to $A_n = \{(x, y) \in A \mid |x| < n, |y| < n\}$ and then let $n \to \infty$. ∎

The following theorem is a generalization of the most primitive, and popular, notion of an integral as the "area under the graph of a nonnegative function."

Theorem 2. *Given a nonnegative function f: $E^1 \to [0, \infty)$, let*

$$A_0 = \{x \mid f(x) = 0\}, \qquad A_+ = \{x \mid f(x) > 0\},$$

and define the **inner ordinate** *set of f by*

$$\underline{0}(f) = \{(x, y) \mid x \in A_+ \text{ and } 0 \leqslant y < f(x)\}.$$

(a) *Then $\underline{0}(f)$ is m_2-measurable iff f is measurable.*
(b) *If f is measurable, then*

$$\int_{E^1} f(x)\, dx = m_2 \underline{0}(f).$$

Proof (a). First let $\underline{0}(f)$ be m_2-measurable, take any number $a \geqslant 0$, and set

$$B = \{(x, y) \mid (x, y) \in \underline{0}(f) \quad \text{and} \quad y > a\} = \underline{0}(f) \cap \{E^1 \times (a, \infty)\}.$$

Then B is m_2-measurable. Hence for almost all y, $0 < y < \infty$, the horizontal section $B(y)$ is m-measurable (by Theorem 1) and, by avoiding a set of measure zero, there is a sequence $y_1 > y_2 > y_3 > \cdots$ where $y_n \to a$ and $B(y_n) \subset E^1$ is m-measurable. Since $B(y_1) \subset B(y_2) \subset B(y_3) \subset \cdots$ and

$$\bigcup B(y_n) = \{x \in A \mid f(x) > a\}$$

is m-measurable, it follows that f is a measurable function.

Conversely, let f be measurable. Let $\{r_n\}$ be the positive rationals. Then for each n,

$$P_n = \{x \in A_+ \mid 0 \leqslant f(x) < r_n\}$$

is m-measurable, $P_n \times [0, r_n)$ is m_2-measurable by Theorem 5–4.2, and

$$\underline{0}(f) = \bigcup \{P_n \times [0, r_n)\} \text{ is } m_2\text{-measurable.}$$

Proof (b). Let s be a nonnegative measurable step function expressed as

$$s = \sum_{i=1}^{n} c_i \chi_{A_i}, \qquad c_i \geqslant 0, \bigcup A_i = E^1$$

with each A_i m-measurable, $A_i A_j = \varnothing$ for $1 \leqslant i < j \leqslant n$. Then

$$\sum_{i=1}^{n} c_i mA_i = \mathscr{I}(s, E^1) = \int_{E^1} s(x)\, dx,$$

in which zero is used in the sum if $c_i = 0$ even if $mA_i = \infty$. Let $I_i = \{y \mid 0 \leqslant y < c_i\}$, which is empty if $c_i = 0$. Then $A \times I_i$ is m_2-measurable (Theorem 5–4.2) and

$$m_2(A_i \times I_i) = \begin{cases} c_i mA_i & \text{if } c_i > 0 \\ 0 & \text{if } c_i = 0. \end{cases}$$

Since $(A_i \times I_i) \cap (A_j \times I_j) = \varnothing$ for $1 \leqslant i < j \leqslant n$ or $I_i = \varnothing$, then

$$m_2 \underline{0}(s) = \sum_{i=1}^{n} m_2(A_i \times I_i) = \int_{E^1} s(x)\, dx,$$

which establishes the result for step functions.

Then $\int_{A_0} f(x)\, dx = 0$ and $m_2\{(x, 0) \mid x \in A_0\} = 0$ so it is sufficient to prove that

$$m_2 \underline{0}(f) = \int_{A_+} f(x)\, dx.$$

Let $\{s_n\}$ be a monotonically increasing sequence of nonnegative step functions such that

$$\lim_{n\to\infty} s_n(x) = f(x) \qquad \text{for } x \in A_+,$$

which exists by Theorem 4–3.7. Then

$$\lim_{n\to\infty} \int_{A_+} s_n(x)\, dx = \int_{A_+} f(x)\, dx$$

by Theorem 4–5.2. Also $\underline{0}(s_n)$ is m_2-measurable and

$$m_2 \underline{0}(s_n) = \int_{A_+} s_n(x)\, dx, \qquad n = 1, 2, 3, \cdots$$

by the first part of this proof (b). Furthermore,

$$\underline{0}(s_1) \subset \underline{0}(s_2) \subset \underline{0}(s_3) \subset \cdots \qquad \text{and} \qquad \underline{0}(f = \bigcup \underline{0}(s_n)$$

so that

$$m_2 \underline{0}(f) = \lim_{n\to\infty} m_2 \underline{0}(s_n).$$

Thus the theorem is established. ∎

We have stated Theorems 1 and 2 in as small dimensions as possible. With $\mathbf{x} = (x_1, x_2, \cdots, x_k) \in E^k$ and $\mathbf{y} = (y_1, y_2, \cdots, y_l) \in E^l$, a point in $E^{k+l} = E^k \times E^l$ may be denoted by

$$(\mathbf{x}, \mathbf{y}) = (x_1, \cdots, x_k, y_1, \cdots, y_l).$$

Then, with $A \subset E^{k+l}$, the sections

$$A(\mathbf{x}) = \{\mathbf{y} \mid (\mathbf{x}, \mathbf{y}) \in A\} \qquad \text{and} \qquad A(\mathbf{y}) = \{\mathbf{x} \mid (\mathbf{x}, \mathbf{y}) \in A\}$$

are defined and the proof of Theorem 1 may be reviewed to see that if A is m_{k+l}-measurable, then

$$\int_{E^k} m_l A(\mathbf{x})\, dm_k = m_{k+l}(A) = \int_{E^l} m_k A(\mathbf{y})\, dm_l$$

with integrands defined except on null sets of the respective measures in the respective spaces.

Also if $f\colon E^{k+l} \to [0, \infty)$, the inner ordinate set for f is m_{k+l+1}-measurable iff f is an m_{k+l}-measurable function and, when f is such a function,

$$m_{k+l+1}\underline{0}(f) = \int_{E^{k+l}} f(\mathbf{x}, \mathbf{y})\, d(\mathbf{x}, \mathbf{y}).$$

As the next theorem shows, these facts embody the usual evaluation of a multiple integral by iterated integrals. We again revert to as small dimensions as possible.

Theorem 3. (Fubini) *If $f\colon E^2 \to E^1$ and $f \in \mathscr{L}(E^2)$, then*

$$\int_{E^2} f(x, y)\, d(x, y) = \int_{E^1} \left[\int_{E^1} f(x, y)\, dy\right] dx$$

$$= \int_{E^1} \left[\int_{E^1} f(x, y)\, dx\right] dy.$$

[*Note:* This means that $g(x) = \int_{E^1} f(x, y)\, dy$ is defined for almost all $x \in E^1$, $g \in \mathscr{L}(E^1)$, and $\int_{E^2} f = \int_{E^1} g$.]

Proof. Since f is measurable, the positive part f^+ and the negative part f^- are m_2-measurable. Thus the inner ordinate set $\underline{0}^+ = \underline{0}(f^+)$ is m_3-measurable and

$$m_3\underline{0}^+ = \int_{E^2} f^+(x, y)\, d(x, y).$$

Also, with $N \subset E^1$ the null set for which the section $\underline{0}^+(x) \subset E^2$ is nonmeasurable,

$$m_3\underline{0}^+ = \int_{E^1} m_2\underline{0}^+(x)\, dx = \int_{E^1 - N} m_2\underline{0}^+(x)\, dx.$$

Then, with $x \in E^1 - N$, the section $\underline{0}(x)(y)$ of $\underline{0}^+(x)$ is an m-measurable set for m-almost all $y \in E^1$ and

$$m_2\underline{0}^+(x) = \int_{E^1} f(x, y)\, dy \qquad \text{for } x \in E^1 - N.$$

Hence

$$\int_{E^2} f^+(x, y)\, d(x, y) = m_3\underline{0}^+ = \int_{E^1 - N} m_2 0^+(x)\, dx$$

$$= \int_{E^1 - N} \left[\int_{E^1} f^+(x, y)\, dy\right] dx = \int_{E^1} \left[\int_{E^1} f^+(x, y)\, dy\right] dx.$$

In the same way f^+ may be replaced by f^-. Then the first stated equality holds, by Theorem 4–5.8, and the second equality follows by symmetry. ∎

The case $f \in \mathscr{L}(A, m_2)$, with $A \subset E^2$ and m_2-measurable, falls under Theorem 3. If we define $F(x, y)$ to be $f(x, y)$ or 0 according as (x, y) is in A or A^c, then we write

$$\int_A f\, d(x, y) = \int_{E^2} F\, d(x, y) = \int_{E^1} \int_{E^1} F(x, y)\, dy\, dx = \int_? \int_{A(x)} f(x, y)\, dy\, dx.$$

The following problem throws light on the question mark.

PROBLEM 1. (a) If $A \subset E^2$ is m_2-measurable, then the projection $P(A) = \{x \mid (x, y) \in A\}$ need not be m-measurable.

(b) If $A \subset E^2$ is m_2-measurable and $\underset{\smile}{A}$ is an F_σ subset of A such that $m_2(A - \underset{\smile}{A}) = 0$, then the projection $P(\underset{\smile}{A})$ is an F_σ subset of $P(A)$.

(c) Upon defining $F(x, y)$ to be $f(x, y)$ or 0 according as (x, y) is in $\underset{\smile}{A}$ or $\underset{\smile}{A}^c$, then

$$\int_A f\, d(x, y) = \int_{\underset{\smile}{A}} f\, d(x, y) = \int_{P(\underset{\smile}{A})} \int_{A(x)} f\, dy\, dx.$$

Fubini's theorem also shows that the Lebesgue iterated integrals exist and are equal, provided $f \in \mathscr{L}(E^2)$. The following example shows that the m_2-measurability of f and the existence of the iterated integrals are not sufficient for equality of the iterated integrals. In this example we use the customary notation for integrals.

EXAMPLE. We shall define a real valued m_2-measurable function such that

$$\int_0^1 \int_0^1 f(x, y)\, dx\, dy \neq \int_0^1 \int_0^1 f(x, y)\, dy\, dx. \tag{1}$$

Divide the open square $S = (0, 1) \times (0, 1)$ into mutually exclusive semiopen subsquares $A_1, B_1, C_1, A_2, B_2, C_2, A_3, \cdots$ following the scheme in Figure 2.

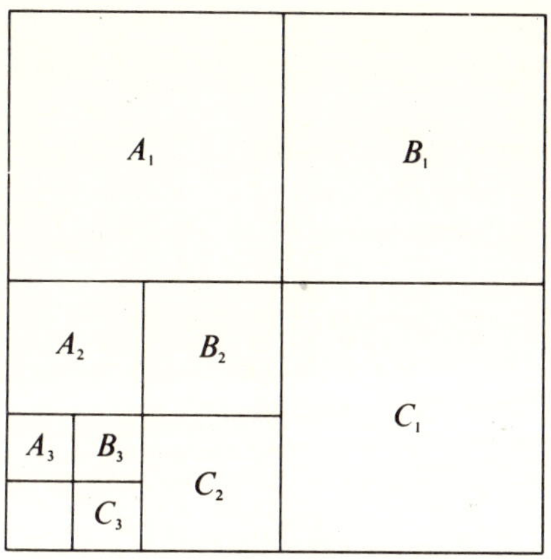

FIG. 2

Specifically,

$$A_n = \left\{(x, y) \mid 0 < x < \frac{1}{2^n}, \frac{1}{2^n} \leqslant y < \frac{1}{2^{n-1}}\right\},$$

$$B_n = \left\{(x, y) \,\middle|\, \frac{1}{2^n} \leqslant x < \frac{1}{2^{n-1}}, \frac{1}{2^n} \leqslant y < \frac{1}{2^{n-1}}\right\},$$

and

$$C_n = \left\{(x, y) \,\middle|\, \frac{1}{2^n} \leqslant x < \frac{1}{2^{n-1}}, 0 < y < \frac{1}{2^n}\right\}.$$

Define $f(x, y) = 0$ for $(x, y) \in \bigcup B_n$ or $(x, y) \in S^c$. Let $\{a_n\}$ and $\{c_n\}$ be real sequences, which will be chosen later to accomplish our purposes. Define

$$f(x, y) = \begin{Bmatrix} a_n & \text{if } (x, y) \in A_n \\ c_n & \text{if } (x, y) \in C_n \end{Bmatrix}, \qquad n = 1, 2, 3, \cdots.$$

Then f is defined on E^2 and, regardless of what the real sequences $\{a_n\}$ and $\{c_n\}$ are, f is seen to be m_2-measurable. Also

$$\text{if } \frac{1}{2} \leqslant y < 1, \qquad \text{then } \int_0^1 f(x, y)\, dx = \frac{a_1}{2},$$

$$\text{if } \frac{1}{2^2} \leqslant y < \frac{1}{2}, \qquad \text{then } \int_0^1 f(x, y)\, dx = \frac{a_2}{4} + \frac{c_1}{2},$$

and upon setting

$$\sigma_n = \int_0^1 f(x, y)\, dx \qquad \text{if } \frac{1}{2^n} \leqslant y < \frac{1}{2^{n-1}},$$

then

$$\sigma_n = \frac{a_n}{2^n} + \frac{c_{n-1}}{2^{n-1}} + \frac{c_{n-2}}{2^{n-2}} + \cdots + \frac{c_2}{2^2} + \frac{c_1}{2}.$$

Hence

$$\sigma_{n+1} - \sigma_n = \frac{a_{n+1}}{2^{n+1}} + \frac{c_n}{2^n} - \frac{a_n}{2^n}, \qquad n = 1, 2, 3, \cdots.$$

We now choose the sequences to make

$$\int_0^1 f(x, y)\, dx \text{ independent of } y \text{ for } 0 < y < 1.$$

This will be true if $\sigma_{n+1} = \sigma_n$, and hence we require that

$$a_{n+1} = 2(a_n - c_n).$$

In addition we specify that $c_n = -a_n$ and thus have

$$a_{n+1} = 4a_n, \qquad n = 1, 2, 3, \cdots.$$

Finally take $a_1 = 4$ and obtain

$$a_n = 4^n \qquad \text{and} \qquad c_n = -4^n, \qquad n = 1, 2, 3, \cdots.$$

Therefore:

$$\text{if } 0 < y < 1, \qquad \text{then } \int_0^1 f(x, y)\, dx = \sigma_1 = \frac{a_1}{2} = 2$$

and

$$\text{if } 0 < x < 1, \qquad \text{then } \int_0^1 f(x, y)\, dy = -2.$$

Thus the first iterated integral in (1) has value 2; and the second, -2.

PROBLEM 2. If $f: E^2 \to [0, \infty)$ is m_2-measurable, then the existence or infinite value of any one of

$$\int_{E^2} f\, d(x, y), \qquad \int_{E^1} \int_{E^1} f\, dy\, dx, \qquad \int_{E^1} \int_{E^1} f\, dx\, dy$$

implies the same for the others and the equality of all three.

[*Hint:* With $I_n = [-n, n]$, define

$$f_n(x, y) = \begin{cases} f(x, y) & \text{if } (x, y) \in I_n^2 \text{ and } f(x, y) \leqslant n \\ n & \text{if } (x, y) \in I_n^2 \text{ and } f(x, y) > n \\ 0 & \text{if } (x, y) \notin I_n^2. \end{cases}$$

Apply Fubini's theorem to f_n (since $f_n \in \mathscr{L}(E^2)$) and check limits as $n \to \infty$.]

5–6 OUTER ORDINATE SETS

We shall show that inner ordinate sets may be replaced in the previous work by outer ordinate sets.

Definition 1. *Given a fvnction $f: A \to [0, \infty]$ with $A \subset E^k$, let*

$$A_\infty = \{\mathbf{x} \in A \mid f(\mathbf{x}) = \infty\}, \qquad A_0 = \{(\mathbf{x}, 0) \mid \mathbf{x} \in A, f(\mathbf{x}) = 0\}$$

and define the **outer ordinate** *set of f on A by*

$$\bar{0}(f, A) = \{(\mathbf{x}, y) \mid \mathbf{x} \in A - A_\infty,\ \ 0 \leqslant y \leqslant f(\mathbf{x})\} \cup \{A_\infty \times [0, \infty)\}.$$

Also, the finite graph of f is defined by

$$\begin{aligned} G(f) &= \bar{0}(f, A) - \underline{0}(f, A) \\ &= \{(\mathbf{x}, y) \mid \mathbf{x} \in A - A_\infty,\ y = f(\mathbf{x})\}. \end{aligned}$$

Lemma 1. *Given a set $B \subset E^1$ and a number $\lambda > 0$, let*

$$\begin{aligned} \lambda B &= \{y \mid y = \lambda x \quad \text{for some } x \in B\} \\ &= \{y \mid \lambda^{-1} y \in B\}. \end{aligned}$$

Then $m^*(\lambda B) = \lambda m^* B$.

Proof. An interval $I = [a, b]$ has $\lambda I = [\lambda a, \lambda b]$ and thus

$$|\lambda I| = \lambda b - \lambda a = \lambda |I|.$$

Given an $\varepsilon > 0$, let $\{I_n\}$ be such that

$$B \subset \bigcup I_n \qquad \text{and} \qquad \sum |I_n| \leqslant m^*B + \varepsilon.$$

Then $\lambda B \subset \bigcup (\lambda I_n)$ so that

$$m^*(\lambda B) \leqslant \sum |\lambda I_n| = \lambda \sum |I_n| \leqslant \lambda(m^*B + \varepsilon)$$

and hence $m^*(\lambda B) \leqslant \lambda m^*B$. Therefore

$$m^*B = m^*[\lambda^{-1}(\lambda B)] \leqslant \lambda^{-1} m^*(\lambda B)$$

so that $m^*(\lambda B) = \lambda m^*B$. ▌

Lemma 2. *The outer and inner ordinate sets of a function $f\colon A \to [0, \infty]$ with $A \subset E^k$ are such that*

$$m^*_{k+1}\,\bar{\text{O}}(f, A) = m^*_{k+1}\,\underline{\text{O}}(f, A).$$

Proof. By extension of the proof used in Lemma 1, it follows that if $\lambda > 0$, then

$$m^*_{k+1}\,\underline{\text{O}}(\lambda f, A) = \lambda m^*_{k+1}\,\underline{\text{O}}(f, A).$$

Let $\lambda_n = 1 + 1/n$, $n = 1, 2, 3, \cdots$. Then $\{\underline{\text{O}}(\lambda_n f, A)\}$ is a monotonically decreasing sequence of sets. Also

$$\bigcap \underline{\text{O}}(\lambda_n f, A) = \bar{\text{O}}(f, A) - A_0$$

where $A_0 \subset E^{k+1}$ (as given in Definition 1) has $m^*_{k+1} A_0 = 0$. Thus

$$\begin{aligned} m^*_{k+1}\,\bar{\text{O}}(f, A) &\leqslant \lim_{n\to\infty} m^*_{k+1}\,\underline{\text{O}}(\lambda_n f, A) \\ &= \lim_{n\to\infty} [\lambda_n m^*_{k+1}\,\underline{\text{O}}(f, A)] \\ &= m^*_{k+1}\underline{\text{O}}(f, A) \leqslant m^*_{k+1}\bar{\text{O}}(f, A). \end{aligned}$$ ▌

Theorem 1. *Given an m_k-measurable set $A \subset E^k$ and a function $f\colon E^k \to [0, \infty]$, then $\bar{\text{O}}(f, A)$ is m_{k+1}-measurable iff $\underline{\text{O}}(f, A)$ is m_{k+1}-measurable* [*and thus, by Theorem* 5–5.2(a), *iff f is an m_k-measurable function*].

Proof. Let $J_p = [0, p]$, $p = 1, 2, 3, \cdots$. Then $A \times J_p$ is m_{k+1}-measurable, with

$$m_{k+1}(A \times J_p) = |J_p|\, m_k A$$

by Theorem 5–4.2. Let f_p be the function defined by

$$f_p(\mathbf{x}) = \begin{cases} f(\mathbf{x}) & \text{if } 0 \leqslant f(\mathbf{x}) < p \\ p & \text{if } f(\mathbf{x}) \geqslant p. \end{cases}$$

Then $m^*_{k+1}\underline{O}(f_p, A) = m^*_{k+1}\bar{O}(f_p, A)$ by Lemma 2. Let

$$\bar{P}(f_p, A) = (A \times J_p) - \underline{O}(f_p, A) \qquad \text{and}$$
$$\underline{P}(f_p, A) = (A \times J_p) - \bar{O}(f_p, A).$$

The proof of Lemma 2 may be altered to show that

$$m^*_{k+1}\,\bar{P}(f_p, A) = m^*_{k+1}\,\underline{P}(f_p, A).$$

Thus

$$\begin{aligned} m_{k+1}(A \times J_p) &\leqslant m^*_{k+1}\,\underline{O}(f_p, A) + m^*_{k+1}\bar{P}(f_p, A) \\ &= m^*_{k+1}\,\bar{O}(f_p, A) + m^*_{k+1}\,\underline{P}(f_p, A). \end{aligned}$$

In case $m_k A < \infty$, then $m_{k+1}(A \times J_p) < \infty$ and the equality holds (by Theorem 5–1.8) iff $\underline{O}(f_p, A)$ is m_{k+1}-measurable and then (by the same theorem) iff $\bar{O}(f_p, A)$ is m_{k+1}-measurable. But if $\underline{O}(f, A)$ is m_{k+1}-measurable, then

$$\underline{O}(f_p, A) = \underline{O}(f, A) \cap (E^k \times J_p)$$

is m_{k+1}-measurable and also

$$\bar{O}(f, A) = \bigcup_{p=1}^{\infty} \bar{O}(f_p, A) \quad \text{is } m_{k+1}\text{-measurable.}$$

The same reasoning holds with upper and lower bars interchanged in case $m_k A < \infty$.

In case $m_k A = \infty$, let $I_n = [-n, n]$, $n = 1, 2, 3, \cdots$, apply the above result to $A_n = A \bigcap I_n{}^k$ and then use

$$\bar{O}(f, A) = \bigcup_{n=1}^{\infty} \bar{O}(f, A_n) \qquad \text{and} \qquad \underline{O}(f, A) = \bigcup_{n=1}^{\infty} \underline{O}(f, A_n). \blacksquare$$

Theorem 2. *If f is a nonnegative m_k-measurable function on E^k, then the finite graph has*

$$m_{k+1}G(f) = 0.$$

[*Note:* This may not be surprising, since we usually visualize the graph of a function as a curve in E^2 or a surface in E^3. There is, however, an example† of a real-valued function of a real variable whose graph $G(f)$ is m_2-nonmeasurable and thus has $m^*_2\,G(f) > 0$.]

Proof. The outer and inner ordinate sets are m_{k+1}-measurable and

$$m_{k+1}\bar{O}(f, A) = m_{k+1}\underline{O}(f, A) = \int_{E^k} f\,dm_k.$$

† Bernard R. Gelbaum and John M. H. Olmsted, *Counter Examples in Analysis*, Chap. 10, example 23, Holden-Day, San Francisco, 1964.

If the integral is finite, then (since $\underline{0}$ is an m_{k+1}-measurable subset of $\bar{0}$)

$$\begin{aligned} 0 &= m_{k+1}\bar{0} - m_{k+1}\underline{0} \\ &= m_{k+1}(\bar{0} - \underline{0}) = m_{k+1}G(f). \end{aligned}$$

In case the integral is not finite, let n and p be positive integers, set

$$R_n = I_n^{\,k} \qquad \text{where } I_n = [-n, n],$$

and define f_{np} by

$$f_{np}(\mathbf{x}) = \begin{cases} f(\mathbf{x}) & \text{if } \mathbf{x} \in R_n \text{ and } 0 \leqslant f(\mathbf{x}) \leqslant p \\ p & \text{if } \mathbf{x} \in R_n \text{ and } f(\mathbf{x}) > p \\ 0 & \text{if } \mathbf{x} \in E^k - R_n. \end{cases}$$

Then f_{np} is an m_k-measurable function for which

$$\int_{E^k} f_{np}\, dm_k < \infty,$$

and thus $m_{k+1}G(f_{np}) = 0$. Then, also, $m_{k+1}G(f) = 0$, since

$$G(f) \subset \bigcup_{n=1} \left[\bigcup_{p=1} G(f_{np}) \right].$$

Corollary. *If $f\colon E^k \to E^*$ is an m_{k+1}-measurable function, then $m_{k+1}G(f) = 0$, where*

$$G(f) = \{(\mathbf{x}, y) \mid -\infty < f(\mathbf{x}) < \infty \quad \textit{and} \quad y = f(x)\}.$$

5-7 ERGODIC THEORY

In the following example, T is a function, but we use Tx rather than $T(x)$ to avoid nesting parentheses and in conformity with notation usually used for transformations.

EXAMPLE 1. Let $I = [0, 1)$ and let T be the function (transformation) defined by

$$Tx = \begin{cases} 2x & \text{if } 0 \leqslant x < \frac{1}{2} \\ 2x - 1 & \text{if } \frac{1}{2} \leqslant x < 1; \end{cases}$$

or, briefly, $Tx = 2x (\text{mod } 1)$. For example, $T\frac{1}{8} = \frac{1}{4}$ and $T\frac{5}{8} = \frac{1}{4}$ so that both $\frac{1}{8}$ and $\frac{5}{8}$ are transformed into $\frac{1}{4}$. Also, any $y \in (0, 1)$ is the transform of both

$$x_1 = \tfrac{1}{2}y \qquad \text{and} \qquad x_2 = \tfrac{1}{2}(y + 1).$$

T^2 is defined by setting $T^2x = T(Tx)$ for $0 \leqslant x < 1$ and, with T^n defined, $T^{n+1} = T^nT$. For example,

$$T\tfrac{1}{10} = \tfrac{1}{5}, \qquad T^2\tfrac{1}{10} = \tfrac{2}{5}, \qquad T^3\tfrac{1}{10} = \tfrac{4}{5}, \qquad T^4\tfrac{1}{10} = \tfrac{3}{5}, \qquad T^5\tfrac{1}{10} = \tfrac{1}{5}$$

so that $T^5\frac{1}{10} = T\frac{1}{10}$. Then $T^n\frac{1}{10}$, as a function of n, is periodic of period 4. Moreover, none of these transforms is in [0.1, 0.2); that is, a particle starting at 0.1 $\in$ [0.1, 0.2) "never returns" to this interval. However, 0.11 $\in$ [0.1, 0.2) has none of the transforms $T\,0.11 = 0.22$, $T^2\,0.11 = 0.44, \cdots$, $T^7\,0.11 = 0.08$ in [0.1, 0.2), but

$$T^8\,0.11 = 0.16 \in [0.1, 0.2).$$

By continuing to successive transforms it will be found that $T^{13}\,0.11 = 0.12 \in [0.1, 0.2)$, with none of the intermediate transforms in [0.1, 0.2). Thus, if the transforms are performed once per millisecond, a particle starting at 0.11 spends an average of $\frac{3}{13}$ milliseconds in the interval [0.1, 0.2) during the first 13 milliseconds.

A question inspired by statistical mechanics is:

If a particle starts in a set A and is transformed once every time unit, will the average time of its sojourn in A approach a limit as the number n of time units increases indefinitely?

With χ_A the characteristic function of A, a way of asking the question is:

$$\text{Does } \lim_{n\to\infty} \frac{1}{n} \sum_{n=0}^{n-1} \chi_A(T^n x) \text{ exist for } x \in A?$$

This is the central question of ergodic† theory. It will follow (at the end of this section) that if A is measurable, then for almost all $x \in A$ the limit exists and is equal to

$$mA = \int_A \chi_A(x)\,dx.$$

As further motivation, we point out another property of the transformation of this example. With $A \subset [0, 1)$,

$$T^{-1}A = \{x \mid Tx \in A\}.$$

† The ergodic hypothesis of Boltzmann and Maxwell is described by G. D. Birkhoff and B. O. Koopman in *Proc. National Academy of Science*, 1932, p. 279.

E. T. Whittaker (*J. London Math. Soc.*, **20** pp. 121–128) (1945) in reviewing the work of G. D. Birkhoff, stated:

"The name *ergodic* was given more than seventy years ago by Boltzmann to those mechanical systems which had the property that each particular motion, when continued indefinitely, passes through every configuration and state of motion of the system which is compatible with the value of total energy; the assumption that the systems considered in the kinetic theory of gases are ergodic was made by Maxwell and Boltzmann and is known as the *ergodic hypothesis*. It was shown by Lord Kelvin in 1891 that the hypothesis could scarcely be rigorously true.... Now, Poincaré in his discussion of stability in the *Nouvelles méthodes*, had pointed out that the occurrence of certain properties of a dynamical system might have probability unity...the exceptions forming, in the later-introduced terminology of Lebesgue theory, a set of measure zero. The importance of this observation was realized by Birkhoff, who...published in 1931 a fundamental result which he called the *ergodic theorem*,....The theorem...justifies the physical intuitions of Maxwell and Boltzmann in the ergodic hypothesis, and goes far towards supplying a rigorous foundation for statistical mechanics."

Thus, for $0 \leqslant a < b < 1$, (a, b) has length $b - a$ and

$$T^{-1}(a, b) = \left(\frac{a}{2}, \frac{b}{2}\right) \cup \left(\frac{a+1}{2}, \frac{b+1}{2}\right),$$

where the sum of the lengths of the two intervals on the right is also $b - a$. If $0 \leqslant a_1 < b_1 \leqslant a_2 < b_2 < 1$, then

$$T^{-1}(a_1, b_1) \cap T^{-1}(a_2, b_2) = \varnothing.$$

Hence an open set $G \subset I$ is such that

$$m(T^{-1}G) = m(G),$$

as seen by considering the component intervals of G. Given a set $A \subset I$, take a set $G \subset I$, G open relative to I and $A \subset G$. Since $T^{-1}A \subset T^{-1}G$, then

$$m^*T^{-1}A \leqslant mT^{-1}G = mG.$$

Consequently

$$m^*T^{-1}A \leqslant m^*A \qquad \text{and} \qquad m^*T^{-1}A^c \leqslant m^*A^c,$$

where $A^c = I - A$. Hence if A is measurable, then

$$|I| \leqslant m^*T^{-1}A + m^*T^{-1}A^c \leqslant mA + mA^c = |I|.$$

Therefore, it follows that:

If $A \subset I$ is measurable, then $T^{-1}A$ is measurable, and

$$mT^{-1}A = mA,$$

for the T of this example.

Note, however, that $m(TA)$ need not equal mA. For example, $m[0, \frac{1}{2}) = \frac{1}{2}$, but

$$mT[0, \tfrac{1}{2}) = m[0, 1) = 1.$$

Nevertheless, if B is a set such that $A = TB$, then

$$m(T^{-1}TB) = mTB,$$

but $T^{-1}TB \neq B$. This nonequality is illustrated by

$$T^{-1}T[0, \tfrac{1}{4}) = T^{-1}[0, \tfrac{1}{2}) = [0, \tfrac{1}{4}) \cup [\tfrac{1}{2}, \tfrac{3}{4}) \neq [0, \tfrac{1}{4}).$$

PROBLEM 1. Express $x \in (0, 1)$ in binary form

$$x = \underset{2}{0.}\alpha_1\alpha_2\alpha_3\cdots,$$

using the terminating expansion if x is a binary rational. Then for the transformation T of Example 1,

$$Tx = \underset{2}{0.}\alpha_2\alpha_3\alpha_4\cdots \qquad \text{and} \qquad T^{-1}x = \{\underset{2}{0.}0\alpha_1\alpha_2\alpha_3\cdots, \underset{2}{0.}1\alpha_1\alpha_2\alpha_3\cdots\}.$$

Further discussion will be for an abstract (universal) space X on which an outer measure μ^* is defined such that

$$(1) \qquad 0 < \mu X < \infty.$$

Thus any set A has $\mu^* A < \infty$, it being understood that all $A \subset X$.

Transformations on X into X are also involved. Hence if T is a transformation, then

$$Tx \in X \qquad \text{for all } x \quad (\text{again } x \in X).$$

T is on X into X, not necessarily onto X. Thus there may be a $y \in X$ such that $Tx \neq y$ for all x. On the other hand, T as a function need not have an inverse function; that is, it is possible to have $x_1 \neq x_2$ but $Tx_1 = Tx_2$.

Nevertheless, $T^{-1}A$ with A a set is defined by

$$T^{-1}A = \{x \mid Tx \in A\}.$$

For any two sets A and B it follows that both

$$(T^{-1}A) \cap (T^{-1}B) = T^{-1}(A \cap B),$$

and

$$(2) \qquad (T^{-1}A) \cup (T^{-1}B) = T^{-1}(A \cup B).$$

As indicated in the example, T^n is defined for n a positive integer, and with m also a positive integer,

$$T^{m+n} = T^m T^n.$$

The inverse transformation T^{-1} is defined and then, inductively, $T^{-(n+1)} = T^{-n}T^{-1}$ so that

$$T^{-(m+n)} = T^{-m}T^{-n}.$$

However, $T^{-(m+n)}T^n \neq T^{-m}$ is possible. [In the example

$$T^{-2}T[0, \tfrac{1}{4}) = T^{-2}[0, \tfrac{1}{2}) = T^{-1}\{[0, \tfrac{1}{4}) \cup [\tfrac{1}{2}, \tfrac{3}{4})\} \neq T^{-1}[0, \tfrac{1}{4}).]$$

Then T^0 means that $T^0 A = A$ so that

$$TT^{-1} = T^0, \qquad T^n T^{-n} = T^0,$$

but

$$T^{-1}T \neq T^0 \quad \text{may hold.}$$

[Again from the example, $T^{-1}T\{\tfrac{1}{8}\} = T^{-1}\{\tfrac{1}{4}\} = \{\tfrac{1}{8}, \tfrac{5}{8}\}$.]

With $x \in X$ we use $T^{-1}x$ rather than $T^{-1}\{x\}$. Even with x an element, $T^{-1}x$ is a set which may be empty or have a single element or have many elements, whereas Tx is a single element.

Definition 1. *A transformation T on X into X is said to be*

(i) *A* **measurable transformation** *if $B = T^{-1}A$ is a measurable set whenever A is a measurable set.*

(ii) *A* **measure-preserving transformation** *if it is measurable and, for every measurable set A,*

$$\mu(T^{-1}A) = \mu A.$$

PROBLEM 2. If T is a measure-preserving transformation and A is a measurable set, then

$$\mu(A - T^{-1}A) = \mu(T^{-1}A - A).$$

[Intuitively, X is a sealed jar, of volume $\mu X = 1$, containing material being moved about by a law of nature T, and the locations of all particles are recorded simultaneously once a day. A region A of volume μA is selected. Since

$$\frac{\mu A}{\mu X} = \mu A,$$

the probability that any designated particle is in A on a specified recording is therefore μA. The laws of nature are (ideally, at any rate) time-independent and probability is a law of nature (in a toss of a coin the probability of a head yesterday was the same as today and will be the same tomorrow). Let $T^{-1}A$ be the set which yesterday contained exactly the locations of particles in A today. The probability that a particle was in $T^{-1}A$ yesterday is thus the same as the probability that the same particle is in A today; that is, $\mu(T^{-1}A) = \mu A$. On the other hand, the particles in A today will be in $TA = B$ tomorrow, but T may also send some particles not in A today into B tomorrow so that $\mu(TA)$ and μA need not be equal.]

The first lemma is not concerned with either transformations or measures, but includes a definition and result which will be a key in proving Lemma 3.

Lemma 1. *Given a real succession $c_0, c_1, c_2, \cdots, c_{n-1}$ of length n, a term c_j is called a vit (very important term) if at least one of the sums*

$$\begin{array}{l} c_j \\ c_j + c_{j+1} \\ \cdots \\ c_j + c_{j+1} + \cdots + c_{n-1} \end{array}$$

is positive. [For $1, -\frac{1}{2}, -\frac{4}{5}, \frac{6}{5}$, the vits are $1, -\frac{4}{5}, \frac{6}{5}$.] *Let v be the sum of the vits if there are any, but $v = 0$ if there are none.*

The result to be proved is that $v \geqslant 0$.

Proof. Induction on the lengths n of successions will be used. Any succession of length $n = 1$ consists of a single term c and $v = c$ if $c \geqslant 0$, but $v = 0$ if $c < 0$. Let $n \geqslant 2$ be an integer such that the result holds for *all* successions of lengths $\leqslant (n-1)$. Take any n-length succession

$$c_0, c_1, c_2, \cdots, c_{n-1}. \tag{3}$$

Form the sums

$$s_0 = c_0$$
$$s_1 = c_0 + c_1$$
$$\cdots$$
$$s_{n-1} = c_0 + c_1 + \cdots + c_{n-1}.$$

Case 1. All $s_k \leqslant 0$. Then c_0 is not a vit of (3). Hence all vits of (3), if there are any, are vits of the $(n-1)$-length succession $c_1, c_2, \cdots, c_{n-1}$ so that $v \geqslant 0$ by the induction hypothesis.

Case 2. $s_0 > 0$. Then $c_0 > 0$, c_0 is a vit of (3) and any other vit of (3) is also a vit of $c_1, c_2, \cdots, c_{n-1}$. Hence $v \geqslant c_0 > 0$.

Case 3. $s_0 \leqslant 0, s_1 \leqslant 0, \cdots, s_{k-1} \leqslant 0$, but $s_k > 0$ with $1 \leqslant k \leqslant n-1$. Then

$$0 < s_k = s_{k-1} + c_k$$
$$0 < s_k = s_{k-2} + (c_k + c_{k-1})$$
$$\cdots$$
$$0 < s_k = s_0 + (c_k + c_{k-1} + \cdots + c_1).$$

Hence $c_1 + c_2 + \cdots + c_k, c_2 + c_3 + \cdots + c_k, \cdots, c_{k-1} + c_k, c_k$ are all positive, making $c_1, c_2, \cdots, c_k$ all vits of (3). Also, c_0 is a vit (since $s_k > 0$), but the vits $c_0, c_1, \cdots, c_k$ have s_k as their positive sum. If there are any vits of (3) other than these, then the others are also vits of the shorter succession $c_{k+1}, \cdots, c_{n-1}$, which (inductive hypothesis) has sum $\geqslant 0$. Hence $v \geqslant 0$. ∎

The following lemma gives change of variable of integration formulas.

Lemma 2. *If f: $X \to E^*$ is summable (on X), if T is a measure-preserving transformation, and if M is a measurable set, then*

$$\int_M f(x)\,d\mu = \int_{T^{-1}M} f(Tx)\,d\mu = \int_{T^{-n}M} f(T^n x)\,d\mu. \tag{4}$$

In particular, when M is the whole space X,

$$\int_X f(x)\,d\mu = \int_X f(T^i x)\,d\mu, \qquad i = 0, 1, 2, 3, \cdots. \tag{5}$$

Proof. Let $N = T^{-1}M$ and, given any number a, let

$$N_a = \{y \mid f(Ty) \geqslant a\} \qquad \text{and} \qquad M_a = \{x \mid f(x) \geqslant a\}.$$

Then M_a is measurable (since f is measurable) and

$$y \in N_a \Leftrightarrow [f(Ty) \geqslant a] \Leftrightarrow [Ty \in M_a] \Leftrightarrow y \in T^{-1}M_a.$$

Hence $N_a = T^{-1}M_a$. Since T is a measurable transformation, then N_a (and also N) is a measurable set. But T is also measure-preserving so that

$$\begin{aligned}\mu(NN_a) &= \mu(T^{-1}M \cap T^{-1}N_a)\\ &= \mu(T^{-1}M) + \mu T^{-1}N_a - \mu(T^{-1}M \cup T^{-1}M_a)\\ &= \mu M + \mu N_a - \mu[T^{-1}(M \cup M_a)] \qquad \text{by (2)}\\ &= \mu M + \mu M_a - \mu(M \cup M_a) = \mu(MM_a).\end{aligned}$$

All of the above holds as well with $\geqslant a$ replaced by $\leqslant a$. Hence the characteristic functions χ_M and χ_N are such that

$$\int_M \chi_M(x)\,d\mu = \int_N \chi_N(x)\,d\mu = \int_{T^{-1}M} \chi_M(Tx)\,d\mu.$$

Thus the first equation in (4) follows, since any step function approximating either integral has a corresponding step function approximating the other integral.

In the first equation of (4) replace M by $T^{-1}M$ and x by Tx; than use induction to see that (4) holds.

From the first equation of (4) we obtain the first equation in

$$\int_X f(x)\,d\mu = \int_{T^{-1}X} f(Tx)\,d\mu = \int_X f(Tx)\,d\mu$$

and the second equation holds, since $T^{-1}X$ is almost all of X, as $\mu(T^{-1}X) = \mu(X)$ states. Then induction establishes (5). ∎

The following lemmas and theorem are for a function $f\colon X \to E^*$ which is summable on X and thus is defined and finite a.e. By avoiding a set of measure zero where any $T^i x$, $i = 0, 1, 2, \cdots, n$ is such that $f(T^i x)$ is undefined or has an infinite value, the function σ_n, where

(6) $$\sigma_n(x) = \frac{1}{n}\sum_{i=0}^{n-1} f(T^i x), \qquad n = 1, 2, 3, \cdots,$$

is defined a.e. (in particular no two terms have opposite infinite values). The sequence $\{\sigma_n\}$ of these "averaging functions" plays a central role.

For later use we note that

$$\begin{aligned}\int_X \sigma_n(x)\,d\mu &= \frac{1}{n}\sum_{i=0}^{n-1}\int_X f(T^i x)\,d\mu \\ &= \frac{1}{n}\sum_{i=0}^{n-1}\int_X f(x)\,d\mu \qquad \text{by (5)} \\ &= \int_X f(x)\,d\mu,\end{aligned}$$

and therefore

(7) $$\lim_{n\to\infty}\int_X \sigma_n(x)\,d\mu = \int_X f(x)\,d\mu.$$

As a guide for what is coming, the goal is to prove that $\lim_{n\to\infty}\sigma_n(x)$ exists a.e. and that the limit and integral in (7) may be interchanged.

Lemma 3. *If $f\colon X \to E^*$ is summable and T is a measure preserving transformation, then the set*

$$M = \left\{x \,\middle|\, \sup_{n\geqslant 1}\sigma_n(x) > 0\right\}$$

is measurable and

$$\int_M f(x)\, d\mu \geqslant 0.$$

Also, if λ *is a number and* $M^\lambda = \{x \mid \sup_{n\geqslant 1} \sigma_n(x) > \lambda\}$, *then*

$$\int_{M^\lambda} f(x)\, d\mu \geqslant \lambda\mu(M^\lambda).$$

Proof. For $k = 1, 2, 3, \cdots$ let M_k be defined by the first equation

$$\begin{aligned} M_k &= \left\{x \,\middle|\, \sup_{1\leqslant n\leqslant k} \sigma_n(x) > 0\right\} \\ &= \left\{x \,\middle|\, \sup_{1\leqslant n\leqslant k} \sum_{i=0}^{n-1} f(T^i x) > 0\right\} \end{aligned}$$

and note why the second equation then holds (even though there is no $1/n$ before the summation sign). From the second form we observe, for use presently, that $x \in M_k$ iff at least one of

(8)
$$\begin{aligned} &f(x) \\ &f(x) + f(Tx) \\ &\cdots \\ &f(x) + f(Tx) + \cdots + f(T^{k-1}x) \end{aligned}$$

is positive. The set M_k is measurable (by Theorem 4–3.6), since $f, f(T\ \ ), f(T^2\ \ )$, $\cdots$ are measurable functions. Also,

$$M_1 \subset M_2 \subset M_3 \subset \cdots \qquad \text{and} \qquad M = \bigcup M_n$$

so that M is a measurable set and then

$$\lim_{k\to\infty} \int_{M_k} f(x)\, d\mu = \int_M f(x)\, d\mu,$$

by the corollary, page 204, applied to the positive and negative parts of the summable function f. The convergence of this sequence implies the convergence of its arithmetic means (Corollary, Theorem 3–7.12); that is,

$$\lim_{n\to\infty} \frac{1}{n} \sum_{k=1}^{n} \int_{M_k} f(x)\, d\mu = \int_M f(x)\, d\mu.$$

The first part of the lemma will therefore be proved upon showing that

(9)
$$\sum_{k=1}^{n} \int_{M_k} f(x)\, d\mu \geqslant 0, \qquad n = 1, 2, 3, \cdots,$$

which we proceed to do.

The sum in (9) is equal to

$$\int_{M_n} f(x)\,d\mu + \int_{M_{n-1}} f(x)\,d\mu + \cdots + \int_{M_1} f(x)\,d\mu$$

$$= \int_{T^0 M_n} f(T^0 x)\,d\mu + \int_{T^{-1}M_{n-1}} f(T^1 x)\,d\mu + \cdots + \int_{T^{-(n-1)}M_1} f(T^{n-1}x)\,d\mu$$

by Lemma 2. By setting $N^j = T^{-j}M_{n-j}$ and using the characteristic function χ_{N_j},

$$\sum_{k=1}^{n} \int_{M_k} f(x)\,d\mu = \sum_{j=0}^{n-1} \int_{N_j} f(T^j x)\,d\mu = \sum_{j=0}^{n-1} \int_X f(T^j x)\chi_{N_j}(x)\,d\mu \tag{10}$$

$$= \int_X \sum_{j=0}^{n-1} f(T^j x)\chi_{N_j}(x)\,d\mu.$$

We now establish (9) by showing that the integrand in (10) is nonnegative for all $x \in X$.

Since $x \in M_k$ iff at least one of the sums in (8) is positive, then $x \in M_{n-j}$ iff at least one of

$$\begin{aligned} &f(x)\\ &f(x) + f(Tx)\\ &\cdots\\ &f(x) + f(Tx) + \cdots + f(T^{n-j-1}x) \end{aligned}$$

is positive. From $N_j = T^{-j}M_{n-j}$, then $T^j N_j = M_{n-j}$. Thus $x \in N_j$ iff $T^j x \in M_{n-j}$ and hence iff at least one of

$$\begin{aligned} &f(T^j x)\\ &f(T^j x) + f(T^{j+1}x)\\ &\cdots\\ &f(T^j x) + f(T^{j+1}x) + \cdots + f(T^{n-1}x) \end{aligned}$$

is positive. Stated in terms of vits (defined in Lemma 1) $x \in N_j$ iff $f(T^j x)$ is a vit in the succession

$$f(x), \quad f(Tx), \quad f(T^2 x), \quad \cdots, \quad f(T^{n-1}x). \tag{11}$$

Consider any $x \in X$ and any j among $0, 1, 2, \cdots, n-1$. Either $x \notin N_j$, in which case $\chi_{N_j}(x) = 0$, or else $x \in N_j$, in which case $f(T^j x)\chi_{N_j}(x) = f(T^j x)$ is a vit of (11). Hence, for any $x \in X$, the succession (11) has its $v(x)$, as defined in Lemma 1, given by

$$v(x) = \sum_{j=0}^{n-1} f(T^j x)\chi_{N_j}(x),$$

which is thus nonnegative, by Lemma 1. Since this sum is the integrand in (10), we have established (9) and proved the first part of the lemma.

The second part of the lemma follows upon replacing f in the first part by $f - \lambda$. ∎

The symmetric difference $A \triangle B$ of sets A and B is used in the next definition. Hence

$$\begin{aligned} A \triangle B &= (A - B) \cup (B - A) \\ &= AB^c \cup BA^c. \end{aligned}$$

Definition 2. *With respect to a transformation T, we say that*

(i) *A set A is* **invariant** *if*

$$\mu(A \triangle T^{-1}A) = 0.$$

(ii) *A function f is* **invariant**† *if*

$$f(Tx) = f(x) \quad a.e.$$

Hence A is invariant if A and $T^{-1}A$ differ at most by a set of measure zero.

EXAMPLE 2. Let T: $[0, 1) \to (0, 1)$ be defined by

$$Tx = \begin{cases} \frac{1}{2} - x & \text{if } 0 \leqslant x < \frac{1}{2} \\ x & \text{if } \frac{1}{2} \leqslant x < 1. \end{cases}$$

Then $A = (0, \frac{1}{2})$ and $B = [\frac{1}{2}, 1)$ are both invariant, but

$$(x \in A, x \neq \tfrac{1}{4}) \Rightarrow T^{-1}x \neq x.$$

PROBLEM 3. Let T: $X \to X$ be a measure-preserving transformation and let A be a measurable set.

(a) If either of the sets

$$A - T^{-1}A \qquad \text{or} \qquad T^{-1}A - A$$

has measure zero, then so does the other and A is invariant.

(b) If A is invariant, then its characteristic function χ_A is invariant.

(c) If $\mu A = 0$ or if $\mu A^c = 0$, then A is invariant.

Lemma 4. *Let f: $X \to E^*$ be a summable function, T a measure-preserving transformation, and λ a number. If A is a measurable invariant set such that for each $x \in A$:*

(a) $$\sup_{n \geqslant 1} \sigma_n(x) > \lambda, \qquad \textit{then} \qquad \int_A f(x)\, d\mu \geqslant \lambda\mu(A)$$

(b) $$\inf_{n \geqslant 1} \sigma_n(x) < \lambda, \qquad \textit{then} \qquad \int_A f(x)\, d\mu \leqslant \lambda\mu(A).$$

Proof. First consider the special case $\lambda = 0$. With $g = f\chi_A$ and

$$M = \left\{x \,\middle|\, \sup_{n \geqslant 1} \frac{1}{n} \sum_{i=0}^{n-1} g(T^i x) > 0\right\}, \qquad \text{then} \qquad \int_M g(x)\, d\mu \geqslant 0$$

† "Almost invariant" would be better in both (i) and (ii).

by Lemma 3. The characteristic function χ_A is invariant by Problem 2(b). Thus $\chi_A(T^i x) = \chi_A(x)$ a.e. and then $g(T^i x) = f(T^i x)\chi_A(x)$ a.e. so that

$$g(T^i x) = \begin{cases} f(T^i x) & \text{a.e. in } A \\ 0 & \text{a.e. in } A^c. \end{cases}$$

Hence

$$\frac{1}{n}\sum_{i=0}^{n-1} g(T^i x) = \begin{cases} \sigma_n(x) & \text{a.e. in } A \\ 0 & \text{a.e. in } A^c. \end{cases}$$

Therefore M differs from A by a set of measure zero and

$$0 \leqslant \int_M g(x)\, d\mu = \int_A g(x)\, d\mu = \int_A f(x)\, d\mu,$$

which proves (a) under the special case $\lambda = 0$.

In case λ is not zero, then under the hypothesis in (a),

$$\sup_{n \geqslant 1} \frac{1}{n}\sum_{i=0}^{n-1} [f(T^i x) - \lambda] = \sup_{n \geqslant 0} [\sigma_n(x) - \lambda] > 0$$

and

$$0 \leqslant \int_A [f(x) - \lambda]\, d\mu = \int_A f(x)\, d\mu - \lambda\mu(A)$$

by the above special case, with f replaced by $f - \lambda$. Hence (a) holds.

Now apply (a) with f replaced by $-f$ and obtain (b). ∎

In the following theorem a function $\hat{f}$ is defined in terms of a function f.

Theorem 1. (Ergodic) *If $f\colon X \to E^*$ is a summable function and T is a measure-preserving transformation, then the following limit exists a.e. and the function $\hat{f}$ defined by*

$$\lim_{n\to\infty} \frac{1}{n}\sum_{i=0}^{n-1} f(T^i x) = \hat{f}(x) \quad \text{a.e.} \tag{12}$$

is invariant, measurable, and summable with

$$\int_X \hat{f}(x)\, d\mu = \int_X f(x)\, d\mu. \tag{13}$$

Proof. As in the previous work, σ_n is defined by

$$\sigma_n(x) = \frac{1}{n}\sum_{i=0}^{n-1} f(T^i x)$$

for all x where $f(x)$ is defined and $f(T^i x) \neq \pm\infty$. For all such x, let

$$\underline{\sigma}(x) = \varliminf_{n\to\infty} \sigma_n(x) \qquad \text{and} \qquad \bar{\sigma}(x) = \varlimsup_{n\to\infty} \sigma_n(x).$$

Hence $\underline{\sigma}$ and $\bar{\sigma}$ are defined a.e., but either or both may have infinite values.

Both $\underline{\sigma}$ and $\bar{\sigma}$ are measurable functions by Theorem 4–3.6, since each σ_n is a measurable function. Also

$$\underline{\sigma}(Tx) = \varliminf_{n\to\infty} \sigma_n(Tx) = \varliminf_{n\to\infty} \left[-\frac{f(x)}{n} + \frac{n+1}{n} \frac{1}{n+1} \sum_{i=0}^{n} f(T^i x) \right]$$

$$= 0 + 1 \cdot \varliminf_{n\to\infty} \sigma_{n+1}(x) = \varliminf_{n\to\infty} \sigma_n(x) = \underline{\sigma}(x)$$

where the zero follows, since we avoid any x where $f(x) = \pm\infty$. Thus $\underline{\sigma}$ is invariant and in the same way $\bar{\sigma}$ is invariant. With $a < b$ any constants, the set defined by

$$A_{ab} = \{x \mid \underline{\sigma}(x) < a < b < \bar{\sigma}(x)\}$$

is a measurable set. Also

$$T^{-1}A_{ab} = \{x \mid Tx \in A_{ab}\}$$
$$= \{x \mid \underline{\sigma}(Tx) < a < b < \bar{\sigma}(Tx)\}$$

in which, except for a set of measure zero, we may replace Tx by x because $\underline{\sigma}$ and $\bar{\sigma}$ are invariant. Thus A_{ab} is an invariant measurable set and moreover

$$A_{ab} \subset \left\{ x \,\middle|\, \inf_{n\geqslant 1} \sigma_n(x) < a < b < \sup_{n\geqslant 1} \sigma_n(x) \right\}.$$

Hence, by Lemma 3,

$$\int_{A_{ab}} f(x)\, d\mu \leqslant a\mu A_{ab} \leqslant b\mu A_{ab} \leqslant \int_{A_{ab}} f(x)\, d\mu.$$

Consequently $\mu A_{ab} = 0$ for all $a < b$. Thus $\mu A = 0$, where

$$A = \{x \mid \underline{\sigma}(x) < \bar{\sigma}(x)\} = \bigcup \{x \mid \underline{\sigma}(x) < a < b < \bar{\sigma}(x)\}$$

with the countable union taken over all rational pairs $a < b$. Hence, upon setting

$$\lim_{n\to\infty} \sigma_n(x) = \hat{f}(x)$$

whenever the limit exists*, we have $\hat{f}$ defined a.e. as in (12). Moreover, $\hat{f}$ is invariant and measurable, since $\hat{f} = \underline{\sigma}$ a.e.

To see that $\hat{f}$ is summable, the same method used in obtaining (7) yields

$$\int |\sigma_n(x)|\, d\mu \leqslant \frac{1}{n} \sum_{i=0}^{n-1} \int |f(T^i x)|\, d\mu$$
$$= \frac{1}{n} \sum_{i=0}^{n-1} \int |f(x)|\, d\mu$$
$$= \int |f(x)|\, d\mu.$$

Hence, by Fatou's lemma (Theorem 4–5.3),

$$\int \underline{\lim}\, |\sigma_n(x)|\, d\mu \leqslant \underline{\lim} \int |\sigma_n(x)|\, d\mu \leqslant \int |f(x)|\, d\mu < \infty.$$

But $\underline{\lim}\, |\sigma_n(x)| = \lim |\sigma_n(x)| = |\hat{f}(x)|$ a.e. and we have that $\hat{f}$ is summable, since

$$\int |\hat{f}(x)|\, d\mu < \infty.$$

The properties proved so far and (7) give

$$\text{(14)} \quad \int \hat{f}(x)\, d\mu = \int \lim_{n\to\infty} \sigma_n(x)\, d\mu \qquad \text{and} \qquad \lim_{n\to\infty} \int \sigma_n(x)\, d\mu = \int f(x)\, d\mu.$$

We shall show that in the first equation the integral and limit may be interchanged, from which the equality (13) will follow.

Let $\varepsilon > 0$ be given. Fix a positive number λ such that

$$\text{(15)} \qquad \int_N |f(x)|\, d\mu < \varepsilon, \qquad \text{where } N = \{x \mid |f(x)| > \lambda\}$$

this being possible, by Theorem 4–5.10, since f is summable.

Set

$$M_m = \left\{x \,\middle|\, \sup_{n\geqslant 1} \frac{1}{n} \sum_{i=0}^{n-1} |f(T^i x)| > m\right\}, \qquad m = 1, 2, 3, \cdots.$$

With m and n any positive integers,

$$\left|\int \hat{f} - \int \sigma_n\right| \leqslant \int |\hat{f} - \sigma_n|$$

$$\text{(16)} \qquad \leqslant \int_{(M_m)^c} |\hat{f} - \sigma_n| + \int_{M_m} |\hat{f}| + \int_{M_m} |\sigma_n|.$$

Since $x \in M_m{}^c \Rightarrow |\sigma_n(x)| \leqslant m$, it follows that

$$|\hat{f}(x)| = \left|\lim_{n\to\infty} \sigma_n(x)\right| \leqslant m \qquad \text{for almost all } x \in M_m{}^c.$$

Hence, for each m, the integrand of the first integral in (16) is dominated by $2m$ for almost all $x \in M_m{}^c$, converges to zero as $n \to \infty$ for almost all x and thus, by the dominated convergence Theorem 4–5.9,

$$\lim_{n\to\infty} \int_{(M_m)^c} |\hat{f} - \sigma_n| = 0, \qquad m = 1, 2, 3, \cdots.$$

Hence for $m = 1, 2, 3, \cdots$, we have, from (16), that

$$\text{(17)} \qquad \overline{\lim_{n\to\infty}} \left|\int \hat{f} - \int \sigma_n\right| \leqslant 0 + \int_{M_m} |\hat{f}| + \overline{\lim_{n\to\infty}} \int_{M_m} |\sigma_n|.$$

The left side is independent of m and we proceed to show that the right side may be made arbitrarily small for some choice of m. First consider the last integral on the right. From the definition of N,

$$x \in T^{-i}N \Rightarrow T^i x \in N \Rightarrow |f(T^i x)| > \lambda,$$

and

$$\begin{aligned}\int_{M_m} |f(T^i x)|\,d\mu &= \int_{M_m(T^{-i}N)} |f(T^i x)|\,d\mu + \int_{M_m - T^{-i}N} |f(T^i x)|\,d\mu \\ &\leqslant \int_{T^{-i}N} |f(T^i x)|\,d\mu + \int_{M_m} \lambda\,d\mu \\ &= \int_N |f(x)|\,d\mu + \lambda\mu(M_m) \qquad \text{(by Lemma 2)} \\ &< \varepsilon + \lambda\mu(M_m) \qquad \text{by (15).}\end{aligned}$$

Thus, for any integer n,

$$\int_{M_m} |\sigma_n| \leqslant \frac{1}{n}\sum_{i=0}^{n-1} \int_{M_m} |f(T^i x)|\,d\mu < \varepsilon + \lambda\mu(M_m).$$

Hence, from (17),

$$\overline{\lim_{n\to\infty}} \left| \int \hat{f} - \int \sigma_n \right| \leqslant \int_{M_m} |\hat{f}| + \varepsilon + \lambda\mu(M_m). \tag{18}$$

From the definition of M_m and the second part of Lemma 3 (with $|f|$ in place of f),

$$\mu(M_m) \leqslant \frac{1}{m}\int_{M_m} |f| \leqslant \frac{1}{m}\int |f| \to 0 \qquad \text{as } m \to \infty.$$

Since $|\hat{f}|$ is summable, we now have

$$\int_{M_m} |\hat{f}| \to 0 \qquad \text{as } m \to \infty$$

by Theorem 4–5.10. Thus, by letting $m \to \infty$ in (18) we have (since λ is fixed)

$$\overline{\lim_{n\to\infty}} \left| \int \hat{f} - \int \sigma_n \right| \leqslant \varepsilon.$$

Hence, the existence of the limit and the equality

$$\lim_{n\to\infty} \int \sigma_n(x)\,d\mu = \int \hat{f}(x)\,d\mu$$

is obtained. This, together with the second equation in (14), establishes (13) and thus finishes the proof of the ergodic theorem. ∎

In Problem 3(c) it was seen that if T is a measure-preserving transformation, then sets of measure zero and their complements are invariant sets.

Definition 3. *A measure-preserving transformation T is said to be an* **ergodic transformation** *if the only invariant measurable sets have either measure zero or measure equal to μX.*

Corollary. *If, in addition to the conditions of the ergodic theorem, T is ergodic, then $\hat{f}$ is constant a.e. and*

$$\hat{f}(x) = \frac{1}{\mu(X)} \int_X f(t)\, d\mu \quad a.e.$$

Proof. With $N = \{x \mid \hat{f}(Tx) \neq \hat{f}(x)\}$ and y any number, let

$$A_y = \{x \notin N \mid \hat{f}(x) > y\}.$$

From the invariance of $\hat{f}$, both $\mu N = 0$ and

$$TA_y = \{x \notin N \mid \hat{f}(Tx) > y\} = \{x \notin N \mid \hat{f}(x) > y\} = A_y.$$

Thus, for each y, the set A_y is invariant and, since T is ergodic, either $\mu(A_y) = 0$ or $\mu(A_y) = \mu(X)$. Since $\hat{f}$ is finite a.e. there are numbers a and b such that $\mu(A_a) < \frac{1}{2}\mu(X)$ and $0 < \mu(A_b)$. Thus $\mu(A_a) = 0$ and $\mu(A_b) = \mu(X)$ so that

$$c \doteq \sup\{y \mid \mu(A_y) = \mu(X)\}$$

is finite. With $y_1 < y_2 < y_3 < \cdots$ and $y_n \to c$ as $n \to \infty$, then

$$A_{y_1} \supset A_{y_2} \supset A_{y_3} \supset \cdots \qquad \text{and} \qquad A = \bigcap A_{y_n}$$

are such that, by Theorem 4–2.10,

$$\mu A = \lim_{n \to \infty} \mu A_{y_n} = \mu X.$$

Also, $x \in A \Rightarrow x \in A_{y_n} \Rightarrow \hat{f}(x) > y_n \Rightarrow \hat{f}(x) \geqslant c$.

On the other hand, let $B = \{x \in A \mid f(x) > c\}$. Then, with $z_1 > z_2 > z_3 > \cdots$, $z_n \to c$,

$$B = \bigcup A_{z_n}, \qquad \mu B = \lim_{n \to \infty} \mu A_{z_n} = 0.$$

Thus $\hat{f}(x) = c$ a.e. in the set A having $\mu A = \mu X$. Hence

$$\int \hat{f}(t)\, d\mu = \int_A c\, d\mu = c\mu(A) = \hat{f}(x)\mu(X) \quad \text{a.e.}$$

and the conclusion of the corollary follows from (13). ∎

Research has not discovered general tests for ergodicity of transformations comparable with tests for convergence of series or summability of functions. Several transformations of practical or theoretical interest have been shown to be ergodic, but each such transformation has required a special method. The following definition and theorem are sometimes useful.

Definition 4. *With respect to a transformation T, a set A is said to be* **strictly invariant** *if*

$$T^{-1}A = A.$$

PROBLEM 4. For any transformation T and any set A, the set

$$B = \overline{\lim_{n\to\infty}}\ T^{-n}A$$

is strictly invariant. [*Hint:* From Problem 2–3.3

$$B = \{x \mid x \in T^{-n}A \text{ for infinitely many } n\}.]$$

Observation. If T is a measure-preserving transformation and A is an invariant measurable set, then the strictly invariant set B of Problem 4 has $\mu B = \mu A$.

Test. A sufficient condition for ergodicity of a measure-preserving transformation T is that each strictly invariant measurable set A has either $\mu A = 0$ or $\mu A = \mu X$.

PROOF OF OBSERVATION. Since T is a measure-preserving transformation, then

$$\mu A = \mu(T^{-1}A) = \mu(T^{-2}A) = \cdots = \mu(T^{-n}A) = \cdots.$$

Thus, by Theorem 4–2.11,

$$\mu B = \mu\Big(\overline{\lim_{n\to\infty}}\ T^{-n}A\Big) \geqslant \overline{\lim_{n\to\infty}}\ \mu(T^{-n}A) = \mu A.$$

But A is invariant and hence the set

$$N = (T^{-1}A - A) \text{ has } \mu N = 0.$$

Thus $0 = \mu N = \mu(T^{-1}N) = \mu(T^{-2}N) = \cdots$ and

$$B \subset A \cup N \cup T^{-1}N \cup T^{-2}N \cup \cdots.$$

Consequently, $\mu B \leqslant \mu A$ and hence $\mu B = \mu A$. ∎

EXAMPLE I (continued). Hence $Tx = 2x(\text{mod } 1)$, $X = [0, 1)$, $\mu = m =$ Lebesgue measure and $m[0, 1) = 1$. We shall show that T is ergodic by proving that any strictly invariant measurable set A has either $mA = 0$ or $mA = 1$.

Accordingly, let A be such that

$$T^{-1}A = A \qquad \text{and} \qquad mA > 0.$$

Take any $x \in A$. Then $T^{-1}x \subset A$ and $T^{-1}x$ consists of the two points

$$x_1 = \tfrac{1}{2}x \qquad \text{and} \qquad x_2 = \tfrac{1}{2} + \tfrac{1}{2}x.$$

Since x was any point of A, the sets $A_1 = A[0, \frac{1}{2})$ and $A_2 = A[\frac{1}{2}, 1)$ are translations by $\frac{1}{2}$ of one another so that, by Theorem 5–2.3, $mA_1 = mA_2$. Also,

$$mA = m(A_1 \cup A_2) = mA_1 + mA_2 = 2mA_1 = 2mA_2.$$

Then $T^{-1}x_1$ and $T^{-1}x_2$ are

$$\begin{aligned} x_{11} &= \tfrac{1}{2}x_1 = \tfrac{1}{4}x, & x_{12} &= \tfrac{1}{2} + \tfrac{1}{2}x_1 = \tfrac{1}{2} + \tfrac{1}{4}x, \quad \text{and} \\ x_{21} &= \tfrac{1}{2}x_2 = \tfrac{1}{4} + \tfrac{1}{4}x, & x_{22} &= \tfrac{1}{2} + \tfrac{1}{2}x_2 = \tfrac{3}{4} + \tfrac{1}{4}x, \end{aligned}$$

respectively, and all are points of A. Hence

$$A_{11} = A[0, \tfrac{1}{4}), \qquad A_{12} = A[\tfrac{1}{2}, \tfrac{3}{4}), \qquad A_{21} = A[\tfrac{1}{4}, \tfrac{1}{2}), \qquad A_{22} = A[\tfrac{3}{4}, 1)$$

are congruent to one another and

$$mA = 2^2 mA_{ij}; \qquad i = 1, 2, \quad j = 1, 2.$$

By induction, for n a positive integer, the portion of A in an interval

$$I_{nk} = \left(\frac{k}{2^n}, \frac{k+1}{2^n}\right), \qquad k = 0, 1, 2, \cdots, 2^n - 1$$

is a translation of the portion of A in any other of these intervals and

$$mA = 2^n m(AI_{nk}) = \frac{m(AI_{nk})}{|I_{nk}|}.$$

The density of A exists and is equal to 1 at almost all points of A (Problem 5–2.1(b)). Select an $x_0 \in A$ such that $D(A, x_0) = 1$. For each n, let I_n be the interval I_{nk} containing x_0. Then

$$mA = \frac{m(AI_n)}{|I_n|} \qquad \text{and} \qquad mA = \lim_{n\to\infty} \frac{m(AI_n)}{|I_n|} = D(A, x_0) = 1.$$

Thus if A is strictly invariant and measurable, then either $mA = 0$ or $mA > 0$, in which case $mA = 1$ and this, by the test (p. 275), is sufficient to guarantee that T is ergodic.

Hence, with A a measurable set and $f = \chi_A$,

$$\hat{f}(x) = \lim_{n\to\infty} \frac{1}{n} \sum_{i=0}^{n-1} \chi_A(T^i x) \quad \text{a.e.} \qquad \text{and}$$

$$\hat{f}(x) = \int_X \chi_A(x)\, d\mu = mA \qquad \text{(corollary, Theorem 1)}$$

which confirms the statement about this specific transformation made on page 261.

PROBLEMS

1. (Converse of Theorem 5–1.4.) Let S be a metric space on which an outer measure μ^* is defined for which all open sets are μ-measurable. Then μ^* is a metric outer measure.

   [*Hint:* Take $d(A, B) > 0$, G open, $A \subset G$, and $BG = \varnothing$. Use $A \cup B$ in the criterion for μ-measurability of G.]

2. (Another nonmeasurable set.) Given $x \in [-1, 1]$, let

$$B(x) = \{y \in [-1, 1] \mid x - y \ \text{ is rational}\}.$$

   (a) Show that if $x', x'' \in [-1, 1]$, then either $B(x') = B(x'')$ or $B(x') \cap B(x'') = \varnothing$.

   We thus have a mutually exclusive family of nonempty sets whose union is $[-1, 1]$. Select one and only one point from each member of this family and let

A be the set so selected. Hence $A \subset [-1, 1]$. Let $r_0 = 0, r_1, r_2, \cdots$ be an ordering, without duplication, of the rationals on $[-2, 2]$ and set

$$A_n = r_n + A, \qquad n = 0, 1, 2, \cdots.$$

(b) Show that $A_n \subset [-3, 3]$, $m \neq n \Rightarrow A_m A_n = \varnothing$, $[-1, 1] \subset \bigcup A_n$, and therefore $2 \leqslant m^*(\bigcup A_n) \leqslant 6$.

Since $m^*A_n = m^*A$ and A_n is measurable iff A is measurable (by Theorem 5–2.3), the assumption that A is measurable leads to

$$2 \leqslant \sum_{n=1}^{\infty} mA_n = \lim_{n\to\infty} \sum_{k=1}^{n} m(A_k) = \lim_{n\to\infty} nm(A) \leqslant 6$$

and $m(A) > 0$ from the left inequality, but $m(\mathrm{A}) = 0$ from the right inequality.

3. Let $A_1, A_2, \cdots, A_n$ and A be measurable sets on an interval $I = [a, b]$. If every point of A belongs to at least k of the sets $A_1, A_2, \cdots, A_n$, then

$$kmA \leqslant mA_1 + mA_2 + \cdots + mA_n.$$

[*Hint:* Integrate the sum of characteristic functions. For an application see the third reference under 1, at the end of this chapter.]

4. Given an interval I and a number $0 < r < 1$, let $0(r)$ be the operation of removing an open interval J with the same midpoint as I and such that $|J| = r|I|$.

(a) Perform the operation $0(1/4)$ on the interval $[0, 1]$, then on the remaining intervals, then on the remaining intervals, etc. With E the set removed, find the measure of $[0, 1] - E$.

(b) Perform the operation $0(1/2)$ on $[0, 1]$, then $0(1/2^2)$ on the remaining intervals, then $0(1/2^3)$ on the remaining intervals, etc. Show that the set removed has measure <1 and is dense on $[0, 1]$.

5. With $A_n = (1/2n, 1/n)$ let $f_n = 2n\chi_{A_n}$, $n = 1, 2, \cdots$. Calculate

$$\int_0^1 [\lim f_n]\, dm \qquad \text{and} \qquad \lim \int_0^1 f_n\, dm.$$

6. Construct a function $f\colon (0, 1] \to E^1$ which is summable on each of the sets $A_n = (1/2^n, 1/2^{n-1}]$, $n = 1, 2, 3, \cdots$ and

$$\int_{A_n} f \text{ converges,}$$

but such that f is not summable on $A = (0, 1] = \bigcup A_n$.

7. If $f \in \mathscr{L}$ and $A_n = \{x \mid |f(x)| \geqslant n\}$, then $\sum mA_n$ converges.

8. Given that f is measurable and nondecreasing on $[a, b]$, let

$$F(x) = \int_a^x f(t)\, dt, \qquad a \leqslant x \leqslant b.$$

If $a \leqslant x_1 < x_2 \leqslant b$, then

$$F\left(\frac{x_1 + x_2}{2}\right) \leqslant \frac{1}{2}\,[F(x_1) + F(x_2)].$$

9. If $A, B \subset E^1$ are measurable sets, each having density 1 at all of its points, then the difference set $\Delta(A, B) = \{a - b \mid a \in A, b \in B\}$ is open.

10. (a) Given a set $A \subset E^1$, then the functions $\underline{f}$ and $\bar{f}$ defined by

$$\underline{f}(x) = \underline{D}(A, x) \qquad \text{and} \qquad \bar{f}(x) = \bar{D}(A, x), \qquad x \in E^1,$$

are measurable.

[*Hint:* Problem 10, page 221.]

(b) $A^D = \{x \mid D(A, x) = 1\}$ is a measurable set, $mA^D = m^*A$, and $D(A^D, x) = 1$ at all points of A^D.
(c) A is measurable iff $m(A^D - A) = 0$.
(d) If also $B \subset E^1$, then the difference set $\Delta(A^D, B^D)$ is open.

11. Let g be a nonnegative function such that $g \in \mathscr{L}[a, b]$. If $\{f_n\}$ is such that

$$f_n \in \mathscr{L}[a, b], \qquad f_n(x) \leqslant g(x), \qquad \text{and} \qquad \lim f_n(x) = f(x),$$

then for any bounded measurable function h on $[a, b]$,

$$\lim_{n \to \infty} \int_a^b hf_n = \int_a^b hf.$$

Problems on Inner Measure

Definition. *Given a space S and an outer measure μ^*, the corresponding inner measure μ_* is defined for each set $A \subset S$ by*

$$\mu_* A = \sup\{\mu M \mid M \textit{ measurable and } M \subset A\}.$$

The following problems are theorems about inner measure.

PROBLEMS

1. For each set A, $\mu_* A \leqslant \mu^* A$.
2. For each set A there is a μ-measurable set $\underset{\vee}{A} \subset A$ such that $m\underset{\vee}{A} = m_* A$.
3. If $B \subset A$, then $\mu_* B \leqslant \mu_* A$.
4. If $A \cap B = \varnothing$, then $\mu_* A + \mu_* B \leqslant \mu_*(A \cup B)$.
5. Given a pairwise disjoint sequence $\{A_n\}$ of sets, then

$$\sum \mu_* A \leqslant \mu_* \left(\bigcup A_n\right) \leqslant \mu^* \left(\bigcup A_n\right) \leqslant \sum \mu^* A_n .$$

In the next eight problems we make the specialization $S = E^1$, $\mu^* = m^*$, and $\mu_* = m_*$. Hence, for each set $A \subset E$, there is a measurable set $\hat{A} \supset A$ with $m\hat{A} = m^* A$. The only reason we do not stay with the general μ^* is that the corresponding "equi-measurable superset property" has not been hypothesized. For the sake of simplicity we are not taking $S = E^k$, $\mu^* = m_k^*$, and $\mu_* = m_{*k}$, but it should be seen that we could have.

6. $m_* A = \sup\{mF \mid F \text{ closed and } F \subset A\}$.
7. There is an F_σ set $B \subset A$ with $mB = m_* A$.
8. Given that $A \cap B = \varnothing$, then

$$m_* A + m_* B \leqslant m_*(A \cup B) \leqslant \begin{Bmatrix} m_* A + m^* B \\ m^* A + m_* B \end{Bmatrix} \leqslant m^*(A \cup B) \leqslant m^* A + m^* B.$$

If $m^*(A \cup B) = m^* A + m^* B < \infty$, then $m(\hat{A}\hat{B}) = m(\underset{\vee}{A}\underset{\vee}{B}) = 0$ and

$$m_*(A \cup B) = m_* A + m_* B.$$

[*Hint:* $m^*(A \cup B) = \cdots = m(\hat{A} \cup \hat{B}) + m(\hat{A}\,\hat{B}) \geqslant m^*(A \cup B) + m(\hat{A}\hat{B})$.

Take a measurable set $M \subset A \cup B$ having $mM = m_*(A \cup B)$.

$$m^*(A \cup B) = m(M - \hat{A}) + m(M - \hat{B}) + m(M\hat{A}\hat{B})$$
$$\leqslant m_* B + m_* A + 0 = m\underset{\vee}{A} + m\underset{\vee}{B} = \cdots .]$$

9. (a) A set A having $m^* A < \infty$ is measurable iff $m_* A = m^* A$.
(b) Give an example of a nonmeasurable set A having $m_* A = m^* A$.

10. The nonmeasurable set A of Theorem 5–2.6 has $m_*A = 0$.

11. If a set A has $m_*A > 0$, then the set $\Delta(A)$ (of differences of points in A) contains an interval.

12. Let $\{B_n\}$ be a pairwise disjoint sequence of measurable sets. Then for any sequence $\{A_n\}$ such that $A_n \subset B_n$, $n = 1, 2, 3, \cdots$, it follows that

$$m_*(\bigcup A_n) = \sum m_*A_n \qquad \text{and}$$
$$m^*(\bigcup A_n) = \sum m^*A_n.$$

[*Hint:* Take a measurable set $\underset{\vee}{A} \subset \bigcup A_n$ such that $m\underset{\vee}{A} = m^*(\bigcup A_n)$ and obtain the inequality reverse of the first one in Problem 5.]

13. (a) Given a measurable set A, then for any set W,

$$m_*W = m_*WA + m_*WA^c.$$

[*Hint:* Use $\underset{\vee}{W}$ and obtain an inequality reverse to one known to hold by Problem 8.]

(b) Let A be a set. If $mW = m_*WA + m_*WA^c$ for every measurable set W with $mW < \infty$, then A is measurable.

Outer Measure and Measurability (Alternative)

Let S be a space in which there is a ring $\mathscr{E}$ of subsets such that

$$S = \bigcup_{A \in \mathscr{E}} A$$

(but S need not be a member of $\mathscr{E}$, since $\mathscr{E}$ need not be a σ-ring). Also let a real-valued function γ with domain $\mathscr{E}$ be given such that:

(i) $A \in \mathscr{E} \Rightarrow 0 \leqslant \gamma(A) < \infty$.
(ii) $\gamma(A) > 0$ for at least one $A \in \mathscr{E}$ (to avoid triviality).
(iii) γ is additive on $\mathscr{E}$.

For $A \in \mathscr{E}$, $\gamma(A)$ is called the **content** of A.

[*Note:* For a realization, consider $S = E^k$, call A an **elementary** set if A is the union of a finite number of rectangles, and let $\mathscr{E}$ be the family of all elementary sets. A rectangle

$$R = I_1 \times I_2 \times \cdots \times I_n$$

may have any of its edges I_i an open, closed, or open-closed interval. Check that:

(a) $\mathscr{E}$ is a ring (but not a σ-ring).

(b) If $A \in \mathscr{E}$, then A is expressible as the union of a finite number of mutually disjoint rectangles (called a **decomposition** of A).

Define $\gamma(A)$ for $A \in \mathscr{E}$ by

$$\gamma(A) = \sum_{i=1}^{n} |R_i|,$$

where $R_1, R_2, \cdots, R_n$ is a decomposition of A, and check that if $R_1', R_2', \cdots, R_m'$ is another decomposition of A, then

$$\sum_{j=1}^{m} |R'_j| = \sum_{i=1}^{n} |R_i|.$$

Finally see that (i), (ii), and (iii) hold for γ so defined.

For $A \in \mathscr{E}$, $\gamma(A)$ is called the **Jordan content** of A.]

Going back to the general situation, the following definition is given.

Definition 1. *Given any $A \subset S$ with $A \neq \varnothing$, μ^*A is defined by*

$$\mu^*A = \inf \sum \gamma(A_n)$$

where the infimum is taken of all sums such that

$$A \subset \bigcup A_n \qquad \text{with each } A_n \in \mathscr{E}.$$

Also, $\mu^*\varnothing = 0$.

It is then a matter of routine to check that this μ^* satisfies Definition 4–2.1 (i), (ii), and (iii) (but not necessarily (iv) of Definition 5–1.1, since S might not even be a metric space). The big difference between the previous development of measure theory and this one is in the definition of measurability. Recall that the symmetric difference of two sets A and B is

$$(A \triangle B) = (A - B) \cup (B - A).$$

Definition 2.

(i) *The* **distance** *between two sets A and B is*

$$d(A, B) = \mu^*[(A \triangle B)],$$

(which might better be called the μ^* distance). *For a sequence* $\{A_n\}$ *and a set A we write*

$$A_n \to A \qquad \text{if } \lim_{n\to\infty} d(A_n, A) = 0.$$

(Sometimes $l.i.\mu_{n\to\infty} A_n = A$ is used instead of $A_n \to A$.)

(ii) *A set A is said to be* **finitely** μ**-measurable** *if there is a sequence* $\{A_n\}$ *such that each*

$$A_n \in \mathscr{E} \qquad \text{and} \qquad A_n \to A.$$

The family of all finitely μ-measurable sets is denoted by $\mathfrak{M}_F(\mu)$.

(iii) *A set A is* μ**-measurable** *if A is the union of a countable collection of finitely μ-measurable sets.* $\mathfrak{M}(\mu)$ *denotes the family of all such sets.*

(iv) *If* $A \in \mathfrak{M}_F(\mu)$ *or* $A \in \mathfrak{M}(\mu)$ *the μ-measure of A is defined by* $\mu A = \mu^*A$.

The following problems furnish a guide for an alternate route to obtaining that the μ-measurable sets form a σ-ring.

PROBLEMS

1. $A \in \mathscr{E} \Rightarrow A \in \mathfrak{M}_F(\mu)$ with $\mu(A) = \gamma(A)$.
2. $d(A, B) = d(B, A)$, $d(A, A) = 0$.
   $d(A, B) \leqslant d(A, C) + d(C, B)$.
   $$\left.\begin{array}{l} d(A_1 \cup A_2, B_1 \cup B_2) \\ d(A_1 \cap A_2, B_1 \cap B_2) \\ d(A_1 A_2^c, B_1 B_2^c) \end{array}\right\} \leqslant d(A_1, B_1) + dA_2, B_2).$$
   [*Hint:* Problem 2, page 49.]
3. $\mu^*A \leqslant d(A, B) + \mu^*B$ and if at least one of μ^*A or μ^*B is finite, then
   $$|\mu^*A - \mu^*B| \leqslant d(A, B).$$
4. Given $A \in \mathfrak{M}_F(\mu)$, $B \in \mathfrak{M}_F(\mu)$, $\{A_n\}$ with $A_n \in \mathscr{E}$ and $A_n \to A$, and $\{B_n\}$ with $B_n \in \mathscr{E}$ and $B_n \to B$, then
   (a) $A_n \cup B_n \to A \cup B$.
   (b) $A_n \cap B_n \to A \cap B$.

(c) $A_n - B_n \to A - B$.
(d) $\mathfrak{M}_F(\mu)$ is a ring.
(e) $\mu(A_n) \to \mu A < \infty$.
(f) $\mu(A_n) + \mu(B_n) = \mu(A_n \cup B_n) + \mu(A_n \cap B_n)$.
(g) $\mu(A) + \mu(B) = \mu(A \cup B) + \mu(A \cap B)$.
(h) μ is additive on $\mathfrak{M}_F(\mu)$.

5. Let $A \in \mathfrak{M}(\mu)$.
(a) Then there is a sequence $\{A_n\}$ of mutually exclusive sets in $\mathfrak{M}_F(\mu)$ such that $A = \bigcup A_n$, and for this sequence

$$\mu(A) = \sum_{n=1}^{\infty} \mu(A_n).$$

(b) If $\mu(A) < \infty$, then $A \in \mathfrak{M}_F(\mu)$.
[*Hint for* (b). Set $B_n = A_1 \cup A_2 \cup \cdots \cup A_n$ for the sequence $\{A_n\}$ of part (a) and show that $d(A, B_n) \to 0$.]

6. If $A_n \in \mathfrak{M}(\mu)$, $n = 1, 2, 3, \cdots$, then $\bigcup A_n \in \mathfrak{M}(\mu)$.

7. Let $A \in \mathfrak{M}(\mu)$, $B \in \mathfrak{M}(\mu)$, $A = \bigcup A_n$ with $A_n \in \mathfrak{M}_F(\mu)$ and $B = \bigcup B_n$ with $B_n \in \mathfrak{M}_F(\mu)$. Then
(a) $A_n \cap B = \bigcup_{i=1}^{\infty} (A_n \cap B_i) \in \mathfrak{M}(\mu)$.
(b) $A_n \cap B \in \mathfrak{M}_F(\mu)$. [*Hint:* Problem 5(b).]
(c) $A_n - B \in \mathfrak{M}_F(\mu)$.
(d) $A - B \in \mathfrak{M}(\mu)$.

Problems 6 and 7 then show that $\mathfrak{M}(\mu)$ is a σ-ring.

Problems on Outer Measures from Measures

Experience with specfic outer measures is a medium for seeing how the spirit of measure theory could have evolved in the following way with more subterranean digging and, possibly, some economy in overall effort. The setting is in an abstract space X.

Definition 1. *Given a ring $\mathscr{R}$ of sets, a function*

$$\mu : \mathscr{R} \to [0, \infty]$$

is said to be a **measure** *on $\mathscr{R}$ if $\mu(\varnothing) = 0$ and μ is countably additive on $\mathscr{R}$* [that is, whenever $\{R_n\}$ is a disjoint sequence of members of $\mathscr{R}$ such that

$$\bigcup R_n \in \mathscr{R}, \qquad \text{then } \mu(\bigcup R_n) = \sum \mu R_n].$$

For μ a measure on a ring $\mathscr{R}$, an $A \in \mathscr{R}$ is said to be σ-finite *if there is a sequence $\{R_n\}$ of elements of $\mathscr{R}$ such that*

$$A \subset \bigcup R_n \qquad \text{with } \mu R_n < \infty \quad \text{for } n = 1, 2, 3, \cdots;$$

the measure μ is said to be σ-**finite** *if every $A \in \mathscr{R}$ is σ-finite; and if the whole space $X \in \mathscr{R}$* (so $\mathscr{R}$ is an algebra) *and X is σ-finite, then μ is said to be* **totally** σ-**finite**. *Also, μ is said to be* **complete** *if*

$$(A \in \mathscr{R},\ \mu A = 0 \text{ and } B \subset A) \Rightarrow B \in \mathscr{R}.$$

For example, Lebesgue measure m_k is totally σ-finite and complete. A way of maintaining contact with reality is to visualize $\mathscr{R}$ as the Borel sets; then what evolves is a completion of the Borel sets to obtain the Lebesgue measurable sets, very much in analogy with the reals being the completion of the rationals.

"The real numbers are uncountable" may be stated as: If $\mathscr{R}$ is the algebra of all subsets of E^1 and $\mu A =$ (number of points in A) for each $A \in \mathscr{R}$, then μ is not totally σ-finite. A significant example of a nontotally σ-finite measure is started on page 283.

A gauge of the grasp of previous work on measure is the frequency of back glances when proving the statements in the following problems. In this fresh approach to measure theory, be cautious about accepting a familiar statement before its validity is established under the present ground rules.

PROBLEM 1. With μ a measure on a ring $\mathscr{R}$:

(a) If $A, B \in \mathscr{R}$, and $A \subset B$, then

$$\mu B = \mu A + \mu(B - A) \qquad [\mu \text{ is } \textbf{subtractive}]$$

and

$$\mu A \leqslant \mu B \qquad [\mu \text{ is } \textbf{monotone}].$$

(b) $A, B \in \mathscr{R} \Rightarrow \mu(A \cup B) \leqslant \mu(A) + \mu(B) \qquad [\mu$ is **subadditive**].

PROBLEM 2. Given a measure μ on a ring $\mathscr{R}$ and a sequence (finite or infinite) $\{A_n\}$ of members of $\mathscr{R}$ [but $\bigcup A_n$ need not be in $\mathscr{R}$]:

(a) Then $(A \in \mathscr{R}$ and $A \subset \bigcup A_n) \Rightarrow \mu A \leqslant \sum \mu A_n \qquad [\mu$ is **countably subadditive**].

(b) If $\bigcup A_n \subset A$ with $A \in \mathscr{R}$ and $\{A_n\}$ is disjoint, then

$$\sum \mu A_n \leqslant \mu A.$$

(c) If $\{A_n\}$ is monotonically increasing and $\lim A_n \in \mathscr{R}$, then

$$\mu\,(\lim A_n) = \lim \mu A_n.$$

(d) If $\{A_n\}$ is monotonically decreasing, $\lim A_n \in \mathscr{R}$, and $\mu A_n < \infty$ for at least one n, then

$$\mu\,(\lim A_n) = \lim \mu A_n.$$

PROBLEM 3. Given a ring $\mathscr{R}$, there is a smallest σ-ring containing $\mathscr{R}$ [denoted by $\sigma(\mathscr{R})$ and called the **σ-ring generated** by $\mathscr{R}$].

[*Hint:* In case Section 5–3 was skipped, look at Theorem 5–3.1.]

Definition 2. *A nonempty family $\mathscr{E}$ of sets is said to be* **hereditary** *if whenever $E \in \mathscr{E}$ and $F \subset E$, then $F \in \mathscr{E}$.*

PROBLEM 4. A hereditary family $\mathscr{E}$ is a σ-ring iff $\mathscr{E}$ is closed under formation of countable unions; that is, whenever $\{E_n\}$ is a sequence of members of $\mathscr{E}$, then $\bigcup E_n \in \mathscr{E}$.

PROBLEM 5. If $\mathscr{E}$ is a nonempty family (not necessarily hereditary), then there is a smallest hereditary σ-ring containing $\mathscr{E}$ [denoted by $\mathscr{H}(\mathscr{E})$ and called the **hereditary σ-ring** generated by $\mathscr{E}$].

PROBLEM 6. If $\mathscr{E}$ is a nonempty family of sets, then $\mathscr{H}(\mathscr{E})$ is the family of

(a) All sets which may be covered by countably many sets in $\mathscr{E}$.

(b) All subsets of sets in $\mathscr{E}$.

Definition 3. *Given a hereditary σ-ring $\mathscr{H}$, a function*

$$\varphi\colon \mathscr{H} \to [0, \infty],$$

which is countably subadditive and $\varphi(\varnothing) = 0$ is said to be an **outer measure** *on $\mathscr{H}$.*

PROBLEM 7. With μ a measure on a ring $\mathscr{R}$, define μ^* on $\mathscr{H}(\mathscr{R})$ by

$$\mu^*(A) = \inf\{\textstyle\sum \mu(R_n) \mid R_n \in \mathscr{R} \text{ and } A \subset \bigcup R_n\}.$$

Then μ^* is an outer measure on $\mathscr{H}(\mathscr{R})$ [called the outer measure **generated** by μ].

Definition 4. *With* μ^* *an outer measure on a hereditary* σ*-ring* $\mathscr{H}$, *a set* $A \in \mathscr{H}$ *is said to be*† μ^**-measurable if for every* $W \in \mathscr{H}$,

$$\mu^* W = \mu^* WA + \mu^* WA^c.$$

[Note that all this is in a space X so that A^c is defined and $WA^c \in \mathscr{H}$ even though A^c may not be.]

PROBLEM 8. Let μ^* be an outer measure on a hereditary σ-ring $\mathscr{H}$ and let $\mathfrak{M}$ be the family of all μ^*-measurable sets.

(a) $A \in \mathfrak{M}$ iff $\mu^* W \geqslant \mu^* WA + \mu^* WA^c$ for every $W \in \mathscr{H}$.
(b) $\mathfrak{M}$ is a ring.
(c) $\mathfrak{M}$ is a σ-ring.
(d) If $A \in \mathscr{H}$ and $\{M_n\}$ is a disjoint sequence of sets in $\mathfrak{M}$, then

$$\mu^*[A(\bigcup M_n)] = \sum \mu^*(AM_n).$$

(e) If $A \in \mathscr{H}$ and $\mu^* A = 0$, then $A \in \mathfrak{M}$.
(f) Upon defining the set function $\bar{\mu}$ by setting

$$\bar{\mu}A = \mu^* A \qquad \text{for all } A \in \mathfrak{M},$$

then $\bar{\mu}$ is a measure on $\mathfrak{M}$ and moreover is complete. [$\bar{\mu}$ is called the measure **induced** by the outer measure μ^*.]

In the following problem, μ is a measure on a ring $\mathscr{R}$, μ^* is the generated outer measure on $\mathscr{H}(\mathscr{R})$, and $\bar{\mu}$ is the measure induced by μ^* on the σ-ring $\mathfrak{M}$ of all μ^*-measurable sets of $\mathscr{H}(\mathscr{R})$. Recall the definition of $\sigma(\mathscr{R})$ in Problem 3.

PROBLEM 9. (a). If $R \in \mathscr{R}$, then R is μ^*-measurable.

[*Hint:* Take any $W \in \mathscr{H}(\mathscr{R})$ and $\varepsilon > 0$, and see that there is a sequence $\{E_n\}$ in $\mathscr{R}$ such that $W \subset \bigcup E_n$ and

$$\begin{aligned}\mu^* W + \varepsilon \geqslant \sum \mu E_n &= \sum \{\mu E_n R + \mu(E_n - R)\}\\ &\geqslant \mu^* WR + \mu^* WR^c.]\end{aligned}$$

(b) Every set in $\sigma(\mathscr{R})$ is μ^*-measurable. [*Hint:* By part (a), $\mathscr{R} \subset \mathfrak{M}$.]
(c) If $A \in \mathscr{H}(\mathscr{R})$, then

$$\begin{aligned}\mu^* A &= \inf\{\bar{\mu}(M) \mid A \subset M \in \mathfrak{M}\}\\ &= \inf\{\bar{\mu}(M) \mid A \subset M \in \sigma(\mathscr{R})\}.\end{aligned}$$

(d) If $A \in \mathscr{H}(\mathscr{R})$ and A is σ-finite for μ^*, then there is an $\hat{A} \in \sigma(\mathscr{R})$ such that

$$A \subset \hat{A} \qquad \text{and} \qquad \mu^* A = \bar{\mu}(\hat{A}).$$

Problems on Linear Measure of Sets in E^2

The diameter $d(U)$ of a set $U \subset E^2$ is defined by

$$d(U) = \sup\{|u_1 - u_2| \mid u_1 \in U, u_2 \in U\}.$$

Given a nonempty set $A \subset E^2$ and a number $\rho > 0$, a sequence $\{U_n\}$ of sets is said to be a ρ-covering of A if

$$A \subset \bigcup U_n \qquad \text{and} \qquad d(U_n) < \rho \qquad \text{for } n = 1, 2, 3, \cdots.$$

† We say μ^*-measurable now because μ-measurable would not carry the proper meaning.

Let

$$l_\rho(A) = \inf \sum d(U_n),$$

where the infimum is taken over all ρ-coverings of A. Then

$$\rho_1 < \rho_2 \Rightarrow l_{\rho_1}(A) \geqslant l_{\rho_2}(A).$$

The **outer linear measure** of A is defined by

$$l^*(A) = \lim_{\rho \to 0} l_\rho(A).$$

Also, by definition, $l^*(\varnothing) = 0$.

Then A is said to be l-measurable (linearly measurable) if

$$l^* W = l^* WA + l^* WA^c$$

for each set W. As noted on page 186 this was the first use of this definition of measurability.

PROBLEMS

1. The same number $l^*(A)$ is obtained if every member of every ρ-covering used is
   (a) A convex set.
   (b) An open convex set.
   (c) A closed convex set.
2. l^* is a metric outer measure. Hence, from the definition of measurability, every open set, closed set, G_δ set, F_σ set, etc., is linearly measurable.
3. Every nonempty open set has infinite linear measure.
4. Given any set $A \subset E^2$,
   (a) There is a G_δ set $\hat{A}$ such that
   $$\hat{A} \subset A \qquad \text{and} \qquad l(\hat{A}) = l^*(A).$$
   (b) Any set $\hat{A}$ as in (a) can be expressed as
   $$\hat{A} = \bigcap H_n, \ H_n \text{ open}, \qquad \text{and } H_1 \supset H_2 \supset H_3 \supset \cdots.$$
5. With $A \subset E^2$ and P the projection of A on the x-axis, then $m^*P \leqslant l^*(A)$.
6. Let A_0 be the interior and boundary of an equilateral triangle of side 1. Let A_1 be the union of the three triangles of side 1/3 at each vertex of A_0. Let A_2 be the union of the nine triangles of side 1/9 shown below. Continue in this way to obtain

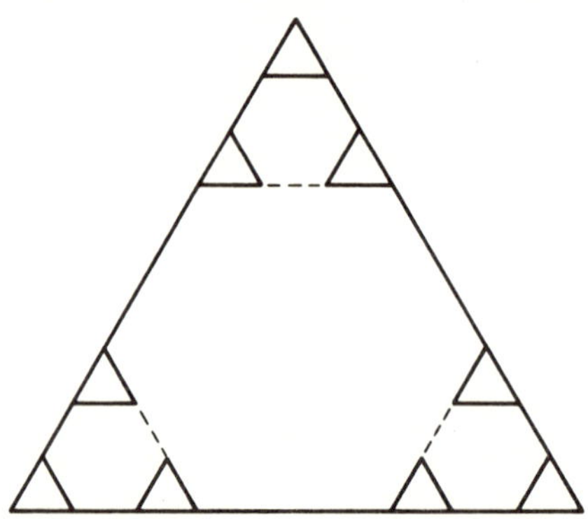

$A_3, A_4, \cdots$. Then the set

$$A = A_0 \bigcap A_1 \bigcap A_2 \bigcap \cdots \quad \text{has } l(A) = 1.$$

7. A set B with $l^*(B) < \infty$ and a number $\varepsilon > 0$ is given. Show that there is a number $\delta > 0$ such that if $\{V_n\}$ is any sequence of closed convex sets with each $d(V_n) < \delta$, then

$$\sum d(V_n) > l^*[(B(\bigcup V_n)] - \varepsilon.$$

[*Hint:* Let $\delta > 0$ be such that $l_\delta(B) > l^*(B) - (\varepsilon/2)$. Take any $\{V_n\}$ as described. Cover $B - \bigcup V_n$ by a sequence $\{W_n\}$ having $d(W_n) < \delta$ and

$$\sum d(W_n) < l^*(B - \bigcup V_n) + \frac{\varepsilon}{2}.]$$

8. Let $B \subset A$ both be linearly measurable, with $l(A) < \infty$. Then corresponding to each $\varepsilon > 0$, there is an open set H such that

$$B \subset H \qquad \text{and} \qquad l(AH) \subset l(B) + \varepsilon.$$

[*Hint:* Use Problem 4(b) on the set B.]

Definition. *Let A be linearly measurable. Denote the closed circular disk with center z and radius $r > 0$ by $c(z, r)$. The lower density $\underline{D}(A, z)$ and upper density $\overline{D}(A, z)$ of A at z are defined by the limits inferior and superior as $r \to 0$ of*

$$\frac{l[Ac(z, r)]}{2r}.$$

9. The set of Problem 6 has $\underline{D}(A, z) \leqslant \frac{1}{2}$ for all $z \in E^2$.

10. Let A be linearly measurable, with $l(A) < \infty$.
(a) With $r > 0$ fixed, $\{z \mid lAc(z, r) > a\}$ is closed for each real a.
(b) With r fixed, the function f defined by

$$f(z) = \frac{l[Ac(z, r)]}{2r}$$

is a measurable function.

(c)
$$\overline{\lim_{n\to\infty}} \frac{l[Ac(z, 1/n)]}{2/n} = \overline{D}(A, z).$$

[*Hint:* With $r > 0$ and n such that $1/(n+1) < r \leqslant 1/n$,

$$\frac{l[Ac(z, r)]}{2r} \leqslant \frac{n+1}{n} \frac{l[Ac(z, 1/n)]}{2/n}\Bigg].$$

(d) The function f defined by

$$f(z) = \overline{D}(A, z)$$

is a measurable function.

11. If A is linearly measurable and $l(A) < \infty$, then

$$\overline{D}(A, z) \leqslant 1 \qquad \text{for almost all } z \in A.$$

[*Hint:* With

$$B_n = \{z \in A \mid \overline{D}(A, z) > 1 + 1/n\}, \qquad n = 1, 2, \cdots,$$

show that $l(B_n) = 0$. For the set B_n use the open set H as in Problem 8 and $\delta > 0$ as in Problem 7. B_n can be covered in the Vitali sense by using circles satisfying appropriate conditions.]

Theorem 1. *If A is linearly measurable with $l(A) < \infty$, then*

$$\bar{D}(A, z) \geq \tfrac{1}{2} \quad \text{for } l\text{-almost all } z \in A.$$

Proof. Assume $\{z \in A \mid \bar{D}(A, z) < \frac{1}{2}\}$ has positive linear measure. Choose $n \geq 3$ such that

$$B = \left\{z \in A \,\middle|\, \bar{D}(A, z) < \frac{1}{2} - \frac{1}{n}\right\} \text{ has } l(B) > 0.$$

Choose $m \geq 1$ such that $l(C) > 0$, where

$$C = \left\{z \in B \,\middle|\, \frac{l[Ac(z, r)]}{2r} < \frac{1}{2} - \frac{1}{n} \text{ for all } r < \frac{1}{m}\right\}.$$

Given a number $\varepsilon > 0$, take $\{U_p\}$ such that

$$C \subset \bigcup U_p, \qquad CU_p \neq \varnothing, \qquad 0 < d(U_p) < \frac{1}{m},$$

and

$$\sum d(U_p) < l(C) + \varepsilon.$$

Set $d_p = d(U_p)$ and select $z_p \in CU_p$. Then

$$U_p \subset c(z_p, d_p) \qquad \text{and} \qquad CU_p \subset Cc(z_p, d_p).$$

Hence

$$\begin{aligned} l(C) &\leq \sum l(CU_p) \leq \sum l[Cc(z_p, d_p)] \\ &\leq \sum \frac{l[Ac(z_p, d_p)]}{2d_p} \cdot 2d_p < \left(\frac{1}{2} - \frac{1}{n}\right) 2 \sum d_p \\ &< \left(1 - \frac{2}{n}\right)[l(C) + \varepsilon]. \end{aligned}$$

Therefore $l(C) \leq (1 - 2/n)l(C)$ which is a contradiction to $0 < l(C) \leq l(A) < \infty$.

Comments. One policy I have followed, with few exceptions, is to omit a proof of a theorem or stated fact only if I felt confident the reader could supply a proof by generalizing or amalgamating previously included material. I depart from this policy in these comments about one of my early interests. The numbers in parentheses refer to the Bibliography at the end of these comments.

In Carathéodory's original paper on linear measure (cited on page 186) he showed that if A is the set of points forming a simple plane rectifiable arc, then $l(A) =$ length of the arc. Thus Carathéodory linear measure is indeed a generalization of arc length.

Carathéodory did not consider density properties; this phase was started mainly by Besicovitch (1). Unlike sets on the line, a plane linearly measurable set need not have density 1 at almost all of its points (as seen in the above problems). If $l(A) < \infty$, then

$$0 \leq \underline{D}(A, a) \leq 1 \qquad \text{and} \qquad \tfrac{1}{2} \leq \bar{D}(A, a) \leq 1$$

for almost all points of A, and these are the best possible [(1) and (11)] in the sense that each equality holds for some set with $l(A) > 0$.

It is shown in (1) and by different methods in (11), that a set with $l(A) < \infty$ differs in linear measure from a set on the union of countably many rectifiable arcs iff the density of A is 1 at almost all points of A. Hence the set of Problem 6 (see Problem 9) intersects every rectifiable arc in a set of linear measure zero. Stated another way:

Following the method of defining an outer measure by covering a set with elementary sets (page 279), we might have considered rectifiable arcs as "elementary" sets. An outer linear measure would be obtained, but the extremely sparse set of Problem 6 would have this linear measure infinite.

Another fact is that if $A \subset B$ and $l(B) < \infty$, then "locally" the set A is very much the same as B. By this is meant (14) that at almost all points of A,

$$\underline{D}(A, a) = \underline{D}(B, a) \qquad \text{and} \qquad \bar{D}(A, a) = \bar{D}(B, a).$$

Hence if $A \subset E^1$, so that $l(A) = m(A)$, then $D(A, a) = 1$ at almost all points of A, since $D(E^1, a) = 1$ for all $a \in E^1$.

There are definitions of linear measure of plane sets other than Carathéodory's. See Refs. (5), (6), (7), (9), (10), and (16). Each of these has some advantages over the others. For example, with $A \subset E^2$, let A_x and A_y be the projections of A on the x-axis and y-axis. Then certainly $l(A_x) \leqslant l(A)$ and $l(A_y) \leqslant l(A)$, and it might be expected that

$$l(A) \text{ would be} \geqslant \sqrt{[l(A_x)]^2 + [l(A_y)]^2}.$$

There is, however, an example (10) of a set for which

$$l(A) = l(A_x) = l(A_y) = 1.$$

Gillespie linear measure $g(A)$ was named after my Ph.D thesis director who casually proposed the definition. This linear measure has (10) the properties

$$g(A) \geqslant \sqrt{[g(A_x)]^2 + [g(A_y)]^2}, \qquad g(A_x) \leqslant g(A), \qquad \text{and } g(A_y) \leqslant g(A).$$

It is known (10) that $l(A) \leqslant g(A) \leqslant (\pi/2)l(A)$, and these are the best possible (8).

Each of the linear measures has all the requisite properties to support a theory of Lebesgue integration. Thus if $A \subset E^2$ and $f: A \to E^*$ is an l or g linearly measurable function, then

$$\int_A f(x, y)\, dl \qquad \text{or} \qquad \int_A f(x, y)\, dg$$

has meaning as a generalized line integral. Then a plane set G with $m_2 G > 0$ may have a boundary that is not a curve but a set A with $l(A) < \infty$. Theorems (of Green's variety) relating Lebesgue m_2-integrals over G and Lebesgue l-integrals over the boundary may then be considered (13). For extensions to higher dimensions, see Ref. (3).

Each of the linear measures has an extension to a "surface" measure of sets in E^3. It is natural to visualize an "ordinate" set

$$H = \{(x, y, z) \mid (x, y) \in A, \qquad 0 \leqslant x \leqslant f(x, y)\}$$

as a 2-dimensional set in E^3 and to wonder if

$$l_2 H \qquad \text{and} \qquad \int_A f(x, y)\, dl$$

are equal. Simpler (but equivalently), if a set $A \subset E^2$ and an interval $I \subset E^1$ are taken, is

$$l_2(A \times I) \quad \text{equal to} \quad |I|\, l(A)?$$

Examples, (2) and (4), show they need not be equal. For Gillespie g and g_2, however, (10),

$$g_2(A \times I) = |I|\, g(A).$$

None of these 2-measures for sets in E^3, however, satisfactorily answer the perplexing question of surface area (12).

Bibliography

1, A. S. Besicovitch, "On the Fundamental Geometric Properties of Linearly Measurable Sets of Points," *Mathematische Annalen*, **98** (1928), pp. 422–464; **115** (1938), pp. 296–329; **116** (1939), pp. 347–357.
2. A. S. Besicovitch and P. A. P. Moran, "The Measure of Product and Cylinder Sets," *J. London Math. Soc.* **20** (1945), pp. 110–120.
3. H. Federer, "Surface Area I, II," *Trans. Am. Math. Soc.*, **55** (1944), pp. 420–456.
4. G. Freilich, "Carathéodory Measure of Cylinders," *Trans. Am. Math. Soc.*, **122** (1966), pp. 153–162.
5. W. Gross, "Über Flächenmass von Punktmengen," *Monatshefte für Mathematik und Physik*, **29** (1918) pp. 145–176; and "Über das lineare Mass von Punktmengen," *ibid.*, pp. 177–193.
6. O. Jansen, "Über einige stetige Kurven, über Bogenlangen, linearen Inhalt und Flächeninhalt," Inaugural Dissertation, Königsberg, 1907.
7. R. L. Jeffrey, "Sets of k-Extent in n-Dimensional Space," *Trans. Am. Math. Soc.* **35** (1933), pp. 629–647.
8. D. S. Miller, "Some Properties of Carathéodory and Gillespie Linear Measure," *Duke Math. J.*, **11** (1944), pp. 139–145.
9. N. Minkowski, "Über die Begriffe Länge, Oberfläche, und Volumen," *Jahresbericht der Deutschen Mathematiker-Vereinigung*, **9** (1900), pp. 115–121.
10. A. P. Morse and J. F. Randolph, "Gillespie Measure," *Duke Math. J.*, **6** (1940), pp. 408–419.
11. ———, "The φ Rectifiable Subsets of the Plane," *Trans. Am. Math. Soc.*, **55** (1944), pp. 236–305.
12. T. Radó, "Length and Area," *Am. Math. Soc. Colloquium Publications*, **30** (1948).
13. J. F. Randolph, "Carathéodory Measure and a Generalization of the Gauss-Green Lemma," *Trans. Am. Math. Soc.*, **38** (1935), pp. 531–548.
14. ———, "Some Density Properties of Point Sets," *Annals of Mathematics*, **37** (1936), pp. 336–344.
15. ———, "On Generalizations of Length and Area," *Bull. of the Am. Math. Soc.*, **42** (1936), pp. 268–274.
16. J. P. Schauder, "The Theory of Surface Measure," *Fundamenta Mathematicae*, **8** (1926), pp. 1–48.

Chapter 6

CONTINUITY

6–1 LIMITS AND CONTINUITY OF FUNCTIONS

For the sake of simplicity, abstract metric spaces X and Y will be considered until properties indigenous to specific realizations (such as E^1, E^k, or $\mathscr{L}^2$) are discussed. The distance functions of these spaces will be denoted by d_X and d_Y. Hence if $x_1, x_2, x_3 \in X$, then

$$d_X(x_1, x_2) = d_X(x_2, x_1) \geqslant 0, \qquad \text{with} \qquad = 0 \text{ iff } x_1 = x_2,$$

$$d_X(x_1, x_3) \leqslant d_X(x_1, x_2) + d_X(x_2, x_3),$$

and corresponding relations hold for d_Y and points of Y. Subscripts will also designate the space in which a neighborhood lies:

$$N_X(x_0, r) = \{x \in X \mid d_X(x, x_0) < r\},$$

$$N_Y(y_0, r) = \{y \in Y \mid d_Y(y, y_0) < r\}, \qquad r > 0,$$

which may be visualized as open balls in X and Y. Also

$$N_X{}^\circ(x_0, r) = N_X(x_0, r) - \{x_0\}$$

is a punctured neighborhood of x_0.

Definition 1. *A function f with domain $D \subset X$ and range in Y; that is,*

$$f: D \to Y,$$

is said to have limit $y_0 \in Y$ at $x_0 \in X$, and we write $f(x) \to y_0$ as $x \to x_0$ if both
(i). *$x_0 \in D'$ (D' is the derived set of D) and*
(ii). *corresponding to x_0 and each number $\varepsilon > 0$ there is a number $\delta > 0$ such that*

$$f(D \cap N_X{}^\circ(x_0, \delta)) \subset N_Y(y_0, \varepsilon).$$

The set notation in (ii) may, of course, be replaced by

$$[x \in D \text{ and } 0 < d_X(x, x_0) < \delta] \Rightarrow d_Y(f(x), y_0) < \varepsilon.$$

The appearance of $0 < d_X(x, x_0)$ here, or the punctured neighborhood $N_X{}^\circ$ in (ii), shows that $f(x_0)$ need not be defined, and even if $f(x_0)$ is defined its value is not used in determining whether $f(x) \to y_0$ as $x \to x_0$.

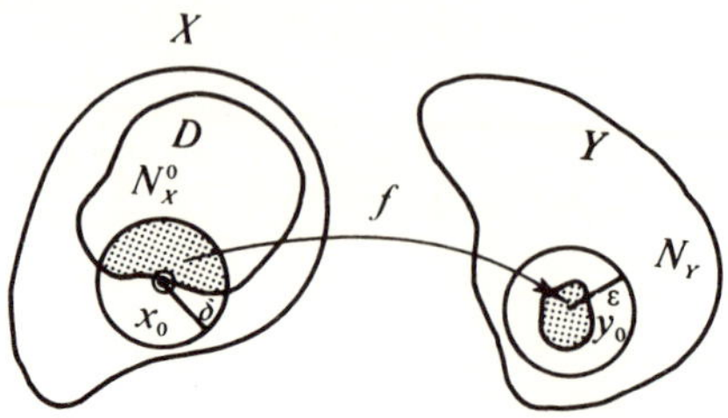

FIG. 1

Theorem 1. *If $f: D \to Y$ has a limit at $x_0 \in D'$, then this limit is unique.*

Proof. Let $f(x) \to y_0$ and $f(x) \to y_1$ as $x \to x_0 \in D'$ and select $\varepsilon > 0$ arbitrarily. Determine $\delta > 0$ such that $x \in D$ and $0 < d_X(x, x_0) < \delta$ imply both

$$d_Y(f(x), y_0) < \varepsilon \qquad \text{and} \qquad d_Y(f(x), y_1) < \varepsilon$$

and therefore imply

$$d_Y(y_1, y_0) \leqslant d_Y(f(x), y_1) + d_Y(f(x), y_0) < 2\varepsilon.$$

Hence $d_Y(y_1, y_0) = 0$, so that $y_1 = y_0$. ∎

Because of Theorem 1, we may use

$$\lim_{x \to x_0} f(x) = y_0$$

as an alternative for $f(x) \to y_0$ as $x \to x_0$. We shall never write $\lim_{x \to x_0}$ unless $x_0 \in D'$, but with $x_0 \in D'$ we use

$$\lim_{x \to x_0} f(x) \neq y_0,$$

provided either $\lim_{x \to x_0} f(x)$ does not exist or exists and is not y_0; precisely,

$$\exists\, \varepsilon > 0 \text{ such that with } \delta > 0 \text{ arbitrary,}$$

$$f(D \cap N_X{}^\circ(x_0, \delta)) - N_Y(y_0, \varepsilon) \neq \varnothing.$$

Theorem 2. *If $f: D \to Y$ and $x_0 \in D'$, then the two statements*
(a) $\lim_{x \to x_0} f(x) = y_0$; *and*
(b) $\lim_{n \to \infty} f(x_n) = y_0$ *for every sequence* $\{x_n\}$ *such that* $x_n \in D$, $x_n \neq x_0$, *and* $\lim_{n \to \infty} x_n = x_0$,

are equivalent.

Proof. Let (a) hold. Given a number $\varepsilon > 0$, determine $\delta > 0$ such that

$$f(D \cap N_X{}^\circ(x_0, \delta)) \subset N_Y(y_0, \varepsilon).$$

Select any sequence as specified in (b). Corresponding to the number δ determined above, let N be such that

$$n \geqslant N \Rightarrow x_n \in N_X{}^\circ(x_0, \delta).$$

Hence $n \geqslant N \Rightarrow f(x_n) \in f(D \cap N_X{}^\circ(x_0, \delta)) \subset N_Y(y_0, \varepsilon)$; that is, $\lim_{n\to\infty} f(x_n) = y_0$. Thus (a) $\Rightarrow$ (b).

Conversely, take an f and y_0 for which $\lim_{x\to x_0} f(x) \neq y_0$, and accordingly determine $\varepsilon > 0$ such that for $\delta > 0$ arbitrary,

$$f(D \cap N_X{}^\circ(x_0, \delta)) - N_Y(y_0, \varepsilon) \neq \varnothing.$$

Hence, for each positive integer n, an $x_n \in D \cap N_X{}^\circ(x_0, 1/n)$ may be selected such that $f(x_n) \notin N_Y(y_0, \varepsilon)$. Then

$$\lim_{n\to\infty} f(x_n) \neq y_0$$

even though $\lim_{n\to\infty} x_n = x_0$, $x_n \in D$ and $x_n \neq x_0$. Thus (b) does not hold. Hence

$$\text{not (a)} \Rightarrow \text{not (b)}; \qquad \text{that is, (b)} \Rightarrow \text{(a)}. \ \blacksquare$$

Theorem 3. *With X a metric space and $D \subset X$, let functions*

$$f : D \to E^1 \qquad \textit{and} \qquad g : D \to E^1$$

both have limits at some $x_0 \in D'$ denoted by

$$\lim_{x\to x_0} f(x) = y_1 \qquad \textit{and} \qquad \lim_{x\to x_0} g(x) = y_2 .$$

Then the following limits exist and equalities hold:

(a) $\lim_{x\to x_0} (f + g)(x) = y_1 + y_2$,
(b) $\lim_{x\to x_0} (fg)(x) = y_1 y_2$, *and*
(c) $\lim_{x\to x_0} (f/g)(x) = y_1/y_2$, *provided* $y_2 \neq 0$.

The same is true with E^1 replaced by E^2 interpreted as the complex plane.

Upon replacing E^1 by E^k and using vector notation (including the inner product), then

(a) $\lim_{x\to x_0} (\mathbf{f} + \mathbf{g})(x) = \mathbf{y}_1 + \mathbf{y}_2$ and
(b) $\lim_{x\to x_0} (\mathbf{f} \cdot \mathbf{g})(x) = \mathbf{y}_1 \cdot \mathbf{y}_2$

but (c) *has no analog in E^k with $k > 2$.*

Proof. Because of Theorem 2, these results follow from corresponding results for sequences. $\blacksquare$

In accordance with previous usage (Sec. 2–1), if

$$f : D \to Y \qquad \text{and} \qquad A \subset X,$$

then

$$f(A) = \{y \in Y \mid \exists\, x \in A \cap D \text{ and } y = f(x)\}.$$

Thus $f(A) = f(AD)$ and, in particular if $AD = \varnothing$, then $f(AD) = \varnothing$; it being unnecessary to distinguish empty sets of X and Y. Hence

$$f(N_X{}^\circ(x_0, r)) = f(D \cap N_X{}^\circ(x_0, r))$$

and we shall henceforth so simplify the notation. Also, if $B \subset Y$, then

$$f^{-1}(B) = \{x \in D \mid f(x) \in B\}.$$

Hence, again, if $B \cap f(D) = \varnothing$, then $f^{-1}(B) = \varnothing$. Even if $B \cap f(D) \neq \varnothing$, the set $f(f^{-1}(B))$ may be a proper subset of B, so all that can be said is

$$f(f^{-1}(B)) \subset B.$$

Definition 2. *A function* $f: D \to Y$ *with* $D \subset X$ *is said to be* **continuous** *at* $x \in D$ *if corresponding to this* x *and each* $\varepsilon > 0$ *there is a* $\delta > 0$ *such that*

$$f(N_X(x, \delta)) \subset N_Y(f(x), \varepsilon).$$

A function which is continuous at each point of its domain is said to be a **continuous function.**

Hence a positive way of stating the negative assertion

"f: $D \to Y$ is not continuous"

is: "There exists a point $x_0 \in D$ and a number $\varepsilon > 0$ such that corresponding to each arbitrary number $\delta > 0$ there is an $x \in D \cap N_X(x_0, \delta)$ for which $\{f(x)\} \cap N_Y(f(x_0), \varepsilon) = \varnothing$.

In Definition 2, $x \in D$ need not be a cluster point of D and the neighborhood $N_X(x, \delta)$ is not punctured. Hence f is continuous at $x_0 \in D$ iff either

1. $x_0 \in D - D'$; that is, $x_0 \in D$ is an isolated point of D, or
2. $x_0 \in D \cap D'$ and the limit exists and the equality holds in

$$\lim_{x \to x_0} f(x) = f(x_0).$$

Either "f is discontinuous at x" or "x is a point of discontinuity of f" may be used in place of "$x \in X$ and f is not continuous at x." In particular, any point of $X - D = D^c$ is a point of discontinuity of $f: D \to Y$.

EXAMPLE 1. Let D be the set of all rationals of E^1 and define f on D (and nowhere else) by $f(x) = 1$ iff $x \in D$. Then $f: D \to E^1$ is a continuous function. For this function, x is a point of discontinuity iff x is irrational.

EXAMPLE 2. Let $D = E^1$ and define f by setting

$$f(x) = \begin{cases} 1 & \text{if } x \text{ is rational} \\ 0 & \text{if } x \text{ is irrational.} \end{cases}$$

Then $f: E^1 \to E^1$ has every $x \in E^1$ as a point of discontinuity.

PROBLEM 1. Express each rational on $(0, 1) \subset E^1$ in simplest form; that is, as p/q with p and q positive integers with no common factor other than 1. Define $f: (0, 1) \to E^1$ by

$$f(x) = \begin{cases} 1/q & \text{if } x = p/q \text{ in simplest form} \\ 0 & \text{if } x \text{ is irrational.} \end{cases}$$

Then f is discontinuous at every rational and continuous at every irrational of its domain.

PROBLEM 2. The function $f: (0, 1] \to E^1$, defined by

$$f(x) = \frac{1}{x}, \qquad 0 < x \leqslant 1,$$

is a continuous function.

EXAMPLE 3. The Cantor function $\psi: [0, 1] \to E^1$ is a continuous function transforming a set of measure zero (the Cantor set Γ itself) into a set of positive measure.

To see that ψ is continuous, first take points

$$\left.\begin{aligned} c' &= 0_{\underset{3}{.}}(2\alpha_1)(2\alpha_2)(2\alpha_3)\cdots \\ c'' &= 0_{\underset{3}{.}}(2\beta_1)(2\beta_2)(2\beta_3)\cdots \end{aligned}\right\} \qquad \begin{aligned} &\text{each } \alpha_i \text{ and } \beta_i \\ &\text{either 0 or 1} \end{aligned}$$

of the Cantor set and note that if $|c' - c''| < 1/3^n$, then $\alpha_i = \beta_i$ for $1 \leqslant i \leqslant n+1$ and therefore

$$|\psi(c') - \psi(c'')| < \frac{1}{2^n}.$$

Hence, if $c_0 \in \Gamma$ and $\{c_n\}$ is such that $c_n \in \Gamma$ and $c_n \to c_0$ as $n \to \infty$, then

$$\psi(c_n) \to \psi(c_0) \qquad \text{as } n \to \infty.$$

Next let $x_0 \in [0, 1]$ and $\{x_n\}$ be such that $x_n \in [0, 1]$, $x_n \neq x_0$ and $x_n \to x_0$ as $n \to \infty$. Consider two cases.

Case 1. $x_0 \notin \Gamma$. Then $x_0 \in I \subset (0, 1)$ with I an open interval complementary to Γ. Then $x_n \in I$ and $\psi(x_n) = \psi(x_0)$ for all except a finite number of n, and thus

$$\psi(x_n) \to \psi(x_0) \qquad \text{as } n \to \infty.$$

Case 2. $x_0 \in \Gamma$. For each n such that $x_n \in \Gamma$, set $c_n = x_n$. If $x_n \notin \Gamma$ then, with I the complementary interval of Γ containing x_n,

(a) If $x_n < x_0$, set c_n the upper end point of I.
(b) If $x_0 < x_n$, set c_n the lower end point of I.

Hence, in any case, $\psi(x_n) = \psi(c_n) \to \psi(x_0)$ as $n \to \infty$. Since $\{x_n\}$ was any sequence satisfying the stated conditions, then (by using Theorem 2) ψ is continuous.

We have already shown (Theorem 2–2.10) that ψ carries Γ onto $[0, 1]$; that is, a set of measure zero into a set of positive measure.

Theorem 4. *Given metric spaces X, Y, Z and functions*

$$f: A \to Y \qquad \text{and} \qquad g: B \to Z, \qquad \text{with } A \subset X \text{ and } f(A) \subset B \subset Y,$$

let $h = g \circ f: A \to Z$ be the **composition** *function defined by*

$$h(x) = g(f(x)) \qquad \text{iff } x \in A.$$

If f and g are continuous, then h is continuous.

Proof. Select any $x \in A$ and set $y = f(x)$ and $z = g(y)$. Given a number $\varepsilon > 0$, determine $\eta > 0$ such that

$$g(N_Y(y, \eta)) \subset N_Z(z, \varepsilon).$$

Corresponding to this number $\eta > 0$, determine $\delta > 0$ such that

$$f(N_X(x, \delta)) \subset N_Y(y, \eta).$$

Then

$$h(N_X(x, \delta)) = g[f(N_X(x, \delta))] \subset g(N_Y(y, \eta)) \subset N_Z(z, \varepsilon)$$

which says that h is continuous at x. ▮

Definition 3. *Given a function $f: D \to Y$ and a set $A \subset D$, we denote by f_A the function having domain A defined by*

$$f_A(x) = f(x) \qquad \text{iff } x \in A$$

and shall call $f_A: A \to Y$ the **restriction** *of f to A.*

With f the function of Example 2, let A be the set of all rational numbers. Then the function f, which is discontinuous everywhere (in E^1), has the restriction f_A continuous.

PROBLEM 3. If $f: D \to Y$ is continuous and $A \subset D$, then the restriction f_A is continuous.

Theorem 5. *Let f and g be continuous complex-valued functions having common domain D in a metric space X. Then $f + g$ and fg are continuous and f_A/g_A is continuous where these are the restrictions of f and g to*

$$A = \{x \in D \mid g(x) \neq 0\}.$$

Proof. At points of $D \cap D'$ or $A \cap A'$ the results follow from Theorem 3. At any isolated point x of D there is a $\delta > 0$ such that

$$(f + g)(N_X(x, \delta)) = (f + g)(x) \subset N_{E^2}((f + g)(x), \varepsilon)$$

for all $\varepsilon > 0$, with similar results for fg and f_A/g_A. ▮

Since X in Theorem 5 may be taken as E^2 (or E^1), then with $a_0, a_1, \cdots, a_n$ complex numbers, the polynomial

$$P(x) = a_0 + a_1 x + a_2 x^2 + \cdots + a_n x^n, \qquad x \in E^2$$

is continuous. Then any complex rational function P/Q, with P and Q polynomials, is continuous at each point of $\{x \in E^2 | Q(x) \neq 0\}$.

Also with $\mathbf{x} = (x_1, x_2, \cdots, x_k) \in E^k = X$, continuity is obtained for polynomials of the form

$$P(\mathbf{x}) = \sum a_{n_1 n_2 \ldots n_k} {x_1}^{n_1} {x_2}^{n_2} \cdots {x_k}^{n_k},$$

where $n_1, n_2, \cdots, n_k$ are nonnegative integers, the a's are complex numbers, and there is a finite number of terms in the sum.

Theorem 6. *Let $f_1, f_2, \cdots, f_k$ be complex-valued functions with common domain D in a metric space X and let*

$$\mathbf{f}: D \to E^k$$

be defined by

$$\mathbf{f}(x) = (f_1(x), f_2(x), \cdots, f_k(x)), \qquad x \in D.$$

(a) *Then* $\mathbf{f}$ *is continuous iff each component function f_i is continuous.*
(b) *With* $\mathbf{g}: D \to E^k$ *also such a function, if* $\mathbf{f}$ *and* $\mathbf{g}$ *are continuous, then*

$$\mathbf{f} + \mathbf{g}: D \to E^k \qquad \textit{and} \qquad \mathbf{f} \cdot \mathbf{g}: D \to E^2$$

are continuous.

Proof. At points of $D \cap D'$ all limits exist and have proper values from Theorem 3 and appropriate theorems about sequences. At isolated points of D the method in the proof of Theorem 5 may be used. ∎

6–2 RELATIVE OPENNESS AND CONTINUITY

A point $a \in A \subset D \subset X$, with X a metric space, is said to be an **interior point of** A **relative to** D, provided $\exists\, \delta > 0$ such that

$$D \cap N_X(a, \delta) \subset A.$$

Then A is said to be **open relative** to D if each point of A is interior to A relative to D. These notations were given in Definition 2–11.5 and will be further illustrated and then related to continuity of functions.

Let D inherit the distance function of X; that is, think of d_D defined on $D \times D$ (and nowhere else) by

$$d_D(x_1, x_2) = d_X(x_1, x_2) \qquad \text{iff } x_1, x_2 \in D.$$

Then D is itself a metric space (a submetric space of X as defined in Definition 2–11.4). Thus a set $A \subset D$ is open relative to D if a D-lander would see A as an open set. The dictator of D decrees that any comrade may approach the wall surrounding D, but must not touch it unless he is the sole occupant of an isolated

1-point province. As any metric space is both open and closed to its inhabitants, D is open and closed relative to D. For example, if $X = E^1$ and $D = (0, 1] \subset E^1$, then the "lebensraum" of D is $(0, 1)$ and an analogy to the extended real number system E^* is to consider $D^* = [0, 1]$. Then $A = (\frac{1}{2}, 1]$ is not open in E^1 but is open relative to D, whereas $(\frac{1}{2}, 0.99)$ is open in D as well as in E^1. Since $[0, \frac{1}{2})$ is not a subset of D, any question of its openness or closedness relative to D is irrelevant.

Given $A \subset D \subset X$, then A is open relative to D iff there is an open set $U \subset X$ such that

$$A = U \cap D.$$

We shall use this fact several times throughout the section, but will not point out each time that it is the statement of Theorem 2–11.5.

Theorem 1. *Let X be a metric space having a metric outer measure μ^* and let a function*

$$f : D \to E^1, \qquad D \subset X$$

be given. If f is continuous, then for each $c \in E^1$ the set $\{x \in D \mid f(x) > c\}$ is open relative to D. If, in addition, D is a measurable set, then f is a measurable function.

Proof. Given a number $c \in E^1$, let

$$D_c = \{x \in D \mid f(x) > c\}.$$

Consider that $D_c \neq \varnothing$, select any $x_0 \in D_c$, set $y_0 = f(x_0)$ and $\varepsilon = y_0 - c$. Then $\varepsilon > 0$. Corresponding to $\varepsilon/2$, determine a $\delta > 0$ such that

$$f(N_X(x_0, \delta)) \subset N_{E^1}(y_0, \tfrac{1}{2}\varepsilon).$$

Hence

$$\begin{aligned} x \in D \cap N_X(x_0, \delta) &\Rightarrow y_0 - \tfrac{1}{2}\varepsilon < f(x) < y_0 + \tfrac{1}{2}\varepsilon \\ &\Rightarrow f(x) > y_0 - \tfrac{1}{2}(y_0 - c) = \tfrac{1}{2}(y_0 + c) > c. \end{aligned}$$

Thus $D \cap N_X(x_0, \delta) \subset D_c$ so that x_0 is interior to D_c relative to D. Since this is true for every $x_0 \in D_c$, the set D_c is open relative to D. Accordingly, take an open set $U \subset X$ such that

$$D_c = U \cap D.$$

D is given as measurable and U is measurable (Theorem 5–1.4), since μ^* is a metric outer measure. Thus D_c is measurable by Theorem 4–2.4. This being true for every $c \in E^1$, even if $D_c = \varnothing$, the function f satisfies Definition 4–3.1 of measurability of functions. ▮

As preparation for the next theorem, let $f : (0, 1] \to E^1$ be the continuous function of Problem 6–1.2. Then the open set

$$V = (\tfrac{1}{4}, 2) \subset E^1$$

is such that $f^{-1}(V) = (\frac{1}{2}, 1]$, which is open relative to the domain $D = (0, 1]$ of f.

For the same V, but the continuous function of Example 6–1.1, then

$$f^{-1}(V) = \{x \mid x \text{ is rational}\}$$

is open relative to D, since it is D. However, with f the discontinuous function of Example 6–1.2 and the same V, then the same $f^{-1}(V)$ is obtained, but this time is not open relative to the domain $D = E^1$ of this function.

It is expected that an appreciation of the power of the next theorem will grow as proofs of later theorems in the section are read.

Theorem 2. *A function $f: D \to Y$ with $D \subset X$ is continuous iff $f^{-1}(V)$ is open relative to D for every open set $V \subset Y$.*

Proof. First let f be continuous and take any open set $V \subset Y$. If $f^{-1}(V) = \varnothing$, then it is open, and hence consider $f^{-1}(V) \neq \varnothing$. Take any $x \in f^{-1}(V) \subset D$ and set $y = f(x)$. Then $y \in V$, hence is an interior point of V, and we choose $\eta > 0$ such that $N_Y(y, \eta) \subset V$. Hence, corresponding to this y and any ε such that $0 < \varepsilon < \eta$, there is a $\delta > 0$ for which

$$f(N_X(x, \delta)) \subset N_Y(y, \varepsilon) \subset V,$$

and therefore $\varnothing \neq D \cap N_X(x, \delta) \subset f^{-1}(V)$. This says that x is interior to $f^{-1}(V)$ relative to D. Since x was any point of $f^{-1}(V)$, then $f^{-1}(V)$ is open relative to D.

Conversely, let f be such that $f^{-1}(V)$ is open relative to D for every open $V \subset Y$. Take any $x \in D$, any $\varepsilon > 0$, and set $y = f(x)$. Then $N_Y(y, \varepsilon)$ is an open subset of Y so that

$$f^{-1}(N_Y(y, \varepsilon))$$

is open relative to D and we take an open subset U of X such that

$$f^{-1}(N_Y(y, \varepsilon)) = U \cap D.$$

Since $U \subset X$ is open and $x \in U$, take $\delta > 0$ such that

$$N_X(x, \delta) \subset U.$$

[*Recall:* $f(A) = f(DA)$ for any $A \subset X$ and $f(f^{-1}(B)) \subset B$ for any $B \subset Y$.] Hence

$$f(N_X(x, \delta)) \subset f(U) = f(f^{-1}(N_Y(y, \varepsilon))) \subset N_Y(y, \varepsilon),$$

which says that f is continuous at x. Thus f is continuous at each point of its domain D; that is, f is continuous. ∎

Recall that a subset A of a metric space is said to be compact if every open covering of A contains a finite subcovering of A.

Theorem 3. *If $f: D \to Y$ is continuous and the domain D is compact, then the range $f(D)$ is compact. If, in addition, a closed set $A \subset D$ is taken, then $f(AD) = f_A(D)$ is compact.*

Proof. Given f and D satisfying the hypotheses in the first sentence, let $\mathscr{V}$ be an open covering of $f(D)$. Then $V \in \mathscr{V} \Rightarrow f^{-1}(V)$ is open relative to D, by Theorem 2. Associate with each $V \in \mathscr{V}$ an open set $U \subset X$ such that

$$f^{-1}(V) = U \cap D.$$

The resulting family is an open covering of the compact set D and hence contains a finite subcovering $U_1, U_2, \cdots, U_n$. Therefore

$$D \subset \bigcup_{k=1}^{n} U_k, \qquad \text{and} \qquad D = \bigcup_{k=1}^{n} f^{-1}(V_k),$$

where V_k is a member of $\mathscr{V}$ with which U_k was associated. Hence

$$f(D) = \bigcup_{k=1}^{n} f(f^{-1}(V_k)) \subset \bigcup_{k=1}^{n} V_k$$

which shows that $f(D)$ is compact.

The second part of the theorem follows from Problem 6–1.3 and the fact that AD is compact (Theorem 2–6.2), since A is closed and D is compact. ∎

By definition, $f: D \to Y$ is a **bounded function** if its range $f(D)$ is a bounded set.

In the following two theorems the domain space X may be any metric space, but the range spaces are specific metric spaces.

Theorem 4. *If $f: D \to E^k$ is continuous and D is compact, then $f(D)$ is closed and bounded (so f is bounded).*

Proof. In E^k a set is compact iff it is closed and bounded (Theorem 2–7.5). ∎

Theorem 5. *If $f: D \to E^1$ is continuous and D is compact, then $\exists\, \underline{x}, \bar{x} \in D$ such that*

$$f(\underline{x}) \leqslant f(x) \leqslant f(\bar{x}) \qquad \textit{for each } x \in D;$$

concisely, f has a maximum and a minimum on D.

Proof. See Theorem 2–6.6. ∎

In the next theorem the domain space is E^k and Lebesgue measure m_k is used.

Theorem 6. *If $f: D \to E^1$, with $D \subset E^k$, is continuous and D is compact, then $f \in \mathscr{L}(D, m_k)$ and*

$$f(\underline{x}) m_k D \leqslant \int_D f\, dm_k \leqslant f(\bar{x}) m_k D,$$

where $\underline{x}, \bar{x} \in D$ are as in Theorem 5.

Proof. D is a closed bounded set by Theorem 2–7.5 (since the space is E^k) and thus is not only m_k-measurable but also $m_k D < \infty$. Then f is a bounded measurable function and this is sufficient (Problem 4–5.4) for the stated conclusions. ∎

The next theorem requires that D be connected (Definition 2–10.1), but allows the range space Y to be any metric space.

Theorem 7. *If $f: D \to Y$ is continuous and the domain D is connected, then the range $f(D) \subset Y$ is connected.*

Proof. Assume that the hypotheses hold but that $f(D)$ is disconnected; accordingly let $V, W \subset Y$ be open sets such that

$$V \cap f\ D) \neq \varnothing, \qquad W \cap f(D) \neq \varnothing, \qquad V \cap W = \varnothing, \qquad \text{and}$$

$$f(D) \subset V \cup W.$$

Then $f^{-1}(V)$ and $f^{-1}(W)$ are nonempty disjoint subsets of D whose union is D, but they are also open relative to D (by Theorem 2), which they cannot be because D is connected. ∎

In the next theorem the result in (a) is referred to as "the intermediate value property of continuous functions," and (b) is called "the second law of the mean for integrals," it being understood that

$$\int_a^b f(x)\,dx \qquad \text{means} \qquad \int_{[a,b]} f\,dm.$$

Theorem 8. *Let $f: [a, b] \to E^1$, $-\infty < a < b < \infty$, be a continuous function.*

(a) *If $f(a) < f(b)$ and c is such that $f(a) < c < f(b)$, or if $f(a) > f(b)$ and c is such that $f(a) > c > f(b)$, then there is an $x_0 \in (a, b)$ for which $f(x_0) = c$.*

(b) *If g is a function such that $g \in \mathscr{L}[a, b]$ and either $g(x) \geqslant 0$ or else $g(x) \leqslant 0$ for almost all $x \in (a, b)$, then there is a $\xi \in (a, b)$ for which*

$$f(\xi)\int_a^b g(x)\,dx = \int_a^b f(x)g(x)\,dx. \tag{1}$$

Proof (a). Since $[a, b]$ is connected, then $f([a, b])$ is connected, by Theorem 7, and for c as described (in either case),

$$c \in f([a, b]),$$

again by Theorem 7. Hence there is an $x_0 \in [a, b]$ such that $f(x_0) = c$. But $f(a) \neq c$ and $f(b) \neq c$ so that $x_0 \in (a, b)$.

Proof (b). First take $g \geqslant 0$ a.e. in (a, b). If $g = 0$ a.e. in (a, b), then both sides of (1) are zero for all $\xi \in (a, b)$. Thus consider $g > 0$ a.e. in (a, b) so that

$$\int_a^b g(x)\,dx > 0.$$

Select a $\xi_1 \in (a, b)$ such that

$$f(\xi_1)\int_a^b g(x)\,dx \leqslant \int_a^b f(x)g(x)\,dx.$$

There is such a ξ_1, since otherwise a.e. in (a, b):

$$f(t)g(t)\int_a^b g(x)\,dx > g(t)\int_a^b f(x)g(x)\,dx \qquad \text{and}$$

$$\left(\int_a^b f(t)g(t)\,dt\right)\int_a^b g(x)\,dx > \left(\int_a^b g(t)\,dt\right)\int_a^b f(x)g(x)\,dx.$$

In the same way there is a $\xi_2 \in (a, b)$ such that

$$f(\xi_2)\int_a^b g(x)\,dx \geqslant \int_a^b f(x)g(x)\,dx.$$

If the equality does not hold for either ξ_1 or ξ_2, there is a ξ between them for which (1) holds, by part (a). If $g \leqslant 0$ a.e. in (a, b), apply the above result to $-g$. ∎

The following corollary is called "the first mean value theorem for integrals"; it is obtained by taking $g(x) = 1$, $a \leqslant x \leqslant b$.

Corollary. *If* $-\infty < a < b < \infty$ *and* $f\colon [a, b] \to E^1$ *is continuous, then there is a* $\xi \in (a, b)$ *such that*

$$f(\xi) = \frac{1}{b-a}\int_a^b f(x)\,dx.$$

As previously used

$$f\colon D \leftrightarrow R$$

means that f is 1–1 on D onto $R = f(D)$ and thus the inverse function f^{-1} not only exists with domain R and range D but also

$$f^{-1}\colon R \leftrightarrow D.$$

In the next theorem, X and Y may be any metric spaces.

Theorem 9. *If* $f\colon D \leftrightarrow R$ *is continuous and* D *is compact, then* $f^{-1}\colon R \leftrightarrow D$ *is also continuous.*

Proof. In Theorem 2 interchange the pairs f and f^{-1}, D and R, and X and Y to see from the "if" part that f^{-1} will be continuous "if $f(U)$ turns out to be open relative to R for every open set $U \subset X$." Accordingly, select any open set $U \subset X$.

Then, f being 1–1, we have $f(U) \cap f(U^c) = \varnothing$ and thus

$$f(U) = R - f(U^c) = R \cap [f(U^c)]^c.$$

But U^c is closed so that DU^c is compact (Problem 2–6.1). Hence $f(U^c)$, which is really $f(DU^c)$, is compact by Theorem 3, second part. Therefore $[f(U^c)]^c$ is open, and it follows that $R \cap [f(U^c)]^c = f(U)$ is open relative to R as we hoped. ∎

The essence of the next theorem is that a continuous function of a measurable function is measurable.

Theorem 10. *Let $g: E^1 \to E^1$ be continuous and*

$$f: D \to E^1$$

be measurable, where D is a measurable subset of a metric space X having a metric outer measure μ^*. Then the composition function $h = g \circ f: D \to E^1$ is a measurable function.

Proof. Given a number c, let

$$D_c = \{x \in D \mid h(x) > c\}.$$

To help keep things straight, denote the domain and range spaces of g by $Y = E^1$ and $Z = E^1$, respectively. Define $G_c \subset Y$ by

$$G_c = \{y \mid g(y) > c\}.$$

Since g is continuous on E^1, the set G_c is open (Theorem 1). G_c being an open subset of E^1, we let $\{I_n\}$ be its open component intervals (Theorem 2–5.5) so that

$$G_c = \bigcup I_n \qquad \text{and} \qquad I_n \cap I_k = \varnothing \qquad \text{for } n \neq k.$$

With $I_n = (a_n, b_n)$, the sets A_n and A defined by

$$A_n = \{x \in D \mid a_n < f(x) < b_n\} \qquad \text{and} \qquad A = \bigcup A_n$$

are measurable. We now check that $x \in D_c$ iff $x \in A_n$ for some n and hence that $A = D_c$. Therefore h is a measurable function. ∎

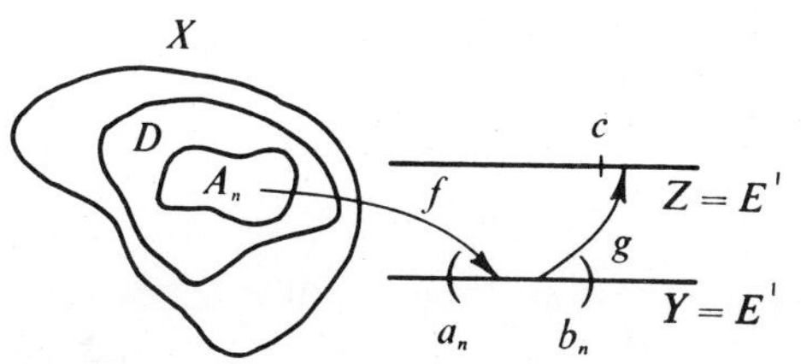

It would seem from many previous results that a measurable function is obtained by any reasonable operations using measurable functions or sequences of measurable functions. Thus the alliteration "a measurable function of a measurable function is a measurable function" sounds reasonable, but unfortunately the statement is not true. The next example shows that even "measurable" and "continuous" cannot be interchanged in Theorem 10.

EXAMPLE 1. With $\psi: [0, 1] \to [0, 1]$ the Cantor function, let $\varphi: [0, 1] \to E^1$ be defined by

$$\varphi(x) = \psi(x) + x \qquad \text{for } 0 \leqslant x \leqslant 1.$$

Then φ is a continuous function, since ψ is continuous (Example 6–1.3), and thus the range of φ is connected (Theorem 7). We know the first implication below and check the others,

$$\begin{aligned}0 \leqslant x_1 < x_2 \leqslant 1 &\Rightarrow 0 = \psi(0) \leqslant \psi(x_1) \leqslant \psi(x_2) \leqslant \psi(1) = 1\\ &\Rightarrow 0 = \psi(0) + 0 \leqslant \psi(x_1) + x_1 < \psi(x_2) + x_2 \leqslant \psi(1) + 1 = 2\\ &\Rightarrow 0 = \varphi(0) \leqslant \varphi(x_1) < \varphi(x_2) \leqslant \varphi(1) = 2\end{aligned}$$

to see that φ is 1–1 on [0, 1] onto [0, 2]. Thus φ^{-1} exists and

(2) $$\varphi^{-1}\colon [0, 2] \leftrightarrow [0, 1] \text{ is continuous,}$$

by Theorem 9. Another fact we need is

(3) $$m\varphi(\Gamma) > 0,$$

where Γ is the Cantor set and m is Lebesgue measure. To see this, let $\{I_n\}$, $I_n = (a_n, b_n)$, be the component open intervals of $[0, 1] - \Gamma$. Thus $\varphi(I_n)$ is also an interval with end points $\varphi(a_n)$ and $\varphi(b_n)$ so that

$$m[\varphi(I_n)] = \psi(b_n) + b_n - \psi(a_n) - a_n = b_n - a_n,$$

since $\psi(a_n) = \psi(b_n)$. Hence

$$m[\varphi(\bigcup I_n)] = m[\bigcup \varphi(I_n)] = \sum m\varphi(I_n) = \sum (b_n - a_n) = 1,$$

and thus $m\varphi(\Gamma) = m\{[0, 2] - \varphi(\bigcup I_n)\} = m[0, 2] - 1 = 1 > 0$.

For the end we have in mind, it is now better to switch orientation by thinking of the Cantor set on the y-axis and considering φ as mapping

$$\{y \mid 0 \leqslant y \leqslant 1\} \qquad \text{onto} \qquad \{x \mid 0 \leqslant x \leqslant 2\}.$$

Hence φ^{-1} maps the second set 1–1 continuously onto the first set. This φ^{-1} is one of the functions we want, and we set

$$(x) = \varphi^{-1}(x) \qquad \text{for } 0 \leqslant x \leqslant 2.$$

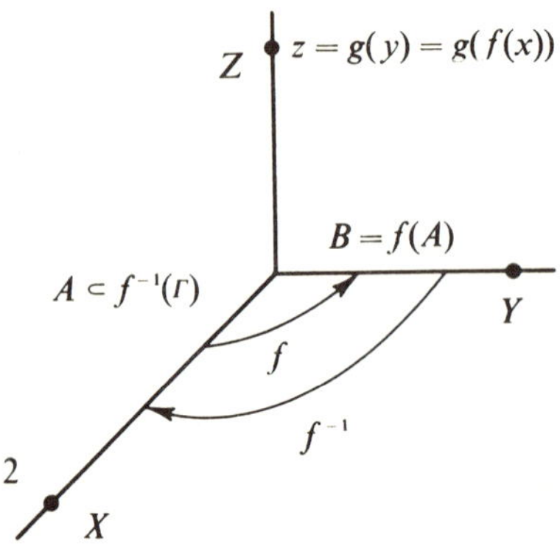

We now have from (2) that

$$f\colon [0, 2] \leftrightarrow [0, 1] \text{ is continuous,}$$

and we translate (3) into

$$mf^{-1}(\Gamma) > 0.$$

In the set $f^{-1}(\Gamma)$ we now select a nonmeasurable set A, this being possible by the corollary, page 243. Set $B = f(A)$. Not only is

$$B \subset \{y \mid 0 \leqslant y \leqslant 1\}, \qquad \text{but also } B \subset \Gamma.$$

Hence B is measurable, since $0 \leqslant m^*B \leqslant m\Gamma = 0$.

The second function g we want is defined by

$$g(y) = \begin{cases} 1 & \text{if } y \in B \\ 0 & \text{if } 0 \leqslant y \leqslant 1 \quad \text{and} \quad y \notin B. \end{cases}$$

Hence g is on $\{y \mid 0 \leqslant y \leqslant 1\}$ into $Z = E^1$. Also g is a measurable function, since

$$\{y \mid g(y) \geqslant c\}$$

is $[0, 1]$, B, or $\varnothing$ according as $c \leqslant 0$, $0 < c \leqslant 1$, or $c > 1$.

Now the composition function

$$h = g \circ f\colon \{x \mid 0 \leqslant x \leqslant 2\} \to Z$$

is such that $\{x \mid h(x) > \frac{1}{2}\} = A$. Since A is nonmeasurable by choice, the measurable function g of the continuous function f is a nonmeasurable function.

6–3 UNIFORMITY

Again X and Y are abstract metric spaces.

Definition 1. *A function $f\colon D \to Y$, $D \subset X$, is said to be* **uniformly continuous** *on D if corresponding to each given number $\varepsilon > 0$ there is a number $\delta > 0$ such that*

$$d_Y(f(x), f(t)) < \varepsilon; \qquad x, t \in D,$$

provided only that $d_X(x, t) < \delta$.

A careful dichotomy of the concepts of continuity and uniform continuity should be made. Continuity is a "local" property in that continuity "at a point" is first defined and subsequently a function is said to be continuous on D if it is continuous **at each point** of D. Uniform continuity, however, is defined first and only "throughout" D, and thus may be perceived as a "global" property or a "property in the large." A query about uniform continuity at a point would show clear lack of understanding.† Another way of expressing the distinction is that the definition of continuity mentions

first a point, then ε, and then δ,

whereas in the definition of uniform continuity the order is (and has to be)

ε first, δ second, and then points.

PROBLEM 1. Make a positive statement which says that a function $f: D \to Y$ is not uniformly continuous on D.

PROBLEM 2. If $f: D \to Y$ is uniformly continuous on D, then f is continuous on D.

The sensation that it is "easier" for a function to be continuous than uniformly continuous is confirmed by Problem 2. A condition on D which, together with continuity, implies uniform continuity is given in the next theorem. The proud history of compactness began by distilling the essence of why a continuous function on a bounded closed interval is uniformly continuous.

Theorem 1. *If $f: D \to Y$ is continuous and D is compact, then f is uniformly continuous on D.*

Proof. Let $\varepsilon > 0$ be given. Corresponding to $x \in D$ and the number $\frac{1}{2}\varepsilon$, let $r(x)$ denote a positive number such that, from the continuity of f at x,

$$f[N_X(x, r(x))] \subset N_Y(f(x), \tfrac{1}{2}\varepsilon).$$

Denote by $C(x)$ the neighborhood centered at x, but having radius $\frac{1}{2}r(x)$, so that

$$C(x) = N_X(x, \tfrac{1}{2}r(x)).$$

This being done for each $x \in D$, the family of these smaller neighborhoods is an open covering of D; in fact, each point of D is the center of a member. Since D is compact, there is a finite number of points $x_1, x_2, \cdots, x_n$ of D such that

$$D \subset \bigcup_{i=1}^{n} C(x_i).$$

Choose $\delta = \frac{1}{2}\min[r(x_1), r(x_2), \cdots, r(x_n)]$. Hence $\delta > 0$.

In order to check the uniformity condition, choose any pair of points $x, t \in D$ subject only to

$$d_X(x, t) < \delta.$$

At least one of $C(x_1), C(x_2), \cdots, C(x_n)$ contains x. We let k be such that $x \in C(x_k)$. Thus

$$d_X(x, x_k) < \tfrac{1}{2}r(x_k) \qquad \text{and also} \qquad x \in N_X(x_k, r(x_k)).$$

† Even a function defined at a point p and nowhere else is not (to split hairs) uniformly continuous at p, but is uniformly continuous on the set $\{p\}$.

Hence $d_X(t, x_k) \leqslant d_X(t, x) + d_X(x, x_k) < \delta + \frac{1}{2}r(x_k) \leqslant r(x_k)$. Therefore $N_X(x_k, r(x_k))$ includes both t and x. Hence f carries both t and x into the neighborhood $N_Y(f(x_k), r(x_k))$. We thus have both

$$d_Y(f(x), f(x_k)) < \tfrac{1}{2}\varepsilon \qquad \text{and} \qquad d_Y(f(t), f(x_k)) < \tfrac{1}{2}\varepsilon$$

and then, by the triangular inequality, that

$$d_Y(f(x), f(t)) < \varepsilon.$$

Consequently, f is uniformly continuous on D. ▮

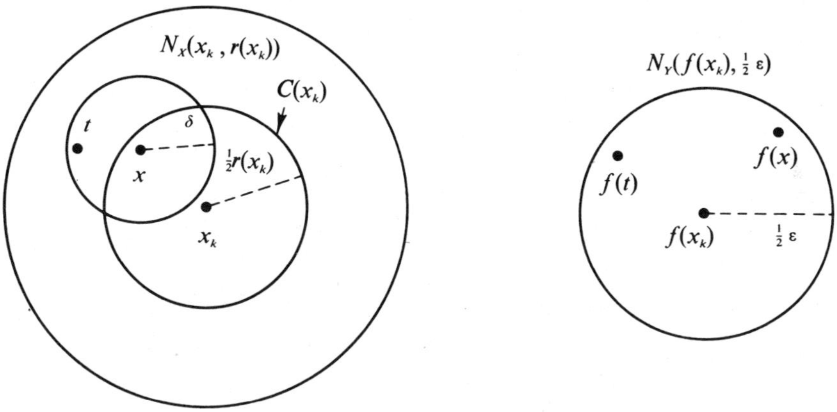

PROBLEM 3. If $f: D \to E^1$ is continuous and D is compact, then corresponding to each number $\varepsilon > 0$ there is a number $\delta > 0$ such that

$$|f(x_1)f(x_2) - f^2(x)| < \varepsilon$$

for all triples x, x_1, x_2 in D satisfying $d_X(x, x_1) < \delta$ and $d_X(x, x_2) < \delta$.

Another line of development was to define convergence of a sequence $\{a_n\}$ of constants and then convergence on a set D of a sequence $\{f_n\}$ of functions. This latter concept is again a local one, since what is meant is that **for each** $x \in D$, the sequence $\{f_n(x)\}$ of constants converges.

EXAMPLE 1. Let $f: [0, 1] \to E^1$ and $f_n: [0, 1] \to E^1$ for $n = 1, 2, 3, \cdots$ be defined by

$$f(x) = \begin{cases} 0 & \text{if } x = 0 \\ 1 & \text{if } 0 < x \leqslant 1; \end{cases} \qquad f_n(x) = \begin{cases} nx & \text{if } 0 \leqslant x \leqslant 1/n \\ 1 & \text{if } 1/n \leqslant x \leqslant 1. \end{cases}$$

Then, as should be checked, no matter what x, $0 \leqslant x \leqslant 1$, is given,

$$\lim_{n \to \infty} f_n(x) = f(x).$$

Note, however, that each f_n is continuous, but the limit function f is not continuous.

The next definition gives a global convergence property.

Definition 2. *Given* $f_n \colon D \to Y, n = 1, 2, 3, \cdots$ *and* $f \colon D \to Y$, *then* $\{f_n\}$ *is said to* **converge uniformly** *to* f **on** D, *and we write*

$$f_n \rightrightarrows f \qquad \text{on } D$$

if corresponding to each given number $\varepsilon > 0$ *there is a positive integer* N *such that*

$$n \geqslant N \Rightarrow d_Y(f_n(x), f(x)) < \varepsilon \qquad \text{for all } x \in D.$$

[A geometric interpretation of uniform convergence is that an ε-band about the graph of $y = f(x)$ includes the whole graph over D of $y = f_n(x)$ for all sufficiently large n.]

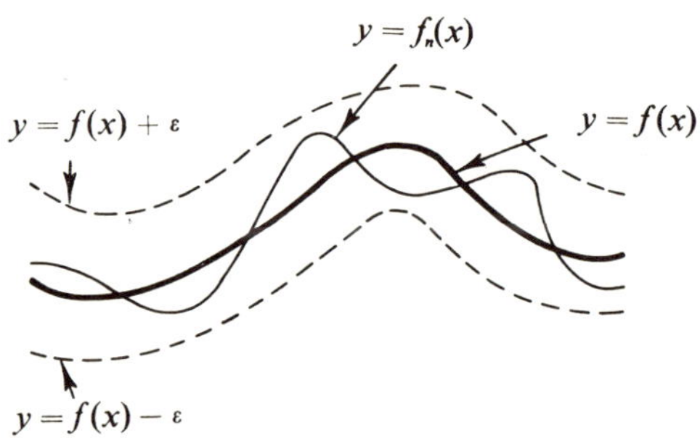

Clearly, if $f_n \rightrightarrows f$ on D, then the sequence $\{f_n\}$ converges pointwise; i.e., for each given $x \in D$,

$$\lim_{n\to\infty} f_n(x) = f(x).$$

In Example 1 the convergent sequence does not converge uniformly. For there is an $\varepsilon > 0$ (namely $\varepsilon = \frac{1}{2}$) such that if N is arbitrary, then corresponding to each $n \geqslant N$ there is an x (for example, $x = \frac{1}{3}n$) such that $|f_n(x) - f(x)| > \varepsilon$.

Definition 3. *A sequence* $\{f_n\}$ *with each* $f_n \colon D \to Y$ *is said to be a* **uniform Cauchy sequence** *on* D *if corresponding to each given number* $\varepsilon > 0$ *there is an* N *such that*

(1) $$n, m \geqslant N \Rightarrow d_Y[f_n(x), f_m(x)] < \varepsilon \qquad \text{for all } x \in D.$$

Theorem 2. *Let* Y *be a complete metric space and*

$$f_n \colon D \to Y, \qquad n = 1, 2, 3, \cdots.$$

The sequence $\{f_n\}$ *converges uniformly on* D *iff it is a uniform Cauchy sequence on* D.

Proof. First let $f_n \rightrightarrows f$ on D and let $\varepsilon > 0$ be given. Corresponding to $\varepsilon/2$, let N be such that

$$n \geqslant N \Rightarrow d_Y[f_n(x), f(x)] < \frac{\varepsilon}{2} \qquad \text{for all } x \in D.$$

Then $n, m \geqslant N \Rightarrow$

$$d_Y[f_n(x), f_m(x)] \leqslant d_Y[f_n(x), f(x)] + d_Y[f_m(x), f(x)] < \varepsilon$$

for all $x \in D$. Hence $\{f_n\}$ is uniform Cauchy on D.

Conversely, let $\{f_n\}$ be a uniform Cauchy sequence on D. Then for each $x \in D$, picked in advance, the sequence of constants $\{f_n(x)\}$ is Cauchy and hence converges (since Y is complete), and we designate its limiting value by $f(x)$. Considering this done for each $x \in D$, we have $f: D \to Y$ such that $\{f_n\}$ converges on D to f, but so far we have only convergence. We know, however, that for n any positive integer,

$$\lim_{m \to \infty} d_Y[f_n(x), f_m(x)] = d_Y[f_n(x), f(x)] \qquad \text{for all } x \in D,$$

since d_Y is a continuous function.

To establish uniform convergence, let $\varepsilon > 0$ be given. Determine an N such that (1) holds and then, with $n \geqslant N$ held fixed, let $m \to \infty$ to obtain

$$n \geqslant N \Rightarrow d_Y[f_n(x), f(x)] \leqslant \varepsilon \qquad \text{for all } x \in D. \blacksquare$$

The following test may be easier to apply than trying to establish whether a sequence of functions is uniformly Cauchy.

Theorem 3. *Given that $f_n(x) \to f(x)$ for each $x \in D$, set*

$$M_n = \sup_{x \in D} d_Y[f_n(x), f(x)].$$

Then $f_n \rightrightarrows f$ on D iff $M_n \to 0$ as $n \to \infty$.

Proof. First let $f_n \rightrightarrows f$ on D. Given an $\varepsilon > 0$, let N be such that

$$n \geqslant N \Rightarrow d_Y[f_n(x), f(x)] < \varepsilon \qquad \text{for } all \ x \in D.$$

Then

$$n \geqslant N \Rightarrow \sup_{x \in D} d_Y[f_n(x), f(x)] = M_n \leqslant \varepsilon,$$

which says that $M_n \to 0$ as $n \to \infty$.

Conversely, let $M_n \to 0$ as $n \to \infty$. Then with $\varepsilon > 0$ given and N such that $n \geqslant N \Rightarrow M_n < \varepsilon$, we have

$$n \geqslant N \Rightarrow d_Y[f_n(x), f(x)] \leqslant M_n < \varepsilon \qquad \text{for all } x \in D,$$

and hence that $f_n \rightrightarrows f$ on D. $\blacksquare$

The following definition and theorem specialize Y to E^2 (which, of course, may be E^1) so that arithmetic may be done on elements in the range space and the range space is complete.

Definition 4. *Given that $u_n(x) \in E^2$ is defined for each $x \in D$, then the series $\sum u_n$ is said to* **converge uniformly** *on D if the sequence $\{s_n\}$ of partial sums converges uniformly on D.*

Theorem 4. (Weierstrass M-test) *If*

$$|u_n(x)| \leqslant M_n \qquad \text{for each } x \in D,$$

and $\sum M_n$ converges, then $\sum u_n$ converges uniformly on D.

Proof. Given an $\varepsilon > 0$, let N be such that

$$N \leqslant n \leqslant m \Rightarrow \sum_{k=n}^{m} M_k < \varepsilon,$$

which is possible because the partial sums of a convergent sequence of constants in E^1 form a Cauchy sequence. Thus

$$N \leqslant n \leqslant m \Rightarrow \left|\sum_{k=n}^{m} u_k(x)\right| \leqslant \sum_{k=n}^{m} |u_k(x)| \leqslant \sum_{k=n}^{m} M_k < \varepsilon$$

for each $x \in D$. Hence the partial sums $\{s_n\}$ of $\sum u_n$ form a uniform Cauchy sequence on D. Thus $\{s_n\}$ converges uniformly on D by Theorem 3. Hence, by Definition 4, $\sum u_n$ converges uniformly on D. ∎

The next theorem goes back to having range space Y, but in this theorem Y is complete.

Theorem 5. *If $f_n \rightrightarrows f$ on D, if $t \in D'$ at which each $\lim_{x\to t} f_n(x)$ exists, and if the range space Y is complete, then all limits exist and the equality holds in*

$$\lim_{x\to t} \lim_{n\to\infty} f_n(x) = \lim_{n\to\infty} \lim_{x\to t} f_n(x). \tag{2}$$

Proof. Set $a_n = \lim_{x\to t} f_n(x)$. Then a_n depends on t, but t is fixed. Let $\varepsilon > 0$ be given, determine N such that

$$n, m \geqslant N \Rightarrow d_Y[f_n(x), f_m(x)] < \varepsilon \qquad \text{for all } x \in D,$$

and now let $x \to t$. This yields

$$n, m \geqslant N \Rightarrow d_Y(a_n, a_m) \leqslant \varepsilon.$$

Hence (Y being complete) $\{a_n\}$ converges and we set

$$\lim_{n\to\infty} a_n = a.$$

We now know the existence of the iterated limit on the right of (2) and may write

$$\lim_{n\to\infty} \lim_{x\to t} f_n(x) = \lim_{n\to\infty} a_n = a.$$

On the left of (2) the inside limit exists and is $f(x)$, since $f_n \rightrightarrows f$ on D. Thus we ask:

$$\text{Does } \lim_{x\to t} f(x) \text{ exist and } = a\,?$$

For each $x \in D$ and each n we know that

$$(3) \qquad d_Y[f(x), a] \leqslant d_Y[f(x), f_n(x)] + d_Y[f_n(x), a_n] + d_Y[a_n, a].$$

Ignoring the middle term on the right for the moment, we pick a specific n such that both

$$d_Y[f(x), f_n(x)] < \frac{\varepsilon}{3} \qquad \text{for all } x \in D, \text{ and } d_Y[a_n, a] < \frac{\varepsilon}{3},$$

this being possible because the sequence of functions converges uniformly and the sequence of constants converges. Holding this n fixed, we choose a $\delta > 0$ such that

$$x \in N_Y(t, \delta) \Rightarrow d_Y[f_n(x), a_n] < \frac{\varepsilon}{3}.$$

Hence, from (3), for the n we are now holding,

$$x \in N_Y(t, \delta) \Rightarrow d_Y[f(x), a] < \varepsilon.$$

But n does not appear in this expression and therefore

$$\lim_{x \to t} f(x) = a.$$

This answers the above question affirmatively and thus (2) holds. ∎

The next result is considered very important because of its many applications.

Theorem 6. *If $f_n \rightrightarrows f$ on D, if each f_n is continuous on D, and if Y is complete, then f is also continuous on D.*

Proof. This result is a corollary of Theorem 5 in that we may now replace the inside limits on the left and right in (2) by $f(x)$ and $f_n(t)$, respectively, to obtain

$$\lim_{x \to t} f(x) = \lim_{n \to \infty} f_n(t)$$

$$= f(t) \qquad \text{for } t \in D \cap D',$$

and there is no difficulty at isolated points of D. ∎

The converse of Theorem 6 is not true; i.e., it is possible to have a continuous limit of a sequence of continuous functions, but the approach not uniform.

PROBLEM 4. Such a counterexample is $\{f_n\}$, where

$$f_n(x) = \frac{x^2}{x^2 + (1 - nx)^2}, \qquad 0 \leqslant x \leqslant 1,\ n = 1, 2, 3, \cdots.$$

[*Hint:* Use $f_n(1/n)$ to obtain the nonuniform approach.]

Lemma 1. *Given a metric space X and a closed nonempty set $C \subset X$, then the function φ defined by $\varphi(x) = d(x, C)$ is continuous on X and strictly positive on C^c.*

Proof. Take $x, y \in X$ and then any $c \in C$. Hence

$$\varphi(x) = d(x, C) \leqslant d(x, c) \leqslant d(x, y) + d(y, c).$$

Since this holds for any $c \in C$, then

$$\varphi(x) \leqslant d(x, y) + \inf_{c \in C} d(y, c) = d(x, y) + \varphi(y).$$

In the same way, $\varphi(y) \leqslant d(x, y) + \varphi(x)$ so that

$$|\varphi(y) - \varphi(x)| \leqslant d(x, y).$$

Thus φ is continuous on X (actually uniformly continuous whether C is closed or not).

Select any $x \in X$ for which $\varphi(x) = 0$. By definition, $\varphi(x) = d(x, C)$ and thus there is a sequence $\{c_n\}$ in C such that

$$\lim d(x, c_n) = \varphi(x) = 0.$$

If the set $\{c_1, c_2, c_3, \cdots\}$ is finite, then $x \in C$, but if it is infinite, then $x \in C'$ so that again $x \in C$, since C is closed. Thus

$$x \notin C \Rightarrow \varphi(x) > 0. \quad \blacksquare$$

Theorem 7. (Tietze Extension Theorem) *Let X be a metric space and let $F \subset X$ be a closed set. If $f: F \to E^1$ is continuous on F (relative to F) and if f is bounded with*

$$\sup\{|f(x)| \mid x \in F\} = M < \infty,$$

then there is a continuous (on X) function $g: X \to E^1$ such that

$$x \in F \Rightarrow g(x) = f(x)$$

and $|g(x)| \leqslant M$ for all $x \in X$.

Proof. If $M = 0$, then $g \equiv 0$ satisfies the conclusion, so we consider $M > 0$ and, of course, $F \neq \varnothing$ and $F \neq X$. Let

$$A = \{x \in F \mid f(x) \leqslant -\tfrac{1}{3}M\} \qquad \text{and} \qquad B = \{x \in F \mid f(x) \geqslant \tfrac{1}{3}M\}.$$

Then $A \cap B = \varnothing$, A and B are both closed, and at least one of them is nonempty so that if both $A \neq \varnothing$ and $B \neq \varnothing$, then $d(x, A) + d(x, B) > 0$ for each $x \in X$ by Lemma 1. Define $\varphi_1: X \to E^1$ by

$$\varphi_1(x) = \begin{cases} -\dfrac{1}{3}M + \dfrac{2}{3}M \dfrac{d(x, A)}{d(x, A) + d(x, B)} & \text{if } A \neq \varnothing \text{ and } B \neq \varnothing \\ \tfrac{1}{3}M & \text{if } A = \varnothing \quad (\text{and hence } B \neq \varnothing) \\ -\tfrac{1}{3}M & \text{if } B = \varnothing \quad (\text{and hence } A \neq \varnothing). \end{cases}$$

In any case φ_1 is continuous on X and

$$|\varphi_1(x)| \leqslant \tfrac{1}{3}M, \qquad x \in X.$$

In case both $A \neq \varnothing$ and $B \neq \varnothing$, then

$$x \in A \Rightarrow \varphi_1(x) = -\tfrac{1}{3}M \Rightarrow |f(x) - \varphi_1(x)| \leqslant \tfrac{2}{3}M,$$

$$x \in B \Rightarrow \varphi_1(x) = \tfrac{1}{3}M \Rightarrow |f(x) - \varphi_1(x)| \leqslant \tfrac{2}{3}M,$$

and, by also checking $x \in F - (A \cup B)$,

(4) $$x \in F \Rightarrow |f(x) - \varphi_1(x)| \leqslant \tfrac{2}{3}M.$$

Note, furthermore, that (4) holds either if $A = \varnothing$ or else if $B = \varnothing$.

We now apply the same reasoning to $f - \varphi_1$ and obtain a function $\varphi_2\colon X \to E^1$ which is continuous,

$$|\varphi_2(x)| \leqslant \tfrac{1}{3}\tfrac{2}{3}M, \qquad x \in X \qquad \text{and}$$

$$x \in F \Rightarrow |f(x) - \varphi_1(x) - \varphi_2(x)| \leqslant (\tfrac{2}{3})^2 M.$$

By this method, a sequence $\{\varphi_n\}$ of continuous functions is obtained such that for $n = 1, 2, 3, \cdots$,

$$|\varphi_n(x)| \leqslant \tfrac{1}{3}(\tfrac{2}{3})^{n-1}M, \qquad x \in X, \qquad \text{and}$$

$$x \in F \Rightarrow |f(x) - \varphi_1(x) - \cdots - \varphi_n(x)| \leqslant (\tfrac{2}{3})^n M.$$

Let $g_n = \varphi_1 + \varphi_2 + \cdots + \varphi_n$ and note that

$$|g_n(x)| \leqslant \tfrac{1}{3}M[1 + \tfrac{2}{3} + \cdots + (\tfrac{2}{3})^{n-1}] \leqslant M, \qquad x \in X;$$

if $m > n$

$$|g_m(x) - g_n(x)| = |\varphi_{n+1}(x) + \cdots + \varphi_m(x)|$$
$$\leqslant \tfrac{1}{3}(\tfrac{2}{3})^n M[1 + \tfrac{2}{3} + (\tfrac{2}{3})^2 + \cdots] = (\tfrac{2}{3})^n M, \qquad x \in X,$$

and

$$x \in F \Rightarrow |f(x) - g_n(x)| \leqslant (\tfrac{2}{3})^n M.$$

Thus $\{g_n\}$ is a uniformly convergent sequence of continuous functions whose limit function g,

$$g(x) = \lim g_n(x), \qquad x \in X,$$

is continuous (by Theorem 6, since E^1 is complete) and

$$x \in F \Rightarrow g(x) = f(x). \; \blacksquare$$

In elementary calculus, $y = f(x)$ may have been referred to as a curve, but also curves for parametric equations

(5) $$x = f(t), \qquad y = g(t)$$

were studied. Such parametric curves may cross themselves, as shown in Fig. 1, whereas a "curve" $y = f(x)$ cannot. It is natural to say that the curve of (5) is continuous if both f and g are continuous, since $|t - t_0| < \delta$, implying both $|f(t) - f(t_0)| < \varepsilon/\sqrt{2}$ and $|g(t) - g(t_0)| < \varepsilon/\sqrt{2}$ yields

$$|t - t_0| < \delta \Rightarrow (f(t), g(t)) \in N_{E^2}[(f(t_0), g(t_0)), \varepsilon].$$

It came as quite a surprise to the mathematical world when G. Peano (1858–1932) in 1890 obtained a continuous curve having positive area. The following example defines a continuous curve passing through every point on and interior to a square. This is one example of a space-filling curve.

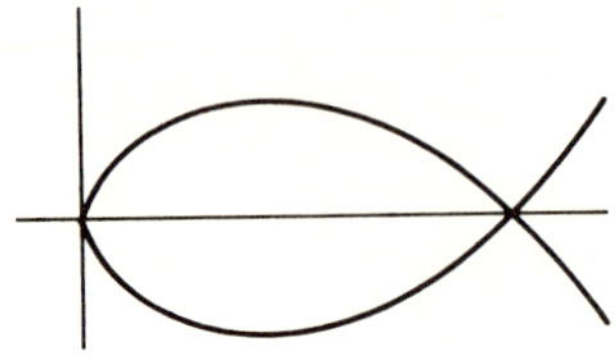

$x = 2t^2, \ y = t(t^2-1)$

FIG. 1

EXAMPLE 2. Let φ be the periodic function of period 2 whose graph for $0 \leqslant t \leqslant 2$ is given in Fig. 2. Thus φ is a continuous function, and for each integer K,

$$\varphi(2K + t) = \varphi(t) \qquad \text{for all } t \in E^1. \tag{6}$$

Let f and g be defined for $0 \leqslant t \leqslant 1$ by

$$f(t) = \frac{\varphi(3^0 t)}{2} + \frac{\varphi(3^2 t)}{2^2} + \frac{\varphi(3^4 t)}{2^3} + \cdots = \sum_{n=0}^{\infty} \frac{\varphi(3^{2n} t)}{2^{n+1}},$$

$$g(t) = \frac{\varphi(3t)}{2} + \frac{\varphi(3^3 t)}{2^2} + \frac{\varphi(3^5 t)}{2^3} + \cdots = \sum_{n=0}^{\infty} \frac{\varphi(3^{2n+1} t)}{2^{n+1}}.$$

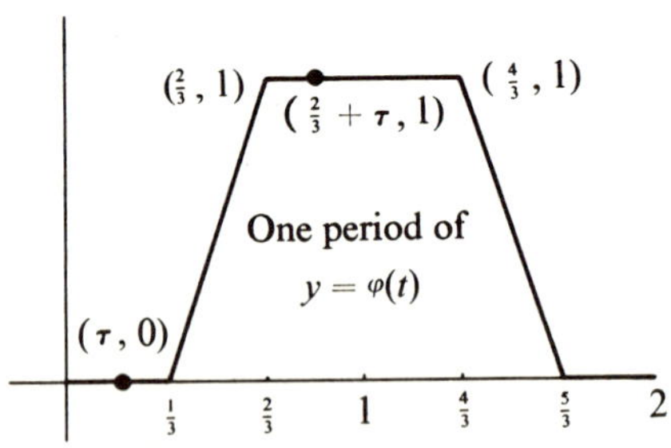

FIG. 2

These definitions are valid and, in addition, f and g are continuous as we now show. The nth term of each series is nonnegative and less than or equal to

$$M_n = \frac{1}{2^n}, \qquad \text{with } \sum M_n \text{ converging,}$$

so the partial sums converge uniformly on [0, 1] by the Weierstrass M-test. But each partial sum is continuous and thus the limit functions f and g are continuous by Theorem 6. Also $0 \leqslant f(t) \leqslant 1$ and $0 \leqslant g(t) \leqslant 1$ for $0 \leqslant t \leqslant 1$.

To see that the continuous parametric curve

$$x = f(t), \qquad y = g(t), \qquad 0 \leqslant t \leqslant 1,$$

passes through each point of a square, pick $\bar{x}$ and $\bar{y}$ arbitrarily except that

$$0 \leqslant \bar{x} \leqslant 1, \qquad 0 \leqslant \bar{y} \leqslant 1.$$

Both $\bar{x}$ and $\bar{y}$ have binary expansions and we let $\{\alpha_n\}$ be a sequence of 0's and 1's such that

$$\bar{x} = \frac{\alpha_0}{2} + \frac{\alpha_2}{2^2} + \frac{\alpha_4}{2^3} + \cdots = \sum_{n=0}^{\infty} \frac{\alpha_{2n}}{2^{n+1}} \qquad \text{and}$$

$$\bar{y} = \frac{\alpha_1}{2} + \frac{\alpha_3}{2^2} + \frac{\alpha_5}{2^3} + \cdots = \sum_{n=0}^{\infty} \frac{\alpha_{2n+1}}{2^{n+1}}.$$

Now select the number $\bar{t}$ defined by

$$\bar{t} = \frac{2\alpha_0}{3} + \frac{2\alpha_1}{3^2} + \frac{2\alpha_2}{3^3} + \frac{2\alpha_3}{3^4} + \cdots = \sum_{k=0}^{\infty} \frac{2\alpha_k}{3^{k+1}}.$$

Hence $0 \leqslant \bar{t} \leqslant 1$ and we claim that both

$$f(\bar{t}) = \bar{x} \qquad \text{and} \qquad g(\bar{t}) = \bar{y}.$$

We must compute $\varphi(3^k \bar{t})$ for $k = 0, 1, 2, \cdots$. To do so, first write

$$\begin{aligned} 3^k \bar{t} &= 2 \cdot 3^k \left(\frac{\alpha_0}{3} + \frac{\alpha_1}{3^2} + \cdots + \frac{\alpha_{k-1}}{3^{k-1}} \right) + 3^k \frac{2\alpha_k}{3^{k+1}} \\ &\qquad + 2 \cdot 3^k \left(\frac{\alpha_{k+1}}{3^{k+2}} + \frac{\alpha_{k+2}}{3^{k+3}} + \frac{\alpha_{k+3}}{3^{k+4}} + \cdots \right) \\ &= 2K + \frac{2\alpha_k}{3} + \tau, \end{aligned}$$

where it should be seen that what we replaced by K is indeed an integer and that $0 \leqslant \tau \leqslant \frac{1}{3}$. Hence, by using (6),

$$\varphi(3^k \bar{t}) = \varphi\left(\frac{2\alpha_k}{3} + \tau \right).$$

The α_k is either 0 or 1, and accordingly we consider two cases.

Case 1. $\alpha_k = 0$. Then $\varphi(3^k \bar{t}) = \varphi(\tau) = 0$, from Fig. 2.

Case 2. $\alpha_k = 1$. Then $\varphi(3^k \bar{t}) = \varphi(\frac{2}{3} + \tau) = 1$.

Hence, in either case, $\varphi(3^k \bar{t}) = \alpha_k$. Thus

$$f(\bar{t}) = \frac{\alpha_0}{2} + \frac{\alpha_2}{2^2} + \frac{\alpha_4}{2^3} + \cdots = \bar{x} \qquad \text{and}$$

$$g(\bar{t}) = \frac{\alpha_1}{2} + \frac{\alpha_3}{2^2} + \frac{\alpha_5}{2^3} + \cdots = \bar{y}.$$

Therefore the curve passes through the arbitrary selected point $(\bar{x}, \bar{y})$ of $\{(x, y) \mid 0 \leqslant x \leqslant 1, 0 \leqslant y \leqslant 1\}$. Thus the curve covers this square and does it continuously.

PROBLEM 5. For the parametric curve of Example 3, show that the curve

(a) Passes through $(\frac{1}{2}, 1)$ when $t = \frac{1}{3}$.

(b) Passes through $(\frac{1}{2}, \frac{3}{4})$ when $t = \frac{74}{81}$, and find three other values of t for which the curve passes through the same point.

(c) Maps the Cantor set onto the whole square.

With f and g the functions of Example 2, the function F: $[0, 1] \to [0, 1] \times [0, 1]$, defined by

$$F(t) = (f(t), g(t)), \qquad 0 \leqslant t \leqslant 1,$$

is a continuous mapping of $[0, 1]$ onto $[0, 1] \times [0, 1]$; it was seen in Problem 5 that this mapping is not 1–1. In fact:

There is no continuous 1–1 mapping of $[0, 1]$ onto $[0, 1] \times [0, 1]$.

For assume there is such a function and call it h. Then h^{-1} is a function on $[0, 1] \times [0, 1]$ onto $[0, 1]$, is 1–1, and is continuous by Theorem 6–2.9. Then h^{-1} sends the connected T of the figure into a closed connected subset (i.e., subinterval) of $[0, 1]$ by Theorem 6–2.7. But the inverse of the upright and the inverse of the crossbar of T are also subintervals which overlap, since $h^{-}(c)$ is interior to both. Then any interior point of the overlap other than $h^{-1}(c)$ is sent by h to both the upright and to the crossbar. Hence we have the contradiction that h is not a function.

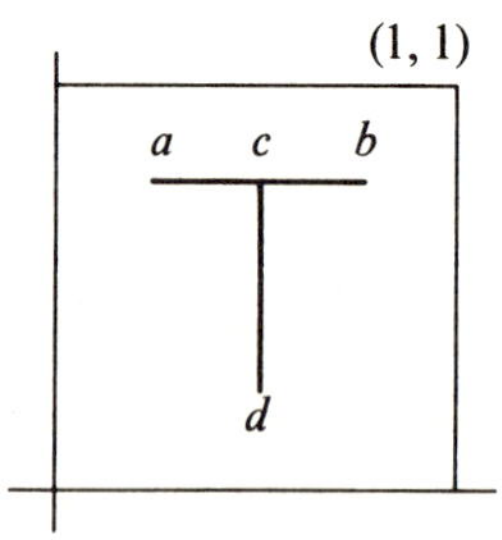

[Anyone interested may find in the first book of the bibliography, by H. Hahn, that any continuous space-filling curve has its double points dense in the range, this set has cardinal $\mathfrak{c}$ of the continuum, its triple points are dense in the range, and there is a continuous space-filling curve with no quadruple points.]

Recall that the function G defined in the proof of Theorem 2–2.9 is 1–1 on $[0, 1]$ onto $[0, 1] \times [0, 1]$, and hence we now know that this function G is not continuous.

6–4 WEIERSTRASS APPROXIMATION THEOREM

A real-valued continuous function on an interval may be approximated uniformly by a polynomial. A more precise statement than this will be given after some properties of binomial coefficients $\binom{n}{k}$ are obtained by using elementary calculus.

The binomial expansion of $(x + c)^n$ is

$$\sum_{k=0}^{n} \binom{n}{k} x^k c^{n-k} = (x + c)^n.$$

Take the derivative of both sides with respect to x and then multiply by x to obtain

$$\sum_{k=0}^{n} \binom{n}{k} k x^k c^{n-k} = nx(x + c)^{n-1}.$$

Another derivative and multiplication by x yields

$$\sum_{k=0}^{n} \binom{n}{k} k^2 x^k c^{n-k} = n(n-1)x^2(x + c)^{n-2} + nx(x + c)^{n-1}.$$

Lemma 1

(1) $$\sum_{k=0}^{n} (nx - k)^2 \binom{n}{k} x^k (1 - x)^{n-k} \leqslant \frac{1}{4} n \qquad \textit{for } 0 \leqslant x \leqslant 1.$$

Proof. In the above formulas replace c by $1 - x$ and obtain

(2) $$\sum_{k=0}^{n} \binom{n}{k} x^k (1 - x)^{n-k} = 1,$$

$$\sum_{k=0}^{n} k \binom{n}{k} x^k (1 - x)^{n-k} = nx, \qquad \text{and}$$

$$\sum_{k=0}^{n} k^2 \binom{n}{k} x^k (1 - x)^{n-k} = n(n-1)x^2 + nx.$$

Multiply both sides of these, respectively, by n^2x^2, $-2nx$, and 1; then add the results and collect terms to see that

$$\sum_{k=0}^{n} (nx - k)^2 \binom{n}{k} x^k (1 - x)^{n-k} = nx(1 - x).$$

Since $x(1 - x) \leqslant \frac{1}{4}$ for $0 \leqslant x \leqslant 1$, the inequality (1) follows. ∎

The proof of the Weierstrass approximation theorem we shall give is due to Serge N. Bernstein (1880–).

Definition 1. Given a function g: $[0, 1] \to E^1$, the polynomial B_n defined by

$$B_n(x) = \sum_{k=0}^{n} g\left(\frac{k}{n}\right)\left[\binom{n}{k}x^k(1-x)^{n-k}\right]$$

is called the nth **Bernstein polynomial** for g.

Lemma 2. *If* g: $[0, 1] \to E^1$ *is continuous, then*

$$B_n \rightrightarrows g \quad \text{on } [0, 1].$$

Proof. Given a number $\varepsilon > 0$, let $\delta > 0$ be such that

$$(s, t \in [0, 1], |s - t| < \delta) \Rightarrow |g(s) - g(t)| < \frac{\varepsilon}{2},$$

which is possible because g is uniformly continuous on $[0, 1]$ by Theorem 6–3.1. Also, $|g|$ has a maximum value (Theorem 6–2.5), which we denote by M so that

$$0 \leqslant x \leqslant 1 \Rightarrow |g(x)| \leqslant M.$$

Take a positive integer N such that

$$n \geqslant N \Rightarrow \frac{M}{2n\delta^2} < \frac{\varepsilon}{2}$$

and note that N does not depend on x in any way.

With δ and N determined, we now take any $n \geqslant N$ and only after this select an $x \in [0, 1]$ arbitrarily. Then, by using (2),

$$g(x) = \sum_{k=0}^{n} g(x)\binom{n}{k}x^k(1-x)^{n-k}$$

and hence

$$|B_n(x) - g(x)| \leqslant \sum_{k=0}^{n} \left|g\left(\frac{k}{n}\right) - g(x)\right|\binom{n}{k}x^k(1-x)^{n-k}.$$

Next divide the integers $k = 0, 1, 2, \cdots, n$ into two groups A and B by the rule

$$k \in A \quad \text{if} \left|\frac{k}{n} - x\right| < \delta, \qquad k \in B \quad \text{if} \left|\frac{k}{n} - x\right| \geqslant \delta.$$

Hence $k \in A \Rightarrow |g(k/n) - g(x)| < \varepsilon/2$ because of the choice of δ, so that

$$\sum_{k \in A} \left|g\left(\frac{k}{n}\right) - g(x)\right|\binom{n}{k}x^k(1-x)^{n-k}$$

$$< \frac{\varepsilon}{2}\sum_{k \in A}\binom{n}{k}x^k(1-x)^{n-k} \leqslant \frac{\varepsilon}{2}\sum_{k=0}^{n}\binom{n}{k}x^k(1-x)^{n-k} = \frac{\varepsilon}{2}$$

by using (2) again. On the other hand,

$$k \in B \Rightarrow \left[\frac{(k-nx)^2}{n^2} \geqslant \delta^2\right] \Rightarrow \left[1 \leqslant \frac{(k-nx)^2}{n^2\delta^2}\right]$$

and hence

$$\begin{aligned}
\sum_{k \in B} \left|g\left(\frac{k}{n}\right) - g(x)\right| \binom{n}{k} x^k(1-x)^{n-k} &\leqslant 2M \sum_{k \in B} (1) \binom{n}{k} x^k(1-x)^{n-k} \\
&\leqslant 2M \sum_{k \in B} \frac{(k-nx)^2}{n^2\delta^2} \binom{n}{k} x^k(1-x)^{n-k} \\
&\leqslant \frac{2M}{n^2\delta^2} \sum_{k=0}^{n} (k-nx)^2 \binom{n}{k} x^k(1-x)^{n-k} \leqslant \frac{2M}{n^2\delta^2}\frac{n}{4} \qquad \text{by (1)} \\
&< \frac{\varepsilon}{2}.
\end{aligned}$$

We thus have that

$$n \geqslant N \Rightarrow |B_n(x) - g(x)| < \varepsilon \qquad \text{for all } x \in [0, 1];$$

that is, the sequence $\{B_n\}$ of Bernstein polynomials for g converges uniformly on $[0, 1]$ to g. ∎

Theorem 1. (Weierstrass Approximation) *If*

$$f\colon [a, b] \to E^1, \qquad -\infty < a < b < \infty,$$

is continuous and a number $\varepsilon > 0$ is given, then there is a polynomial P such that

$$|P(x) - f(x)| < \varepsilon \qquad \textit{for all } x \in [a, b]. \tag{3}$$

Proof.† Let g be the function defined by

$$g(t) = f(a + t(b-a)), \qquad 0 \leqslant t \leqslant 1.$$

Then g is continuous on its domain $\{t \mid 0 \leqslant t \leqslant 1\}$ by Theorem 6–1.4. Corresponding to the given number ε, let B be a Bernstein polynomial for g such that

$$|B(t) - g(t)| < \varepsilon \qquad \text{for all } t, \quad 0 \leqslant t \leqslant 1.$$

Then the function P defined by

$$P(x) = B\left(\frac{x-a}{b-a}\right)$$

is a polynomial and (3) holds for this polynomial, since

$$g\left(\frac{x-a}{b-a}\right) = f(x), \qquad x \in [a, b]. \ ∎$$

† Among the problems at the end of Chapter 7 there is an alternate proof using elementary properties of integrals. Neither of these is the original proof by Weierstrass.

Definition 2. *A polynomial is said to be a* **rational polynomial** *if all its coefficients are rational numbers.*

PROBLEM 1. If f: $[a, b] \to E^1$ is continuous and $\varepsilon > 0$ is given, then there is a rational polynomial Q such that

$$|Q(x) - f(x)| < \varepsilon \qquad \text{for all } x \in [a, b].$$

Definition 3. *With* $-\infty < a < b < \infty$, *we use* $\mathscr{C}$ $[a, b]$ (*or simply* $\mathscr{C}$) *for the space of all continuous real-valued functions on* $[a, b]$, *and for* $f \in \mathscr{C}$ *call*

$$\|f\|_{\mathscr{C}} = \max_{a \leqslant x \leqslant b} |f(x)|$$

the $\mathscr{C}$**-norm** *of* f. *The usual addition of functions and multiplication of a function by a real number c are used to define*

$$f + g \qquad \text{and} \qquad cf \qquad \text{for} \quad f, g \in \mathscr{C}$$

and $\|f - g\|_{\mathscr{C}}$ *is called the* $\mathscr{C}$**-distance** *between* f *and* g.

Hence $f \in \mathscr{C}$ *iff* f: $[a, b] \to E^1$ *is uniformly continuous, by Theorem 6–3.1.*

PROBLEM 2. $\mathscr{C}$ is a

(a) Normed vector space.
(b) Metric space.
(c) Banach space.
(d) Separable space.

Hence circle, open set, closed set, etc., have meaning in $\mathscr{C}$.

PROBLEM 3. Given numbers $-\infty < c < d < \infty$, let

$$\mathscr{A} = \{f \in \mathscr{C} \mid c \leqslant f(x) \leqslant d,\ a \leqslant x \leqslant b\}.$$

(a) The set $\mathscr{A}$ is a closed subset of $\mathscr{C}$.
(b) $\mathscr{A}$ is not a subvector space of $\mathscr{C}$.
(c) $\mathscr{A}$ is a sub-Banach space of $\mathscr{C}$.
(d) Is $\mathscr{A}$ a separable space in the $\mathscr{C}$-norm?

PROBLEM 4. In $\mathscr{C}$ [0, 1] let $c \in \mathscr{C}$ be defined by $c(x) = x^2$, $0 \leqslant x \leqslant 1$. Describe the "open circular disk"

$$\{f \in \mathscr{C} \mid \|f - c\|_{\mathscr{C}} < 1\}.$$

Show that this set is open.

6–5 ABSOLUTE CONTINUITY

Throughout this section a closed interval $[a, b]$ is fixed and all functions are real-valued, each with domain $[a, b]$.

Definition 1. *A function f is said to be* **absolutely continuous**† *if, corresponding to each given* $\varepsilon > 0$*, there is a* $\delta > 0$ *such that*

$$\sum |f(\beta_i) - f(\alpha_i)| < \varepsilon \tag{1}$$

whenever $[\alpha_1, \beta_1], [\alpha_2, \beta_2], \cdots, [\alpha_n, \beta_n]$ *is a finite collection of nonoverlapping subintervals of* $[a, b]$ *such that*

$$\sum (\beta_i - \alpha_i) < \delta.$$

[It is all right for two intervals to have end points in common, and n is not fixed.]

Hence if f is absolutely continuous on $[a, b]$, then it is uniformly continuous on $[a, b]$ as seen by placing a single interval of length $<\delta$ anyplace on $[a, b]$.

Theorem 1. *An absolutely continuous function* $f\colon [a, b] \to E^1$ *transforms:*

(a) *Sets of measure zero into sets of measure zero, and*
(b) *Measurable sets into measurable sets.*

Proof (a). Take a set $Z \subset [a, b]$ having $mZ = 0$. With $\varepsilon > 0$ given and δ as in the definition, select an open set $G \supset Z$ such that $mG < \delta$. The component intervals $\{I_n\}$ of G are mutually exclusive. With $I_n = (a_n, b_n)$ let

$$\bar{I}_n = [a_n, b_n],$$

where $a_n = a$ in case $a \in I_n$ and $b_n = b$ in case $b \in I_n$. Then $Z \subset \bigcup \bar{I}_n$, $\{\bar{I}_n\}$ are pairwise nonoverlapping and, even if they are countably infinite,

$$\sum_{i=1}^{n} (b_i - a_i) < \delta, \qquad n = 1, 2, 3, \cdots.$$

Let $\underline{x}_i$ and $\bar{x}_i$ be points of $[a_i, b_i]$ such that $f(\underline{x}_i)$ is the minimum and $f(\bar{x}_i)$ the maximum of f on $[a_i, b_i]$. Since $f(Z) \subset \bigcup f(\bar{I}_i)$, then

$$m^*f(Z) \leqslant \sum mf(\bar{I}_i) = \sum [f(\bar{x}_i) - f(\underline{x}_i)] \leqslant \varepsilon,$$

since $\sum_{i=1}^{n} |\underline{x}_i - {}_i\bar{x}| \leqslant \sum_{i=1}^{n} (b_i - a_i) < \delta$, $n = 1, 2, 3, \cdots$. Hence $mf(Z) = 0$.

Proof (b). Take any measurable set $A \subset [a, b]$ and then a monotonically increasing sequence $\{F_n\}$ of closed subsets of A such that

$$m(A - \bigcup F_n) = 0.$$

Then $\{f(F_n)\}$ is a monotonically increasing sequence of closed subsets of $f(A)$ and $\bigcup f(F_n)$ is a measurable subset of $f(A)$. Since

$$f(A - \bigcup F_n) \supset f(A) - \bigcup f(F_n)$$

† Sometimes called completely continuous.

and $mf(A - \bigcup F_n) = 0$ by part (a), we have

$$m[f(A) - \bigcup f(F_n)] = 0.$$

Hence $f(A)$ is a measurable set (Theorem 5–1.10) and (b) is proved. ∎

EXAMPLE 1. The Cantor function ψ is not absolutely continuous on [0, 1]. In fact the Cantor function was designed to show that a uniformly continuous function need not be absolutely continuous.

PROBLEM 1. If f and g are absolutely continuous on $[a, b]$, then so are $f + g$, fg, and cf for any constant c, and so is $1/f$ in case $f(x) \neq 0$, $a \leqslant x \leqslant b$.

[*Hint* for fg. Take M the larger of $\max|f|$ and $\max|g|$. Then

$$|f(\beta_i)g(\beta_i) - f(\alpha_i)g(\alpha_i)| \leqslant M\{|f(\beta_i) - f(\alpha_i)| + |g(\beta_i) - g(\alpha_i)|\}.]$$

PROBLEM 2. Replace (1) by $|\sum [f(\beta_i) - f(\alpha_i)]| < \varepsilon$. If a function satisfies this altered definition, then it is absolutely continuous.

The following theorem shows how intimately absolute continuity and Lebesgue integrability are related. We state the theorem now although half of the proof must be delayed until the next chapter.

Theorem 2. *A function g is absolutely continuous on $[a, b]$ iff there is a summable function f and a constant c such that*

$$g(x) = \int_{[a,x]} f\,dm + c \qquad \text{for all } x \in [a, b].$$

Proof (if). Let f be summable on $[a, b]$, let c be any constant, and [in the more familiar integral notation] define g by

$$g(x) = \int_a^x f(t)\,dt + c, \qquad a \leqslant x \leqslant b.$$

Corresponding to $\varepsilon > 0$ given, let $\delta > 0$ be such that (by Theorem 4–5.10) if $B \subset [a, b]$ is measurable and $mB < \delta$, then

$$\int_B |f|\,dm < \varepsilon.$$

Select $[a_1, b_1], [a_2, b_2], \cdots, [a_n, b_n]$ nonoverlapping intervals on $[a, b]$ having $\sum (b_i - a_i) < \delta$ and set $B = \bigcup [a_i, b_i]$. Then

$$\sum |g(b_i) - g(a_i)| = \sum \left| \int_a^{b_i} f - \int_a^{a_i} f \right|$$

$$= \sum \left| \int_{a_i}^{b_i} f \right| \leqslant \sum \int_{a_i}^{b_i} |f|$$

$$= \int_B |f| < \varepsilon.$$

Thus the function g so defined is absolutely continuous. ∎

[The "only if" part will follow by showing (in Sec. 7-5) that for any absolutely continuous function g, the derived function g' exists almost everywhere, is summable, and

$$g(x) = g(a) + \int_a^x g'(t)\, dt.$$

Note that the Cantor function ψ is not equal to the integral of its derivative. This follows, since $\psi'(x) = 0$ at each x in a complementary interval of Γ. Thus ψ' exists and is zero at almost all points of [0, 1], and hence is summable on [0, 1] with

$$\psi(1) - \psi(0) = 1 \neq \int_0^1 \psi'(t)\, dt = 0.$$

But, as we know, ψ is not absolutely continuous.]

The concept in the next definition is often useful as a test for absolute continuity; and it is used in other connections as well.

Definition 2. *A function f is said to be* **Lipschitzian** *on $[a, b]$ (or satisfy a* **Lipschitz condition***) if there is a constant M such that*

$$|f(x) - f(t)| \leqslant M|x - t| \qquad \text{for all } x, t \in [a, b].$$

Theorem 3. *If f is Lipschitzian on $[a, b]$, then it is absolutely continuous on $[a, b]$.*

Proof. With $\varepsilon > 0$ given, take $\delta = \varepsilon/M$. ∎

6-6 EQUICONTINUITY

The sequences $\{f_n\}$ of this section are such that

$$f_n\colon D \to E^2, \qquad D \subset X, \qquad X \text{ a metric space}$$

except for the specializations in, and related to, the example. Also we do not repeat $x, t \in D$.

Definition 1. *$\{f_n\}$ is said to be* **equicontinuous** *on D if corresponding to each given $\varepsilon > 0$, there is a $\delta > 0$ such that*

$$|f_n(x) - f_n(t)| < \varepsilon \qquad \text{whenever } d_X(x, t) < \varepsilon, \quad n = 1, 2, 3, \cdots.$$

In this definition, δ does not depend on a point, or points, and furthermore δ does not depend on n. Thus an equicontinuous sequence of functions satisfies a doubly uniform condition in that the uniform continuity condition is satisfied uniformly by all functions of the sequence.

Theorem 1. *If $\{f_n\}$ converges uniformly, if each f_n is continuous, and if D is compact, then $\{f_n\}$ is equicontinuous.*

Proof. First note that $\{f_n\}$ satisfies the uniform Cauchy condition (Theorem 6–3.2). Given an $\varepsilon > 0$, let N be such that

$$n \geqslant N \Rightarrow |f_n(x) - f_N(x)| < \varepsilon \qquad \text{for all } x \in D.$$

Hence there is a $\delta > 0$ (the smallest of N positive numbers known to exist) such that

$$d_X(x, t) < \delta \Rightarrow |f_i(x) - f_i(t)| < \varepsilon, \qquad i = 1, 2, \cdots, N,$$

this from the uniform continuity (since D is compact) of each individual f_n. Now if both $d_X(x, t) < \delta$ and $n \geqslant N$, then

$$|f_n(x) - f_n(t)| \leqslant |f_n(x) - f_N(x)| + |f_N(x) - f_N(t)| + |f_N(t) - f_n(t)| < 3\varepsilon.$$

By using this expression if $n > N$ and the one immediately above if $1 \leqslant n \leqslant N$, we have

$$|f_n(x) - f_n(t)| < 3\varepsilon \qquad \text{whenever } d_X(x, t) < \delta, \qquad n = 1, 2, 3, \cdots.$$

Thus $\{f_n\}$ is equicontinuous on D. ▌

Any bounded sequence of real or complex numbers has a convergent subsequence (from the Weierstrass Theorem 2–7.3). The following example, however, displays a sequence $\{f_n\}$ of functions

$$f_n\colon [0, 2\pi] \to E^1, \qquad \text{with } |f_n(x)| \leqslant 1 \text{ for } 0 \leqslant x \leqslant 2\pi,$$

such that no subsequence converges. This example will use the fact (which will be established later) that any bounded function on a bounded closed interval whose integral was evaluated in elementary calculus is also summable and its Lebesgue integral has the same value. Thus, for any two different integers p and q,

(1)
$$\int_0^{2\pi} (\sin px - \sin qx)^2 \, dx = 2\pi.$$

EXAMPLE. $f_n(x) = \sin nx$, $0 \leqslant x \leqslant 2\pi$, $n = 1, 2, 3, \cdots$. Suppose there is a subsequence $n_1 < n_2 < n_3 \cdots$ of positive integers such that $\{\sin n_k x\}$ converges for each $x \in [0, 2\pi]$. Then

$$\lim_{k\to\infty} \sin n_k x = \lim_{k\to\infty} \sin n_{k+1} x$$

and $\lim_{k\to\infty}(\sin n_k x - \sin n_{k+1}x)^2 = 0$. Hence, from Theorem 4–5.9,

$$\lim_{k\to\infty} \int_0^{2\pi} (\sin n_k x - \sin n_{k+1} x) \, dx = 0,$$

which contradicts (1).

The next definition is given to avoid any possible misunderstanding.

Definition 2. *A sequence $\{f_n\}$ is said to be*

(i) **Pointwise bounded** *on D if for each $x \in D$ the sequence of complex numbers $\{f_n(x)\}$ is bounded.*

(ii) **Uniformly bounded** *on D if there is a positive number M such that*

$$|f_n(x)| < M \qquad \textit{for all } x \in D \textit{ and } n = 1, 2, 3, \cdots.$$

Theorem 2. *If $\{f_n\}$ is pointwise bounded and equicontinuous and if D is compact, then*

(a) *$\{f_n\}$ contains a uniformly convergent subsequence, and*
(b) *$\{f_n\}$ is uniformly bounded on D.*

Proof (a). Every compact subset of a metric space is separable (Theorem 2–6.5). We let

$$A = \{a_1, a_2, a_3, \cdots\}$$

be a countable subset of D such that

$$A \cup A' = D.$$

Then $\{f_n(a_1)\}$ is a bounded sequence of complex numbers and thus has a convergent subsequence. We denote such a subsequence by

$$f_{11}(a_1), f_{12}(a_1), f_{13}(a_1), \cdots$$

so that f_{1n} is some f_m with $n \leqslant m$ and

$$\lim_{n\to\infty} f_{1n}(a_1) \text{ exists.}$$

Then $\{f_{1n}(a_2)\}$ is a bounded sequence of complex numbers and we denote a convergent subsequence of it by

$$f_{21}(a_2), f_{22}(a_2), f_{23}(a_2), \cdots,$$

each f_{2n} being some f_{1m} with $n \leqslant m$, and

$$\lim_{n\to\infty} f_{2n}(a_2) \text{ existing.}$$

Continuing in this way with $a_3, a_4, a_5, \cdots$, we display the successive subsequences of functions as

$$\begin{array}{llllll} f_{11}, & f_{12}, & f_{13}, & \cdots, & f_{1n}, & \cdots \\ f_{21}, & f_{22}, & f_{23}, & \cdots, & f_{2n}, & \cdots \\ \cdots & & & & & \\ f_{k1}, & f_{k2}, & f_{k3}, & \cdots, & f_{kn}, & \cdots \\ \cdots & & & & & \end{array}$$

We next consider the main diagonal sequence and set

$$g_n = f_{nn}.$$

The infinite matrix is such that any function in the $(k + 1)$st row is a function directly above or farther to the right in the kth row; that is,

$$f_{k+1,n} \text{ is some } f_{k,m} \text{ with } n \leqslant m.$$

Thus for $n \geqslant k$, we have that $g_n, g_{n+1}, g_{n+2}, \cdots$ is a subsequence of

$$f_{k,1}, f_{k,2}, f_{k,3}, \cdots.$$

Hence for each k, the sequence $\{g_n(a_k)\}$ converges and is therefore a Cauchy sequence.

We next show that

(2) $\{g_n\}$ is a uniform Cauchy sequence on D.

To do so let $\varepsilon > 0$ be given and consider $\delta > 0$ determined (from the equicontinuity of $\{f_n\}$) such that

$$d_X(x, t) < \delta \Rightarrow |f_n(x) - f_n(t)| < \varepsilon, \qquad n = 1, 2, 3, \cdots.$$

The family of neighborhoods $N_X(a_k, \delta)$, $k = 1, 2, 3, \cdots$ is an open covering of D, since $A \subset D$ is dense in D. Since D is compact, there is an integer p such that

$$D \subset \bigcup_{k=1}^{p} N_X(a_k, \delta).$$

Let N (the largest of p determinable integers) be such that

$$m, n \geqslant N \Rightarrow |g_n(a_k) - g_m(a_k)| < \varepsilon, \qquad k = 1, 2, \cdots, p.$$

The δ and N having been determined, we next select an arbitrary $x \in D$. There is at least one k among $1, 2, \cdots, p$ such that $d_X(x, a_k) < \delta$, and for such a k and $m, n \geqslant N$ we have

$$\begin{aligned} |g_n(x) - g_m(x)| &\leqslant |g_n(x) - g_n(a_k)| + |g_n(a_k) - g_m(a_k)| + |g_m(a_k) - g_m(x)| \\ &< 3\varepsilon. \end{aligned}$$

Hence

$$m, n \geqslant N \Rightarrow |g_n(x) - g_m(x)| < 3\varepsilon \qquad \text{for all } x \in D,$$

thus establishing the statement (2).

Therefore $\{g_n\}$ converges uniformly on D by Theorem 6–3.2. Since $\{g_n\}$ is a subsequence of $\{f_n\}$, part (a) has been proved.

Proof (b). From pointwise boundedness of $\{f_n\}$ the function φ defined for each $x \in D$ by

$$\varphi(x) = \sup\{|f_n(x)|\}$$

is finite. Given an $\varepsilon > 0$, let $\delta > 0$ be such that

$$d_X(x, t) < \delta \Rightarrow |f_n(x) - f_n(t)| < \varepsilon, \qquad n = 1, 2, 3, \cdots.$$

With x and t fixed for the moment, but $d_X(x, t) < \delta$, it follows that

$$|f_n(x)| < |f_n(t)| + \varepsilon \qquad \text{and} \qquad |f_n(t)| < |f_n(x)| + \varepsilon,$$

so that

$$\varphi(x) \leqslant \varphi(t) + \varepsilon \qquad \text{and} \qquad \varphi(t) \leqslant \varphi(x) + \varepsilon.$$

Thus $|\varphi(x) - \varphi(t)| \leqslant \varepsilon$ for any $x, t \in D$ such that $d_X(x, t) < \delta$. Hence $\varphi: D \to E^1$ is continuous and thus has a maximum value (Theorem 6–2.5), since D is compact. This maximum value (necessarily finite) is a uniform bound of $\{|f_n|\}$ on D. ∎

6–7 SEMICONTINUITY

Limits inferior and superior for sequences of sets and sequences of real numbers will now be extended to real-valued functions. The definition is given with domains in a metric space X, although most of the applications will specialize X to E^1. An r-neighborhood of $a \in X$ will be denoted by $N(a, r)$ rather than $N_X(a, r)$. Also $N^\circ(a, r)$ is a deleted neighborhood.

Definition 1. *Given $f: D \to E^*$, let $a \in D'$. For each $\lambda > 0$, let $A(\lambda) = \{f(x) \mid x \in N^\circ(a, \lambda) \cap D\}$ and set*

$$\underline{f}(a, \lambda) = \inf A_\lambda \qquad \text{and} \qquad \bar{f}(a, \lambda) = \sup A_\lambda .$$

By definition,

$$\underline{f}(a) = \sup_{\lambda > 0} \underline{f}(a, \lambda) \qquad \text{and} \qquad \bar{f}(a) = \inf_{\lambda > 0} \bar{f}(a, \lambda)$$

are called the **limit inferior** *and* **limit superior**, *respectively, of f at a, and the notations*

$$\varliminf_{x \to a} f(x) = \underline{f}(a) \qquad \text{and} \qquad \varlimsup_{x \to a} f(x) = \bar{f}(a)$$

are used.

If $0 < \lambda$ and $0 < \mu$, we let $\delta = \min(\lambda, \mu)$ to see that $A(\delta) \subset A(\lambda)$, $A(\delta) \subset A(\mu)$, and

$$\underline{f}(a, \lambda) \leqslant \underline{f}(a, \delta) \leqslant \bar{f}(a, \delta) \leqslant \bar{f}(a, \mu).$$

Hence

$$\underline{f}(a) = \sup_{\lambda > 0} \underline{f}(a, \lambda) \leqslant \bar{f}(a, \mu)$$

and then

$$\underline{f}(a) \leqslant \inf_{\mu > 0} \bar{f}(a, \mu) = \bar{f}(a);$$

that is,

(1) $$\varliminf_{x \to a} \underline{f}(x) \leqslant \varlimsup_{x \to a} \bar{f}(x).$$

The following theorem may be established by the method used for the corresponding theorem about real sequences.

Theorem 1. *Given $f: D \to E^*$ and $a \in D'$, then*

(a) *$\underline{f}(a) = -\infty$ iff $G > 0$ and $\delta > 0$ imply that*

$$\{x \mid x \in N^\circ(a, \delta) \cap D \text{ and } f(x) < -G\} \text{ is infinite.}$$

(b) $\underline{f}(a) = \infty$ *iff corresponding to each given* $G > 0$ *there is a* $\delta > 0$ *such that*

$$x \in N^\circ(a, \delta) \cap D \Rightarrow f(x) > G.$$

(c) $-\infty < \underline{f}(a) < \infty$ *iff with* $\varepsilon > 0$ *given, then both*
(i) $\{x \mid x \in N^\circ(a, \delta) \cap D$ *and* $f(x) < \underline{f}(a) + \varepsilon\}$ *is infinite for all* $\delta > 0$, *and*
(ii) *there exists a* $\delta > 0$ *such that*

$$x \in N^\circ(a, \delta) \cap D \Rightarrow \underline{f}(a) - \varepsilon < f(x).$$

We shall, of course, also use a companion Theorem $\bar{1}$ for $\overline{\lim}$. It then follows in view of (1), that

$$\lim_{x \to a} f(x) \text{ exists* iff } \varliminf_{x \to a} f(x) \geqslant \varlimsup_{x \to a} f(x). \tag{2}$$

In the following problems, $f: D \to E^*$ and $a \in D'$.

PROBLEM 1. Let $\{x_n\}$ be a sequence such that $x_n \in D - \{a\}$ and $x_n \to a$ as $n \to \infty$.

(a) Then

$$\varliminf_{x \to a} f(x) \leqslant \varliminf_{n \to \infty} f(x_n) \leqslant \varlimsup_{n \to \infty} f(x_n) \leqslant \varlimsup_{x \to a} f(x).$$

(b) There is such a sequence $\{x_n\}$ for which $\{f(x_n)\}$ converges* and

$$\lim_{n \to \infty} f(x_n) = \varliminf_{x \to a} f(x)$$

and similarly, perhaps a different $\{x_n\}$, for the limit superior.

PROBLEM 2. $\varlimsup_{x \to a} f(x) = - \varliminf_{x \to a} [-f(x)].$

PROBLEM 3. If also $g: D \to E^*$, then

$$\underline{f}(a) + \underline{g}(a) \leqslant \varliminf_{x \to a} [f(x) + g(x)] \leqslant \begin{Bmatrix} \underline{f}(a) + \bar{g}(a) \\ \bar{f}(a) + \underline{g}(a) \end{Bmatrix}$$
$$\leqslant \varlimsup_{x \to a} [f(x) + g(x)] \leqslant \bar{f}(a) + \bar{g}(a),$$

provided the outside and middle terms have meaning.

Definition 2. *A function* $f: D \to E^*$ *is said to be* **lower (upper) semicontinuous** *at* $a \in D \cap D'$ *if*

$$f(a) \leqslant \varliminf_{x \to a} f(x) \qquad \left(\varlimsup_{x \to a} f(x) \leqslant f(a)\right)$$

and is said to be both lower and upper semicontinuous (and continuous) at any isolated point of D.

We shall use the abbreviations lsc and usc.

For example, the function f which is 0 at all rationals and 1 at all irrationals, is lsc at each rational and usc at each irrational. Also, the characteristic function χ_A of a set A:

$$\chi_A = \begin{cases} 1 & \text{if } x \in A \\ 0 & \text{if } x \in A^c \end{cases}$$

is lsc at each point of A^c and usc at each point of A. If A is open, then χ_A is lsc everywhere, but if A is closed, then χ_A is usc everywhere.

Theorem 2. *$f: D \to E^*$ is continuous at a point $a \in D \cap D'$ where $f(a)$ is finite iff f is both lsc and usc.*

Proof. If f is both lsc and usc, then

$$\varliminf_{x \to a} f(x) \geqslant f(a) \geqslant \varlimsup_{x \to a} f(x)$$

and f is continuous at a, since, from (2),

$$\lim_{x \to a} f(x) \text{ exists and } = f(a).$$

The converse is trivial. ∎

Because of Problem 2 and the symmetry in Definition 2, we shall state the following theorems, whose numbers are underlined, for lsc functions. Corresponding theorems (appropriate inequalities reversed, etc.) hold for usc functions.

Theorem $\underline{3}$. *If $f: D \to E^*$ and $g: D \to E^*$ are both lsc at $a \in DD'$ and $\underline{f}(a) + \underline{g}(a)$ has meaning, then $f + g$ is also lsc at a.*

Proof. $f(a) + g(a)$ has meaning, since $f(a) \leqslant \underline{f}(a)$ and $g(a) \leqslant \underline{g}(a)$. Thus, from Problem 3,

$$(f + g)(a) = f(a) + g(a) \leqslant \underline{f}(a) + \underline{g}(a) \leqslant \varliminf_{x \to a} [f(x) + g(x)],$$

which says that $f + g$ is lsc at a. ∎

PROBLEM $\underline{4}$. The converse of Theorem $\underline{3}$ is not true.

Theorem $\underline{4}$. *Let $f: D \to E^*$ and $a \in D \cap D'$ be given at which $f(a) > -\infty$. Then f is lsc at a iff corresponding to each number $c < f(a)$ there is a $\delta > 0$ such that*

(3) $$x \in N(a, \delta) \cap D \Rightarrow f(x) \geqslant c.$$

[f is automatically lsc at a if $f(a) = -\infty$.]

Proof. Consider the (c, δ)-condition holding. Given a $c < f(a)$, determine $\delta > 0$ for which (3) follows. In the notation of Definition 1, we then have that

$$A(\delta) = \{f(x) \mid x \in N^\circ(a, \delta) \cap D\}$$

is such that $c \leqslant \inf A(\delta) = \underline{f}(a, \delta)$ and consequently

$$c \leqslant \sup_{\lambda > 0} \underline{f}(a, \lambda) = \underline{f}(a).$$

Since this holds for every $c < f(a)$, then $\underline{f}(a) \geqslant f(a)$, which says that f is lsc at a.

Conversely, let f be lsc at a (so $f(a) \leqslant \underline{f}(a)$) and let $c < f(a)$ be given. The existence of a $\delta > 0$ for which (3) holds then follows from Theorem 1; (b) in case $f(a) = \infty$ and (c) (ii) in case $f(a) < \infty$. ∎

Theorem 5. *If $f\colon D \to E^*$ and $g\colon D \to E^*$ are both lsc at $a \in DD'$, then the function h defined by*

$$h(x) = min[f(x), g(x)], \qquad x \in D$$

is also lsc at a.

Proof. Consider $-\infty < h(a)$ and choose any number c such that $-\infty < c < h(a)$. Then $c < f(a)$, $c < g(a)$, and there is a $\delta > 0$ such that

$$\begin{aligned} x \in N(a, \delta) &\Rightarrow [f(x) > c \text{ and } g(x) > c] \\ &\Rightarrow h(x) > c, \end{aligned}$$

which, by Theorem 4, means that h is lsc at a. ∎

The next theorem involves limits inferior of both sequences and functions.

Theorem 6. *Let $\{f_n\}$ be a sequence of functions such that for each n,*

$$f_n\colon D \to E^* \text{ is lsc at an } a \in D \cap D'.$$

If the sequence is monotonically increasing,

$$f_1(x) \leqslant f_2(x) \leqslant f_3(x) \leqslant \cdots, \qquad x \in D,$$

then the limit function (which exists)*

$$f(x) = \lim_{n \to \infty} f_n(x), \qquad x \in D$$

is also lsc at a.

Proof. If $f(a) = -\infty$ there is nothing to prove. Hence consider $f(a) > -\infty$ and select any number

$$c < f(a).$$

There is then an integer N such that $f_N(a) > c$, since $\lim f_n(a) = f(a)$. From f_N being lsc at a, select a $\delta > 0$ such that for the neighborhood $N(a, \delta)$,

$$x \in N(a, \delta) \Rightarrow f_N(x) > c.$$

Then $[n \geqslant N, x \in N(a, \delta)] \Rightarrow f_n(x) > c$ from monotonicity of $\{f_n\}$. Thus

$$x \in N(a, \delta) \Rightarrow f(x) = \underline{\lim}_{n \to \infty} f_n(x) \geqslant c.$$

Therefore, by Theorem 4, f is lsc at a. ∎

Definition 3. *A function f: $D \to E^*$ is said to be lower semicontinuous on D if it is lsc at each $x \in D$.*

Theorem 7. *f: $D \to E^*$ is lsc on D iff for each real c the set*

$$B(c) = \{x \in D \mid f(x) \leqslant c\}$$

is closed relative to D.

Proof. Consider that $B(c)$ is closed relative to D for all $c \in E^1$. Assume there is an $a \in D$ at which f is not lsc. Then $\underline{f}(a) < f(a)$ so $f(a) \neq -\infty$. Let c be any number such that

$$\underline{f}(a) < c < f(a).$$

Take $\{x_n\}$ such that $x_n \in D - \{a\}$, $x_n \to a$ and $f(x_n) \to \underline{f}(a)$ as $n \to \infty$, which is possible by Problem 1(b), and a is not an isolated point of D. Hence there is an integer N such that

$$n \geqslant N \Rightarrow x_n \in B(c).$$

Thus $a \in B(c)$, since $B(c)$ is closed relative to D and a is the (one and only) cluster point of the subset $x_N, x_{N+1}, x_{N+2}, \cdots$ of $B(c)$. Hence $f(a) \leqslant c$, contradicting $c < f(a)$.

Conversely, let f be lsc on D. Let c be such that $B(c)$ has at least one cluster point in D and let $a \in D$ be any such cluster point. Select $\{x_n\}$, $x_n \in B(c)$, $x_n \neq a$, and $x_n \to a$ as $n \to \infty$. Then, from f being lsc and Problem 1(a),

$$f(a) \leqslant \underline{\lim}_{x \to a} f(x) \leqslant \underline{\lim}_{n \to \infty} f(x_n) \leqslant c.$$

Thus $a \in B(c)$ and hence $B(c)$ is closed relative to D. ∎

The following theorem is a corollary of Theorem 7.

Theorem 8. *If X has a metric outer measure μ^*, then every lsc function on a measurable set D is measurable on D.*

Theorem 9. *If f: $D \to E^*$ is lsc and if D is compact, then f has a minimum value* [which may be $-\infty$, but not $+\infty$, since we reject the uninteresting possibility that $f(x) = \infty$ for all $x \in D$].

Proof. Let $\underline{m} = \inf\{f(x)\}$ for all $x \in D$. Select $\{x_n\}$, $x_n \in D$, such that

$$\lim_{n \to \infty} f(x_n) = \underline{m}.$$

If any $f(x_n) = \underline{m}$ we are through, so consider that the set $\{x_1, x_2, x_3, \cdots\}$ is infinite. This set has at least one cluster point in D (since D is compact) and we let $a \in D$ be such a cluster point. Then $\{x_n\}$ has a subsequence $\{t_n\}$ such that $t_n \to a$ as $n \to \infty$. Hence

$$\underline{m} \leqslant f(a) \leqslant \varliminf_{x \to a} f(x) \leqslant \lim_{n \to \infty} f(t_n) = \lim_{n \to \infty} f(x_n) = \underline{m},$$

so that $f(a) = \underline{m}$ with $a \in D$. ∎

Example 6–3.1 displays a monotonically increasing sequence of continuous functions whose limit function is not continuous. It follows, however, from Theorem $\underline{6}$, that:

Theorem $\underline{10}$. *If $\{f_n\}$ is a monotonically increasing sequence of functions each continuous on D, then the limit function f defined by*

$$f(x) = \lim_{n \to \infty} f_n(x), \qquad x \in D,$$

is lower semicontinuous.

The converse of this theorem, with slight restrictions on D and the range space, is given in the following theorem.

Theorem $\underline{11}$. *Given that $f: D \to (-\infty, \infty]$ is lsc and that D is compact. Then there is a monotonically increasing sequence $\{f_n\}$ of functions, each continuous on D, such that*

$$\lim_{n \to \infty} f_n(x) = f(x) \qquad \textit{for all } x \in D.$$

Proof. Since the range space is $(-\infty, \infty]$, it follows that $-\infty < f(x)$ for all $x \in D$. Hence the minimum value $\underline{m}$ of f, which exists by Theorem $\underline{9}$, is such that

$$-\infty < \underline{m} \leqslant f(x) \qquad \text{for all } x \in D.$$

We shall first prove the theorem under the additional restriction that

$$f(x) < \infty \qquad \text{for all } x \in D. \tag{4}$$

Even with this restriction, f may be unbounded above.

Select an $x \in D$ and an integer n, both held fixed for the moment, and define φ_n by

$$\varphi_n(t) = f(t) + nd(x, t), \qquad t \in D.$$

Then φ_n is lsc (Theorem $\underline{3}$) on the compact set D and thus has a minimum $\geqslant \underline{m} > -\infty$. Let $X_n(x) \in D$ be a point where this minimum is assumed. Thus

$$\varphi_n[X_n(x)] \leqslant \varphi_n(t) \qquad \text{for all } t \in D.$$

We now define $f_n(x)$ for each $x \in D$ by setting

$$\begin{aligned} f_n(x) &= \varphi_n[X_n(x)] \\ &= f[X_n(x)] + nd[x, X_n(x)]. \end{aligned} \tag{5}$$

Hence $f_n(x) \leqslant \varphi_n(t)$ for all $t \in D$ and this may be written as

$$f_n(x) \leqslant f(t) + nd(x, t), \qquad t \in D. \tag{6}$$

This inequality with $t = x$ yields

$$f_n(x) \leqslant f(x). \tag{7}$$

Incidentally, all these terms are finite, since we have restriction (4) and

$$\begin{aligned} -\infty < \underline{m} \leqslant f[X_n(x)] &\leqslant f[X_n(x)] + d[x, X_n(x)] \\ &= f_n(x) \leqslant f(x) < \infty. \end{aligned}$$

Now with x freed, we have the functions

$$X_n\colon D \to D \qquad \text{and} \qquad f_n\colon D \to E^1, \qquad n = 1, 2, 3, \cdots,$$

for which (5) and (7) hold for all $x \in D$ and (6) holds for all pairs $x, t \in D$. Hence, upon replacing t in (6) by $X_n(t)$, we have

$$\begin{aligned} f_n(x) &\leqslant f_n[X_n(t)] + nd[x, X_n(t)] \\ &\leqslant f_n[X_n(t)] + nd[t, X_n(t)] + nd(t, x) \\ &= f_n(t) + nd(x, t) \qquad \text{from (5).} \end{aligned}$$

Then $f_n(t) \leqslant f_n(x) + nd(t, x)$ by interchanging x and t. Thus

$$|f_n(t) - f_n(x)| \leqslant nd(x, t)$$

for all $x, t \in D$. Consequently:

$$f_n \text{ is continuous on } D.$$

Furthermore, for $m > n$,

$$\begin{aligned} f_n(x) &\leqslant f[X_m(x)] + nd[x, X_m(x)] \qquad \text{from (6)} \\ &\leqslant f[X_m(x)] + md[x, X_m(x)] = f_m(x) \end{aligned}$$

that is, the sequence

$$\{f_n(x)\} \text{ is monotonically increasing for each } x \in D.$$

Hence, because of (7), the sequence $\{f_n(x)\}$ is convergent and

$$\lim_{n\to\infty} f_n(x) \leqslant f(x), \qquad x \in D. \tag{8}$$

We next show, for each $x \in D$, that:

$$\text{the sequence } \{X_n(x)\} \text{ converges to } x. \tag{9}$$

Since D is compact, any convergent subsequence of $\{X_n(x)\}$ converges to a point in D. Let $\{X_{n_k}(x)\}$ be any convergent subsequence of $\{X_n(x)\}$ and denote by $x_0 \in D$ its limit. Hence $n_1 < n_2 < n_3 < \cdots$ and

$$\lim_{k\to\infty} X_{n_k}(x) = x_0 .$$

Since f is lsc at x_0, we have

$$\begin{aligned}
-\infty < f(x_0) &\leqslant \varliminf_{t\to x_0} f(t) \leqslant \varliminf_{k\to\infty} f[X_{n_k}(x)] \qquad \text{by Problem 1}\\
&\leqslant \varliminf_{k\to\infty} f[X_{n_k}(x)] + \varlimsup_{k\to\infty} n_k\, d[x, X_{n_k}(x)]\\
&\leqslant \varlimsup_{k\to\infty} \{f[X_{n_k}(x)] + n_k\, d[x, X_{n_k}(x)]\}\\
&\leqslant \varlimsup_{n\to\infty} \{f[X_n(x)] + nd[x, X_n(x)]\}\\
&= \lim_{n\to\infty} f_n(x) \leqslant f(x) < \infty.
\end{aligned}$$

Thus all terms are finite, and in particular

$$0 \leqslant \varlimsup_{k\to\infty} n_k\, d[x, X_{n_k}(x)] < \infty.$$

Since $n_k \to \infty$ as $k \to \infty$, it follows that

$$\lim_{k\to\infty} X_{n_k}(x) = x.$$

Hence $x_0 = x$. Thus every convergent subsequence of $\{X_n(x)\}$ converges to x and therefore (Problem 3–4.5) the sequence $\{X_n(x)\}$ itself converges to x; that is, (9) is established.

Consequently, for each $x \in D$,

$$\begin{aligned}
f(x) &\leqslant \varliminf_{t\to x} f(t) \leqslant \varliminf_{n\to\infty} f[X_n(x)]\\
&\leqslant \lim_{n\to\infty} \{f[X_n(x)] + nd[x, X_n(x)]\}\\
&= \lim_{n\to\infty} f_n(x) \leqslant f(x) \qquad \text{by (8).}
\end{aligned}$$

Hence $\{f_n\}$ satisfies all properties demanded in the conclusion of the theorem, but under restriction (4).

Restriction (4) is removed by first defining

$$g_m(x) = \begin{cases} f(x) & \text{if } f(x) \leqslant m \\ m & \text{if } f(x) > m \end{cases} = \min[f(x), m], \qquad m = 1, 2, 3, \cdots.$$

Thus g_m is lsc by Theorem 5 and satisfies (4). Hence, for each m, we take a monotonically increasing sequence $\{g_{mn}\}$:

$$g_{m1}(x) \leqslant g_{m2}(x) \leqslant \cdots \leqslant g_{mn}(x) \leqslant \cdots,$$

each continuous on D and such that

$$\lim_{n\to\infty} g_{mn}(x) = g_m(x), \qquad x \in D.$$

Now, for each integer n define h_n by

$$h_n(x) = \max[g_{1n}(x), g_{2n}(x), \cdots, g_{mn}(x)].$$

Then each h_n is continuous on D and it is easily checked that $\{h_n\}$ is monotonically increasing with

$$\lim_{n\to\infty} h_n(x) = f(x), \qquad x \in D. \quad \blacksquare$$

PROBLEM 5. If both f and g are upper semicontinuous at a point x_0, then $f+g$ is upper semicontinuous at x_0, but not conversely.

PROBLEM 6. A function f is usc at a point x_0 iff $-f$ is lsc at x_0.

PROBLEM 7. Given $f: D \to E^1$, then

(a) f is usc on D iff for each number λ the set $\{x \mid f(x) \geqslant \lambda\}$ is closed relative to D (or equivalently $\{x \mid f(x) < \lambda\}$ is open relative to D).

(b) f is lsc on D iff $\{x \mid f(x) \leqslant \lambda\}$ is closed (or $\{x \mid f(x) > \lambda\}$ is open) relative to D for all λ.

Theorem 12. *If $f: A \to [0, \infty)$ is summable on the measurable set A, then there is a sequence $\{A_n\}$ of measurable sets and a sequence $\{c_n\}$ of nonnegative constants such that*

$$A = \bigcup A_n, \qquad f(x) = \sum c_n \chi_{A_n}(x) \qquad \textit{for } x \in A, \text{ and}$$

$$\int_A f(x)\, dx = \sum_{n=1}^{\infty} c_n\, mA_n.$$

[*Note:* $f(x)$ is nonnegative and finite for each $x \in A$, but f may be unbounded, provided $\int_A |f| < \infty$. Also, $\{A_n\}$ need not be a mutually exclusive sequence.]

Proof. The inner and outer ordinate sets of f,

$$\underline{0} = \{(x, y) \mid x \in A, 0 \leqslant y < f(x)\}$$

$$\bar{0} = \{(x, y) \mid x \in A, 0 \leqslant y \leqslant f(x)\}$$

are both m_2-measurable and (Theorem 5–5.2 and Section 5–6)

$$\int_A f(x)\, dx = m_2 \underline{0} = m_2 \bar{0}.$$

We first obtain a set B such that $\underline{0} \subset B \subset \bar{0}$ together with a sequence $\{E_k\}$ of subsets of A and a sequence of nonnegative constants satisfying $m_2 B = \sum c_k m E_k$.

Let $E_0 = \{x \in A \mid f(x) = 0\}$, $B_0 = \{(x, 0) \mid x \in E_0\}$ and $c_0 = 0$. Then $0 = m_2 B_0 = c_0\, mA_0$ even if $mA_0 = \infty$.

For $k = 1, 2, 3, \cdots$ let

$$E_k = \{x \in A \mid f(x) > k\}, \qquad B_k = \{(x, y) \mid x \in E_k, k-1 \leqslant y < k\},$$

and $c_k = 1$. Hence $B_i \cap B_k = \varnothing$ for $0 \leqslant i < k$,

$$B_0 \cup B_1 \cup B_2 \cup \cdots \subset \bar{0} \qquad \text{and} \qquad m_2 B_k = c_k\, mE_k,$$

since $c_k = 1$ for $k \geqslant 1$. Also, $A - E_0 \supset E_1 \supset E_2 \supset \cdots$ and

$$x \in A \Rightarrow \left| f(x) - \sum_{k=0}^{\infty} c_k \chi_{E_k}(x) \right| < 1.$$

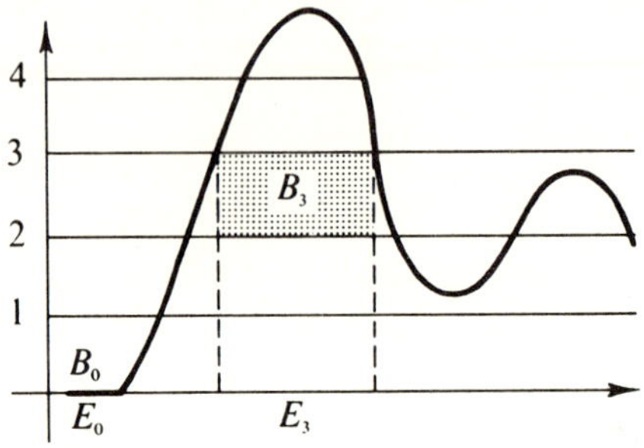

FIG. 1

The method of proceeding is simple, but the notation becomes somewhat involved. For $k = 0, 1, 2, \cdots$ let

$$E_{k+\frac{1}{2}} = \{x \in A - E_{k+1} | f(x) > k + \tfrac{1}{2}\},$$
$$B_{k+\frac{1}{2}} = \{(x, y) \,|\, x \in E_{k+\frac{1}{2}},\, k \leqslant y < k + \tfrac{1}{2}\} \subset \bar{0}$$

and $c_{k+\frac{1}{2}} = \frac{1}{2}$. Hence $m_2(B_{k+\frac{1}{2}}) = c_{k+\frac{1}{2}}\, m(E_{k+\frac{1}{2}})$ and

$$x \in A \Rightarrow \left| f(x) - \sum_{k=0}^{\infty} [c_k \chi_{E_k}(x) + c_{k+\frac{1}{2}} \chi_{E_{k+\frac{1}{2}}}(x)] \right| < \tfrac{1}{2}.$$

For $k = 0, 1, 2, \cdots$, let

$$E_{k+\frac{1}{4}} = \{x \in A - E_{k+\frac{1}{2}} \,|\, f(x) > k + \tfrac{1}{4}\}, \qquad c_{k+\frac{1}{4}} = \tfrac{1}{4},$$
$$E_{k+\frac{3}{4}} = \{x \in A - E_{k+1} \,|\, f(x) > k + \tfrac{3}{4}\}, \qquad c_{k+\frac{3}{4}} = \tfrac{1}{4},$$

and define B_q with $q = k + \frac{1}{4}$ and $q = k + \frac{3}{4}$.

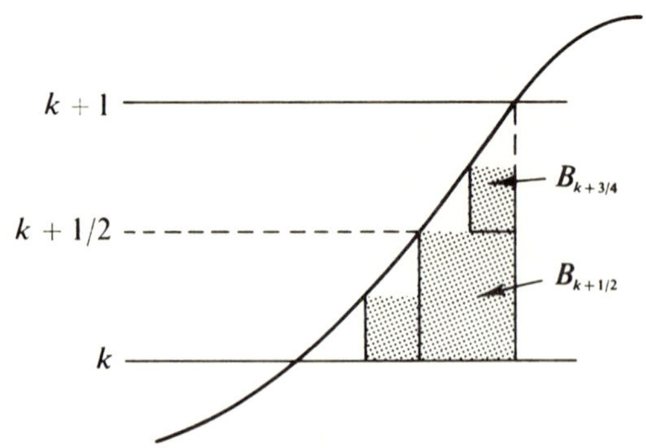

FIG. 2

By continuing in this way we obtain a countable mutually exclusive family of B's whose union we call B is such that $B \subset \bar{0}$. Also $\underline{0} \subset B$, since if $(x, y) \in \underline{0}$, then $x \in A$, $0 \leqslant y < f(x)$, and there are positive integers n, i such that

$$y < \frac{i}{2^n} < f(x)$$

so that one of the B's includes (x, y). Hence

$$m_2 B = m_2 \underline{0} = m_2 \bar{0}.$$

The countable family of E's we now rename and arrange in a sequence $A_1, A_2, A_3, \cdots$, each carrying its c. Thus

$$A_1 \cup A_2 \cup A_3 \cup \cdots = A, \qquad f(x) = \sum_{n=1}^{\infty} c_n \chi_{A_n}(x) \qquad \text{for } x \in A,$$

and

$$\sum_{n=1}^{\infty} c_n mA_n = m_2 B = \int_A f(x)\, dx. \quad \blacksquare$$

A finite summable function f may have many discontinuities, but its integral may be approximated by squeezing f between two semicontinuous functions.

Theorem 13. (Vitali-Carathéodory) *If $f: A \to E^1$ is summable on the measurable set A and $\varepsilon > 0$ is given, then there is an usc function u and a lsc function l such that*

(10) $$u(x) \leqslant f(x) \leqslant l(x) \qquad \textit{for } x \in A \qquad \textit{and}$$

$$\int_A (l - u) < \varepsilon.$$

Proof. Corresponding to the positive part f^+ of f, select sequences $\{A_n\}$ and $\{c_n\}$ as in Theorem 12. For $n = 1, 2, 3, \cdots$ let F_n be closed, G_n open,

$$F_n \subset A_n \subset G_n \qquad \text{and} \qquad c_n m(G_n - F_n) < \frac{\varepsilon}{4} 2^{-n}.$$

Using the convergence of $\sum c_n mA_n$, choose N such that

$$\sum_{N+1}^{\infty} c_n mA_n < \frac{\varepsilon}{4}.$$

Since $\bigcup_{n=1}^{N} F_n$ is closed, the function

$$u^+ = \sum_{n=1}^{N} c_n \chi_{F_n}$$

is usc by Problem 7. Similarly the function

$$l^+ = \sum_{n=1}^{\infty} c_n \chi_{G_n}$$

is lsc. Moreover, for each $x \in A$

$$0 \leqslant u^+(x) \leqslant \sum_{n=1}^{\infty} c_n \chi_{A_n}(x) = f(x)$$

$$\leqslant \sum_{n=1}^{\infty} c_n \chi_{G_n}(x) = l^+(x) \qquad \text{and}$$

$$l^+ - u^+ = \sum_{n=1}^{N} c_n(\chi_{G_n} - \chi_{F_n}) + \sum_{N+1}^{\infty} c_n \chi_{F_n}$$

$$\leqslant \sum_{n=1}^{\infty} c_n(\chi_{G_n} - \chi_{F_n}) + \sum_{N+1}^{\infty} c_n \chi_{A_n}.$$

Thus

$$\int_A (l^+ - u^+) \leqslant \sum_{n=1}^{\infty} c_n m(G_n - F_n) + \sum_{N+1}^{\infty} c_n m(A_n) < \frac{\varepsilon}{2}.$$

The same reasoning applied to f^- yields u^- usc, l^- lsc,

$$0 \leqslant u^- \leqslant f^- \leqslant l^-, \qquad \text{and} \qquad \int_A (l^- - u^-) < \frac{\varepsilon}{2}.$$

Consequently

$$u^+ - l^- \leqslant f^+ - f^- = f \leqslant l^+ - u^-,$$

$u = u^+ + (-l^-)$ is usc, $l = l^+ - u^-$ is lsc, and (10) holds. ∎

In the proof of the important Theorem 7–7.5 we shall need both strict inequalities in the

Corollary. *If in addition $mA < \infty$, then corresponding to each $\varepsilon > 0$ there is a lsc function g such that*

$$f(x) < g(x) \qquad \textit{for } x \in A \textit{ and} \int_A (g - f) < \varepsilon.$$

Proof. Take l lsc such that

$$f(x) \leqslant l(x) \qquad \text{and} \qquad \int_A (l - f) < \frac{\varepsilon}{2}.$$

This l increased by $\varepsilon/(1 + 2mA)$ is a satisfactory g. ∎

6–8 DISCONTINUITIES

Given that $f: (a, b) \to E^1$, then for $a < x < b$ we use

$$\underline{f}(x-) = \underline{\lim}_{t \to x-} f(t) \qquad \text{and} \qquad \bar{f}(x-) = \overline{\lim}_{t \to x-} f(t)$$

for the limit inferior and limit superior of the restriction of f to (a, x). If $\underline{f}(x-) = \bar{f}(x-)$, this common value is denoted by $f(x-)$ and called the **left-hand limit** of f at x. Similarly, right-hand limits inferior $\underline{f}(x+)$, superior $\bar{f}(x+)$, and limit $f(x+)$ are defined.

If f is discontinuous at $x \in (a, b)$ and if $f(x-)$ and $f(x+)$ exist, then f is said to have a **discontinuity of the first kind** (or a **simple** discontinuity) at x. A simple discontinuity of the first kind at x is said to be a **removable discontinuity** if $f(x-) = f(x+) \neq f(x)$ and is said to be a **jump discontinuity** if $f(x-) \neq f(x+)$, regardless of the value $f(x)$.

Any discontinuity not of the first kind is said to be of the **second kind**. For example, the function f defined by

$$f(x) = \begin{cases} \sin(1/x) & \text{if } x \neq 0 \\ 0 & \text{if } x = 0 \end{cases}$$

has a discontinuity of the second kind at $x = 0$.

Theorem 1. *If* $f: (a, b) \to E^1$ *is nondecreasing on* (a, b); that is, $a < x < y < b \Rightarrow f(x) \leqslant f(y)$, *then*

(a) $a < x < b \Rightarrow f(x-) \leqslant f(x) \leqslant f(x+)$
  [*and thus all discontinuities of f are of the first kind*].
(b) $a < x < y < b \Rightarrow f(x+) \leqslant f(y-)$.
(c) *The set of points of discontinuity of f on* (a, b) *is countable.*

Proof (a). Given $a < x < b$, then $a < t < x \Rightarrow f(t) \leqslant f(x)$ and thus

$$\underline{f}(x-) \leqslant \bar{f}(x-) \leqslant f(x).$$

Let $\varepsilon > 0$ be given. Then there is a number c such that $a < c < x$ and

$$\bar{f}(x-) - \varepsilon < f(c).$$

Since f is nondecreasing,

$$c < t < x \Rightarrow [f(c) \leqslant f(t)] \Rightarrow f(c) \leqslant \underline{f}(x-).$$

Thus $\bar{f}(x-) - \varepsilon \leqslant \underline{f}(x-)$ for each $\varepsilon > 0$ so that $\bar{f}(x-) \leqslant \underline{f}(x-)$. Since $\underline{f}(x-) \leqslant \bar{f}(x-)$ we have that $f(x-)$ exists and

$$f(x-) = \bar{f}(x-) \leqslant f(x).$$

In a similar way, $f(x+)$ exists and $f(x) \leqslant f(x+)$.

Proof (b). Given x and y such that $a < x < y < b$, then

$$\begin{aligned} x < t < s < y &\Rightarrow f(x) \leqslant f(t) \leqslant f(s) \leqslant f(y) \\ &\Rightarrow f(x) \leqslant f(x+) \leqslant f(s) \leqslant f(y) \\ &\Rightarrow f(x) \leqslant f(x+) \leqslant f(y-) \leqslant f(y). \end{aligned}$$

Proof (c). Let S be the set of discontinuities of f on (a, b). Hence $s \in S$ implies (from part (a)) that $f(s-) < f(s+)$. Corresponding to each $s \in S$ select a rational number $r(s)$ such that

$$f(s-) < r(s) < f(s+).$$

Then, by part (b) for the $\leqslant$ below,

$$(s, t \in S, s < t) \Rightarrow r(s) < f(s+) \leqslant f(t-) < r(t)$$

from which it follows that $r(s) \neq r(t)$. Hence we have a 1–1 correspondence between the points of S and a subset of the rational numbers. Thus S is countable. ∎

The following problem shows that any countable set on (a, b) is the set of discontinuities of a monotonically increasing function on (a, b).

PROBLEM 1. Let $S = \{s_1, s_2, s_3, \cdots\}$ be a countable set on (a, b). Take any sequence $\{c_n\}$ of positive constants such that $\Sigma\, c_n$ converges. For each $x \in (a, b)$ let

$$A(x) = \{n \mid a < s_n < x\}.$$

Define the function f by

$$f(x) = \sum_{n \in A(x)} c_n \qquad \text{if } A(x) \neq \varnothing, \quad f(x) = 0 \quad \text{if } A(x) = \varnothing.$$

Then

(a) f is monotonically increasing on (a, b).
(b) $f(s_n+) - f(s_n-) = c_n$; that is, f is discontinuous at each point of S.
(c) $f(x-) = f(x)$ for each $x \in (a, b)$.
(d) f is continuous at each $x \in (a, b) - S$.

Definition 1. *Given a function $f: E^1 \to E^1$, we define:*

(i) *The* **oscillation** *of f* **over an open interval** *I by*

$$\omega[f, I] = \underset{x \in I}{\text{l.u.b.}}\, f(x) - \text{g.l.b.} f(x), \qquad \textit{and}$$

(ii) *The* **oscillation** *of f* **at** *$a \in E^1$ by*

$$\omega(f, a) = \underset{a \in I}{\text{g.l.b.}}\ \omega[f, I].$$

PROBLEM 2. Let $f: E^1 \to E^1$ be given.

(a) Then f is continuous at $a \in E^1$ iff $\omega(f, a) = 0$.
(b) For each number $r > 0$, $\{x \mid \omega(f, x) \geqslant r\}$ is closed.
(c) If D is the set of points of discontinuity of f, then D is an F_σ set.

Recall that a set $A \subset E^1$ is nowhere dense (Definition 2–4.3) if each open interval contains an open interval free of points of A; equivalently, $\bar{A}$ contains no open interval. All intervals mentioned are, of course, nonempty. Thus $A \subset E^1$ is nowhere dense iff $\bar{A}$ is nowhere dense.

Definition 2. *The subsets of E^1 are divided into two mutually exclusive classes by:*

(i) *A is of* **first category** *if A is expressible as $A = \bigcup A_i$ with each A_i nowhere dense.*
(ii) *A is of* **second category** *if it is not of first category.*

PROBLEM 3. (a) $A \cup B$ is of first category iff A and B are both of first category.
(b) The set of rationals is of first category.
(c) E^1 is of second category.
(d) The set of irrationals is of second category.
(e) The set of irrationals is not an F_σ set.

[*Hint* for *c:* Assume $E^1 = \bigcup A_n$, A_n nowhere dense. Take a bounded open interval I_1 free of points of A_1, $J_1 \subset I_1$ and closed, $I_2 \subset J_1$ free of points of $A_2, \cdots$. $\bigcap J_n$ is nonempty and empty.]

Recall that the function of Problem 6–1.1 is continuous at each irrational and discontinuous at each rational.

PROBLEM 4. There is no real-valued function of a real variable which is continuous at each rational and discontinuous at each irrational.

PROBLEM 5. (Category Theorem of Baire) Any complete metric space is of second category.

6–9 APPROXIMATE CONTINUITY

All continuous functions with measurable domains are measurable functions (Theorem 6–2.1). The main goal of this section is to show how nearly the converse comes to being true.

In order to focus attention on the central theme, real-valued functions of a real variable and Lebesgue measure are used. Without exception, domains could be in E^k with the measure† m_k, but ranges must remain in E^1. Also the sets are bounded, although some results hold without this condition, as could be seen by the technique, amply illustrated in Chapter 4, of first restricting consideration to $I_n = (-n, n)$ and then expanding.

† There are treatments with further ejections.

Theorem 1. (Egoroff). *Let $\{f_n\}$ be a sequence of measurable functions all having domains including a measurable set M. If $\{f_n\}$ converges at almost all points of M to a function f, then corresponding to each number $\eta > 0$ there is a measurable set $A \subset M$ such that $m(M - A) < \eta$ and $\{f_{nA}\}$ converges uniformly on A, where f_{nA} is the restriction of f_n to A.*

Proof. By neglecting a set of measure zero we consider that $\{f_n\}$ converges to f at all points of M. Since f is the limit of measurable functions, it is itself measurable. Thus for each pair n, k of positive integers, the set

$$B_{nk} = \bigcup_{i=n}^{\infty} \left\{ x \in M \,\middle|\, |f_i(x) - f(x)| \geqslant \frac{1}{k} \right\}$$

is a measurable subset of M. Moreover

$$(i \geqslant n \text{ and } x \in M - B_{nk}) \Rightarrow |f_i(x) - f(x)| < \frac{1}{k}.$$

Also for each k,

$$M \supset B_{1k} \supset B_{2k} \supset B_{3k} \cdots \qquad \text{and} \qquad \bigcap_{n=1}^{\infty} B_{nk} = \varnothing,$$

since $\{f_i(x)\}$ converges to $f(x)$ for each $x \in M$. Since $mM < \infty$ (M is bounded, as agreed of all sets to be used), then

$$\lim_{n \to \infty} m(B_{nk}) = 0, \qquad k = 1, 2, 3, \cdots$$

by Theorem 4–2.10. For each k let n_k be such that

$$m(B_{n_k k}) < \frac{\eta}{2^k}.$$

Hence, for each k,

$$(i \geqslant n_k \text{ and } x \in M - B_{n_k k}) \Rightarrow |f_i(x) - f(x)| < \frac{1}{k}.$$

Let B be defined by $B = \bigcup_{k=1}^{\infty} B_{n_k k}$ so that

$$mB \leqslant \sum_{k=1}^{\infty} mB_{n_k k} < \eta.$$

Given a number $\varepsilon > 0$, choose K such that $1/K < \varepsilon$. Hence n_K is determined and

$$\begin{aligned}(i \geqslant n_K \text{ and } x \in M - B) &\Rightarrow (i \geqslant n_K \text{ and } x \in M - B_{n_K K}) \\ &\Rightarrow |f_i(x) - f(x)| < \frac{1}{K} < \varepsilon.\end{aligned}$$

Upon setting $M - B = A$ we thus have that $\{f_{nA}\}$ converges uniformly on A and also

$$m(M - A) = mB < \eta. \quad \blacksquare$$

Lemma 1. *Given a step function*

$$s = \sum_{i=1}^{n} c_i \chi_{A_i}, \qquad A_i \cap A_j = \varnothing$$

for $1 \leqslant i < j \leqslant n$, c_i *real, and each* A_i *measurable, let* $A = \bigcup_{i=1}^n A_i$. *Then corresponding to each number* $\eta > 0$, *there is a closed set* $F \subset A$ *such that* $m(A - F) < \eta$ *and the restriction* s_F *of* s *to* F *is continuous on* F.

Proof. For each $i = 1, 2, \cdots, n$ let C_i be a closed subset of A_i such that $m(A_i - C_i) < \eta/n$, this being possible by Theorem 5–1.9. Set $F = \bigcup_{i=1}^n C_i$. Then s_F is constant on each of the closed sets $F_1, F_2, \cdots, F_n$, no one containing a point nor a cluster point of the others, there being only a finite number of them. Thus s_F is continuous on F. Moreover

$$A - F = \bigcup_{i=1}^{n} (A_i - C_i)$$

and $m(A - F) = \sum_{i=1}^n m(A_i - C_i) < \eta$. ∎

Theorem 2. (Lusin) *Let* f *be a measurable function on a measurable set* M *and let* $\varepsilon > 0$ *be given. Then*

(a) *There is a closed set* $F \subset M$ *such that* $m(M - F) < \varepsilon$ *and the restriction* f_F *is continuous on* F.

(b) *There is a bounded*† *closed set* $F \subset M$ *and a continuous function* $g: E^1 \to E^1$ *such that*

$$m(M - F) < \varepsilon \qquad \text{and} \qquad g(x) = f(x) \qquad \text{for all } x \in F.$$

Proof. (a). Let $\{s_n\}$ be a sequence of measurable step functions converging to f on M. This would first be done for f^+ and f^- separately (by Theorem 4–3.7) and the difference taken to form $\{s_n\}$. By using Egoroff's theorem with $\eta = \varepsilon/2$, take a measurable set $A \subset M$ such that

$$m(M - A) < \frac{\varepsilon}{2}$$

and $\{s_{nA}\}$ converges uniformly to f on A.

With n a positive integer, apply Lemma 1 to obtain a closed set $F_n \subset A$ such that

$$m(A - F_n) < \frac{\varepsilon}{2^{n+1}}$$

and the restriction of s_{nA} to F_n, which we denote by σ_n, is continuous on F_n. Hence the set

$$F = \bigcap_{n=1}^{\infty} F_n$$

† "Bounded" is inserted here, since in referring back to this theorem it might be overlooked that early in the section we pointed out that the blanket consideration of "boundedness" of all sets in this section is not a significant restriction.

is closed and $A - F \subset \bigcup (A - F_n)$ so that

$$m(A - F) \leqslant \sum_{n=1}^{\infty} m(A - F_n) < \frac{\varepsilon}{2}.$$

But $M - F = (M - A) \cup (A - F)$ and hence

$$m(M - F) < \varepsilon.$$

Moreover, each function σ_n is continuous at each point of F, since it is so on $F_n \supset F$. We now let f_n be the restriction of σ_n to F. Hence not only is each f_n continuous on F, but in addition the sequence $\{f_n\}$ converges uniformly on F, since it inherits this uniformity from $\{s_{nA}\}$ on the larger set A. Hence $\{f_n\}$ converges on F to a continuous function having domain F, by Theorem 6–3.6. Since this limit function has to be the restriction f_F, part (a) is proved.

Proof. (b). Since f_F is continuous on the bounded closed set F, then f_F is bounded on F. We then apply the Tietze extension theorem (6–3.7) to obtain a function g which is continuous on all of E^1 and agrees with f_F (and hence with f) on F. ∎

PROBLEM 1. If f is a measurable function on a measurable set M, then there is a sequence $\{f_n\}$ of continuous functions such that

$$\lim_{n \to \infty} f_n(x) = f(x) \quad \text{a.e.} \qquad \text{in } M.$$

In spite of Lusin's theorem, a measurable function can be wildly discontinuous. For example, the function equal to 1 at all irrationals and 0 otherwise is measurable and discontinuous everywhere.

The Lusin theorem has many applications, but for now we give an easy (and curious) example of its use.

EXAMPLE 1. Let $f: E^1 \to E^1$ satisfy

$$f(x + y)f(x - y) = f^2(x) - f^2(y) \qquad \text{for all } x, y \in E^1.$$

If there is a measurable set A with $mA > 0$ on which f is measurable, then f is continuous at the origin.

Proof. Let F be a bounded closed subset of A with $mF > 0$ and such that the restriction f_F is continuous on F. Choose any $\varepsilon > 0$. Since f_F is uniformly continuous on F, let $\delta_1 > 0$ be such that

$$|f_F(x_1)f_F(x_2) - f_F{}^2(x)| < \varepsilon$$

if x_1, x_2, and x is any triple of points of F satisfying $|x_1 - x| < \delta_1$ and $|x_2 - x| < \delta_1$. See Problem 6–3.3.

Since F is measurable and $mF > 0$, let $\eta > 0$ be such that corresponding to each number h with $|h| < \eta$ there are points for which all three of

$$x \in F, \qquad x + h \in F, \qquad \text{and} \qquad x - h \in F$$

hold (see the corollary of Theorem 5–2.4). Let $\delta = \min(\delta_1, \eta)$. Hence

$$|h| < \delta \Rightarrow |f(h)| < \sqrt{\varepsilon},$$

since corresponding to each such h there is at least one x for which we are justified in writing

$$\begin{aligned} |-f^2(h)| &= |f(x+h)f(x-h) - f^2(x)| \\ &= |f_F(x+h)f_F(x-h) - f_F{}^2(x)| < \varepsilon. \end{aligned}$$

Since $f(0) = 0$ (prove this quickly),

$$\lim_{h\to 0} f(h) = 0 = f(0).$$

[A reason for interest in the above example is that it was known from several independent discoveries,† that

If f, satisfying the functional equation of the example, has at least one point of continuity, then it must have one or another of the forms

$$f(x) = ax, \qquad f(x) = a \sin bx, \qquad \textit{or} \qquad f(x) = a \sinh bx.$$

Note that all these solutions are continuous at all points of E^1.

A related curiosity is that the functional equation

$$f(x+y) = f(x) + f(y)$$

has all its measurable solutions not only continuous but of the form $f(x) = ax$ (see page 363). Thus any discontinuous solution—and there are some (which we state for information only since we do not prove it)—is nonmeasurable. The axiom of choice is, however, used in obtaining discontinuous solutions. The existence of such discontinuous solutions of this special equation may seem of little significance, but a concept introduced to show that they exist led to what is now called a Hamel base‡ for the real numbers, and Hamel bases are used in many different connections in more advanced mathematics.]

The following observation motivates the next definition on which another continuity property of measurable functions is based.

If $f: (a, b) \to E^1$ is continuous at an $x \in (a, b)$, then for each given $\varepsilon > 0$, the set

$$\{t \mid |f(t) - f(x)| < \varepsilon\}$$

contains an interval centered at x. In other words, the values of t such that $f(t)$ differs by less than ε from $f(x)$ are so dense around x that they completely fill an interval.

† A rather complete bibliography is on pp. 114–115 of J. Aczel, *Vorlesungen uber Funktionalgeichungen und Ihre Anwendugnen*, Basel, 1961. The above example was proved by Sanford L. Segal, " On A Sine Functional Equation," *Am. Math. Monthly*, Vol. 70, No. 3, (March, 1963), pp. 306–308.

‡ G. Hamel, "Eine Basis aller Zahlen und die Losungen der Funktionalgleichung $f(x+y) = f(x) + f(y)$," *Mathematische Annalen*, Vol. 60 (1905), pp. 459–462.

Definition. *A function* $f\colon [a, b] \to E^1$ *is said to be* **approximately continuous** *at an* $x \in [a, b]$ *if for each given* $\varepsilon > 0$ *the set*

$$A(x, \varepsilon) = \{t \mid |f(t) - f(x)| < \varepsilon\}$$

has density 1 *at* x; *that is*, (see Definition 5–2.2),

$$\underline{D}[A(x, \varepsilon), x] = \bar{D}[A(x, \varepsilon), x] = 1.$$

Theorem 3. (A. Denjoy) *If* $f\colon [a, b] \to E^1$ *is measurable and finite a.e. on* $[a, b]$, *then* f *is approximately continuous a.e. on* $[a, b]$.

Proof. Given a number $\varepsilon > 0$, let $F \subset [a, b]$ be a closed set and g a continuous function such that

$$mF > (b - a) - \varepsilon \qquad \text{and} \qquad x \in F \Rightarrow g(x) = f(x).$$

Then

$$B = \{x \in F \mid D(F, x) = 1\} \qquad \text{has } mB = mF$$

by Theorem 5–2.2, and in particular B is a measurable set. Fix any $x_0 \in B$. Then $x_0 \in (a, b)$, since both $\bar{D}(F, a) \leqslant \frac{1}{2}$ and $\bar{D}(F, b) \leqslant \frac{1}{2}$. Let $\eta > 0$ be given. Then the set

$$\{t \mid |g(t) - g(x_0)| < \eta\}$$

contains intervals centered at x_0 of arbitrarily small length and within (a, b), since $x_0 \in (a, b)$. Thus the following limit, taken over such x_0-centered intervals, exists and

$$\lim_{|I| \to 0} \frac{m(F \cap I)}{|I|} = 1.$$

Since $x \in F \cap I \Rightarrow f(x) = g(x)$, the set

$$A(x_0, \eta) = \{t \mid |f(t) - f(x_0)| < \eta\}$$

also has density 1 at x_0 so that x_0 is a point of approximate continuity of f. Thus the set

$$H = \{x \in [a, b] \mid f \text{ is approximately continuous at } x\}$$

contains the measurable set B with $mB > (b - a) - \varepsilon$. Hence $mH > (b - a) - \varepsilon$ for all $\varepsilon > 0$, and thus $mH = b - a$. ∎

6–10 CONTINUOUS LINEAR FUNCTIONALS

Except for the continuity properties below and the special case at the very end, the material of this section could have been included in Section 2–12 in which a vector space V over the complex number field Z (Definition 2–12.1) was considered. We shall now use $x, y, \cdots$ (rather than $\mathbf{x}, \mathbf{y}, \cdots$) for elements of V and $\alpha, \beta, \cdots$ for complex numbers. Also the V used here has an abstract inner product (Definition 2–12.4) and the standard norm $\|x\| = \sqrt{x \cdot x}$ is understood.

Hence, for the resulting metric space, the parallelogram law

$$\|x - y\|^2 + \|x + y\|^2 = 2\|x\|^2 + 2\|y\|^2 \tag{1}$$

holds as well as the Schwarz and Minkowski inequalities

$$|x \cdot y| \leqslant \|x\| \, \|y\| \qquad \text{and} \qquad \|x + y\| \leqslant \|x\| + \|y\|.$$

It is common practice to refer to a function with domain V and range in Z as a **functional.**

Definition 1. *A functional $L : V \to Z$ is said to be* **linear** *if for all $x, x_1, x_2 \in V$ and $\alpha \in Z$*

$$L(\alpha x) = \alpha L(x) \qquad \textit{and} \qquad L(x_1 + x_2) = L(x_1) + L(x_2).$$

[Thus $L(0) = 0$ where the first 0 is the zero vector (origin) and the second means $0 + i0$.]

EXAMPLE 1. Given a space X having a σ-algebra of subsets on which a measure μ is defined, let $\mathscr{L}^2(X, \mu)$ be the Hilbert space of square summable complex-valued functions. Then L defined by

$$L(f) = \int_X f \, d\mu, \qquad f \in \mathscr{L}^2$$

is a linear functional.

EXAMPLE 2. With (V, Z) as above, fix a $y \in V$. Then L defined in terms of the inner product by

$$L(x) = x \cdot y \qquad \text{for all } x \in V$$

is a linear functional since

$$L(\alpha x) = (\alpha x) \cdot y = \alpha(x \cdot y) = \alpha L(x) \qquad \text{and}$$

$$L(x_1 + x_2) = (x_1 + x_2) \cdot y = (x_1 \cdot y) + (x_2 \cdot y) = L(x_1) + L(x_2).$$

PROBLEM 1. With y a fixed element of V, the function defined by $f(x) = y \cdot x$ is a functional, but is not linear.

Theorem 1. *A linear functional $L : V \to Z$ is continuous iff it is continuous at the origin.*

Proof. Let L be continuous at the origin. Given an $\varepsilon > 0$ let $\delta > 0$ be such that

$$\|y\| < \delta \Rightarrow |L(y) - L(0)| = |L(y)| < \varepsilon.$$

Take any $x \in V$. Then

$$z \in V \qquad \text{and} \qquad \|z - x\| < \delta \Rightarrow |L(z) - L(x)| = |L(z - x)| < \varepsilon$$

showing that L is continuous at x [and, moreover, that L is uniformly continuous on V since δ did not depend upon x].

Conversely, let L be continuous at some $x \in V$. Take any $\varepsilon > 0$, and, corresponding to this pair (x, ε), let $\delta > 0$ be such that

$$|L(z) - L(x)| < \varepsilon \qquad \text{whenever } z \in V \text{ and } \|z - x\| < \delta.$$

Select any y having $\|y\| < \delta$ and set $z = x + y$. Then $\|z - x\| < \delta$ and

$$|L(y) - L(0)| = |L(z - x) - 0| = |L(z) - L(x)| < \varepsilon$$

showing that L is continuous at the origin. ∎

A subset W of a vector space V is said to be a **subspace** of V if W is itself a vector space using the same addition and scalar multiplication as defined on V. Hence all subspaces are nonempty (each contains the origin) and the set $\{0\}$ is a subspace.

PROBLEM 2. A nonempty set $W \subset V$ is a subspace of V iff for every $x, y \in W$ and $\alpha \in Z$ both $\alpha x \in W$ and $(x + y) \in W$.

PROBLEM 3. If L is a linear functional on V, then

$$S = \{x \mid L(x) = 0\}$$

is a subspace of V.

As defined in Sec. 2–12, a set $C \subset V$ is **convex** if for each pair $x, y \in C$ also

$$tx + (1 - t)y \in C \qquad \text{for all } 0 \leqslant t \leqslant 1.$$

PROBLEM 4. Any vector space is convex, and hence any subspace of a vector space is convex.

Elements $x, y \in r$ are said to be **orthogonal**, and $x \perp y$ or $y \perp x$ is written, if $x \cdot y = 0$. Now with $y \in V$ we set

$$y^{\perp} = \{x \mid x \perp y\}$$

and with S any nonempty subset of V

$$S^{\perp} = \{x \mid x \perp s \qquad \text{for every } s \in S\} = \bigcap_{s \in S} s^{\perp}.$$

PROBLEM 5. Given any $x \in V$, the set $x^{\perp}$ is a subspace of V.

[*Hint:* Use Problem 2.]

We now consider what further can be said if V is complete; that is, if V is a Hilbert space, still over the field Z of scalars.

Theorem 2. *A nonempty, closed, convex set C of a Hilbert space $\mathscr{H}$ has a unique element $x_0 \in C$ such that*

$$\|x_0\| \leqslant \|x\| \qquad \textit{for all } x \in C;$$

that is, C contains a unique element of minimum norm.

Proof. Let $u = \inf\{\|x\| \mid x \in C\}$. Hence $0 \leqslant u < \infty$ and we take a sequence $\{x_n\}$ in C such that $\lim \|x_n\| = u$. In (1) replace x by $\frac{1}{2}x_n$ and y by $\frac{1}{2}x_m$ to obtain

$$\tfrac{1}{4}\|x_n - x_m\|^2 + \|\tfrac{1}{2}(x_n + x_m)\|^2 = \tfrac{1}{2}\|x_n\|^2 + \tfrac{1}{2}\|x_m\|^2.$$

But $\frac{1}{2}(x_n + x_m) \in C$, since C is convex, so that

$$u \leqslant \|\tfrac{1}{2}(x_n + x_m)\| \qquad \text{and} \qquad \|x_n - x_m\|^2 \leqslant 2\|x_n\|^2 + 2\|x_m\|^2 - 4u^2.$$

As $n \to \infty$ and $m \to \infty$ the right side approaches zero. Thus $\{x_n\}$ is a Cauchy sequence which, therefore, converges to some $x_0 \in \mathscr{H}$ since $\mathscr{H}$ is complete. Thus

$$\|x_0\| \leqslant \lim \|x_n - x_0\| + \lim \|x_n\| = 0 + u.$$

Also $x_0 \in C$, since C is closed, so that $u \leqslant \|x_0\|$ and then $\|x_0\| = u$.

To establish uniqueness, take any $y_0 \in C$ such that $\|y_0\| = u$. Convexity of C gives $\frac{1}{2}(x_0 + y_0) \in C$ and then (1), used much as before, gives

$$\|x_0 - y_0\|^2 = 2\|x_0\|^2 + 2\|y_0\|^2 - 4\|\tfrac{1}{2}(x_0 + y_0)\|^2 \leqslant 2u^2 + 2u^2 - 4u^2 = 0.$$

Hence $\|x_0 - y_0\| = 0$ so that $y_0 = x_0$. ∎

PROBLEM 6. In a Hilbert space $\mathscr{H}$:

(a) If $x \in \mathscr{H}$, then $x^\perp$ is a closed subspace of $\mathscr{H}$.
(b) If S is a subspace of $\mathscr{H}$, then $S^\perp$ is a closed subspace of $\mathscr{H}$.

Definition 2. *Given a vector space (V, Z), a set $S \subset V$, and a $z \in V$, then the translate of S by z is defined by*

$$\begin{aligned} z + S &= \{z + s \mid s \in S\} \\ &= \{x \mid x - z \in S\}. \end{aligned}$$

PROBLEM 7. Let S be a subspace of V. Then:

(a) $z + S = S$ iff $z \in S$.
(b) Either $(z + S) \cap S = \varnothing$ or else $z + S = S$.
(c) If $z_1 \in z + S$, then $z_1 + S = z + S$.

PROBLEM 8. If S is a subspace of a Hilbert space $\mathscr{H}$ and $z \in \mathscr{H}$, then $z + S$ is a nonempty, closed, convex set in $\mathscr{H}$, but need not be a subspace of $\mathscr{H}$.

[The intuitive notion for the proof of the next theorem is that, in a plane, if S is a line through the origin, then the foot of the perpendicular from the origin to a line parallel to S is a positive distance from the origin.]

Theorem 3. *If S is a subspace of a Hilbert space $\mathscr{H}$ and if*

$$S \neq \mathscr{H}, \qquad \text{then } S^\perp \neq \{0\}.$$

Proof. In case $S = \{0\}$, then $y \in S^\perp$ for all y and we take any $y \in \mathscr{H}$ with $y \neq 0$ to see that $S^\perp \neq \{0\}$.

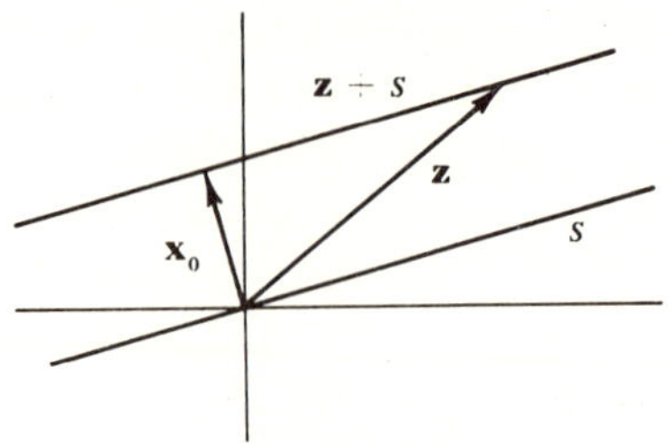

FIG. 1

Consider therefore that $\{0\}$ is a proper subset of S (which itself is a proper subset of $\mathscr{H}$). Take any $z \in \mathscr{H} - S$ and note, in addition to $z \neq 0$, that the translate $z + S$ and S do not intersect. In this translate let x_0 be the vector of smallest norm (by Theorem 2 and Problem 8). Since $x_0 \in z + S$, then

$$x_0 + S = z + S \qquad \text{and} \qquad (x_0 + S) \cap S = \varnothing,$$

so that $x_0 \neq 0$. Take any $y \in S - \{0\}$. Hence $y \neq 0$ and

$$y_0 = \frac{y}{\|y\|} \qquad \text{has} \qquad \|y_0\| = 1.$$

For every $\alpha \in Z$ we have $\alpha y_0 \in S$ and $x_0 - \alpha y_0 \in x_0 + S$. By the minimum norm property of x_0 it follows that

$$x_0 \cdot x_0 \leqslant (x_0 - \alpha y_0) \cdot (x_0 - \alpha y_0) = x_0 \cdot x_0 - \bar{\alpha}(x_0 \cdot y_0) - \alpha(y_0 \cdot x_0) + \alpha\bar{\alpha}(y_0 \cdot y_0)$$

and

$$0 \leqslant -\bar{\alpha}(x_0 \cdot y_0) - \alpha(y_0 \cdot x_0) + |\alpha|^2.$$

The special choice $\alpha = (x_0 \cdot y_0)$ then gives

$$0 \leqslant -\bar{\alpha}\alpha - \alpha\bar{\alpha} + |\alpha|^2 = -|\alpha|^2.$$

Thus $\alpha = 0$; that is, $x_0 \cdot y_0 = 0$. Hence $x_0 \perp y_0$. Therefore

$$x_0 \perp y \qquad \text{for all } y \in S.$$

Consequently $x_0 \in S^\perp$ and $S^\perp \neq \{0\}$ since $x_0 \neq 0$. ∎

PROBLEM 9. [The intuitive idea is that in the plane any vector may be resolved into perpendicular components, one of them in any given direction.] Let S be a closed subspace of a Hilbert space $\mathscr{H}$. Given any $x \in \mathscr{H}$, let $Q(x)$ denote the element of minimum norm in $x + S$ and set $P(x) = x - Q(x)$. Then:

(a) $Q(x) \in S^\perp$ and $P(x) \in S$.
(b) $\|x\|^2 = \|P(x)\|^2 + \|Q(x)\|^2$.
(c) $Q(Q(x)) = Q(x)$ and $P(P(x)) = P(x)$.

[*Note:* $P(x)$ and $Q(x)$ are called the **projections** of x on S and $S^\perp$, respectively.]

One way of obtaining linear functionals on any vector space V was given in Example 2. The next theorem shows that all linear functionals may be obtained in this way if V is a Hilbert space $\mathscr{H}$. The special case following the theorem plays a central role in proving a most significant pair of theorems in the last section of this book.

Theorem 4. *If L is a continuous linear functional on a Hilbert space $\mathscr{H}$, then there is a unique $y \in \mathscr{H}$ such that*

$$L(x) = x \cdot y \qquad \textit{for all } x \in \mathscr{H}. \tag{2}$$

Proof. In case $L(x) = 0$ for all $x \in \mathscr{H}$, then $y = 0$ satisfies (2).
Consider therefore that $L(x) \neq 0$ for at least one $x \in \mathscr{H}$. Set

$$S = \{x \mid L(x) = 0\}.$$

Then S is a subspace of $\mathscr{H}$ by Problem 3, it is closed since L is continuous, $S \neq \mathscr{H}$ since $L(x) \neq 0$ for some x, and therefore $S^\perp \neq \{0\}$ by Theorem 2. Take any $y' \in S^\perp - \{0\}$. Then $L(y') \neq 0$ and $y_0 = \|y'\|^{-1} y'$ has $\|y_0\| = 1$ but also $L(y_0) \neq 0$ since $y_0 \in S^\perp - \{0\}$. Set

$$\alpha = \overline{L(y_0)} \qquad \text{and} \qquad y = \alpha y_0 .$$

Then $\alpha \neq 0$ and we claim that y satisfies (2). Toward seeing that it does, let x be any element of $\mathscr{H}$ and set

$$x_0 = x - \frac{L(x)}{\bar{\alpha}} y_0 .$$

First compute

$$L(x_0) = L\left(x - \frac{L(x)}{\bar{\alpha}} y_0\right) = L(x) - \frac{L(x)}{\bar{\alpha}} L(y_0) = 0.$$

Therefore $x_0 \in S$. But $y_0 \in S^\perp$ so that

$$x_0 \cdot y_0 = 0.$$

Next use $x_0 \cdot y_0 = 0$ in computing

$$x \cdot y = \left(x_0 + \frac{L(x)}{\bar{\alpha}} y_0\right) \cdot (\alpha y_0)$$

$$= \bar{\alpha}(x_0 \cdot y_0) + \frac{L(x)}{\bar{\alpha}} \bar{\alpha}(y_0 \cdot y_0) = 0 + L(x) = L(x).$$

Hence (2) holds for the y as defined above.
To establish uniqueness, take any $y_1 \in \mathscr{H}$ such that

$$L(x) = x \cdot y_1 \qquad \text{for all } x \in \mathscr{H}.$$

Hence $x \cdot y = x \cdot y_1$ and $x \cdot (y - y_1) = 0$ for all $x \in \mathscr{H}$. In particular

$$(y - y_1) \cdot (y - y_1) = 0; \qquad \text{that is, } \|y - y_1\| = 0.$$

Consequently $y_1 = y$ so the above method of obtaining a y satisfying (2) produces the only such y. ∎

The special case mentioned above is stated as a theorem for ready reference in Section 8–7.

Theorem 5. *Let X be a space in which there is a σ-algebra $\mathfrak{M}$ of subsets on which a measure μ is defined. If L is a continuous linear functional on the*† *Hilbert space $\mathscr{L}^2(X, \mu)$, then there is a $g \in \mathscr{L}^2(X, \mu)$ such that*

$$(3) \qquad L(f) = f \cdot g = \int_X (f\bar{g})\, d\mu \qquad \textit{for all } f \in \mathscr{L}^2(X, \mu)$$

and the same is true with g replaced by any function agreeing with it μ-a.e.

[Note that in the proofs of Theorems 2 and 4 we used the property

$$\|x\| = 0 \text{ iff } x = 0$$

of metric spaces, only half of which is known in a pseudometric space. Thus, as a technicality, we apply Theorem 4 to the space of classes of functions in $\mathscr{L}^2$, each class consisting of functions equal μ-a.e. (as discussed at the end of Section 46). Hence (3) holds for all functions in the same class with g; which is what the last part of Theorem 5 says.]

PROBLEMS

1. If $f: [a, b] \to E^1$ is continuous with $f(a) > a$ and $f(b) < b$, then there is an x such that $a < x < b$ and $f(x) = x$.
2. Given that $f: (a, b) \to E^1$ is continuous. Then f is uniformly continuous on the open interval (a, b) iff both

$$\lim_{x \to a+} f(x) \quad \text{and} \quad \lim_{x \to b-} f(x) \quad \text{exist.}$$

3. If $f: (a, b) \to E^1$ has the intermediate value property, then f has no simple discontinuity on (a, b).
4. A real-valued function of a real variable is **locally decreasing** at a point x_0 if there is a neighborhood $\mathscr{U}$ of x_0 such that

$$x_0 \neq x \in \mathscr{U} \Rightarrow \frac{f(x) - f(x_0)}{x - x_0} < 0.$$

Show that there is a continuous function $f: [0, 1] \to E^1$ which is locally decreasing at each point of a set D dense on $[0, 1]$, but $f(0) < f(1)$.

5. (a) Let f be a continuous function whose domain is a circle in E^2 and whose range is an interval $[a, b]$. Then f does not have an inverse.
   (b) If a function f maps $\{(x, y) \mid x^2 + y^2 \leqslant 1\}$ onto $\{(x, y) \mid x^2 + y^2 = 1\}$, then either f is discontinuous or is not 1–1.
   (c) If $f: E^2 \to E^2$ is continuous and $z_0 \in E^2$, then in every neighborhood of z_0

† Recall that the vector space $\mathscr{L}^2(X, \mu)$ is a Hilbert space because completeness follows from Theorem 4–6.3 and the last paragraph of Sec. 4–7.

there are distinct points z_1 and z_2 such that $|f(z_1)| = |f(z_2)|$.
[*Hint:* Assume not, and consider how u defined by $u(z) = |f(z_2)|$ maps some circular disk centered at z_0.]

6. (a) Give an example of a continuous function having connected range and disconnected domain.
(b) If $f: D \to R$ with $D, R \subset E^1$ is continuous on D and $\{(x, y) \mid x \in D, y = f(x)\}$ is connected, then D is connected.

7. (a) Given $f_n(x) = 1/(nx + 1)$, $0 < x < 1$, $n = 1, 2, 3, \cdots$, then $\{f_n\}$ converges pointwise, but not uniformly on (0, 1).
(b) Given $g_n(x) = x/(nx + 1)$, $0 < x < 1$, $n = 1, 2, 3, \cdots$, then

$$g_n \rightrightarrows 0 \quad \text{on } (0, 1).$$

8. For $n = 1, 2, 3, \cdots$ define f_n and g_n on E^1 by

$$f_n(x) = x\left(1 + \frac{1}{n}\right)$$

$$g_n(x) = \begin{cases} 1/n, & x \text{ irrational or } x = 0 \\ q + 1/n & \text{if } x = p/q \text{ in lowest terms.} \end{cases}$$

Also let $h_n = f_n g_n$. Then
(a) Both $\{f_n\}$ and $\{g_n\}$ converge uniformly on every finite interval.
(b) $\{h_n\}$ does not converge uniformly on any finite interval.

9. Let $f_n \rightrightarrows f$ and $g_n \rightrightarrows g$ on $[a, b]$. If $\{f_n\}$ and $\{g_n\}$ are bounded on $[a, b]$, then $f_n g_n \rightrightarrows fg$ on $[a, b]$.

10. Let $f: [a, b] \to E^1$ be continuous and $a < c < d < b$.
(a) There is a number ξ such that $a < \xi < b$ and $2f(\xi) = f(c) + f(d)$.
(b) If α and β are positive numbers, then there is a number ξ such that $a < \xi < b$ and $(\alpha + \beta)f(\xi) = \alpha f(c) + \beta f(d)$.

11. With F a closed subset of E^1, let $f: F \to E^1$ be continuous on F relative to F. Show there is a continuous function $g: E^1 \to E^1$ such that $g(x) = f(x)$ for $x \in F$ and g linear on each component interval of F^c (thus obtaining Tietze's extension theorem by a much easier method in this special case).

12. Let D be a dense subset of a metric space X and let $f: D \to E^1$ be uniformly continuous on D relative to D.
(a) For each $x \in X$ let $C(x, n)$ be the closure of the set

$$f\left[DN\left(x, \frac{1}{n}\right)\right], \qquad n = 1, 2, 3, \cdots.$$

Show that $C(x, 1) \cap C(x, 2) \cap C(x, 3) \cap \cdots$ is a single point of E^1.
(b) For each $x \in X$ let $g(x)$ denote the value obtained in part (a). Show that the function g so defined is the continuous extension of f to X.

13. Let X be a metric space.
(a) If $f: X \to E^1$ is continuous, then $\{x \mid f(x) = 0\}$ and $\{x \mid f(x) = 1\}$ are closed sets.
(b) Conversely, if A and B are disjoint nonempty closed sets, then there is a continuous real-valued function for which A is the 0-set and B is the 1-set.

$$[\textit{Hint: } f(x) = \frac{d(A, x)}{d(A, x) + d(B, x)}, \qquad x \in X.]$$

(c) If A and B are disjoint closed sets, then there are disjoint open sets $G \supset B$ and $H \supset A$.
[*Hint:* For f as in part (b) examine the inverse images of $[0, \frac{1}{2})$ and $(\frac{1}{2}, 1]$.]

14. Every connected metric space having at least two points has cardinality at least $\mathfrak{c}$ (the cardinal of the continuum).

15. Let X and Y be topological space (or metric space if the problems on topological spaces (pp. 117–118) were not worked). Then $f: X \to Y$ is said to be an **open mapping** if $f(G)$ is open whenever G is open.

If $f: E^1 \to E^1$ is a continuous open mapping, then f is monotonic.

16. Let f_1 and g_1 be functions each linear on $[k/9, (k+1)/9]$ and such that $x = f_1(t)$, $y = g_1(t)$ maps this interval onto the heavy vector marked large k in Fig. P.16

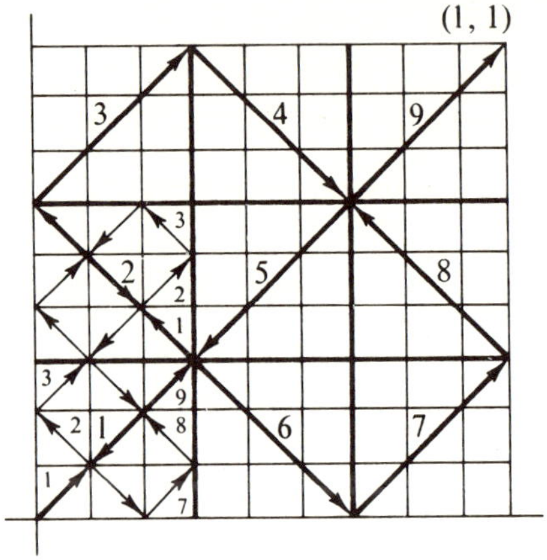

FIG. P.16

for $k = 0, 1, 2, \cdots, 9$. Let f_2 and g_2 be linear on $[l/9^2, (l+1)/9^2]$, $l = 0, 1, \cdots, 80$ and such that $(f_2(t), g_2(t))$ follows the scheme indicated by the light vectors for $0 \leqslant t \leqslant 1$. In particular if

$$\frac{k}{9} \leqslant t \leqslant \frac{k+1}{9}, \text{ then } |f_2(t) - f_1(t)| < \frac{1}{3}, \qquad |g_2(t) - g_1(t)| < \frac{1}{3}.$$

Continue in this way to obtain sequences $\{f_n\}$ and $\{g_n\}$.

(a) The sequences of functions converge to functions f and g such that $x = f(t)$, $y = g(t)$ for $0 \leqslant t \leqslant 1$ is a continuous space-filling curve.

(b) If A is an m_1-measurable set on $[0, 1]$ and B is its image, then B is m_2-measurable and $m_2 B = m_1 A$.

17. If f is a nonnegative continuous function on $[a, b]$ having maximum value M, then

$$\lim_{n\to\infty} \left\{ \int_a^b f^n(x)\, dx \right\}^{1/n} = M.$$

Find

$$\lim_{n\to\infty} \left\{ \int_0^{\pi/2} \left(\frac{\sin x}{x} \right)^n dx \right\}^{1/n}.$$

[*Hint:* $\int_a^b f^n \leqslant M^n(b-a)$. Take $\overline{\lim}$ of nth root. For any $0 < \lambda < M$ there are $a \leqslant c < d \leqslant b$ such that $\int_a^b f^n \geqslant \lambda^n(d-c)$. Take $\underline{\lim}$ of nth root.]

18. Let V be a real vector space on which a real nonnegative nontrivial function p is defined which is continuous in the sense

$$\lim_{t\to 1} p(t\mathbf{x}) = p(\mathbf{x}) \qquad \text{for each } \mathbf{x} \in V,\ t \text{ real.}$$

If p also satisfies the parallelogram law,

$$p^2(\mathbf{x}+\mathbf{y}) + p^2(\mathbf{x}-\mathbf{y}) = 2p^2(\mathbf{x}) + 2p^2(\mathbf{y}), \qquad \mathbf{x}, \mathbf{y} \in V,$$

then p may be used to make V into an inner-product space.

[*Hint:* After proving that $p(\mathbf{0}) = 0$ and $p(-\mathbf{x}) = p(\mathbf{x})$, see Problems 21 and 22, page 113.]

19. For this problem a function f: $[a, b] \to E^1$ is said to be two-to-one if for each y in the range of f there are two and only two distinct values $a \leqslant x_1 < x_2 \leqslant b$ such that $f(x_1) = f(x_2) = y$.

(a) Let $A = \{x \mid 0 < x < \frac{1}{2}, x \text{ irrational}\}$ and with $r_1, r_2, r_3, \cdots$ an ordering without duplication of the rational numbers of $(0, 1)$, let

$$B = A \bigcup \{0, r_1, r_3, r_5, \cdots\}.$$

Define φ on B by $\varphi(x) = x + \frac{1}{2}$ if $x \in A$, $\varphi(0) = 1$, and $\varphi(r_{2k-1}) = r_{2k}$ for $k = 1, 2, 3, \cdots$. Define f on $[0, 1]$ by

$$f(x) = \begin{cases} \varphi(x) & \text{if } x \in B \\ x & \text{if } x \in [0, 1] - B. \end{cases}$$

Then f is two-to-one on $[a, b]$.

(b) There is no continuous two-to-one function f: $[a, b] \to E^1$.

(c) If a function f: $[a, b] \to E^1$ is continuous and takes no value more than twice, then f takes on some value exactly once.

[*Hint:* If f takes its maximum twice, then $f(a) = f(b) = \max$.]

20. (a) With f: $[0, 1] \to [0, 1]$ the continuous function, which was defined in Example 6–3.2, what is the cardinal of the set

$$\{t \mid f(t) = x\} \qquad \text{for each } x \in [0, 1]?$$

(b) There is a continuous real-valued function which is countably infinite to 1. [*Hint: Am. Math. Monthly*, Vol. 72 (1965), page 784.] Compare this fact with Problem 19(b).

21. (a) If f: $[0, 1] \to E^1$ is measurable and both

$$\int_0^1 f(x)\,dx = 0, \qquad \int_0^1 xf(x)\,dx = 1,$$

then $m\{x \mid |f(x)| \geqslant 4\} > 0$.

[*Hint:* If $|f(x)| < 4$ a.e., $1 = \left|\int_0^1 (x - \frac{1}{2})f(x)\,dx\right| < \int_0^1 4|x - \frac{1}{2}|dx = 1$.]

(b) If in addition $|f(x)| \leqslant 4$ a.e. in $[0, 1]$, then $|f(x)| = 4$ a.e. in $[0, 1]$.

22. If f: $E^1 \to E^1$ is continuous and there is a number $\lambda > 0$ such that $|f(x) - f(y)| \geqslant \lambda|x - y|$, then $f(E^1) = E^1$.

[*Hint:* f is 1–1 and $f(E^1)$ is both open and closed.]

Problems on a Totally Disconnected Set Which Becomes Connected by Adding a Single Specific Point

Let P be the set of end points of the open intervals of $[0, 1]$ erased in forming the Cantor set Γ, and let $Q = \Gamma - P$. The Cantor function ψ from $\{x \mid 0 \leqslant x \leqslant 1\}$ to $\{y \mid 0 \leqslant y \leqslant 1\}$ was defined on page 64. Recall that ψ is a continuous monotonically increasing function which maps [0, 1] onto [0, 1], is constant on the closure of each interval erased in forming Γ, $\psi(P)$ is the countable set

$$\left\{y \middle| y = \frac{m}{2^n}; \qquad m = 1, 2, \cdots, n \quad \text{for } n = 1, 2, 3, \cdots\right\}$$

dense on [0, 1], and $\psi(\Gamma) = [0, 1]$. Moreover, if $p \in P$ and $q \in Q$, then $\psi(p) \neq \psi(q)$ and ψ maps the part of Γ on the open interval with end points p and q onto the open interval with end points $\psi(p)$ and $\psi(q)$. Also, if $q_1 \in Q$, $q_2 \in Q$, and $q_1 < q_2$ then

$$\psi(\Gamma \cap (q_1, q_2)) = (\psi(q_1), \psi(q_2)).$$

Theorem. *Let G be an open subset of $\{x \mid 0 < x < 1\}$ such that $G \supset P$. Then $\psi(\Gamma - G)$ is nowhere dense on $\{y \mid 0 \leqslant y \leqslant 1\}$.*

Proof. $\psi(G) \supset \psi(P)$ which is dense on [0, 1]. Any $p \in P$ is in an open subinterval of G which contains a $q \in Q$ with either $q < p$ or $q > p$. Hence $\psi(G)$ contains an open set which is dense on $\{y \mid 0 \leqslant y \leqslant 1\}$.

Since $\Gamma - G = Q - G$ if $y \in \psi(\Gamma - G)$ there is one and only one $x \in G$ such that $\psi(x) = y$ and this $x \notin G$. Thus

$$\psi(\Gamma - G) = \psi(Q) - \psi(G).$$

Hence any open subinterval of the y-interval [0, 1] contains a further subinterval in $\psi(G)$ and thus free of points of $\psi(\Gamma - G)$. Consequently, $\psi(\Gamma - G)$ is nowhere dense on [0, 1]. ∎

PROBLEM 1. If N is any subset of Γ such that $\psi(N)$ is nowhere dense on $\{y \mid 0 \leqslant y \leqslant 1\}$ and I is any open subinterval of $\{x \mid 0 \leqslant x \leqslant 1\}$ such that $I\Gamma \neq \varnothing$, then $\psi(I\Gamma - N)$ contains a closed interval free of points of $\psi(N)$.

[*Hint:* $\psi(I\Gamma)$ contains an open interval and some open subinterval of it is free of points of $\psi(N)$, etc.]

The Cantor set is thinly scattered on [0, 1], but the following work uses a criterion for "thinness" in Γ of a subset of Γ. Recall that a set is of first category if it is a countable union of nowhere dense sets.

Definition 1. *A subset S of the Cantor set Γ is said to be* **sparse** *in Γ if $\psi(S)$ is of first category.*

We thus have the corollary to the theorem.

Corollary. *If G is an open set and $G \supset P$, then $\Gamma - G$ is sparse in Γ.*

PROBLEM 2. If S is sparse in Γ, then $(\Gamma - S)' = \Gamma$.

[*Hint:* Problem 1 above and the hint for Problem 3, page 339.]

PROBLEM 3. If $\{S_n\}$ has each S_n sparse in Γ, then $\bigcup S_n$ is sparse in Γ.

Recall that a set A is connected iff U open, V open, $A \subset U \cup V$, and $UV = \varnothing$ imply that either $A \subset U$ or $A \subset V$. See Theorem 2–10.2.

A set having but a single point is connected. Any finite set of $n \geqslant 1$ points has no connected subset of more than one point.

Definition 2. *A set in a metric space is said to be* **totally disconnected** *if its only connected subsets consist of single elements.*

PROBLEM 4. In E^k the set of all points having all k coordinates rational is totally disconnected.

PROBLEM 5. The Cantor set is totally disconnected.

In E^2, with an ordinary (x, y)-coordinate system, let I_1 be the closed interval joining the point $(1, -\frac{1}{2})$ to $(1, \frac{1}{2})$ and let Γ_1 be the Cantor set on I_1. With $0 < x < 1$, let I_x be the closed interval joining $(x, -\frac{1}{2}x)$ to $(x, \frac{1}{2}x)$ and let Γ_x be the scale reduction of the

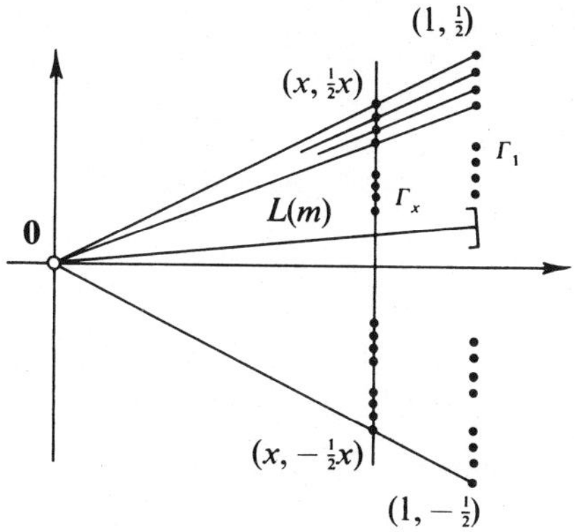

Cantor set on I_x. For $0 < x \leqslant 1$ let P_x be the set of end points of open intervals of I_x erased in forming Γ_x and set $Q_x = \Gamma_x - P_x$.

We now define

$$A_x = \begin{Bmatrix} P_x & \text{if } x \text{ is rational} \\ Q_x & \text{if } x \text{ is irrational} \end{Bmatrix}, \qquad 0 < x \leqslant 1.$$

The set we are interested in is

$$A = \bigcup_{0 < x \leqslant 1} A_x .$$

Note that the origin $\{\mathbf{0}\} = (0, 0)$ is not in A, that $AQ_1 = \varnothing$, since 1 is rational, and that $A\Gamma_1 = P_1$.

As a further useful notation, we set

$$L(m) = \{(x, y) \mid 0 < x \leqslant 1 \quad \text{and} \quad y = mx\}, \qquad -\tfrac{1}{2} \leqslant m \leqslant \tfrac{1}{2}.$$

Thus $L(m)$ is a line segment not including its end point $\{\mathbf{0}\} = (0, 0)$, but including the other end point $(1, m)$. Also

$$A \cap L(m) = \begin{cases} \varnothing & \text{if } (1, m) \notin \Gamma_1 \\ \{(x, mx) \mid 0 < x \leqslant 1 \quad \text{rational}\} & \text{if } (1, m) \in P_1 \\ \{(x, mx) \mid 0 < x < 1 \quad \text{irrational}\} & \text{if } (1, m) \in Q_1 . \end{cases}$$

PROBLEM 6. The set $A \subset E^2$ is totally disconnected.
[*Hint:* Take any $B \subset A$ having at least two points.
Case 1. For some m, $B \subset L(m)$.
Case 2. $m_1 < m_2$ and both $BL(m_1) \neq \varnothing$ and $BL(m_2) \neq \varnothing$.]

The remainder of the work is to show that upon adding the origin, then

The set $A \cup \{\mathbf{0}\}$ is connected.

Let U and V be open sets such that

$$A \cup \{\mathbf{0}\} \subset U \cup V, \qquad U \cap V = \varnothing, \qquad \text{and} \qquad \{\mathbf{0}\} \in U.$$

Hence the origin is an interior point of U and there is a rational number r_0 such that $0 < r_0 \leqslant 1$ and

$$[(x, y) \in A \cup \{\mathbf{0}\} \qquad \text{and} \qquad 0 \leqslant x \leqslant r_0] \Rightarrow (x, y) \in U.$$

For $(1, c) \in \Gamma_1$, we define

$$f(c) = \sup\{0 < x \leqslant 1 \mid \text{if } 0 < t \leqslant x \text{ and } (t, ct) \in A, \text{ then } (t, ct) \in U\}.$$

We thus have that for all $(1, c) \in \Gamma_1$,

$$0 < r_0 \leqslant f(c) \leqslant 1 \qquad \text{and} \qquad (f(c), cf(c)) \in \Gamma_{f(c)}.$$

PROBLEM 7. If $(1, c) \in \Gamma_1$ and $f(c) = 1$, then $AL(c) \subset U$.
[*Hint:* Consider $(1, c) \in P_1$ and $(1, c) \in Q_1$ separately.]

It follows that if $f(c) = 1$ for all $(1, c) \in \Gamma_1$, then $A \cup \{\mathbf{0}\} \subset U$ so that $A \cup \{\mathbf{0}\}$ is connected. We thus consider possibilities if $f(c) < 1$.

PROBLEM 8. If $(1, c) \in \Gamma_1$ and $f(c) < 1$, then

(a) $(f(c), cf(c)) \in (AU)' A^c$.
(b) $(1, c) \in P_1 \Rightarrow f(c)$ is irrational.
(c) $(1, c) \in Q_1 \Rightarrow f(c)$ is rational.

[*Hint for (a):* If $(f(c), cf(c))$ were in A, then this point would either be in U, and $f(c)$ would be greater than it is, or else this point would be in V, etc.]

Note, furthermore, that

$$(1, c) \in Q_1 \Rightarrow (f(c), cf(c)) \in (AU)' A^c,$$

from (a) and (c) if $f(c) < 1$ and if $f(c) = 1$ from $A_1 = P_1$, but if $f(c) = 1$ then $(f(c), cf(c)) = (1, c)$ might be in U.

Let

$$P^f = \{(f(c), cf(c)) \mid (1, c) \in P_1\} \qquad \text{and}$$

$$Q^f = \{(f(c), cf(c)) \mid (1, c) \in Q_1 \text{ and } f(c) < 1\}.$$

We now think of projecting P^f and Q^f into I_1 along lines through the origin and express this by setting

$$P^* = \{(1, c) \mid (f(c), cf(c)) \in P^f\},$$

Q^* by replacing P^f by Q^f, and $S = P^* \cup Q^*$. Since P_1 is countable and $P^* \subset P_1$, it follows that P^* is sparse in Γ_1. We next show that

Q^* is also sparse in Γ_1.

To do so, let r be rational, $r_0 \leqslant r < 1$ and set

$$R_r = Q^f \cap \{(x, y) \mid x = r\}.$$

Thus $R_r \subset \Gamma_r - A_r = \Gamma_r - P_r = Q_r$. Moreover, $A_r \subset U \cup V$ and hence

$$R_r \subset \Gamma_r - (U \cup V).$$

Since $U \cup V$ is open, we have that R_r is sparse in Γ_r (by the corollary). Project R_r into I_1 along lines through the origin. The resulting set, call it S_r, is then sparse in Γ_1. Since $Q^* = \bigcup S_r$, the union taken for all rational r, $r_0 \leqslant r < 1$; we have that Q^* is sparse in Γ_1 (by Problem 3). Consequently $S = P^* \cup Q^*$ is sparse in Γ_1 and, by Problem 2,

$$(\Gamma_1 - S)' = \Gamma_1.$$

Now let $D = \bigcup L(c)$, the union taken for all $(1, c) \in \Gamma_1 - S$. Since $(\Gamma_1 - S)' = \Gamma_1$, we have

$$D' = \Big[\bigcup_{c \in \Gamma_1} L(c)\Big] \cup \{0\} \supset A \cup \{0\}.$$

Recall that if $(1, c) \in \Gamma_1 - S$, then $f(c) = 1$, which means that $AL(c) \subset U$ (by Problem 7) so that $VAL(c) = \varnothing$ and (A being dense on $L(c)$),

$$[UL(c)]' = L(c) \cup \{0\}.$$

Hence UD is dense in D so that $(UD)' = D'$. Thus

$$U' \supset (UD)' = D' \supset A \cup \{0\}.$$

Since $UV = \varnothing$ with U and V open, we have that

$$\varnothing = VU' \supset V(A \cup \{0\}) \qquad \text{and hence} \qquad A \cup \{0\} \subset U$$

because $A \cup \{0\} \subset U \cup V$.

Consequently $A \cup \{0\}$ is connected.†

Problems on Convex Functions and Jensen's Inequalities

For a function $f: E^1 \to E^1$, the value $f(x)$ is excluded from consideration in the Definition 6–7.1 of

$$\underline{f}(x) = \varliminf_{t \to x} f(t) \qquad \text{and} \qquad \overline{f}(x) = \varlimsup_{t \to x} f(t).$$

We temporarily use functions $\check{f}$ and $\hat{f}$ defined by

$$\check{f}(x) = \min[f(x), \underline{f}(x)] \qquad \text{and} \qquad \hat{f}(x) = \max[f(x), \overline{f}(x)].$$

PROBLEM 1. The function f is continuous at x_0 iff

$$\check{f}(x_0) = \hat{f}(x_0).$$

Definition 1. *A function $\varphi: I \to E^1$, where I is a finite or infinite interval in E^1, is said to be* **convex** *on I, provided*

(1) $x, y \in I$ and $0 \leqslant \lambda \leqslant 1 \Rightarrow \varphi((1-\lambda)x + \lambda y) \leqslant (1-\lambda)\varphi(x) + \lambda\varphi(y)$;

† B. Knaster and C. Kuratowski, "Sur les ensembles connexes," *Fundamenta Mathematicae*, Vol. 2 (1921), pp. 206–255, were the first to obtain such a set; their proof used the notion of relative category.

or equivalently, $s < u < t$ all in I implies

$$\frac{\varphi(u)-\varphi(s)}{u-s} \leqslant \frac{\varphi(t)-\varphi(u)}{t-u}. \tag{2}$$

In particular a convex function φ has the midpoint property

$$\varphi\left(\frac{x+y}{2}\right) \leqslant \frac{\varphi(x)+\varphi(y)}{2} \tag{3}$$

for all x, y, in its domain, obtained from setting $\lambda = \frac{1}{2}$ in (1).

For example, the function φ defined by

$$\varphi(x) = \begin{cases} 1 & \text{if } x = 0 \\ 0 & \text{if } 0 < x < 1 \end{cases}$$

is convex on $[0, 1)$ and is discontinuous at the included end point.

PROBLEM 2. If $\varphi : I \to E^1$ is convex and $a < b$ are points of I, then φ is bounded above on $[a, b]$ by $M = \max[|\varphi(a)|, |\varphi(b)|]$.

By adding $-\varphi(t) + \varphi(t)$ to the left numerator in (2), the right inequality in

$$s < u < t \Rightarrow \frac{\varphi(u)-\varphi(s)}{u-s} \leqslant \frac{\varphi(t)-\varphi(s)}{t-s} \leqslant \frac{\varphi(t)-\varphi(u)}{t-u} \tag{4}$$

is obtained. The left inequality in (4) follows by adding $-\varphi(s) + \varphi(s)$ to the right numerator in (2).

Theorem 1. *If $\varphi : I \to E^1$ is convex on I, then φ is continuous on the interior I^0 of I.*

[*Note:* Continuity was probably first seen intuitively from a figure showing the ratios in (4) as slopes. If $s < x < u$, then the point $(x, \varphi(x))$ lies in triangle AUS, but in triangle BUT if $u < x < t$, and in either case approaches U as $x \to u$.]

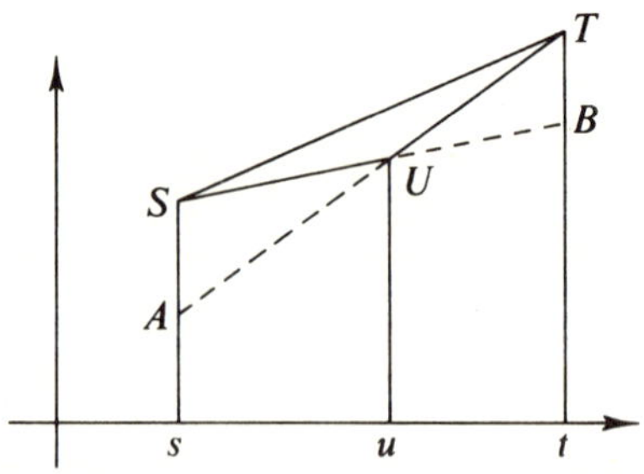

Proof. Take any $x_0 \in I^0$, then a and b also in I^0 such that $a < x_0 < b$. From Problem 2, $\hat{\varphi}(x_0) < \infty$ and since $\varphi(x_0) \leqslant \hat{\varphi}(x_0)$, we have that $\hat{\varphi}(x_0)$ is finite. Determine sequences $\{x_n\}$ and $\{y_n\}$ both in (a, b), both converging to x_0, and such that

$$\lim_{n\to\infty} \varphi(x_n) = \underset{\vee}{\varphi}(x_0) \qquad \text{and} \qquad \lim_{n\to\infty} \varphi(y_n) = \hat{\varphi}(x_0).$$

[For example, if $\underset{\vee}{\varphi}(x_0) = \varphi(x_0)$, choose each $x_n = x_0$, but if $\underset{\vee}{\varphi}(x_0) < \varphi(x_0)$, make $\varphi(x_n) \to \underline{\varphi}(x_0)$.] Let $z_n = 2y_n - x_n$. Hence $z_n \to x_0$ and, by the midpoint property (3) of convex functions,

$$2\varphi(y_n) \leqslant \varphi(x_n) + \varphi(z_n)$$

for all n sufficiently large to make $z_n \in I$. Hence

$$\begin{aligned} 2\hat{\varphi}(x_0) &= \lim_{n\to\infty} 2\varphi(y_n) \leqslant \underline{\lim}_{n\to\infty} \{\varphi(x_n) + \varphi(z_n)\} \\ &\leqslant \underline{\lim}_{n\to\infty} \varphi(x_n) + \overline{\lim}_{n\to\infty} \varphi(z_n) \qquad \text{by Theorem 3–2.4} \\ &\leqslant \lim_{n\to\infty} \varphi(x_n) + \overline{\lim}_{t\to x_0} \varphi(z_n) \leqslant \underset{\vee}{\varphi}(x_0) + \hat{\varphi}(x_0). \end{aligned}$$

Thus $\hat{\varphi}(x_0) \leqslant \underset{\vee}{\varphi}(x_0) \leqslant \hat{\varphi}(x_0)$ and φ is continuous at x_0 by Problem 1. ▌

PROBLEM 3. Let $\varphi : I \to E^1$ be convex, fix a point u on the interior I^0 of I, and define (using only $s \in I^0$)

$$B = \sup_{s<u} \frac{\varphi(u) - \varphi(s)}{u - s}.$$

Then (with u still fixed):

(a) $B \leqslant \inf\limits_{u<t\in I^0} \dfrac{\varphi(t) - \varphi(u)}{t - u}$.

(b) $s, t \in I^0, s < u < t \Rightarrow \dfrac{\varphi(u) - \varphi(s)}{u - s} \leqslant B \leqslant \dfrac{\varphi(t) - \varphi(u)}{t - u}$.

(c) $v \in I^0 \Rightarrow \varphi(v) \geqslant \varphi(u) + B(v - u)$.

In the following statement, E^1 is the space of domains (as well as ranges) and Lebesgue measure m is used for the integrals, although domains (except for φ) could be in any space on which a measure μ is defined.

Theorem 2. *Given a measurable set $A \subset E^1$, let $f : A \to E^1$ and $p : A \to E^1$ be measurable functions such that pf is summable on A, $p(x) \geqslant 0$ for $x \in A$, and p is summable on A with $\int_A p(x)\, dm > 0$. If $\varphi : E^1 \to E^1$ is a convex function, then Jensen's* ***integral inequality***

$$\varphi\left\{\frac{\int_A p(x) f(x)\, dm}{\int_A p(x)\, dm}\right\} \leqslant \frac{\int_A p(x)\varphi(f(x))\, dm}{\int_A p(x)\, dm}$$

holds. [The domain of φ could be any $(c, d) \supset f(A)$.]

Proof. Since φ is continuous (Theorem 1) and f is measurable, the composition function $\varphi(f)$ is measurable, by Theorem 6–2.10, and the integrand in the numerator on the right is measurable.

All integrals are taken over A. Set

$$u = \frac{\int (pf)}{\int p}.$$

Corresponding to this number u, let B be as defined in Problem 3, and then use $v=f(x)$ there. Hence for $x \in A$,

$$\varphi(f(x)) \geqslant \varphi(u) + B\{f(x) - u\} \qquad \text{and}$$

$$\frac{1}{\int p} p(x)\,\varphi(f(x)) \geqslant \frac{\varphi(u)}{\int p} p(x) + B\left\{\frac{1}{\int p} p(x) f(x) - \frac{u}{\int p} p(x)\right\}.$$

The measurable function on the left is thus bounded below by a summable function. Hence, by integrating both sides, the following terms all have meaning; those on the right all finite, the one on the left either finite or $+\infty$:

$$\frac{1}{\int p}\int p\varphi(f) \geqslant \frac{\varphi(u)}{\int p}\int p + B\left\{\frac{1}{\int p}\int (pf) - \frac{u}{\int p}\int p\right\}$$

$$= \varphi(u) + B\{u - u\} = \varphi(u),$$

which is the desired inequality. ∎

In some developments, a function is said to be convex if it has the midpoint property (3), and then additional conditions are obtained to ensure that the function has the properties (1) and (2). There are facts of interest in this approach, some of which are given below. To keep things straight, we shall use the nonstandard terminology in

Definition 2. *A function* $\varphi : I \to E^1$ *is said to be* **midconvex** *on* I, *provided*

$$x, y \in I \Rightarrow \varphi\left(\frac{x+y}{2}\right) \leqslant \frac{\varphi(x) + \varphi(y)}{2}.$$

Hence, if φ is convex, then it is midconvex.

Definition 3. *A function* $f : I \to E^1$ *is said to be* **locally bounded above** *on* I *if each point of the interior* I^0 *is interior to a subinterval on which* f *is bounded above.*

For example, $f(x) = 1/x$ for $x > 0, f(0) = 2$ is locally bounded above on $[0, \infty)$ and midconvex (even convex) on $(0, \infty)$.

PROBLEM 4. Let $\varphi : I \to E^1$ be midconvex on I^0. Then φ is continuous on I^0 iff φ is locally bounded above on I.

[*Hint:* Check that the proof of Theorem 1 has already established this.]

PROBLEM 5. If φ is midconvex on an interval I and $x_k \in I$ for $k = 1, 2, 3, \cdots, n$, then

$$\varphi\left(\frac{x_1 + x_2 + \cdots + x_n}{n}\right) \leqslant \frac{1}{n}\{\varphi(x_1) + \varphi(x_2) + \cdots + \varphi(x\ )\}.$$

[*Hint:* Much as in Hint 1 of Problem 35, page 176, first establish the result with n replaced by $2, 4, \cdots, 2^m$. Then, with $2^m > n$, set

$$\sigma = \frac{1}{n}(x_1 + \cdots + x_n) = \frac{(x_1 + \cdots + x_n) + (2^m - n)\sigma}{2^m}.$$

With the interpretation $x_k = \sigma$ for $k = n+1, n+2, \cdots, 2^m$, obtain, from the formula with n replaced by 2^m,

$$\varphi(\sigma) \leqslant \frac{\varphi(x_1) + \cdots + \varphi(x_n) + (2^m - n)\varphi(\sigma)}{2^m}$$

from which the desired formula follows.]

PROBLEM 6. If the conditions of Problem 5 are satisfied and in addition $r_1, r_2, \cdots, r_n$ are nonnegative rationals such that $r_1 + r_2 + \cdots + r_n > 0$, then

$$\varphi\left(\frac{r_1 x_1 + r_2 x_2 + \cdots + r_n x_n}{r_1 + r_2 + \cdots + r_n}\right) \leqslant \frac{r_1 \varphi(x_1) + r_2 \varphi(x_2) + \cdots + r_n \varphi(x_n)}{r_1 + r_2 + \cdots + r_n}.$$

[*Hint:* As an illustration

$$\varphi\left(\frac{\frac{1}{2}x_1 + \frac{2}{3}x_2}{\frac{1}{2} + \frac{2}{3}}\right) = \varphi\left(\frac{3x_1 + 4x_2}{7}\right) \leqslant \frac{1}{7}\{\varphi(x_1) + \varphi(x_1) + \varphi(x_1) + 4\varphi(x_2)\}$$

$$= \frac{\frac{1}{2}\varphi(x_1) + \frac{2}{3}\varphi(x_2)}{\frac{1}{2} + \frac{2}{3}}.]$$

PROBLEM 7. Let φ be midconvex and also continuous on (a, b), $-\infty \leqslant a < b \leqslant \infty$ [which it will be iff φ is locally bounded above on (a, b)]. If $x_k \in (a, b)$, $k = 1, 2, \cdots, n$, and if $p_1, p_2, \cdots, p_n$ are nonnegative numbers with positive sum, then **Jensen's summation formula,**

$$\varphi\left(\frac{p_1 x_1 + p_2 x_2 + \cdots + p_n x_n}{p_1 + p_2 + \cdots + p_n}\right) \leqslant \frac{p_1 \varphi(x_1) + p_2 \varphi(x_2) + \cdots + p_n \varphi(x_n)}{p_1 + p_2 + \cdots + p_n},$$

holds. In particular,

$$x, y \in (a, b), 0 \leqslant \lambda \leqslant 1 \Rightarrow \varphi((1-\lambda)x + \lambda y) \leqslant (1-\lambda)\varphi(x) + \lambda\varphi(y).$$

[*Hint:* If p_1 is irrational, then in the formula of Problem 6 let $r_1 \to p_1$, etc., valid since φ is continuous.]

It thus follows that

A real-valued function is convex on an open interval iff the function is midconvex and locally bounded above on this open interval.

The geometric-arithmetic mean inequality (Problem 35, page 176) is a special case of Jensen's inequality. To see this, we use elementary properties of logarithmic and exponential functions (taken up again in the next chapter). Note that the function φ defined by $\varphi(x) = e^x = \exp(x)$ is convex on $(-\infty, \infty)$. Set $p_k = 1$. Hence

$$\exp\left(\frac{x_1 + x_2 + \cdots + x_n}{n}\right) \leqslant \frac{e^{x_1} + e^{x_2} + \cdots + e^{x_n}}{n}.$$

Now with $y_1, y_2, \cdots, y_n$ given positive numbers, let

$$x_k = \ln y_k \qquad \text{so that} \qquad e^{x_k} = y_k$$

and $(y_1 y_2 \cdots y_n)^{1/n} \leqslant \dfrac{1}{n}(y_1 + y_2 + \cdots + y_n)$.

Jensen's integral inequality was originally obtained by detailed analysis of limiting processes starting with the summation inequality.

On the other hand, with $x_1, x_2, \cdots, x_n$ given, let μ be the measure defined by

$$\mu\{x_k\}=1, \qquad k=1,2,\cdots,n, \qquad \mu(A)=0 \qquad \text{if } A\cap\{x_1,\cdots,x_n\}=\varnothing.$$

Then Jensen's integral inequality (with μ in place of m) yields

$$\left\{\frac{\sum p(x_k)f(x_k)}{\sum p(x_k)}\right\} \leqslant \frac{\sum p(x_k)\varphi(f(x_k))}{\sum p(x_k)},$$

and then Jensen's summation inequality follows by specializing the functions f and p.

PROBLEM 8. (a) A set $A\subset E^1$ and its negative $-A=\{x\mid -x\in A\}$ have

$$m^*A=m^*(-A).$$

(b) Given a number $c>0$, if $B\subset E^1$ has the property that for each λ, $0\leqslant\lambda\leqslant 1$, at least one of the points $-\lambda c$ or λc is in B, then

$$m^*B\geqslant c.$$

[*Hint:* If $A_1=B\cap[-c,0]$, $A_2=B\cap[0,c]$, then $(-A_1)\cup A_2=[0,c]$.]

Theorem 3. *A function $\varphi:(a,b)\to E^1$ is convex on (a,b) iff φ is midconvex and measurable on (a,b).*

Proof. Since a continuous function is measurable and a midconvex function is continuous on an open interval iff it is locally bounded above on the interval, the proof will be complete upon showing that the assumption

φ is midconvex and measurable on (a, b) but
is not locally bounded above on (a, b)

leads to a contradiction. Under this assumption, let $x_0\in(a,b)$ be such that φ is unbounded above on every open interval containing x_0. Take $c>0$ such that

$$[x_0-2c, x_0+2c]\subset(a,b).$$

With n a positive integer, take $u\in(x_0-c, x_0+c)$ and $\varphi(u)>n$. Then, by midconvexity of φ,

$$n<\varphi(u)\leqslant\tfrac{1}{2}\{\varphi(u-\lambda c)+\varphi(u+\lambda c)\}, \qquad 0\leqslant\lambda\leqslant 1.$$

Hence at least one of $\varphi(u-\lambda c)$ and $\varphi(u+\lambda c)$ is greater than n for every $0\leqslant\lambda\leqslant 1$. Thus upon defining

$$M_n=\{x\mid\varphi(x)>n\}\cap[x_0-2c, x_0+2c],$$

then M_n is measurable and $mM_n\geqslant c$, by Problem 8 applied at the point u instead of the origin. Then $M_1\supset M_2\supset M_3\supset\cdots$ and

$$m\left(\bigcap M_n\right)=\lim_{n\to\infty} mM_n\geqslant c>0$$

by Theorem 4–2.10 (since $mM_1\leqslant 4c<\infty$). Hence $\bigcap M_n\neq\varnothing$. But

$$x\in\bigcap M_n\Rightarrow\varphi(x)>n, \qquad n=1,2,3,\cdots\Rightarrow\varphi(x)=+\infty,$$

contrary to all values of φ being in E^1. ▮

PROBLEM 9. Let $f: E^1 \to E^1$ be such that for all $x, y \in E^1$,

$$f(x+y) = f(x) + f(y).$$

(a) Then f is midconvex on E^1.
(b) For every rational r, $f(rx) = rf(x)$.
(c) If, in addition, f is measurable, then not only is f continuous, but also

$$f(x) = xf(1) \qquad \text{for all } x \in E^1.$$

For comments about nonmeasurable solutions of the functional equation in Problem 9, see page 343.

Chapter 7

DERIVATIVES

7–1 DINI DERIVATIVES

Definition 1. *Given a function $f\colon I \to E^1$, with $I \subset E^1$ an interval, the functions*

$$\underline{f}'\colon I \to E^* \qquad \text{and} \qquad \overline{f}'\colon I \to E^*$$

defined for each $x \in I$ by

$$\underline{f}'(x) = \varliminf_{s \to x} \frac{f(s) - f(x)}{s - x} \qquad \text{and} \qquad \overline{f}'(x) = \varlimsup_{s \to x} \frac{f(s) - f(x)}{s - x}$$

are called the **lower** *and* **upper Dini derivatives** *of f; it being understood that if the lower end point or upper end point of I is included in I, then one-sided limits are taken at these points. At any $x \in I$ where*

$$-\infty < \underline{f}'(x) = \overline{f}'(x) < \infty,$$

the function f is said to be **differentiable** *at x with derivative $f'(x) = \underline{f}'(x) = \overline{f}'(x)$. We also say that $f'(x)$ exists* if $\underline{f}'(x) = \overline{f}'(x)$, finite or infinite.*

EXAMPLE 1. If $f(x) = |x|$, then $\underline{f}'(0) = -1$, $\overline{f}'(0) = 1$, and f is differentiable at $x \neq 0$, with

$$f'(x) = -1 \quad \text{if } x < 0 \qquad \text{and} \qquad f'(x) = 1 \quad \text{if } x > 0.$$

Of course we may use

$$\frac{f(x + h) - f(x)}{h} \qquad \text{instead of} \qquad \frac{f(s) - f(x)}{s - x}.$$

Theorem 1. *If $f\colon I \to E^1$ is differentiable at an $x \in I$, then f is continuous at this x.*

Proof. Let $\delta > 0$ be such that $s \in I$ and $0 < |s - x| < \delta \Rightarrow$

$$\left|\frac{f(s) - f(x)}{s - x} - f'(x)\right| < 1 \Rightarrow |f(s) - f(x)| < |s - x|\{|f'(x)| + 1\}.$$

Hence, as $s \to x$ in I, $f(s) \to f(x)$. ∎

EXAMPLE 2. Let

$$f(x) = \begin{cases} x & \text{if } x \text{ is rational, including } x = 0 \\ x^2 + x & \text{if } x \text{ is irrational.} \end{cases}$$

Then f is discontinuous, and hence nondifferentiable, at each $x \neq 0$, but

$$\frac{f(x) - f(0)}{x - 0} = \begin{cases} 1 & \text{if } x \neq 0 \text{ is rational} \\ x + 1 & \text{if } x \text{ is irrational} \end{cases}$$

so that f is differentiable at 0 with $f'(0) = 1$.

The converse of Theorem 1 is not true; in fact the next theorem will be used in the following example of a function which is continuous on E^1 but is nondifferentiable at each $x \in E^1$. In Theorem 2 we use the result, which follows from previous limit theorems, that if $f'(x)$ exists and $\{x_n\}$ is a sequence in the domain of f such that $x_n \neq x$ with $x_n \to x$, then

$$\lim_{n \to \infty} \frac{f(x_n) - f(x)}{x_n - x} = f'(x).$$

Theorem 2. *Given a function $f: (a, b) \to E^1$ which is differentiable at an $x \in (a, b)$, let $\{\alpha_n\}$ and $\{\beta_n\}$ be any sequences such that $a < \alpha_n < x < \beta_n < b$, $\alpha_n \to x$, and $\beta_n \to x$ as $n \to \infty$. Then*

$$\lim_{n \to \infty} \frac{f(\beta_n) - f(\alpha_n)}{\beta_n - \alpha_n} = f'(x). \tag{1}$$

Proof. Set $\lambda_n = (\beta_n - x)/(\beta_n - \alpha_n)$. Then $0 < \lambda_n < 1$ and

$$\frac{f(\beta_n) - f(\alpha_n)}{\beta_n - \alpha_n} - f'(x) = \lambda_n\left\{\frac{f(\beta_n) - f(x)}{\beta_n - x} - f'(x)\right\} + (1 - \lambda_n)\left\{\frac{f(\alpha_n) - f(x)}{\alpha_n - x} - f'(x)\right\}.$$

The expressions within the braces both $\to 0$ as $n \to \infty$ and the sequences $\{\lambda_n\}$ and $\{1 - \lambda_n\}$ are both bounded. Hence the limit in (1) exists and the equality holds. ∎

Corollary. *If there are sequences $\{\alpha_n\}$ and $\{\beta_n\}$, $a < \alpha_n \leqslant x < \beta_n < b$, $\alpha_n \to x$, $\beta_n \to x$, such that*

$$\lim_{n \to \infty} \frac{f(\beta_n) - f(\alpha_n)}{\beta_n - \alpha_n} \textit{ does not exist,}$$

then f is nondifferentiable at x.

Proof. Eliminate all terms of $\{\alpha_n\}$ and $\{\beta_n\}$ for which $\alpha_n = x$. If infinite subsequences are left, a contrapositive proof using these subsequences establishes the result. If only a finite number of terms are left, then use only $\{\beta_n\}$ in a contrapositive proof. ∎

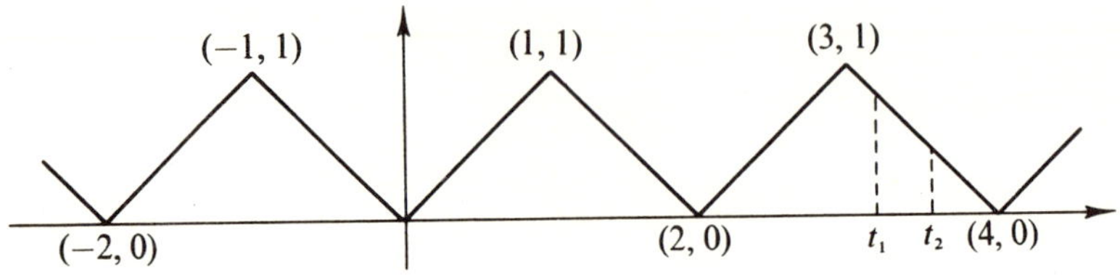

FIG. 1

EXAMPLE 3. We start with the continuous function g of period 2 whose graph is indicated by three periods in Fig. 1. Note, for later use, that if t_1 and t_2 have no integer between them, then

$$|g(t_2) - g(t_1)| = |t_2 - t_1|. \tag{2}$$

Select a number c, $0 < c < 1$, and an even positive integer e. Let f be the function defined for each real number x by

$$f(x) = \sum_{m=0}^{\infty} c^m g(e^m x). \tag{3}$$

Since $0 \leqslant g \leqslant 1$ and $\sum c^n$ converges, the function f is continuous by Theorems 6–3.4 and 6–3.6.

We shall show that c and e may be selected (independent of any x) such that f is nondifferentiable at each $x \in (-\infty, \infty)$.

Select any $x \in (-\infty, \infty)$. In terms of the greatest integer function [ ], we define

$$k_n = [e^n x], \qquad n = 1, 2, 3, \cdots.$$

Hence $k_n \leqslant e^n x < k_n + 1$, and upon setting

$$\alpha_n = \frac{k_n}{e^n} \qquad \text{and} \qquad \beta_n = \frac{k_n + 1}{e^n} = \alpha_n + \frac{1}{e^n},$$

we have $\alpha_n \leqslant x < \beta_n$. It then follows from

$$\overline{\lim_{n\to\infty}}\, \alpha_n \leqslant x \leqslant \underline{\lim_{n\to\infty}}\, \beta_n = \underline{\lim_{n\to\infty}} \left(\alpha_n + \frac{1}{e^n}\right) = \underline{\lim_{n\to\infty}}\, \alpha_n$$

that both $\alpha_n \to x$ and $\beta_n \to x$ as $n \to \infty$. With m a positive integer, we observe that

$$e^m \alpha_n = e^m \frac{k_n}{e^n} \qquad \text{and} \qquad e^m \beta_n = e^m \frac{k_n + 1}{e^n} \qquad \text{are}$$

(a) Both even if $m > n$ (the reason for having e even), and
(b) There is no integer actually between them if $m \leqslant n$.

To see (b), suppose l is an integer such that

$$e^m \frac{k_n}{e^n} < l < e^m \frac{k_n + 1}{e^n}, \qquad m \leqslant n.$$

Then $k_n < l(e^n/e^m) < k_n + 1$, which is impossible, since the middle term is an integer (for $m \leqslant n$) between two consecutive integers.

Hence from (a) and (b),

$$(4) \qquad |g(e^m\beta_n) - g(e^m\alpha_n)| = \begin{cases} |0 - 0| & \text{if } m > n \\ |e^m\beta_n - e^m\alpha_n| = e^m/e^n & \text{if } m \leqslant n, \end{cases}$$

the lower part following from (2). Therefore

$$\begin{aligned} f(\beta_n) - f(\alpha_n) &= \sum_{m=0}^{n} c^m\{g(e^m\beta_n) - g(e^m\alpha_n)\} \\ &= c^n\{g(k_n + 1) - g(k_n)\} + \sum_{m=0}^{n-1} c^m\{g(e^m\beta_n) - g(e^m\alpha_n)\}. \end{aligned}$$

From $|A + B_1 + B_2 + \cdots + B_k| \geqslant |A| - |B_1| - |B_2| - \cdots - |B_k|$ for any numbers $A, B_1, B_2, \cdots, B_k$, and the lower part of (4),

$$\begin{aligned} |f(\beta_n) - f(\alpha_n)| &\geqslant c^n - \sum_{m=0}^{n-1} c^m \frac{e^m}{e^n} \\ &= c^n - \frac{1}{e^n}\frac{(ce)^n - 1}{ce - 1} \qquad \text{if } ce \neq 1. \end{aligned}$$

The first restriction we put on c and e is that $ce > 1$, and then have

$$\begin{aligned} \frac{|f(\beta_n) - f(\alpha_n)|}{\beta_n - \alpha_n} &\geqslant (ce)^n - \frac{(ce)^n}{ce - 1} + \frac{1}{ce - 1} \\ &> (ce)^n \frac{ce - 2}{ce - 1}. \end{aligned}$$

This last term $\to \infty$ as $n \to \infty$ if $ce > 2$. For definiteness we choose $c = \frac{3}{4}$ and $e = 4$, which in no way depends upon the x we are using. For this choice, the function f defined by (3) is not only continuous, but is also nondifferentiable at each $x \in E^1$ by the above corollary. ∎

Now that the relevant limit theorems have been proved, the proof of the following theorem is by the same methods used in elementary calculus.

Theorem 3. *Given that f: $I \to E^1$ and g: $I \to E^1$ are both differentiable at an $x \in I$, then $f + g$ and fg are differentiable at x with*

$$(f + g)'(x) = f'(x) + g'(x) \qquad \text{and}$$
$$(fg)'(x) = f(x)g'(x) + g(x)f'(x).$$

If $g(x) \neq 0$, then also f/g is differentiable at x, with

$$\left(\frac{f}{g}\right)'(x) = \frac{g(x)f'(x) - f(x)g'(x)}{g^2(x)}.$$

Since the function f defined by $f(x) = x$ has $f'(x) = 1$, we use mathematical induction to obtain, in the familiar derivative notation, that

$$\frac{dx^n}{dx} = nx^{n-1}, \qquad n \text{ a positive integer.}$$

Theorem 4. *Let f: $[a, b] \to E^1$ have a point $x \in [a, b]$ at which $f'(x)$ exists* and $f'(x) > 0$. Then there are points a_1 and b_1 such that $x \in (a_1, b_1)$,*

$$s \in (a_1, x) \cap [a, b] \Rightarrow f(s) < f(x), \qquad \textit{and}$$
$$t \in (x, a_2) \cap [a, b] \Rightarrow f(x) < f(t).$$

These inequalities are reversed if $f'(x) < 0$.

[*Note:* This does not say that f is monotonic on (a_1, b_1). A counterexample is to take f the function of Example 4 below and define F by $F(x) = \frac{1}{2}x + f(x)$. Then $F'(0) = \frac{1}{2}$, but F is not monotonic in any interval including the origin.]

Proof. Take ε such that $0 < \varepsilon < f'(x)$ and let $\delta > 0$ be such that for $s \neq x$ and

$$s \in (x - \delta, x + \delta] \cap [a, b] \Rightarrow \begin{cases} \varepsilon < \dfrac{f(s) - f(x)}{s - x} & \text{if } f'(x) = \infty \\[2ex] f'(x) - \varepsilon < \dfrac{f(s) - f(x)}{s - x} & \text{if } f'(x) < \infty. \end{cases}$$

Hence if $(x - \delta, x) \cap [a, b] \neq \varnothing$ and s is in this set, then $s - x < 0$ so that also $f(s) - f(x) < 0$. The other conclusion follows in a similar way. ∎

PROBLEM 1. If f is defined on a neighborhood of a point x_0, if $f(x_0)$ is a local extreme value of f, and if $f'(x_0)$ exists, then $f'(x_0) = 0$.

Definition 2. *We shall use the notation $\underline{f}'_-(x)$ defined by*

$$\underline{f}'_-(x) = \varliminf_{s \to x-} \frac{f(s) - f(x)}{s - x}$$

and† similarly for $\underline{f}'_+(x)$, $\bar{f}'_-(x)$, and $\bar{f}'_+(x)$. Then $f'_-(x)$ exists if $\underline{f}'_-(x) = \bar{f}'_-(x)$ and is equal to this common value. These are called the* **Dini left** *and* **right derivatives.**

† Note that this is not the same as

$$\underline{f}'(x-) = \varliminf_{s \to x-} f'(s).$$

Lemma 1. *If $f\colon I \to E^1$ is differentiable at each point of $(a, b) \subset I$, if $f'_+(a)$ and $f'_-(b)$ exist with $f'_+(a) < 0 < f'_-(b)$, then there exists an $x_0 \in (a, b)$ such that $f'(x_0) = 0$. The same conclusion follows if $f'_+(a) > 0 > f'_-(b)$.*

Proof. From Theorem 1, f is continuous on $[a, b]$ and thus has a minimum value $f(x_0)$ with $x_0 \in [a, b]$, by Theorem 6–2.5. But $s_1, s_2 \in [a, b]$ exist with $f(s_1) < f(a)$ and $f(s_2) < f(b)$ by Theorem 4. Thus $x_0 \in (a, b)$. Then the assumptions $f'(x_0) < 0$ and $f'(x_0) > 0$ both lead to contradictions, by Theorem 4, and hence $f'(x_0) = 0$. ∎

In Example 1 the derived function f' is discontinuous at the one point where it fails to exist. The following example gives a function which is differentiable everywhere, but f' is discontinuous at one point. In this example we use results from elementary calculus, which will be established later.

EXAMPLE 4

$$f(x) = \begin{cases} x^2 \sin(1/x) & \text{if } x \neq 0 \\ 0 & \text{if } x = 0, \end{cases}$$

$$f'(x) = \begin{cases} -\cos(1/x) \neq 2x \sin(1/x) & \text{if } x \neq 0 \\ 0 & \text{if } x = 0, \end{cases}$$

where the bottom line follows, since

$$\left| \frac{f(x) - f(0)}{x - 0} \right| = \left| x \sin \frac{1}{x} \right| \leqslant |x|.$$

But f' is not continuous at $x = 0$; for example, upon setting $x_n = 1/(2n\pi)$, then $x_n \to 0$ as $n \to \infty$, but

$$\lim_{n \to \infty} f'(x_n) = -1 \neq f'(0).$$

This example shows that a function f may be differentiable everywhere, but have f' discontinuous. The next theorem shows, however, that if f' exists on an interval, then f' shares the intermediate value property with continuous functions.

Theorem 5. *If $f\colon [a, b] \to E^1$ is differentiable on $[a, b]$, then f' has the intermediate value property on $[a, b]$.*

Proof. Let $a \leqslant a_1 < b_1 \leqslant b$ be such that $f'(a_1) < f'(b_1)$ and choose any number c such that $f'(a_1) < c < f'(b_1)$. Set

$$g(x) = f(x) - cx, \qquad a_1 \leqslant x \leqslant b_1.$$

Then $g'(x) \doteq f'(x) - c$ for $a_1 < x < b_1$, with

$$g'_+(a_1) = f'(a_1) - c < 0 \qquad \text{and} \qquad g'_-(b_1) = f'(b_1) - c > 0.$$

Hence, by the lemma, there is an $x_0 \in (a_1, b_1)$ such that

$$0 = g'(x_0) = f'(x_0) - c.$$

Thus $f'(x_0) = c$, which means that f' has the intermediate-value property, since if $f'(a_1) > f'(b_1)$, we would apply the above result to $-f$, and if f' is constant on $[a, b]$ there is nothing to prove. ∎

Corollary. *If $f\colon [a, b] \to E^1$ is differentiable on $[a, b]$, then f' has no discontinuity of the first kind on $[a, b]$.*

If f is defined at all points of $[a, b]$ and $f \in \mathscr{L}[a, b]$, it may be that

$$f(t) \neq \frac{1}{b-a}\int_a^b f(x)\,dx \qquad \text{for all } t \in (a, b).$$

Such a function on $[-1, 1]$ is $f(x) = x$ for $x \neq 0$, $f(0) = 1$. The following problem, however, states a second (and hence first) mean-value theorem for integrals of derived functions. This result will be used in Section 7–10.

PROBLEM 2. Let f and g be such that f' exists on (a, b), $f'g \in \mathscr{L}[a, b]$ and $g \in \mathscr{L}[a, b]$. If either $g \geqslant 0$ or $g \leqslant 0$ a.e. in (a, b), then there is a $\xi \in (a, b)$ such that

$$f'(\xi)\int_a^b g(x)\,dx = \int_a^b f'(x)g(x)\,dx.$$

[*Hint:* Proof of Theorem 6–2.8.]

Theorem 6. *With $-\infty \leqslant a < b \leqslant \infty$, let $f\colon (a, b) \to E^1$ be strictly increasing on (a, b) and let $x_0 \in (a, b)$ be such that $f'(x_0)$ exists with $f'(x_0) \neq 0$. Then the inverse function $\varphi = f^{-1}$ is differentiable at $f(x_0)$ and*

$$\varphi'(f(x_0)) = \frac{1}{f'(x_0)}.$$

The same result holds if f is strictly decreasing on (a, b).

Proof. Since f is strictly increasing on (a, b), then f^{-1} exists on (α, β), where $\alpha = f(a+)$ and $\beta = f(b-)$, and is strictly increasing on (α, β). Thus $\alpha < f(x_0) < \beta$ and f^{-1} is continuous at $f(x_0)$, by Theorem 6–2.9. Let $\{y_n\}$ be any sequence such that $\alpha < y_n < \beta$, $y_n \neq f(x_0)$, and $y_n \to f(x_0)$ as $n \to \infty$. Since f^{-1} is continuous at $f(x_0)$, then

$$\lim_{n\to\infty} f^{-1}(y_n) = f^{-1}(f(x_0)) = x_0.$$

Hence, by Theorem 7-1.2, since $f'(x_0)$ exists,

$$f'(x_0) = \lim_{n\to\infty}\frac{f(f^{-1}(y_n)) - f(x_0)}{f^{-1}(y_n) - x_0} = \lim_{n\to\infty}\frac{y_n - f(x_0)}{f^{-1}(y_n) - f^{-1}(f(x_0))}.$$

But, since $f'(x_0) \neq 0$, the limit of the reciprocal also exists and

$$\lim_{n\to\infty}\frac{f^{-1}(y_n) - f^{-1}(f(x_0))}{y_n - f(x_0)} = \frac{1}{f'(x_0)}.$$

Since this is true of any sequence $\{y_n\}$ as described, then by the $(b) \Rightarrow (a)$ part of Theorem 6–1.2,

$$\varphi'(f(x_0)) = \lim_{y \to f(x_0)} \frac{f^{-1}(y) - f^{-1}(f(x_0))}{y - f(x_0)} = \frac{1}{f'(x_0)}.$$

The same method yields the result if f is strictly decreasing on (a, b). ∎

With n a positive integer, the function f defined by

$$f(x) = x^n, \qquad x > 0$$

is such that $f\colon (0, \infty) \to (0, \infty)$ is strictly increasing on its domain and $f'(x) = nx^{n-1} > 0$, $x \in (0, \infty)$. Thus the inverse function

$$\varphi = f^{-1}\colon (0, \infty) \to (0, \infty)$$

is differentiable on its domain, and for x in tne domain of f,

$$\varphi'(f(x)) = \frac{1}{f'(x)} = \frac{1}{nx^{n-1}} = \frac{x}{nx^n}.$$

We now switch notation and think of $y \in (0, \infty)$ in the domain of f. Then $x = y^n = f(y)$ is in the domain of $\varphi = f^{-1}$ and, since $x^{1/n} = y$, we have

$$\varphi'(x) = \varphi'(f(y)) = \frac{y}{ny^n} = \frac{x^{1/n}}{nx} = \frac{1}{n} x^{(1/n)-1}, \qquad x > 0.$$

This formula in differential notation is

$$\frac{dx^{1/n}}{dx} = \frac{1}{n} x^{(1/n)-1}$$

Theorem 7. *Given that $f\colon [a, b] \to E^1$ is continuous on $[a, b]$, let I be an interval containing the range of f and let $g\colon I \to E^1$. If there is an $x \in [a, b]$ such that f is differentiable at x and g is differentiable at $f(x)$, then the composition function $h = g \circ f$ is differentiable at x and*

$$h'(x) = g'(f(x)) \cdot f'(x); \text{ that is,}$$
$$(g \circ f)'(x) = (g' \circ f)(x) \cdot f'(x).$$

Proof. Set $y = f(x)$ and define φ by

$$\varphi(t) = \begin{cases} \dfrac{g(t) - g(y)}{t - y} & \text{for } t \in I, t \neq y \\ g'(y) & \text{for } t = y. \end{cases}$$

From the upper portion we have

$$\lim_{t \to y} \varphi(t) = g'(y) \qquad \text{which} = \varphi(y)$$

by the lower portion. Thus φ is continuous at y. Also

$$g(t) - g(y) = (t - y)\varphi(t), \qquad t \in I$$

whether $t \neq y$ or $t = y$. Hence, upon replacing t by $f(s)$ and using $y = f(x)$,

$$\frac{g(f(s)) - g(f(x))}{s - x} = \frac{f(s) - f(x)}{s - x}\varphi(f(s)); \qquad s \in [a, b], \quad s \neq x.$$

By the continuity of φ at $y = f(x)$, the continuity of f at x, and the existence of $f'(x)$, we have the existence of the following limit and equality:

$$\lim_{s \to x} \frac{h(s) - h(x)}{s - x} = f'(x)g'(f(x)); \text{ that is,}$$

$$h'(x) = g'(f(x)) \cdot f'(x). \quad \blacksquare$$

With p and q positive integers, the functions f and g defined by

$$f(x) = x^{1/q} \qquad \text{and} \qquad g(x) = x^p, \qquad x > 0$$

are differentiable at each $x > 0$,

$$f'(x) = \frac{1}{q}x^{(1/q)-1}, \qquad g'(x) = px^{p-1}, \qquad \text{and } g'(f(x)) = p(x^{1/q})^{p-1}.$$

Hence $h = g \circ f$ is differentiable at each $x > 0$ and

$$h'(x) = p(x^{1/q})^{p-1} \cdot \frac{1}{q}x^{(1/q)-1} = \frac{p}{q}x^{(p/q)-1}.$$

Since $h(x) = (g \circ f)(x) = (x^{1/q})^p = x^{p/q}$, $x > 0$, we have that if r is a positive rational number, then

$$\frac{dx^r}{dx} = rx^{r-1}, \qquad x > 0.$$

In particular, with r a positive rational,

$$\lim_{x \to 1} \frac{x^r - 1}{x - 1} = r. \tag{5}$$

Let α be irrational and $0 < \alpha < 1$. Given a number ε such that $0 < \varepsilon < \alpha < 1 - \varepsilon$, select rational numbers r_1 and r_2 such that

$$\alpha - \varepsilon < r_1 < \alpha < r_2 < \alpha + \varepsilon.$$

Hence

(a) if $0 < x < 1$, then $1 > x^{r_1} > x^\alpha > x^{r_2} > 0$;

(b) if $1 < x$, then $1 < x^{r_1} < x^\alpha < x^{r_2}$;

and in either case

$$\frac{x^{r_1} - 1}{x - 1} < \frac{x^\alpha - 1}{x - 1} < \frac{x^{r_2} - 1}{x - 1}.$$

As $x \to 1$ the limits of the outside members exist by (5) and we have

$$\alpha - \varepsilon < r_1 \leqslant \underline{\lim}_{x \to 1} \frac{x^\alpha - 1}{x - 1} \leqslant \overline{\lim}_{x \to 1} \frac{x^\alpha - 1}{x - 1} \leqslant r_2 < \alpha + \varepsilon.$$

Therefore the existence of the limit and the equality

$$\lim_{x \to 1} \frac{x^\alpha - 1}{x - 1} = \alpha$$

follow, and a similar proof shows this to hold if α is irrational and $1 < \alpha$.

Thus for any positive real number a,

$$\lim_{t \to 1} \frac{t^a - 1}{t - 1} = a,$$

and then for any $x > 0$,

$$\lim_{s \to x} \frac{s^a - x^a}{s - a} = \lim_{s \to x} \frac{(s/x)^a - 1}{(s/x) - 1} x^{a-1} = ax^{a-1}.$$

We thus have, for a any positive number, that

$$\frac{dx^a}{dx} = ax^{a-1}, \qquad x > 0.$$

PROBLEM 2. This formula also holds for any negative real number a.

7–2 MEAN VALUES

A function f is said to have a **local minimum** at a point x_0 of its domain if there is an open interval (α, β) with $x_0 \in (\alpha, \beta)$ such that

$$f(s) \geqslant f(x_0)$$

for each s of the domain of f and $s \in (\alpha, \beta)$. A **local maximum** is defined in the same way with $f(s) \leqslant f(x_0)$.

Theorem 1. (Rolle's Theorem) *If h: $[a, b] \to E^1$ is continuous on $[a, b]$ and differentiable on (a, b), and if $h(a) = h(b) = 0$, then there is an $s \in (a, b)$ such that $h'(s) = 0$.*

Proof. If h is constant on $[a, b]$, then any $s \in (a, b)$ satisfies the conclusion. Hence consider h not constant. Then h has a nonzero maximum or minimum value (by Theorem 6–2.5) $h(s)$ for some $s \in [a, b]$. But $h(s) \neq 0$, while $h(a) = h(b) = 0$ and thus $s \in (a, b)$. Then the assumptions $h'(s) > 0$ and $h'(s) < 0$ both lead to contradictions (by Theorem 7-1.4). Thus $h'(s) = 0$. ▮

Theorem 2. (Generalized Law of the Mean) *Let f and g be continuous real-valued functions on the closed interval $[a, b]$ and differentiable on the open interval (a, b). Then there is an $s \in (a, b)$ such that*

$$\{f(b) - f(a)\}g'(s) = \{g(b) - g(a)\}f'(s).$$

Proof. Define $h(x)$ for $a \leqslant x \leqslant b$ by

$$h(x) = \{f(b) - f(a)\}\{g(x) - g(a)\} - \{g(b) - g(a)\}\{f(x) - f(a)\}.$$

A check will show that the function h satisfies the hypotheses of Theorem 1 and hence there is an $s \in (a, b)$ such that $h'(s) = 0$, which gives the desired result. ∎

Theorem 3. (Law of the Mean) *If f: $[a, b] \to E^1$ is continuous on $[a, b]$ and differentiable on (a, b), then there is an $s \in (a, b)$ such that*

$$f(b) = f(a) + (b - a)f'(s).$$

Proof. Set $g(x) = x$ in Theorem 2. ∎

PROBLEM 1. Let f be continuous at a point c and differentiable at all points of a neighborhood of c except (possibly) at c. If $\lim_{x\to c} f'(x)$ exists, then also $f'(c)$ exists and

$$\lim_{x\to c} f'(x) = f'(c).$$

PROBLEM 2. Let f: $[a, b] \to E^1$ be differentiable on $[a, b]$. If for all $x \in [a, b]$:

(a) $f'(x) \geqslant 0$, then f is nondecreasing.
(b) $f'(x) \leqslant 0$, then f is nonincreasing.
(c) $f'(x) = 0$, then f is constant.
(d) f' is bounded, then f is Lipschitzian.

PROBLEM 3. If $f'(x)$ exists on $[0, 1]$, then

$$f(1) - f(0) = \frac{f'(x)}{2x}$$

has a solution on $(0, 1)$.

PROBLEM 4. (a). If f' is continuous on $[a, b]$, then f is uniformly differentiable on $[a, b]$; that is, given an $\varepsilon > 0$, there is a $\delta > 0$ such that

$$t, x \in [a, b], \qquad 0 < |t - x| < \delta \Rightarrow \left| \frac{f(t) - f(x)}{t - x} - f'(x) \right| < \varepsilon.$$

(b). If f is uniformly differentiable on $[a, b]$, then f' is continuous on $[a, b]$.

The proof of the following theorem is written in such a way that no special consideration need be given if $a = -\infty$ or $b = \infty$.

Theorem 4. (l'Hospital's Rule) *With* $-\infty \leqslant a < b \leqslant \infty$, *let* f *and* g *be differentiable functions on* (a, b) *such that*

(1) $$a < x < b \Rightarrow g'(x) \neq 0.$$

If either

(2) $$\lim_{x \to a+} f(x) = \lim_{x \to a+} g(x) = 0 \qquad or$$

(3) $$\lim_{x \to a+} |g(x)| = \infty,$$

then

(4) $$\underline{\lim}_{x \to a+} \frac{f'(x)}{g'(x)} \leqslant \underline{\lim}_{x \to a+} \frac{f(x)}{g(x)} \leqslant \overline{\lim}_{x \to a+} \frac{f(x)}{g(x)} \leqslant \overline{\lim}_{x \to a+} \frac{f'(x)}{g'(x)}.$$

Also $x \to a+$ *may be replaced by* $x \to b-$ *throughout.*

Proof. Note that

(5) $$a < x < y < b \Rightarrow g(x) \neq g(y),$$

for if the equality held at two points on (a, b), then there would be (by Theorem 3) an s between them where $g'(s) = 0$, contrary to (1). In particular, $g(x) = 0$ has at most one solution on (a, b) and we take c_1 such that $a < c_1 < b$ and

(6) $$a < x < c_1 \Rightarrow g(x) \neq 0.$$

Set

$$\underline{A} = \underline{\lim}_{x \to a+} \frac{f'(x)}{g'(x)}.$$

The first inequality in (4) holds if $\underline{A} = -\infty$, and hence consider

$$-\infty < \underline{A}.$$

Let q be an arbitrary number such that

$$-\infty < q < \underline{A}.$$

Determine a number c, $a < c < c_1$, such that

$$a < s < c \Rightarrow q < \frac{f'(s)}{g'(s)}.$$

With x and y arbitrary except $a < x < y < c$, there is (by Theorem 2) a number s such that

$$x < s < y \qquad \text{and} \qquad \frac{f(y) - f(x)}{g(y) - g(x)} = \frac{f'(s)}{g'(s)}$$

in which the denominators are not zero by (5) and (1). Hence

(7) $$a < x < y < c \Rightarrow q < \frac{f(y) - f(x)}{g(y) - g(x)}.$$

Case 1. (2) holds. Let $x \to a+$ and see, from (2), that

$$a < y < c \Rightarrow q \leqslant \frac{f(y)}{g(y)},$$

in which $g(y) \neq 0$, by (6). Consequently

$$q \leqslant \underline{\lim}_{y \to a+} \frac{f(y)}{g(y)}.$$

Since q was any number $< \underline{A}$, then

$$\underline{A} = \underline{\lim}_{x \to a+} \frac{f'(x)}{g'(x)} \leqslant \underline{\lim}_{x \to a+} \frac{f(x)}{g(x)},$$

and hence the first inequality of (4) holds under condition (2).

Case 2. (3) holds. Fix y such that $a < y < c$. Determine d such that $a < d < y$ and, by using (3), such that

$$a < x < d \Rightarrow 0 < 1 - \frac{g(y)}{g(x)}.$$

Then, from (7), $a < x < d \Rightarrow$

$$q < \frac{f(y) - f(x)}{g(y) - g(x)} = \frac{f(x) - f(y)}{g(x) - g(y)} = \frac{\dfrac{f(x)}{g(x)} - \dfrac{f(y)}{g(x)}}{1 - \dfrac{g(y)}{g(x)}}.$$

Hence, since the last denominator above is positive,

$$q\left\{1 - \frac{g(y)}{g(x)}\right\} + \frac{f(y)}{g(x)} < \frac{f(x)}{g(x)} \qquad \text{for } a < x < d.$$

As $x \to a+$, the left side has q as its limit (and hence limit inferior) because (3) is holding. Therefore, by taking the limit inferior of both sides,

$$q \leqslant \underline{\lim}_{x \to a+} \frac{f(x)}{g(x)}.$$

Again the first inequality in (4) is seen to hold.

With the first inequality of (4) established under either (2) or (3), then (upon replacing f by $-f$, and therefore f' by $-f'$)

$$\underline{\lim}_{x \to a+} \frac{-f'(x)}{g'(x)} \leqslant \underline{\lim}_{x \to a+} \frac{-f(x)}{g(x)}; \qquad \text{that is,}$$

$$-\overline{\lim}_{x \to a+} \frac{f'(x)}{g'(x)} \leqslant -\overline{\lim}_{x \to a+} \frac{f(x)}{g(x)}$$

from which the last inequality of (4) follows. Since the middle inequality of (4) is known, the theorem is proved. ∎

Corollary. *In addition, if*

$$\lim_{x \to a+} \frac{f'(x)}{g'(x)} \text{ exists* and} = L, \text{ then } \lim_{x \to a+} \frac{f(x)}{g(x)} = L.$$

PROBLEM 5. If x is interior to the domain of f and $f''(x)$ exists (but not necessarily anywhere else), then

$$\lim_{h \to 0} \frac{f(x+h) + f(x-h) - 2f(x)}{h^2} = f''(x).$$

Give an example for which the limit exists but $f''(x)$ does not.

Theorem 5. *If F: $[a, b] \to E^1$ and G: $[a, b] \to E^1$ are both differentiable on $[a, b]$ and $F'(x) = G'(x)$ for $x \in [a, b]$, then there is a constant c such that*

$$F(x) = G(x) + c \qquad \text{for } a \leqslant x \leqslant b.$$

Proof. Define the function H by setting

$$H(x) = F(x) - G(x), \qquad a \leqslant x \leqslant b.$$

Then H satisfies the hypotheses of Theorem 3 on $[a, x]$ for each $x \in (a, b]$. Thus for each $x \in (a, b]$ there is an $s_x \in (a, x)$ such that

$$H(x) = H(a) + (x - a)H'(s_x) = H(a),$$

since $H'(s_x) = F'(s_x) - G'(s_x) = 0$. Thus

$$F(x) - G(x) = F(a) - G(a), \qquad a \leqslant x \leqslant b,$$

from which the conclusion follows with $c = F(a) - G(a)$. ∎

We know that if f: $[a, b] \to E^1$ is continuous on $[a, b]$, then $f \in \mathscr{L}([a, b])$, by Theorem 6–2.6, and there is an $s \in (a, b)$ such that

$$\int_a^b f(x)\,dx = (b - a)f(s),$$

by the corollary of Theorem 6–2.8. These facts are used in proving the fundamental theorem of calculus for Lebesgue integrals of continuous functions. This theorem is sufficient for many purposes, including the next section. Conditions under which continuity may be relaxed are given in Sec. 7–7.

Theorem 6. *Given a continuous function f: $[a, b] \to E^1$, there is a differentiable function F: $[a, b] \to E^1$ such that $F'(x) = f(x)$, $a \leqslant x \leqslant b$. Also if G is any function such that $G'(x) = f(x)$, $a \leqslant x \leqslant b$, then*

$$\int_a^b f(x)\,dx = G(b) - G(a). \tag{8}$$

Proof. Let F be the function defined by

$$F(x) = \int_a^x f(u)\,du, \qquad a \leqslant x \leqslant b.$$

Then F is continuous (actually, absolutely continuous) on $[a, b]$ by the "if" part of Theorem 6–5.2. Also, for $x_1, x_2 \in [a, b]$,

$$F(x_2) - F(x_1) = \int_a^{x_2} f(u)\,du - \int_a^{x_1} f(u)\,du = \int_{x_1}^{x_2} f(u)\,du$$

by Theorem 4-5.1.

Let $\varepsilon > 0$ be given. From the uniform continuity of f on $[a, b]$, determine $\delta > 0$ such that

$$(s, t \in [a, b], |s - t| < \delta) \Rightarrow |f(s) - f(t)| < \varepsilon.$$

Take any $x \in [a, b]$ and then $h \neq 0$ such that $|h| < \delta$ and $x + h \in [a, b]$. Then

$$\left|\frac{F(x+h) - F(x)}{h} - f(x)\right| = \left|\frac{1}{h}\int_x^{x+h} f(t)\,dt - f(x)\right|$$
$$= |f(s) - f(x)| < \varepsilon$$

for some s on the interval with end points x and $x + h$. Hence $F'(x)$ exists and

$$F'(x) = f(x), \qquad a \leqslant x \leqslant b.$$

Let $a = x_0 < x_1 < x_2 < \cdots < x_{n-1} < x_n = b$ be such that (for the δ determined above)

$$x_k - x_{k-1} < \delta, \qquad k = 1, 2, \cdots, n.$$

Then there are $s_k, t_k \in (x_{k-1}, x_k)$ such that

$$\int_{x_{k-1}}^{x_k} f(x)\,dx = (x_k - x_{k-1})f(s_k) \qquad \text{and}$$
$$F(x_k) - F(x_{k-1}) = (x_k - x_{k-1})F'(t_k) = (x_k - x_{k-1})f(t_k).$$

Hence

$$\left|\int_a^b f(x)\,dx - [F(b) - F(a)]\right| = \left|\sum_{k=1}^n \left\{\int_{x_{k-1}}^{x_k} f(x)\,dx - [F(x_k) - F(x_{k-1})]\right\}\right|$$
$$= \left|\sum_{k=1}^n (x_k - x_{k-1})[f(s_k) - f(t_k)]\right|$$
$$\leqslant \sum_{k=1}^n (x_k - x_{k-1})\varepsilon = \varepsilon(b - a).$$

Since $\varepsilon > 0$ is arbitrary,

$$\int_a^b f(x)\,dx = F(b) - F(a)$$

and then (8) follows from Theorem 5. ∎

Corollary 1. *If* $\varphi\colon [a, b] \to E^1$ *has* φ' *continuous on* $[a, b]$, *then*

$$\int_a^b \varphi'(t)\, dt = \varphi(b) - \varphi(a).$$

Corollary 2. *If* f' *and* g' *are continuous on* $[a, b]$, *then the integration by parts formula*

$$\int_a^b f(x)g'(x)\, dx + \int_a^b g(x)f'(x)\, dx = f(b)g(b) - f(a)g(a)$$

holds.

Proof. The function $\varphi = fg$ has $\varphi' = (fg)' = fg' + gf'$ continuous on $[a, b]$ and the result follows by Corollary 1. ∎

The change of variable of integration formula of the next theorem will be obtained later (Theorem 7–8.2) under less stringent conditions.

Theorem 7. *Given* $f\colon [a, b] \to E^1$ *continuous on* $[a, b]$ *,then*

$$\int_a^b f(x)\, dx = \int_c^d f(\varphi(t))\varphi'(t)\, dt$$

for any function $\varphi\colon [c, d] \leftrightarrow [a, b]$ *which is strictly increasing and has* φ' *continuous on* $[c, d]$, *where* $c = \varphi^{-1}(a)$ *and* $d = \varphi^{-1}(b)$.

Proof. The first integral exists, since f is continuous on $[a, b]$; and the second integral exists, since $(f \circ \varphi)\varphi'$ is continuous on $[c, d]$.

Given an $\varepsilon > 0$, let $\delta > 0$ be such that

$$(s, x \in [a, b], |s - x| < \delta) \Rightarrow |f(s) - f(x)| < \varepsilon,$$

which is possible because f is uniformly continuous on $[a, b]$. With n a positive integer such that

$$\frac{b-a}{n} < \delta, \qquad \text{set } x_i = a + i\,\frac{(b-a)}{n}, \qquad i = 0, 1, 2, \cdots, n.$$

Let m_i be the minimum and M_i the maximum of f on $[x_{i-1}, x_i]$ and $s_i \in (x_{i-1}, x_i)$ such that

$$\int_{x_{i-1}}^{x_i} f(x)\, dx = f(s_i)(x_i - x_{i-1}), \qquad i = 1, 2, \cdots, n.$$

Then $0 \leqslant M_i - f(s_i) \leqslant \varepsilon$ and $0 \leqslant f(s_i) - m_i \leqslant \varepsilon$. Also set

$$t_i = \varphi^{-1}(x_i), \qquad i = 0, 1, 2, \cdots, n.$$

Then $c = t_0 < t_1 < t_2 < \cdots < t_n = d$ and, by the above Corollary 1,

$$\int_{t_{i-1}}^{t_i} \varphi'(t)\, dt = \varphi(t_i) - \varphi(t_{i-1}) = x_i - x_{i-1}.$$

Furthermore, $t \in [t_{i-1}, t_i] \Rightarrow m_i \leqslant f(\varphi(t)) \leqslant M_i$ and hence

$$[f(s_i) - \varepsilon](x_i - x_{i-1}) \leqslant m_i \int_{t_{i-1}}^{t_i} \varphi'(t)\, dt = \int_{t_{i-1}}^{t_i} m_i \varphi'(t)\, dt$$

$$\leqslant \int_{t_{i-1}}^{t_i} f(\varphi(t)\varphi'(t)\, dt \leqslant [f(s_i) + \varepsilon](x_i - x_{i-1}).$$

Hence by summing from $i = 1$ to n,

$$\int_a^b f(x)\, dx - \varepsilon(b - a) \leqslant \int_c^d f(\varphi(t))\varphi'(t)\, dt \leqslant \int_a^b f(x)\, dx + \varepsilon(b - a).$$

Since $\varepsilon > 0$ is arbitrary, the desired equality holds. ∎

PROBLEM 6. Given f: $[a, b] \to E^1$ continuous on $[a, b]$, then

$$\int_a^b f(x)\, dx = \int_c^d f(\varphi(t))\varphi'(t)\, dt$$

for any function φ: $[d, c] \leftrightarrow [a, b]$ which is strictly decreasing and has φ' continuous on $[d, c]$.

This problem has the following corollary and, of course, there is a similar corollary of Theorem 7.

Corollary 1. *With* $-\infty \leqslant d < c < \infty$, *let* φ: $(d, c] \to E^1$ *satisfy*:

(a) φ *is strictly decreasing on* $(d, c]$.
(b) φ' *exists and is continuous on* $(d, c]$.
(c) $\lim_{y \to d+} \varphi(y) = \infty$.

Upon letting $\varphi(c) = a$, *if* f: $[a, \infty) \to E^1$ *is continuous and* $f \in \mathscr{L}([a, \infty))$, *then*

$$\int_a^\infty f(x)\, dx = \lim_{y \to d+} \int_c^y f(\varphi(t))\varphi'(t)\, dt. \tag{9}$$

[*Note:* The integral on the left in (9) means

$$\int_{[a,\infty)} f\, dm \tag{10}$$

and is not an "improper" Lebesgue integral. Recall, however, that $f \in \mathscr{L}([a, \infty))$ not only means that f is measurable with (10) finite, but also that

$$\int_{[a,\infty)} |f|\, dm < \infty.]$$

Proof. Let $\{d_n\}$ be any sequence such that

$$c > d_1 > d_2 > \cdots > d \qquad \text{and} \qquad d_n \to d. \tag{11}$$

Set $b_n = \varphi(d_n)$. Then, since φ is strictly decreasing,

$$a < b_1 < b_2 < \cdots \qquad \text{and} \qquad b_n \to \infty.$$

For each positive integer n define

$$f_n(x) = \begin{cases} f(x) & \text{for } a \leqslant x \leqslant b_n \\ 0 & \text{for } b_n < x < \infty. \end{cases}$$

Then each function f_n is measurable on $[a, \infty)$ and

$$\int_{[a,\infty)} |f_n|\, dm \leqslant \int_{[a,\infty)} |f|\, dm < \infty,$$

and thus $f_n \in \mathscr{L}([a, \infty))$. Also $\lim_{n\to\infty} f_n(x) = f(x)$ for all $x \in [a, \infty)$. Thus, by Theorem 4–5.9,

$$\lim_{n\to\infty} \int_{[a,\infty)} f_n\, dm = \int_{[a,\infty)} f\, dm.$$

Hence, in the other notation,

$$\int_a^\infty f(x)\, dx = \lim_{n\to\infty} \int_a^\infty f_n(x)\, dx = \lim_{n\to\infty} \int_a^{b_n} f(x)\, dx.$$

By Problem 4, in which the intervals are finite,

$$\lim_{n\to\infty} \int_a^{b_n} f(x)\, dx = \lim_{n\to\infty} \int_c^{\varphi^{-1}(b_n)} f(\varphi(t))\varphi'(t)\, dt$$

and thus, since $d_n = \varphi^{-1}(b_n)$,

$$\int_a^\infty f(x)\, dx = \lim_{n\to\infty} \int_c^{d_n} f(\varphi(t))\varphi'(t)\, dt.$$

Since this is true of any sequence $\{d_n\}$ satisfying (11), then (9) follows. ∎

EXAMPLE 1. Let

$$f(x) = \frac{1}{1 + x^2} \qquad \text{for } x \in E^1.$$

The function g defined by

$$g(x) = \frac{1}{1 + n^2} \qquad \text{for } n \leqslant x < n + 1, \quad n = 1, 2, 3, \cdots$$

is measurable on E^1 and it is seen that

$$g \in \mathscr{L}([1, \infty)), \qquad \text{since } \sum_{n=1}^{\infty} \frac{1}{1 + n^2} \text{ converges.}$$

From $0 < f(x) \leqslant g(x)$, $1 \leqslant x < \infty$, and Theorem 4-5.6, it follows that also $f \in \mathscr{L}([1, \infty))$. The above corollary then justifies the formal substitution $x = 1/t$ to yield

$$\int_1^\infty \frac{1}{1+x^2}\,dx = \lim_{y\to 0+} \int_1^y \frac{1}{1+(1/t)^2}\left(-\frac{1}{t^2}\right)dt$$

$$= \lim_{y\to 0+} \int_y^1 \frac{1}{1+t^2}\,dt = \int_0^1 \frac{1}{1+t^2}\,dt.$$

We therefore have the existence of the integrals and the equalities in

$$\int_0^1 \frac{1}{1+x^2}\,dx = \int_1^\infty \frac{1}{1+x^2}\,dx = \frac{1}{2}\int_0^\infty \frac{1}{1+x^2}\,dx.$$

In these the substitution $x = -t$ then leads to

$$\int_{-1}^0 \frac{1}{1+t^2}\,dt = \int_{-\infty}^{-1} \frac{1}{1+t^2}\,dt = \frac{1}{2}\int_{-\infty}^0 \frac{1}{1+t^2}\,dt.$$

Finally the existence of the integral and equality

$$\int_{-\infty}^\infty \frac{1}{1+x^2}\,dx = \frac{1}{4}\int_0^1 \frac{1}{1+x^2}\,dx$$

is obtained.

These facts will be used in the next section.

We now discuss a few derivative properties of real component vector functions of a real variable represented by

$$\mathbf{f}(x) = (f_1(x), f_2(x), \cdots, f_k(x)).$$

Definition 1. *Given* $\mathbf{f}$: $[a, b] \to E^k$, *where* f_i: $[a, b] \in E^1$, $i = 1, 2, \cdots, k$, *if each* $f_i'(x)$ *exists at some* $x \in [a, b]$, *then* $\mathbf{f}$ *is said to have*

$$\mathbf{f}'(x) = (f_1'(x), f_2'(x), \cdots, f_k'(x))$$

as its **derivative** *at* x.

Hence if $\mathbf{f}$ has a derivative x, then each component function f_i' is continuous at x and thus $\mathbf{f}$ is continuous at x.

The following example, using elementary calculus, shows that the mean-value theorem need not hold for vector functions.

EXAMPLE 2. Given $\mathbf{f}(x) = \cos x + i \sin x$, considered as a function in E^2, then $\mathbf{f}'(x) = -\sin x + i\cos x$. Assume there is an $a < b$ and an s, $a < s < b$, such that

$$\mathbf{f}(b) - \mathbf{f}(a) = (b - a)\mathbf{f}'(s).$$

Then $|\mathbf{f}(b) - \mathbf{f}(a)|^2 = (b-a)^2|\mathbf{f}'(s)|^2 = (b-a)^2$. Since the left side is

$$(\cos b - \cos a)^2 + (\sin b - \sin a)^2 = 4 \sin^2 \tfrac{1}{2}(b - a),$$

we have

$$\sin^2 \tfrac{1}{2}(b - a) = [\tfrac{1}{2}(b - a)]^2,$$

which cannot hold, since $h \neq 0 \Rightarrow \sin h \neq \pm h$.

Vector-valued functions do, however, have a mean-value property, as stated in the following theorem. We will use Theorem 7–1.2 with f replaced by $\mathbf{f}$, trusting that the proof with this replacement is checked.

Theorem 8. *If* $\mathbf{f}$: $[a, b] \to E^k$ *is continuous on* $[a, b]$ *and* $\mathbf{f}'$ *exists on* (a, b), *then there is an* $s \in (a, b)$ *such that*

$$|\mathbf{f}(b) - \mathbf{f}(a)| \leqslant (b - a)|\mathbf{f}'(s)|.$$

Proof. Set $q = |\mathbf{f}(b) - \mathbf{f}(a)|$, $l = \tfrac{1}{3}(b - a)$ and define a function $\mathbf{g}$ by

$$\mathbf{g}(x) = \mathbf{f}(x + l) - \mathbf{f}(x), \qquad a \leqslant x \leqslant a + 2l.$$

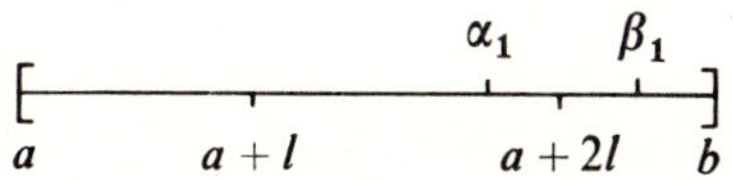

Then $\mathbf{f}(b) - \mathbf{f}(a) = \mathbf{g}(a) + \mathbf{g}(a + l) + \mathbf{g}(a + 2l)$ and hence

(12) $$q \leqslant |\mathbf{g}(a)| + |\mathbf{g}(a + l)| + |\mathbf{g}(a + 2l)|.$$

We now assert that there is an α_1 such that

(13) $$a < \alpha_1 < a + 2l \qquad \text{and} \qquad |\mathbf{g}(\alpha_1)| \geqslant \tfrac{1}{3}q.$$

For assume that $x \in (a, a + 2l) \Rightarrow |\mathbf{g}(x)| < \tfrac{1}{3}q$. Then $|\mathbf{g}(a + l)| < \tfrac{1}{3}q$ and, by the continuity of $\mathbf{g}$ (since $\mathbf{f}$ is continuous),

$$|\mathbf{g}(a)| \leqslant \tfrac{1}{3}q \qquad \text{and} \qquad |\mathbf{g}(a + 2l)| \leqslant \tfrac{1}{3}q,$$

which makes the right side of (12) less than q. Thus (13) holds.

Take α_1 for which (13) holds, set $\beta_1 = \alpha_1 + l$, and then from (13) see that

$$|\mathbf{f}(\beta_1) - \mathbf{f}(\alpha_1)| \geqslant \tfrac{1}{3}q, \qquad a < \alpha_1 < \beta_1 < b, \quad \text{and} \quad \beta_1 - \alpha_1 = \tfrac{1}{3}(b - a).$$

Let $q_1 = |\mathbf{f}(\beta_1) - \mathbf{f}(\alpha_1)|$, apply the above technique to the interval $[\alpha_1, \beta_1]$, and obtain α_2, β_2 such that

$$|\mathbf{f}(\beta_2) - \mathbf{f}(\alpha_2)| \geqslant \tfrac{1}{3}q_1 \geqslant \tfrac{1}{9}q, \qquad \alpha_1 < \alpha_2 < \beta_2 < \beta_1,$$

and

$$\beta_2 - \alpha_2 = \tfrac{1}{3}(\beta_1 - \alpha_1) = \tfrac{1}{9}(b - a).$$

By continuing in this manner, a sequence $\{I_n\}$ of closed intervals

$$(a, b) \supset I_1 \supset I_2 \supset I_3 \supset \cdots$$

is obtained with $I_n = [\alpha_n, \beta_n]$ such that

$$|\mathbf{f}(\beta_n) - \mathbf{f}(\alpha_n)| \geqslant 3^{-n}q \qquad \text{and} \qquad 0 < \beta_n - \alpha_n = 3^{-n}(b - a).$$

By division,

$$\frac{|\mathbf{f}(\beta_n) - \mathbf{f}(\alpha_n)|}{\beta_n - \alpha_n} \geqslant \frac{q}{b-a} = \frac{|\mathbf{f}(b) - \mathbf{f}(a)|}{b-a}.$$

Let $s = \bigcap I_n$, there being one and only one such point. Since $a < \alpha_1 < s < \beta_1 < b$ then $s \in (a, b)$ so that $\mathbf{f}'(s)$ exists. Also, $\alpha_n < s < \beta_n$ with $\alpha_n \to s$ and $\beta_n \to s$ so that, by Theorem 7-1.2 with f replaced by $\mathbf{f}$,

$$|\mathbf{f}'(s)| \geqslant \frac{|\mathbf{f}(b) - \mathbf{f}(a)|}{b-a},$$

from which the desired inequality follows. ∎

Corollary. *Let* $\mathbf{f}\colon [a, b] \to E^k$ *be continuous on* $[a, b]$ *and* $\mathbf{f}'$ *exist on* (a, b) *except possibly at a single point* $c \in (a, b)$. *If*

$$\lim_{x \to c} \mathbf{f}'(x) = \mathbf{u} \text{ exists,}$$

then $\mathbf{f}$ *also has a derivative at* c *with* $\mathbf{f}'(c) = \mathbf{u}$.

Proof. Given a number $\varepsilon > 0$, let $\delta > 0$ be such that $0 < |s| < \delta \Rightarrow c + s \in (a, b)$ and $|\mathbf{f}'(c + s) - \mathbf{u}| < \varepsilon$. Define $\mathbf{g}$ by

$$\mathbf{g}(x) = \mathbf{f}(x) - x\mathbf{u} \qquad \text{for } a \leqslant x \leqslant b.$$

Then $\mathbf{g}$ is continuous on $[c - \delta, c]$ and $\mathbf{g}'$ exists on $(c - \delta, c)$. Hence, by the theorem, corresponding to each h, $0 < h < \delta$, there is an $s \in (c - h, c)$ such that

$$|\mathbf{g}(c - h) - \mathbf{g}(c)| \leqslant h\,|\mathbf{g}'(s)|; \qquad \text{that is,}$$

$$\left|\frac{\mathbf{f}(c - h) - \mathbf{f}(c)}{-h} - \mathbf{u}\right| \leqslant |\mathbf{f}'(s) - \mathbf{u}| < \varepsilon.$$

Similarly, this inequality is obtained with $-h$ replaced by h. Hence $\mathbf{f}'(c)$ exists and is equal to $\mathbf{u}$. ∎

7-3 TRIGONOMETRY

In a few places we have already used the trigonometric function, but only in examples to illustrate concepts and never as a part of the theory. We have followed this policy since the very foundation of trigonometry is not firm, and it could not have been made so in time to have this useful tool available in elementary mathematics. Trigonometry with angles (themselves intuitive) led to the concept of trigonometric functions of a real variable, radians, etc., and in this extension it was assumed that:

1. An arc of a circle has a length.
2. Arcs of any length can be laid off on a circle.
3. Arc length is given by the usual formula.

Although 1. and 3. are demonstrated in some elementary books, 2. is usually overlooked.

We make no claim that the following development† would have been thought of without long experience on the intuitive level. Everything we do, moreover, is with an eye on the properties already attributed to the trigonometric functions.

From this puristic point of view, the ratio of circumference to diameter definition of π cannot be used, since we feel that 1., 2., and 3. have not been established (at least not in this book).

Definition 1

$$\pi = \int_{-\infty}^{\infty} \frac{1}{1+u^2}\, du.$$

Recall that the integral used in this definition was shown to exist in Section 7–2. Also, the equations given there now yield

$$\frac{\pi}{2} = \int_{0}^{\infty} \frac{1}{1+u^2}\, du = \int_{-\infty}^{0} \frac{1}{1+u^2}\, du, \qquad \text{and}$$

$$\frac{\pi}{4} = \int_{0}^{1} \frac{1}{1+u^2}\, du = \int_{1}^{\infty} \frac{1}{1+u^2}\, du.$$

Furthermore, the integrals in the definition of a function p by

$$p(t) = \int_{-t}^{t} \frac{1}{1+u^2}\, du = 2\int_{0}^{t} \frac{1}{1+u^2}\, du, \qquad -\infty < t < \infty$$

exist. We shall use this function p in defining the trigonometric functions, but first note some of its properties.

The absolute continuity of p follows from this property of Lebesgue integrals.

The function p is differentiable and

$$p'(t) = 2\frac{d}{dt}\int_{0}^{t} \frac{1}{1+u^2}\, du = \frac{2}{1+t^2}, \qquad -\infty < t < \infty.$$

† Due in essentially this form to W. F. Eberlein, "The Circular Function(s)," *Mathematics Magazine*, Vol. 39 (1966), pp. 197–201.

Hence p is strictly increasing on $(-\infty, \infty)$ so its inverse p^{-1} exists and is differentiable on the range of p. Since

$$\lim_{t\to-\infty} 2\int_0^t \frac{1}{1+u^2}\,du = 2\int_0^{-\infty} \frac{1}{1+u^2}\,du = -2\int_{-\infty}^0 \frac{1}{1+u^2}\,du = -\pi$$

and

$$\lim_{t\to\infty} p(t) = \pi,$$

we have $p\colon (-\infty, \infty) \leftrightarrow (-\pi, \pi)$ and $p^{-1}\colon (-\pi, \pi) \leftrightarrow (-\infty, \infty)$. Also

$$p(-t) = 2\int_0^{-t} \frac{1}{1+u^2}\,du = -2\int_0^t \frac{1}{1+u^2}\,du = p(t), \qquad -\infty < t < \infty.$$

Thus p is an odd function and hence so is p^{-1}.

Graphs of $y = 1/(1+u^2)$ and $\theta = p(t)$ are given on two superimposed coordinate systems in Fig. 1. It is recommended, however, that no attempt be made to visualize geometric relations at each step of the following analytic development; in fact, the purpose of this development is to avoid intuitive geometry.

In the following definition, E^2 is identified with the complex number field.

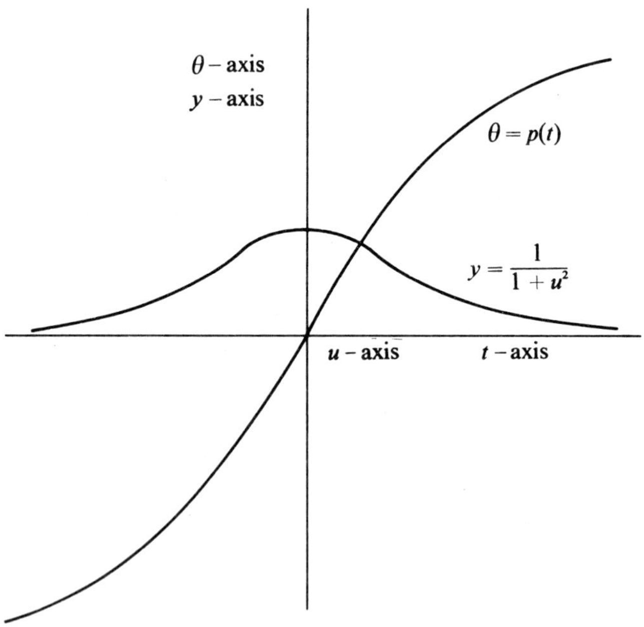

FIG. 1

Definition 2. *The function* cis: $E^1 \to E^2$ *is defined by*

(i) $$\operatorname{cis}\theta = \frac{1+ip^{-1}(\theta)}{1-ip^{-1}(\theta)} \qquad \textit{for } -\pi < \theta < \pi.$$

(ii) $\operatorname{cis} \pi = \operatorname{cis}(-\pi) = -1.$

(iii) cis *is periodic of period* 2π; *that is,* $\operatorname{cis}(\theta + 2\pi) = \operatorname{cis} \theta, \ -\infty < \theta < \infty.$

Upon relating θ and t by $\theta = p(t)$, $t = p^{-1}(\theta)$, then

$$\operatorname{cis} \theta = \frac{1 + it}{1 - it}, \qquad -\infty < \theta < \infty,$$

and since $t \to \infty$ as $\theta \to \pi-$ and $t \to -\infty$ as $\theta \to -(\pi+)$, we have

(1) $$\lim_{\theta \to \pi-} \operatorname{cis} \theta = \lim_{\theta \to -(\pi+)} \operatorname{cis} \theta = -1.$$

This, with (ii) and (iii), shows that cis is continuous on E^1. Also

$$\left| \frac{1 + it}{1 - it} \right| = 1, \qquad -\infty < t < \infty.$$

Consequently

$$|\operatorname{cis} \theta| = 1 \qquad \text{and} \qquad \operatorname{cis} \theta \neq 0 \qquad \text{for all } -\infty < \theta < \infty.$$

Since $p(0) = 0$ and $p(1) = \pi/2$, then $0 = p^{-1}(0)$ and $1 = p^{-1}(\pi/2)$. Thus, from (i),

(2) $$\operatorname{cis} 0 = 1 \qquad \text{and} \qquad \operatorname{cis} \frac{\pi}{2} = \frac{1 + i}{1 - i} = i.$$

Then, from (iii), for $m = 0, \pm 1, \pm 2, \pm 3, \cdots,$

$$\operatorname{cis} 2m\pi = 1 \qquad \text{and} \qquad \operatorname{cis}\left(2m\pi + \frac{\pi}{2}\right) = i.$$

For $-\pi < \theta < \pi$ and $t = p^{-1}(\pi)$, by taking derivatives,

$$\operatorname{cis}' \theta = \frac{\dfrac{d}{dt}\left(\dfrac{1 + it}{1 - it}\right)}{\dfrac{d\theta}{dt}} = \frac{\dfrac{2i}{(1 - it)^2}}{\dfrac{2}{1 + t^2}} = i\,\frac{1 + it}{1 - it} = i \operatorname{cis} \theta.$$

Hence, by the periodicity property (iii), we have

$$\operatorname{cis}' \theta = i \operatorname{cis} \theta \qquad \text{for } \theta \neq m\pi, \qquad m = \pm 1, \pm 2, \cdots.$$

Thus we know that $\operatorname{cis}' \theta$ exists at all points in a neighborhood of π except possibly at π. Note, however, that by using (1),

$$\lim_{\theta \to \pi} \operatorname{cis}' \theta = i \lim_{\theta \to \pi} \operatorname{cis} \theta = -i.$$

Therefore, by the corollary of Theorem 7–2.8, $\operatorname{cis}' \pi$ also exists and $\operatorname{cis}' \pi = -i$. Again by using (iii), the same holds for any positive or negative integer multiple of π, and thus

(3) $$\operatorname{cis}' \theta = i \operatorname{cis} \theta, \qquad -\infty < \theta < \infty.$$

Consequently, in the notation of differential equations, if a is any constant and $y = a \operatorname{cis} \theta$, then

$$\frac{dy}{d\theta} = iy. \tag{4}$$

Theorem 1. *If $f\colon E^1 \to E^2$ is any function satisfying the differential equation* (4), *then*

$$f(\theta) = f(0) \operatorname{cis} \theta, \qquad -\infty < \theta < \infty. \tag{5}$$

Proof. Since $f'(\theta) = i f(\theta)$ and $\operatorname{cis}' \theta = i \operatorname{cis} \theta$,

$$\frac{d}{d\theta} \frac{f(\theta)}{\operatorname{cis} \theta} = \frac{(\operatorname{cis} \theta) \cdot if(\theta) - f(\theta) \cdot (i \operatorname{cis} \theta)}{(\operatorname{cis} \theta)^2} = 0.$$

Thus there is a constant c such that $f(\theta)/\operatorname{cis} \theta = c$; that is,

$$f(\theta) = c \operatorname{cis} \theta.$$

Hence $f(0) = c \operatorname{cis} 0 = c$ and (5) follows. ∎

For any fixed $\varphi \in E^1$, the function f defined by

$$f(\theta) = \operatorname{cis}(\theta + \varphi), \qquad -\infty < \theta < \infty$$

is seen to satisfy (4). Thus, by the theorem,

$$f(\theta) = f(0) \operatorname{cis} \theta = \operatorname{cis} \varphi \operatorname{cis} \theta.$$

Hence we have

$$\operatorname{cis}(\theta + \varphi) = \operatorname{cis} \theta \cdot \operatorname{cis} \varphi; \qquad \theta, \varphi \in E^1. \tag{6}$$

In particular, $\operatorname{cis} \theta \cdot \operatorname{cis}(-\theta) = \operatorname{cis}(\theta - \theta) = \operatorname{cis} 0 = 1$ so that

$$\operatorname{cis}(-\theta) = \frac{1}{\operatorname{cis} \theta} = \overline{\operatorname{cis} \theta} \tag{7}$$

where $\overline{\operatorname{cis} \theta}$ is the complex conjugate of $\operatorname{cis} \theta$.

The following definition names the real and imaginary parts of cis. Notice that

$$\frac{1 - it}{1 + it} = \frac{1 - t^2}{1 + t^2} + i \frac{2t}{1 + t^2}.$$

Definition 3. cos: $E^1 \to E^1$ *and* sin: $E^1 \to E^1$ *are defined by*

$$\operatorname{cis} \theta = \cos \theta + i \sin \theta;$$

that is,

$$\cos \theta = \frac{1 - [p^{-1}(\theta)]^2}{1 + [p^{-1}(\theta)]^2} \quad \text{and} \quad \sin \theta = \frac{2p^{-1}(\theta)}{1 + [p^{-1}(\theta)]^2}. \tag{8}$$

Thus cos is an even function. Since p is an odd function so is p^{-1} and then so is sin.

Equations (2) may now be rewritten as

$$\cos 0 + i \sin 0 = 1 \qquad \text{and} \qquad \cos\frac{\pi}{2} + i \sin\frac{\pi}{2} = i,$$

from which we obtain

$$\cos 0 = 1, \qquad \sin 0 = 0, \qquad \cos\frac{\pi}{2} = 0, \qquad \text{and} \qquad \sin\frac{\pi}{2} = 1.$$

The periodicity relation (iii) yields

$$\cos(\theta + 2\pi) = \cos\theta \qquad \text{and} \qquad \sin(\theta + 2\pi) = \sin\theta$$

so that cos and sin are also periodic of period 2π. Also, (6) becomes

$$\cos(\theta + \varphi) + i \sin(\theta + \varphi) = (\cos\theta + i \sin\theta)(\cos\varphi + i \sin\varphi),$$

and by equating real and imaginary parts,

$$\begin{aligned} \cos(\theta + \varphi) &= \cos\theta \cos\varphi - \sin\theta \sin\varphi \qquad \text{and} \\ \sin(\theta + \varphi) &= \sin\theta \cos\varphi + \cos\theta \sin\varphi. \end{aligned}$$

Since cis satisfies the differential equation (4), we have

$$\frac{d}{d\theta}(\cos\theta + i \sin\theta) = i(\cos\theta + i \sin\theta),$$

$$\cos'\theta + i \sin'\theta = -\sin\theta + i \cos\theta,$$

and therefore

$$\cos'\theta = -\sin\theta, \qquad \sin'\theta = \cos\theta,$$

$$\int_0^\theta \sin s \, ds = 1 - \cos\theta, \qquad \text{and} \qquad \int_0^\theta \cos s \, ds = \sin\theta.$$

From $\sin' 0 = \cos 0 = 1$, it follows that

$$\lim_{\theta \to 0} \frac{\sin\theta - \sin 0}{\theta - 0} = \lim_{\theta \to 0} \frac{\sin\theta}{\theta} = 1,$$

and then by l'Hospital's rule,

$$\lim_{\theta \to 0} \frac{1 - \cos 2\theta}{\theta^2} = \lim_{\theta \to 0} \frac{2 \sin 2\theta}{2\theta} = 2.$$

Also

$$\cos^2\theta + \sin^2\theta = |\text{cis}\,\theta|^2 = 1.$$

The following definition introduces the Euler function as another alternative notation.

Definition 4

$$e^{i\theta} = \text{cis } \theta = \cos \theta + i \sin \theta, \qquad -\infty < \theta < \infty.$$

Then (2), (3), (6), and (iii) take the forms

(2′) $$e^{i0} = 1, \qquad e^{i\pi/2} = i,$$

(3′) $$\frac{de^{i\theta}}{d\theta} = ie^{i\theta},$$

(6′) $$e^{i(\theta+\varphi)} = e^{i\theta}e^{i\varphi}, \qquad e^{i2\theta} = (e^{i\theta})^2,$$

(iii)′ $$e^{i(\theta+2\pi)} = e^{i\theta}.$$

The reciprocal and conjugate relations (7) now yield

$$e^{-i\theta} = \text{cis}(-\theta) = \frac{1}{\text{cis } \theta} = \frac{1}{e^{i\theta}} = \overline{\text{cis } \theta} = \cos \theta - i \sin \theta.$$

From these equations, the definition of $e^{i\theta}$, and (6′),

$$\cos \theta = \frac{1}{2}(e^{i\theta} + e^{-i\theta}) = \frac{1}{2e^{i\theta}}(e^{i2\theta} + 1),$$

$$\sin \theta = \frac{1}{2i}(e^{i\theta} - e^{-i\theta}) = \frac{1}{2ie^{i\theta}}(e^{i2\theta} - 1).$$

Then the function tan is defined by

$$\tan \theta = \frac{\sin \theta}{\cos \theta} \qquad \text{for } \theta \neq \frac{\pi}{2} + m\pi, \qquad m = 0, \pm 1, \pm 2, \cdots,$$

and the other trigonometric functions are defined similarly.

From Definition 2(i) we have, with $\theta = p(t)$, that

$$e^{i\theta} = \frac{1 + it}{1 - it}$$

and by solving this equation for t,

$$t = \frac{e^{i\theta} - 1}{i(e^{i\theta} + 1)} = \frac{(e^{i\theta} - 1)/(2e^{i\theta/2})}{(e^{i\theta} - 1)/(2ie^{i\theta/2})} = \frac{\sin \theta/2}{\cos \theta/2} = \tan \frac{\theta}{2}.$$

Since $-\pi/2 < \theta < \pi/2 \Rightarrow -1 < p^{-1}(\theta) < 1 \Rightarrow \cos \theta > 0$ from (8), we have

$$\sin' \theta = \cos \theta > 0 \qquad \text{for } -\pi/2 < \theta < \pi/2,$$

with $\sin'(-\pi/2) = \sin'(\pi/2) = \cos \pi/2 = 1$. Hence sin is an increasing function on $[-\pi/2, \pi/2]$ onto $[-1, 1]$ and thus has an inverse

$$\sin^{-1}: [-1, 1] \leftrightarrow \left[-\frac{\pi}{2}, \frac{\pi}{2}\right].$$

But sin′ is positive on $(-\pi/2, \pi/2)$ so that $\sin^{-1}$ is differentiable on $(-1, 1)$. Thus for $x \in (-1, 1)$ we have $\cos(\sin^{-1} x) > 0$,

$$x = \sin(\sin^{-1} x), \qquad 1 = \cos(\sin^{-1} x)\frac{d \sin^{-1} x}{dx}, \qquad \text{and}$$

$$\frac{d \sin^{-1} x}{dx} = \frac{1}{\cos(\sin^{-1} x)} = \frac{1}{\sqrt{1 - \sin^2(\sin^{-1} x)}}$$

$$= \frac{1}{\sqrt{1 - x^2}}, \qquad -1 < x < 1.$$

The other inverse trigonometric functions are defined in a similar way (with appropriately restricted domains and ranges) and their derivatives obtained.

From the above development, the function F defined by

$$F(t) = \frac{1 - it}{1 + it} = \frac{1 - t^2}{1 + t^2} + i\frac{2t}{1 + t^2}, \qquad -\infty < t < \infty$$

is a 1–1 mapping of $(-\infty, \infty)$ onto the unit circle minus the point $(-1, 0)$. Since p^{-1} is 1–1 on $(-\pi, \pi)$ onto $(-\infty, \infty)$ such that

$$F(p^{-1}(\theta)) = \operatorname{cis}\theta = e^{i\theta}, \qquad -\pi < \theta < \pi$$

and since $e^{i\pi} = \operatorname{cis}\pi = -1 + i \cdot 0$, we have

If $z = x + iy$ has $|z| = 1$, then there is one and only one value θ such that

$$z = e^{i\theta}, \qquad -\pi < \theta \leqslant \pi.$$

Hence, if $z \neq 0$, the representations

$$z = x + iy = re^{i\theta}; \qquad r > 0, \quad -\pi < \theta \leqslant \pi$$

are unique. This follows from $\big|z/|z|\,\big| = 1$ upon choosing $r = |z|$.

It follows that if $z = re^{i\theta}$, then

$$-z = re^{-i\theta} \qquad \text{and} \qquad z = re^{i(\theta + 2\pi)}.$$

PROBLEM 1. Let $z_k = r_k e^{i\theta_k}$, $r_k \geqslant 0$, $k = 1, 2, \cdots, n$ be given. Then

(a) $\left|\sum_{k=1}^n z_k\right| \geqslant |1/\sqrt{2}\,| \sum_{k=1}^n |z_k|$ if $|\theta_k| \leqslant \pi/4$, $k = 1, 2, \cdots, n$.
(b) The same is true if $|\theta_k - \frac{1}{2}\pi| \leqslant \frac{1}{4}\pi$ or if $|\theta_k + \frac{1}{2}\pi| \leqslant \frac{1}{4}\pi$.
(c) With no restriction on θ_k, there is a set

$$S \subset \{1, 2, \cdots, n\} \qquad \text{such that} \left|\sum_{k \in S} z_k\right| \geqslant \frac{1}{4\sqrt{2}} \sum_{k=1}^n |z_k|.$$

The next theorem is known as the fundamental theorem of algebra. This theorem is important because it shows that the complex number field is algebraically complete in the sense that every nonconstant polynomial with complex coefficients has a complex root.

Theorem 2. *With $a_0, a_1, \cdots, a_n$ complex constants, $n \geqslant 1$ and $a_n \neq 0$, let*

$$P(z) = \sum_{k=0}^{n} a_k z^k, \quad z \text{ complex.}$$

Then this nth-degree polynomial has a root; that is, there is a complex number z_0 such that $P(z_0) = 0$.

Proof. Set

$$u = \inf|P(z)| \qquad \text{for all } z \in E^2.$$

We shall first show that there is a z_0 such that $|P(z_0)| = u$ and then that $u = 0$.

From $a_n z^n = P(z) - \sum_{k=0}^{n-1} a_k z^k$ and $z = re^{i\theta}$, $r > 0$, we obtain

$$|P(re^{i\theta})| \geqslant r^n\left\{|a_n| - \frac{|a_0|}{r^n} - \frac{|a_1|}{r^{n-1}} - \cdots - \frac{|a_{n-1}|}{r}\right\}.$$

As $r \to \infty$ the bracketed terms $\to |a_n| > 0$, and thus there is an $R > 0$ such that

$$|P(Re^{i\theta})| > u \qquad \text{for all } \theta.$$

Hence all z for which $|P(z)| \leqslant u$ must be on the circular disk

$$\{z = re^{i\theta} \mid r \leqslant R\}.$$

Since this disk is a compact set and $|P|$ is continuous, there is a point z_0 of the disk such that

$$|P(z_0)| = u.$$

Next assume that $u > 0$. Now that $P(z_0) \neq 0$, we define another polynomial Q by

$$Q(z) = \frac{P(z + z_0)}{P(z_0)} = \frac{1}{P(z_0)} \sum_{k=0}^{n} a_k(z + z_0)^k = \sum_{k=0}^{n} b_k z^k.$$

We need not determine any of the coefficients b_k except b_0, and for this one,

$$b_0 = Q(0) = \frac{P(z_0)}{P(z_0)} = 1.$$

Some of, but not all, the other coefficients may be zero, and we let b_l be the nonzero coefficient with least positive subscript. Hence

$$Q(z) = 1 + b_l z^l + b_{l+1} z^{l+1} + \cdots + b_n z^n, \qquad b_l \neq 0.$$

Select θ such that $b_l/|b_l| = e^{i\theta}$ and hence

$$-|b_l| = -b_l e^{-i\theta} = b_l e^{i\theta}.$$

Take $r > 0$ sufficiently small to make $1 - |b_l| r^l > 0$. Then, with $\varphi = \theta/l$,

$$Q(re^{i\varphi}) = 1 - |b_l| r^l + b_{l+1} r^{l+1} e^{i(l+1)\varphi} + \cdots + b_n r^n e^{in\varphi}.$$

The sum of the first two terms being positive, we have

$$|Q(re^{i\varphi})| \leqslant 1 - r^l\{|b_l| - |b_{l+1}| r - \cdots - |b_n| r^{n-l}\}.$$

If $r \to 0$, the bracketed terms $\to |b_l| > 0$. Take $r > 0$ smlal enough to make the bracket >0. For this r and φ the complex number $z = re^{i\varphi}$ is such that

$$|Q(z)| < 1.$$

However, $P(z + z_0)$ is a value of P and thus

$$|P(z + z_0)| \geqslant u = |P(z_0)| \quad \text{and} \quad |Q(z)| = \frac{|P(z + z_0)|}{|P(z_0)|} \geqslant 1.$$

Hence $u = 0$ and the theorem is proved. ∎

Since $e^{ikx} = (e^{ix})^k$, we may use the formula for the sum of a geometric progression to obtain

$$\sum_{k=0}^{n} e^{ikx} = \frac{e^{i(n+1)x} - 1}{e^{ix} - 1}, \qquad x \neq 0.$$

Hence by multiplying numerator and denominator by $e^{-ix/2}$, $x \neq 0$,

$$\sum_{k=0}^{n} (\cos kx + i \sin kx) = \frac{e^{i(n+1/2)x} - e^{-ix/2}}{e^{ix/2} - e^{-ix/2}}$$

$$= \frac{\cos(n + \frac{1}{2})x + i \sin(n + \frac{1}{2})x - \cos \frac{1}{2}x + i \sin \frac{1}{2}x}{2i \sin \frac{1}{2}x}.$$

By equating real and imaginary parts,

$$\sum_{k=0}^{n} \cos kx = \frac{1}{2 \sin \frac{1}{2}x} \left\{ \sin\left(n + \frac{1}{2}\right)x + \sin \frac{1}{2} x \right\},$$

$$\sum_{k=0}^{n} \sin kx = \frac{-1}{2 \sin \frac{1}{2}x} \left\{ \cos\left(n + \frac{1}{2}\right)x - \cos \frac{1}{2} x \right\}.$$

The functions given in the following definition will be used in the next section.

Definition 5. *The Dirichlet and Fejér* **kernels** *are defined for* $n = 1, 2, 3, \cdots$ *by*

$$D_n(x) = \sum_{k=-n}^{n} e^{ikx} \quad \text{and} \quad K_n(x) = \frac{1}{n+1} \sum_{m=0}^{n} D_m(x).$$

PROBLEM 2. With $-\pi \leqslant x \leqslant \pi$, then

$$D_n(x) = \frac{\sin(n + \frac{1}{2})x}{\sin \frac{1}{2}x} \quad \text{for } x \neq 0, \quad D_n(0) = 2n + 1,$$

$$K_n(x) = \frac{1 - \cos(n + 1)x}{(n + 1)(1 - \cos x)} \quad \text{for } x \neq 0, \quad K_n(0) = n + 1,$$

and

$$\int_{-\pi}^{\pi} D_n(x)\, dx = \int_{-\pi}^{\pi} K_n(x)\, dx = 2\pi.$$

[*Hint:* To find the expressions for $D_n(x)$ and $K_n(x)$, use the method (not the result) given above. To find the integral of $D_n(x)$, use the definition (not the expression in the problem) and remember there is one term for which $k = 0$.]

7–4 FOURIER SERIES

From facts and formulas of the previous section,

$$\frac{1}{2\pi}\int_{-\pi}^{\pi} e^{ikx}\,dx = \begin{cases} 0 & \text{if } k = \pm 1, \pm 2, \pm 3, \cdots \\ 1 & \text{if } k = 0. \end{cases}$$

Hence for any positive, negative, or zero integers n and k,

$$(1)\qquad \frac{1}{2\pi}\int_{-\pi}^{\pi} e^{i(k-n)x}\,dx = \begin{cases} 0 & \text{if } k \neq n \\ 1 & \text{if } k = n. \end{cases}$$

Definition 1. *With $a_0, a_1, a_2, \cdots, a_n$ and $b_1, b_2, \cdots, b_n$ complex constants,*

$$T_n(x) = \tfrac{1}{2}a_0 + \sum_{k=1}^{n} (a_k \cos kx + b_k \sin kx)$$

$$(2)\qquad = \sum_{k=-n}^{n} c_k e^{ikx}, \qquad x \text{ real}$$

is called a **trigonometric polynomial.** *If c_n, $n = 0, \pm 1, \pm 2, \cdots$, are complex constants, then*

$$\sum_{n=-\infty}^{\infty} c_n e^{inx}$$

is called a **trigonometric series** *having (2) as its partial sum.*

In the definitions of $T_n(x)$, the c's are seen to be related to the a's and b's by writing the first form as

$$T_n(x) = \frac{1}{2}a_0 + \sum_{k=1}^{n} \left\{\frac{a_k}{2}(e^{ikx} + e^{-ikx}) + \frac{b_k}{2i}(e^{ikx} - e^{-ikx})\right\}$$

$$= \frac{1}{2}a_0 + \sum_{k=1}^{n} \left\{\frac{1}{2}(a_k - ib_k)e^{ikx} + \frac{1}{2}(a_k + ib_k)e^{-ikx}\right\}.$$

Hence the second form is obtained in which

$$2c_k = \begin{cases} a_k - ib_k, & k = 1, 2, \cdots, n \\ a_0, & k = 0 \\ a_{-k} + ib_{-k}, & k = -1, -2, \cdots, -n. \end{cases}$$

PROBLEM I

$$a_k = c_k + c_{-k}, \qquad k = 0, 1, 2, \cdots, n$$
$$b_k = i(c_k - c_{-k}), \qquad k = 1, 2, \cdots, n\,.$$

Upon multiplying (2) by e^{-imx}; $m = 0, \pm 1, \cdots, \pm n$, and integrating $1/(2\pi)$ times the product, we obtain

$$\frac{1}{2\pi}\int_{-\pi}^{\pi} T_n(x)e^{-imx}\,dx = \frac{1}{2\pi}\sum_{k=-n}^{n} c_k \int_{-\pi}^{\pi} e^{i(k-m)x}\,dx = c_m\,.$$

Definition 2. *A function* $f\colon [-\pi, \pi] \to E^2$ *belonging to* $\mathscr{L}[-\pi, \pi]$ *is said to have*

$$c_n = \frac{1}{2\pi}\int_{-\pi}^{\pi} f(x)e^{-inx}\,dx; \qquad n = 0, \pm 1, \pm 2, \cdots$$

as its **Fourier coefficients,** *and we write*

$$f(x) \sim \sum_{n=-\infty}^{\infty} c_n e^{inx},$$

which is called the **Fourier series** *of f with partial sums*

$$\sum_{m=-n}^{n} c_m e^{imx}.$$

The symbol $\sim$ is used rather than $=$, since the partial sums may not converge and even if they do they may not converge to $f(x)$.

EXAMPLE 1

$$f(x) = \begin{Bmatrix} 0, & -\pi \leqslant x < 0 \\ 1, & 0 \leqslant x \leqslant \pi \end{Bmatrix}.$$

$$c_n = \frac{1}{2\pi}\int_{-\pi}^{\pi} f(x)e^{-inx}\,dx = \frac{1}{2\pi}\int_0^{\pi} e^{-inx}\,dx$$

$$= \frac{1}{2\pi}\begin{cases} \pi & \text{if } n = 0 \\ \dfrac{i}{n}(\cos n\pi - 1), & n = \pm 1, \pm 2, \pm 3, \cdots. \end{cases}$$

$$a_0 = 1, \qquad a_k = 0 \qquad \text{for } k = 1, 2, 3, \cdots,$$

$$b_k = \frac{1}{k\pi}(1 - \cos k\pi), \qquad k = 1, 2, 3, \cdots.$$

Hence

$$f(x) \sim \frac{1}{2} + \frac{2}{\pi}\left(\frac{\sin x}{1} + \frac{\sin 3x}{3} + \frac{\sin 5x}{5} + \cdots\right).$$

Figure 1 shows the graphs of the first four partial sums.

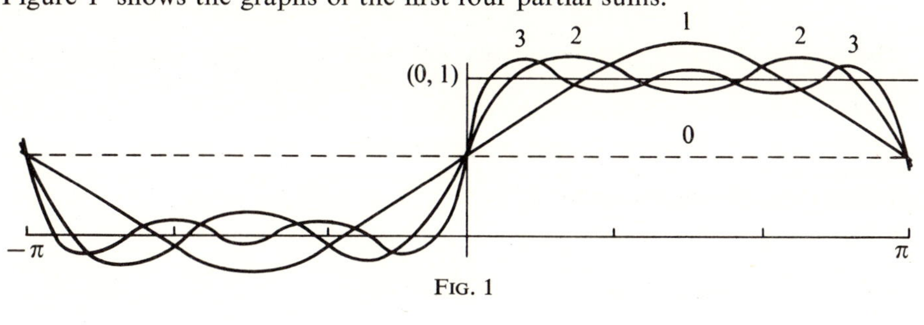

FIG. 1

For this function,

$$\lim_{n\to\infty} T_n(0) = \tfrac{1}{2} \neq f(0).$$

PROBLEM 2

$$|x| \sim \frac{\pi}{2} - \frac{4}{\pi}\left[\cos x + \frac{1}{3^2}\cos 3x + \frac{1}{5^2}\cos 5x + \cdots\right].$$

The following definition is given to indicate one direction generalization may take.

Definition 3. *With φ_n: $[a, b] \to E^2$ and $\varphi_n \in \mathscr{L}^2$ on $[a, b]$ for $n = 1, 2, 3, \cdots$, the sequence $\{\varphi_n\}$ is said to be an* **orthogonal system** *on $[a, b]$ if*

$$\int_a^b \varphi_n(x)\overline{\varphi_m(x)}\, dx = 0 \qquad \text{for } n \neq m$$

and is said to be **orthonormal** *if in addition*

$$\int_a^b |\varphi_n(x)|^2\, dx = 1, \qquad n = 1, 2, 3, \cdots.$$

If $\{\varphi_n\}$ is orthonormal, f: $[a, b] \to E^2$, and $f \in \mathscr{L}^2[a, b]$, then

$$c_n = \int_a^b f(x)\overline{\varphi_n(x)}\, dx \qquad \text{and} \qquad \sum_{n=1}^{\infty} c_n \varphi_n(x)$$

are called the **Fourier coefficients** *and* **series** *of f relative to $\{\varphi_n\}$ on $[a, b]$.*

[*Note:* The domain space E^1 could be replaced by any space X on which an outer measure μ^* is defined and $[a, b]$ replaced by any μ-measurable set $M \subset X$, but we shall not make these further simplifications.]

From the above work it is seen that $\{\varphi_n\}$ defined by

$$\text{(3)} \qquad \varphi_1(x) = \frac{1}{\sqrt{2\pi}}, \qquad \varphi_{2n+1}(x) = \frac{e^{inx}}{\sqrt{\pi}}, \qquad \varphi_{2n}(x) = \frac{e^{-inx}}{\sqrt{\pi}}$$

is orthonormal on $[-\pi, \pi]$.

An example of a real orthonormal system on $[-\pi, \pi]$ is

$$\text{(4)} \qquad \varphi_1(x) = \frac{1}{\sqrt{2\pi}}, \qquad \varphi_{2n+1}(x) = \frac{\cos nx}{\sqrt{\pi}}, \qquad \varphi_{2n}(x) = \frac{\sin nx}{\sqrt{\pi}}.$$

Theorem 1. *In the notation of the definition (with $\{\varphi_n\}$ orthogonal and $\in \mathscr{L}^2[a, b]$) let*

$$s_n(x) = \sum_{k=1}^{n} c_k \varphi_k(x)$$

and let $\{\gamma_n\}$ be any sequence such that

$$t_n(x) = \sum_{k=1}^{n} \gamma_k \varphi_k(x)$$

yields $t_n \in \mathscr{L}^2[a, b]$ *for* $n = 1, 2, 3, \cdots$. *Then*

$$\text{(5)} \qquad \int_a^b |f(x) - s_n(x)|^2\, dx \leqslant \int_a^b |f(x) - t_n(x)|^2\, dx$$

with equality iff $\gamma_n = c_n$, $n = 1, 2, 3, \cdots$. *Also,*

$$\text{(6)} \qquad \sum_{k=1}^{\infty} |c_k|^2 \leqslant \int_a^b |f(x)|^2\, dx \qquad \textbf{(Bessel's inequality).}$$

[*Note:* $\{s_n\}$ furnishes the best mean square approximations of f; that is, the best in the $\mathscr{L}^2$ norm, since (5) means that $\|f - s_n\| \leqslant \|f - t_n\|$.]

Proof. Dropping limits and variables of integration, and indications of sums from 1 to n, we first have

$$\int (f\bar{t}_n) = \int (f \sum \bar{\gamma}_k \bar{\varphi}_k) = \sum \bar{\gamma}_k \int (f\bar{\varphi}_k) = \sum (\bar{\gamma}_k c_k)$$

by the definition of c_k. In the same way

$$\int (\bar{f} t_n) = \sum (\bar{c}_k \gamma_k).$$

Also,

$$\begin{aligned}
\int (t_n \bar{t}_n) &= \int \left[\sum_k (\gamma_k \varphi_k) \ \sum_m (\bar{\gamma}_m \bar{\varphi}_m)\right] = \sum_k \left\{\gamma_k \left[\int \varphi_k \sum_m \bar{\gamma}_m \bar{\varphi}_m\right]\right\} \\
&= \sum_k \left\{\gamma_k \left[\sum_m \bar{\gamma}_m \int (\varphi_k \bar{\varphi}_m)\right]\right\} \\
&= \sum_k \{\gamma_k [\bar{\gamma}_k]\} \qquad \text{by orthonormality of } \{\varphi_n\} \\
&= \sum_k (\gamma_k \bar{\gamma}_k).
\end{aligned}$$

Each of these three equations is used in the calculation

$$\begin{aligned}
\int |f - t_n|^2 &= \int (f - t_n)(\bar{f} - \bar{t}_n) \\
&= \int f\bar{f} - \int f\bar{t}_n - \int \bar{f} t_n + \int t_n \bar{t}_n \\
&= \int |f|^2 - \sum \bar{\gamma}_k c_k - \sum \bar{c}_k \gamma_k + \sum \gamma_k \bar{\gamma}_k \\
&= \int |f|^2 - \sum c_k \bar{c}_k + \sum (c_k \bar{c}_k - \bar{\gamma}_k c_k - \bar{c}_k \gamma_k + \gamma_k \bar{\gamma}_k] \\
&= \int |f|^2 - \sum |c_k|^2 + \sum (c_k - \gamma_k)(\bar{c}_k - \bar{\gamma}_k) \\
&= \int |f|^2 - \sum |c_k|^2 + \sum |c_k - \gamma_k|^2.
\end{aligned}$$

In this leave each c_k, but replace each γ_k by c_k to see that

$$\int |f - s_n|^2 = \int |f|^2 - \sum |c_k|^2 + 0.$$

We thus have (5) and the condition under which equality holds. Also $\int |f - s_n|^2 \geqslant 0$ and the above display gives

$$\sum_{k=1}^{n} |c_k|^2 \leqslant \int_a^b |f(x)|^2 \, dx$$

in which we let $n \to \infty$ to obtain Bessel's inequality (6). ∎

PROBLEM 3. If $h \in \mathscr{L}$ on $[-\pi, \pi]$, then

$$\lim_{n \to \pm\infty} \int_{-\pi}^{\pi} h(t) \cos nt \, dt = \lim_{n \to \pm\infty} \int_{-\pi}^{\pi} h(t) \sin nt \, dt = 0.$$

[*Hint*: Use the orthonormal system (4) and convergence of the series in Bessel's inequality (6).]

Change of variable of integration was given for continuous integrands (Theorem 7-2.11), but the same result also holds for summable integrands, as we shall show in Theorem 7-8.2. For example, if F is summable on $[-\pi, \pi]$ and periodic of period 2π, then for any fixed $x \in E^1$, the substitution $t = s - 2\pi$ gives the first equation in

$$\int_{x-\pi}^{-\pi} F(t) \, dt = \int_{x+\pi}^{\pi} F(s - 2\pi) \, ds = \int_{x+\pi}^{\pi} F(s) \, ds,$$

and the second follows from the periodicity of F. Thus, for such a 2π periodic function,

$$\int_{x-\pi}^{x+\pi} = \int_{x-\pi}^{-\pi} + \int_{-\pi}^{\pi} + \int_{\pi}^{x+\pi} = -\int_{\pi}^{x+\pi} + \int_{-\pi}^{\pi} + \int_{\pi}^{x+\pi} = \int_{-\pi}^{\pi}.$$

We return to the specific orthonormal system of complex exponentials (3).

Theorem 2. *Given $f \in \mathscr{L}[-\pi, \pi]$ and 2π periodic, then the partial sums $\{s_n\}$ of the Fourier series of f are such that*

$$(7) \qquad s_n(x) = \frac{1}{2\pi} \int_{-\pi}^{\pi} f(x - t) D_n(t) \, dt; \qquad x \in E^1, \quad n = 0, 1, 2, 3, \cdots,$$

where D_n is the Dirichlet kernel of Definition 7-3.5.

Proof. Express the mth Fourier coefficient of f as

$$c_m = \frac{1}{2\pi} \int_{-\pi}^{\pi} f(u) e^{-imu} \, du.$$

Then, x being fixed,

$$s_n(x) = \sum_{m=-n}^{n} c_m e^{imx} = \frac{1}{2\pi} \sum_{m=-n}^{n} \int_{-\pi}^{\pi} [f(u)e^{-imu}]e^{imx}\, du$$

$$= \frac{1}{2\pi} \int_{-\pi}^{\pi} f(u) \sum_{m=-n}^{n} e^{im(x-u)}\, du$$

$$= \frac{1}{2\pi} \int_{-\pi}^{\pi} f(u) D_n(x-u)\, du.$$

The substitution $u = x - t$ then gives

$$s_n(x) = \frac{1}{2\pi} \int_{x+\pi}^{x-\pi} f(x-t) D_n(t)(-dt)$$

which, since both f and D_n have period 2π, may be written as (7). ∎

The following "localization" theorem shows that convergence properties of the sequence $\{s_n(x)\}$ for a function f depends only on the values of f in an arbitrarily small neighborhood of x.

Theorem 3. *If $f \in \mathscr{L}[-\pi, \pi]$ and is 2π periodic, then for any number δ such that $0 < \delta < \pi$,*

$$\lim_{n\to\infty} \left\{ \int_{-\pi}^{-\delta} f(x-t) D_n(t)\, dt + \int_{\delta}^{\pi} f(x-t) D_n(t)\, dt \right\} = 0, \qquad x \in E^1.$$

Proof. With x fixed, let $g(t) = 0$ if $|t| < \delta$,

$$g(t) = \frac{f(x-t)}{\sin \frac{1}{2}t} \qquad \text{if } -\pi \leqslant t \leqslant -\delta \quad \text{or} \quad \delta \leqslant t \leqslant \pi$$

and extend g to be periodic of period 2π. By using the expression for $D_n(t)$ in Problem 7-3.2, the sum of the bracketed integrals in the statement of the theorem is equal to

$$\int_{-\pi}^{\pi} g(t) \sin(n + \tfrac{1}{2})t\, dt = \int_{-\pi}^{\pi} g(t) \cos \tfrac{1}{2}t \sin nt\, dt + \int_{-\pi}^{\pi} g(t) \sin \tfrac{1}{2}t \cos nt\, dt.$$

Use $h(t)$ of Problem 3 as $g(t) \cos \frac{1}{2}t$ for the first integral, but as $g(t) \sin \frac{1}{2}t$ for the second integral, and see that both integrals approach zero as $n \to \infty$. ∎

Given a sequence $\{a_n\}$, the sequence $\{s_n\}$ of partial sums is given by

$$s_n = a_0 + a_1 + a_2 + \cdots + a_n$$

and the sequence $\{\sigma_n\}$ of arithmetic means (or first Cesaro means) is defined by

$$\sigma_n = \frac{s_0 + s_1 + s_2 + \cdots + s_n}{n+1}.$$

Recall that if $\{s_n\}$ converges, then so does $\{\sigma_n\}$ and to the same value, but $\{\sigma_n\}$ may converge when $\{s_n\}$ diverges.

Theorem 4. *Given a continuous function f which is periodic of period 2π, let*

$$\sigma_n(x) = \frac{s_0(x) + s_1(x) + s_2(x) + \cdots + s_n(x)}{n+1}, \qquad x \in E^1, \quad n = 1, 2, 3, \cdots,$$

where $\{s_k(x)\}$ is the sequence of partial sums of the Fourier series of f. Then

$$\lim_{n \to \infty} \sigma_n(x) = f(x) \quad \textit{uniformly}.$$

Proof. From (7),

$$(A) \qquad \sigma_n(x) = \frac{1}{n+1} \sum_{k=0}^{n} \frac{1}{2\pi} \int_{-\pi}^{\pi} f(x-t) D_k(t)\, dt = \frac{1}{2\pi} \int_{-\pi}^{\pi} f(x-t) K_n(t)\, dt$$

by Definition 7-3.5 of the Fejér kernel K_n. Hence, by Problem 7-3.2,

$$\sigma_n(x) - f(x) = \frac{1}{2\pi} \int_{-\pi}^{\pi} [f(x-t) - f(x)] K_n(t)\, dt.$$

Given an $\varepsilon > 0$, let δ be such that $0 < \delta < \pi$ and

$$|x - y| < \delta \Rightarrow |f(x) - f(y)| < \varepsilon.$$

Since $K_n \geqslant 0$,

$$\int_{-\delta}^{\delta} |f(x-t) - f(x)|\, K_n(t)\, dt \leqslant \varepsilon \int_{-\delta}^{\delta} K_n(t)\, dt \leqslant \varepsilon \int_{-\pi}^{\pi} K_n(t)\, dt = 2\pi\varepsilon.$$

Problem 7-3.2 also gives

$$0 \leqslant K_n(x) = \frac{1 - \cos(n+1)x}{(n+1)(1 - \cos x)} \leqslant \frac{2}{(n+1)(1 - \cos \delta)} \qquad \text{for } \delta \leqslant |x| \leqslant \pi.$$

Let M be such that $|f| \leqslant M$. Then

$$\int_{-\pi}^{-\delta} |f(x-t) - f(x)|\, K_n(t)\, dt \leqslant \frac{2M(\pi - \delta)}{(n+1)(1 - \cos \delta)} \leqslant \frac{2\pi M}{(n+1)(1 - \cos \delta)}$$

and the same inequality holds for $\int_{\delta}^{\pi}$. Thus

$$\left| \int_{-\pi}^{\pi} [f(x-t) - f(x)] K_n(t)\, dt \right| \leq 2\pi \left\{ \varepsilon + \frac{2\pi M}{(n+1)(1 - \cos \delta)} \right\}.$$

Hence

$$\overline{\lim_{n \to \infty}}\, |\sigma_n(x) - f(x)| \leqslant 2\pi\varepsilon.$$

Therefore there is an integer N such that

$$n \geqslant N \Rightarrow |\sigma_n(x) - f(x)| < 3\pi\varepsilon \qquad \text{for any } x\ E^1,$$

since σ_n inherits 2π periodicity from f and e^{inx}. Consequently, not only does $\sigma_n(x) \to f(x)$ as $n \to \infty$, but it does so uniformly on E^1. ∎

Corollary 1. *If the function f of the theorem is such that*

$$\int_{-\pi}^{\pi} f(x)e^{-inx}\,dx = 0, \qquad n = 0, \pm 1, \pm 2, \cdots,$$

then $f = 0$.

Proof. Every $c_n = 0$, $s_n(x) = 0$, and $\sigma_n(x) = 0$. ∎

Corollary 2. *If f and g are continuous, 2π periodic, and have the same Fourier series, then* $f = g$.

Proof. $\sigma_n(x)$ is the same for both of them, $\sigma_n(x) \to f(x)$ and $\sigma_n(x) \to g(x)$. ∎

PROBLEM 4. Given that f is bounded, measurable, and 2π periodic. If there is an $x \in (-\pi, \pi]$ such that $f(x+)$ and $f(x-)$ exist, then

$$\lim_{n\to\infty} \sigma_n(x) = \tfrac{1}{2}[f(x-) + f(x+)].$$

[*Hints:* (a) From $K_n(-t) = K_n(t)$, use (A), p. 400, to see that

$$\sigma_n(x) = \frac{1}{2\pi}\int_{-\pi}^{\pi} f(x+t)K_n(t)\,dt = \frac{1}{4\pi}\int_{-\pi}^{\pi} [f(x+t) + f(x-t)]K_n(t)\,dt.$$

(b) Define $g(t) = f(t)$ for $t \neq x$ and $g(x) = \frac{1}{2}[f(x+) + f(x-)]$. Obtain

$$\sigma_n(x) - g(x) = \frac{1}{4\pi}\int_{-\pi}^{\pi} [g(x+t) + g(x-t) - 2g(x)]K_n(t)\,dt.$$

(c) Given an $\varepsilon > 0$, see that there is a $\delta > 0$ such that

$$|t| < \delta \Rightarrow |g(x+t) + g(x-t) - 2g(x)| < \varepsilon.$$

(d) Use this δ, as δ was used in the proof of Theorem 4, to obtain an upper bound for $|\sigma_n(x) - g(x)|$.]

Theorem 5. *Given two continuous 2π periodic functions f and g, let c_n and γ_n be the Fourier coefficients of f and g, respectively, and $\{s_n\}$ the partial Fourier sums for f. Then*

$$\lim_{n\to\infty} \int_{-\pi}^{\pi} |f(x) - s_n(x)|^2\,dx = 0, \tag{8}$$

$$\sum_{-\infty}^{\infty} c_n\bar{\gamma}_n = \frac{1}{2\pi}\int_{-\pi}^{\pi} f(x)\overline{g(x)}\,dx, \qquad \textit{and} \tag{9}$$

$$\sum_{\mp\infty}^{\infty} |c_n|^2 = \frac{1}{2\pi}\int_{-\pi}^{\pi} |f(x)|^2\,dx \qquad \textbf{(Parseval's equation)}. \tag{10}$$

Proof. Given $\varepsilon > 0$, take N, by Theorem 4, such that

$$n \geqslant N \Rightarrow |f(x) - \sigma_n(x)| < \varepsilon \qquad \text{for all } x \in E^1.$$

In Theorem 1 replace t_n by σ_n to see that (8) follows from

$$n \geqslant N \Rightarrow \int_{-\pi}^{\pi} |f(x) - s_n(x)|^2\, dx \leqslant \int_{-\pi}^{\pi} |f(x) - \sigma_n(x)|^2\, dx < 2\pi\varepsilon^2.$$

By taking conjugates in the expression for γ_k,

$$\sum_{k=-n}^{n} c_k \bar{\gamma}_k = \sum_{k=-n}^{n} c_k \frac{1}{2\pi} \int_{-\pi}^{\pi} \overline{g(x)e^{ikx}}\, dx = \frac{1}{2\pi} \int_{-\pi}^{\pi} s_n(x)\overline{g(x)}\, dx.$$

Schwarz inequality gives the last step in

$$\left| \int f\bar{g} - 2\pi \sum_{-n}^{n} c_k \bar{\gamma}_k \right| = \left| \int f\bar{g} - \int s_n \bar{g} \right| \leqslant \int \{|f - s_n|\,|\bar{g}|\} \leqslant \left\{ \int |f - s_n|^2 \int |g|^2 \right\}^{1/2}$$

This, with (8), establishes (9). In (9) replace g by f and obtain (10). ∎

We shall prove (in Theorem 7-7.8) that the continuous functions form a dense set in an $\mathscr{L}^2$ space, which is stated in the present context as

If $f \in \mathscr{L}^2[-\pi, \pi]$, then given an $\varepsilon > 0$ there is a continuous function g on $[-\pi, \pi]$ such that

$$\|f - g\|^2 = \int_{-\pi}^{\pi} |f(x) - g(x)|^2\, dx < \varepsilon.$$

The next theorem illustrates how this fact may be used.

Theorem 6. *If $f \in \mathscr{L}^2[-\pi, \pi]$ and $\{c_n\}$ are its Fourier coefficients, then*

(8)
$$\lim_{n\to\infty} \int_{-\pi}^{\pi} |f(x) - s_n(x)|^2\, dx = 0,$$

(10)
$$\sum_{-\infty}^{\infty} |c_n|^2 = \frac{1}{2\pi} \int_{-\pi}^{\pi} |f(x)|^2\, dx \qquad \textbf{(Parseval's equation)},$$

and in particular $\sum |c_n|^2$ converges.

Proof. Given an $\varepsilon > 0$, let $g\colon [-\pi, \pi] \to E^2$ be continuous and such that

(11)
$$\|f - g\| < \varepsilon \qquad \text{and} \qquad g(\pi) = g(-\pi).$$

[To see that this is possible first choose a continuous function h such that

$$\|f - h\| < \frac{\varepsilon}{2}.$$

Take λ, $0 < \lambda < \pi$, and define g by $g(x) = h(x)$ for $-\pi \leqslant x \leqslant \lambda$, but g linear on $[\lambda, \pi]$ joining the points $(\lambda, h(\lambda))$ and $(\pi, h(-\pi))$. Then

$$\|h - g\|^2 = \int_{\lambda}^{\pi} |h(x) - g(x)|^2\, dx \leqslant 4M^2(\pi - \lambda),$$

where $M \geqslant |h(x)|$, $-\pi \leqslant x \leqslant \pi$. Then for λ such that $4M^2(\pi - \lambda) < (\varepsilon/2)^2$, we have $\|h - g\| < \varepsilon/2$ and both parts of (11) hold for this g.]

Extend g to be 2π periodic on all of E^1 so that Theorem 5 may be used. Let the partial sums of the Fourier series of f and g be $\{s_n\}$ and $\{t_n\}$, respectively. Then Theorem 1 inequality (5), in $\mathscr{L}^2$ norm notation, is

$$\|f - s_n\| \leqslant \|f - t_n\|.$$

Hence $\|f - s_n\| \leqslant \|f - g\| + \|g - t_n\| < \varepsilon + \|g - t_n\|$. By using Theorem 5 for the continuous function g, we have

$$\overline{\lim_{n\to\infty}} \|f - s_n\| \leqslant \varepsilon \qquad \text{and then} \qquad \lim_{n\to\infty} \|f - s_n\| = 0,$$

which is (8) for the present function $f \in \mathscr{L}^2[-\pi, \pi]$.

Then (10) follows exactly as in the proof of Theorem 5 (by using a new function g, this time having $g \in \mathscr{L}^2[-\pi, \pi]$). ∎

The following corollary is analogous to Corollary 1 of Theorem 4.

Corollary. *If $f \in \mathscr{L}^2[-\pi, \pi]$ is such that*

$$\int_{-\pi}^{\pi} f(x)e^{-inx}\,dx = 0, \qquad n = 0, \pm 1, \pm 2, \pm 3, \cdots,$$

then $f(x) = 0$ for almost all $x \in [-\pi, \pi]$ so that $\|f\| = 0$.

For the last two theorems of this section, we use the nonspecific orthonormal system $\{\varphi_n\}$ of Definition 3. The orthogonality condition may be stated in inner product notation as

$$\varphi_n \cdot \varphi_m = \int_a^b \varphi_n(x)\overline{\varphi_m(x)}\,dx = 0, \qquad n \neq m,$$

and the normality condition in terms of the $\mathscr{L}^2$ norm,

$$\|\varphi_n\| = 1, \qquad n = 1, 2, 3, \cdots,$$

since $\|\varphi_n\| = (\varphi_n \cdot \overline{\varphi_n})^{1/2}$. Also, the Fourier coefficients $\{c_n\}$ of an $f \in \mathscr{L}^2[a, b]$ are given by

$$c_n = f \cdot \varphi_n = \int_a^b f(x)\overline{\varphi_n(x)}\,dx.$$

Note how nearly the hypothesis and final conclusion of Theorem 6 are interchanged in the following theorem.

Theorem 7. (Riesz-Fischer) *Let $\{c_n\}$ be any sequence of complex constants such that*

(12) $$\sum_{n=1}^{\infty} |c_n|^2 \quad \textit{converges.}$$

Then there is a function $f \in \mathscr{L}^2[a, b]$ such that $\{c_n\}$ is the sequence of Fourier coefficients of f; that is,

(13) $$c_n = f \cdot \varphi_n, \qquad n = 1, 2, \cdots.$$

Furthermore,

(14) $$\lim_{n\to\infty} \left\| f - \sum_{k=1}^{n} c_k \varphi_k \right\| = 0.$$

Proof. Let $s_n = c_1\varphi_1 + c_2\varphi_2 + \cdots + c_n\varphi_n$. For $m < n$,

$$\|s_n - s_m\|^2 = \left(\sum_{k=m+1}^{n} c_k \varphi_k\right) \left(\sum_{l=m+1}^{n} c_l \varphi_l\right)$$

$$= \sum_{k=m+1}^{n} \left\{c_k \sum_{l=m+1}^{n} \bar{c}_l(\varphi_k \cdot \varphi_l)\right\}$$

$$= \sum_{k=m+1}^{n} c_k \bar{c}_k = \sum_{k=1}^{n} |c_k|^2 - \sum_{k=1}^{m} |c_k|^2.$$

Since (12) holds, the partial sums of this series form a Cauchy sequence and thus $\{s_n\}$ is a Cauchy sequence in $\mathscr{L}^2[a, b]$. But every $\mathscr{L}^2$ space is complete (Theorem 4–6.3 and the last paragraph of Sec. 4–7), which translated into the present context says that there is an $f \in \mathscr{L}^2[a, b]$ such that

$$\lim_{n\to\infty} \|f - s_n\| = 0.$$

This states that (14) holds for these constants $\{c_n\}$, and we still want (13) for this same f.

With n a fixed positive integer, choose $m > n$. Then

$$\int_a^b f\bar{\varphi}_n - c_n = \int_a^b f\bar{\varphi}_n - \int_a^b s_m \bar{\varphi}_n = \int_a^b (f - s_m)\bar{\varphi}_n,$$

and hence by Schwarz inequality,

$$\left| \int_a^b f\bar{\varphi}_n - c_n \right| \leqslant \|f - s_m\| \, \|\bar{\varphi}_n\| = \|f - s_m\|.$$

Thus, by letting $m \to \infty$, we obtain (13). ∎

Notice that Theorem 7 does not say that Parseval's equation holds for any function having $\{c_n\}$ as its Fourier coefficients, and for a very good reason as the following problem shows.

PROBLEM 5. Let $\varphi_n = (1/\sqrt{\pi}) \cos nx$, $n = 1, 2, 3, \cdots$.

(a) This sequence $\{\varphi_n\}$ is orthonormal on $[-\pi, \pi]$.

(b) The function $f(x) = x$, $-\pi < x \leqslant \pi$ and 2π periodic, has $c_n = 0$, $n = 1, 2, 3, \cdots$.

(c) $\displaystyle\sum_{n=1}^{\infty} |c_n|^2 \neq \frac{1}{2\pi} \int_{-\pi}^{\pi} |x|^2 \, dx.$

Definition 4. *An orthonormal sequence* $\{\varphi_n\}$ *is said to be* **complete** *if each* $f \in \mathscr{L}^2[a, b]$ *such that*

$$\int_a^b f(x)\overline{\varphi_n(x)}\, dx = 0, \qquad n = 1, 2, 3, \cdots.$$

has $\|f\| = 0$ [that is, only functions equal to 0 almost everywhere in $[a, b]$ can be normal to all members of $\{\varphi_n\}$].

The corollary of Theorem 6 may now be stated:

The complex exponentials (3) *form a complete orthonormal sequence on* $[-\pi, \pi]$.

Theorem 8. *Let* $\{\varphi_n\}$ *be a complete orthonormal sequence on* $[a, b]$. *If* $f \in \mathscr{L}^2[a, b]$, *then its Fourier coefficients* $\{c_n\}$ *are such that*

$$\sum_{n=1}^{\infty} |c_n|^2 = \int_a^b |f(x)|^2\, dx = \|f\|^2. \tag{15}$$

Proof. By the Bessel inequality (6), $\sum |c_n|^2$ converges. Even though $\{c_n\}$ are the Fourier coefficients of f, we use the Riesz-Fischer theorem to see that there is a function $g \in \mathscr{L}^2[a, b]$ also having $\{c_n\}$ as its Fourier coefficients, but in addition such that

$$\lim_{n\to\infty} \|g - s_n\| = 0,$$

where $s_n = \sum_{k=1}^n c_k \varphi_k$. Hence $\|s_n\| \to \|g\|$ and, since

$$\|s_n\|^2 = |c_1|^2 + |c_2|^2 + \cdots + |c_n|^2,$$

$$\sum_{n=1}^{\infty} |c_n|^2 = \int_a^b |g(x)|^2\, dx = \|g\|^2.$$

Let $h = f - g$. Then

$$\int_a^b h\bar{\varphi}_n = \int_a^b f\bar{\varphi}_n - \int_a^b g\bar{\varphi}_n = c_n - c_n = 0,$$

and thus, by the completeness of $\{\varphi_n\}$, we have $\|h\| = 0$. Therefore $\|f\| = \|g\|$ and (15) holds. ∎

Recall that:

(a) Complex l^2 is the space of all absolutely square summable sequences; that is, a sequence $\{a_n\}$ of complex constants is a point of l^2 iff $\sum |a_n|^2 < \infty$.

(b) $\mathscr{L}^2[-\pi, \pi]$ is the space of absolutely square summable functions on $[-\pi, \pi]$; that is, $f\colon [-\pi, \pi] \to E^2$ is a point of $\mathscr{L}^2[-\pi, \pi]$ iff f is measurable on $[-\pi, \pi]$ and

$$\int_{-\pi}^{\pi} |f|^2 < \infty.$$

As different as these spaces are, there is "essentially" a 1–1 correspondence between them—as we now interpret Theorems 7 and 8 to say. With $f \in \mathscr{L}^2[-\pi, \pi]$, let

$$\hat{f} = \{g \mid g(x) = f(x) \quad \text{a.e.} \qquad \text{on } [-\pi, \pi]\}.$$

[Analogously we accept many representations of a rational number; for example, "one-half" is

$$\left\{\frac{m}{n}\,\middle|\, m = 2n \qquad \text{with } n \text{ a nonzero integer}\right\}.\Big]$$

The "hat" is an equivalence relation, since

$$\hat{f} = \hat{g} \qquad \text{iff } \hat{g} = \hat{f}, \qquad \text{and}$$
$$(\hat{f} = \hat{g} \text{ and } \hat{g} = \hat{h}) \Rightarrow \hat{f} = \hat{h}.$$

Let $\hat{\mathscr{L}}^2$ be the set of all $\hat{f}$ for all $f \in \mathscr{L}^2[-\pi, \pi]$.

Let $\{\varphi_n\}$ be a complete orthonormal system in $\mathscr{L}^2[-\pi, \pi]$; for example, use (3). Then

(i) Given $\{c_n\} \in l^2$, associate the $p \in \hat{\mathscr{L}}^2$ by selecting an $f \in \mathscr{L}^2[-\pi, \pi]$ having $\{c_i\}$ as Fourier coefficients (by Theorem 7) and setting $p = \hat{f}$, knowing, by completeness of $\{\varphi_n\}$, that any g also having $\{c_n\}$ as Fourier coefficients is such that $\hat{g} = \hat{f}$.

(ii) Given any $p \in \hat{\mathscr{L}}^2$, take any $f \in p$ and associate the Fourier coefficients $\{c_n\}$ of f with p, knowing that $\{c_n\} \in l^2$ by Theorem 8. Again, any $g \in p$ furnishes the same $\{c_n\}$.

Moreover, the norms of corresponding members of l^2 and $\hat{\mathscr{L}}^2$ are equal by Parseval's equation.

In a sense, then, any point of $\hat{\mathscr{L}}^2$ has a unique set of coordinates in the $\aleph_0$-dimensional Euclidean space l^2.

7-5 DERIVATIVES ALMOST EVERYWHERE

The following theorem is about an arbitrary real-valued function on an open interval and the Dini derivatives.

Theorem 1. *Given any function f: $(a, b) \to E^1$, let*

$$A = \{x \in (a, b) \mid f'_-(x) \textit{ and } f'_+(x) \textit{ exist}^*\} \qquad \textit{and}$$
$$B = \{x \in A \mid f'_-(x) \neq f'_+(x)\}.$$

Then B is a countable set; that is, the set of points on (a, b) where the left and right derivatives exist, but are not equal, is countable.*

Proof. Let $B_1 = \{x \in A \mid f'_-(x) > f'_+(x)\}$. For each $x \in B_1$ choose a rational number r_x such that $f'_-(x) > r_x > f'_+(x)$. Next choose two rational numbers u_x and v_x, $a < u_x < x < v_x < b$ such that

$$s \in (u_x, x) \Rightarrow \frac{f(s) - f(x)}{s - x} > r_x \qquad \text{and}$$

$$s \in (x, v_x) \Rightarrow r_x > \frac{f(s) - f(x)}{s - x}.$$

Hence

$$[s \in (u_x, v_x), s \neq x] \Rightarrow f(s) - f(x) < r_x(s - x). \tag{1}$$

Thus with each $x \in B_1$ we have associated an ordered triple (r_x, u_x, v_x) of rational numbers.

We shall show that:

(2) No two points of B_1 have the same triple.

For assume $x \in B_1$, $y \in B_1$, $x \neq y$, but $r_x = r_y = r$, $u_x = u_y = u$, and $v_x = v_y = v$. Then $x \in (u, v)$, $y \in (u, v)$,

$$[s \in (u, v), s \neq x] \Rightarrow f(s) - f(x) < r(s - x), \qquad \text{and}$$
$$[s \in (u, v), s \neq y] \Rightarrow f(s) - f(y) < r(s - y).$$

In the first of these use $s = y$ and in the second use $s = x$ to obtain

$$f(y) - f(x) < r(y - x) \qquad \text{and} \qquad f(x) - f(y) < r(x - y),$$

which contradict each other and establish the truth of (2). Hence we have a 1–1 correspondence between the points of B_1 and a subset of triples of rational numbers. Since the set of all triples of rational numbers is countable, B_1 is countable.

In the same way we obtain the countability of

$$B - B_1 = \{x \in B \mid f'_-(x) < f'_+(x)\}. \blacksquare$$

Before proceeding it would be well to review the statement of the Vitali covering Theorem 5-2.1.

Theorem 2. *If f: $[a, b] \to E^1$ is monotonic on $[a, b]$, then f is differentiable almost everywhere on $[a, b]$.*

Proof. Consider f nondecreasing. Then $0 \leqslant \underline{f}'(x) \leqslant \bar{f}'(x)$. We first show that the set

$$A = \{x \in [a, b) \mid \underline{f}'_+(x) < \bar{f}'_+(x)\}$$

has $mA = 0$. With u, v rational numbers $0 < u < v$, let

$$A_{uv} = \{x \in A \mid \underline{f}'_+(x) < u < v < \bar{f}'_+(x)\}.$$

Since $x \in A$ iff $x \in A_{uv}$ for some pair u, v, then

$$A = \bigcup A_{uv}$$

with the (countable) union taken over all pairs u, v of rational numbers, $0 < u < v$. Thus

$$m^*A \leqslant \sum_{(u,v)} m^*A_{uv},$$

and we shall have $mA = 0$ if we succeed in showing that

(3) $m^*A_{uv} = 0$ for all pairs (u, v) of rational numbers such that $0 < u < v$.

We fix a pair u, v of rational numbers such that $0 < u < v$ and set

$$m^* A_{uv} = \lambda.$$

Let $\varepsilon > 0$ be arbitrary. Take an open set G such that

$$A_{uv} \subset G \qquad \text{and} \qquad m(G) < m^*(A_{uv}) + \varepsilon.$$

Corresponding to each $x \in A_{uv}$ there exist arbitrarily small positive numbers h and k such that

$$\text{(4)} \quad [x, x+h] \subset G \cap [a, b] \quad \text{and} \quad f(x+h) - f(x) < uh, \quad \text{and}$$

$$\text{(5)} \quad [x, x+k] \subset G \cap [a, b] \quad \text{and} \quad f(x+k) - f(x) > vk.$$

Let $\mathscr{F}_1$ and $\mathscr{F}_2$ be the families of all closed intervals for which (4) and (5) hold, respectively. Each of these families covers A_{uv} in the sense of Vitali. By the corollary of Vitali's theorem, let

$$I_1 = [x_1, x_1 + h_1], \qquad I_2 = [x_2, x_2 + h_2], \cdots, \qquad I_n = [x_n, x_n + h_n]$$

be a finite disjoint subfamily of $\mathscr{F}_1$ such that, with

$$U = \bigcup_{i=1}^{n} I_i, \qquad \text{then } m^*(A_{uv} - U) < \varepsilon.$$

Since $U \subset G$, $|I_i| = h_i$, and the subfamily is disjoint

$$\sum_{i=1}^{n} h_i = \sum_{i=1}^{n} |I_i| = m\left(\bigcup_{i=1}^{n} I_i\right) = mU \leqslant mG < \lambda + \varepsilon.$$

Also, from (4),

$$\sum_{i=1}^{n} \{f(x_i + h_i) - f(x_i)\} < u \sum_{i=1}^{n} h_i < u(\lambda + \varepsilon).$$

Then the interior U^o of U differs from U only by the finite number of end points of $I_1, I_2, \cdots, I_n$, and thus

$$m^*(A_{uv} - U^o) = m^*(A_{uv} - U) < \varepsilon.$$

The family of all intervals of $\mathscr{F}_2$ which lie in the open set U^o is a covering of $A_{uv} \cap U^o$ in the sense of Vitali, and thus contains a finite disjoint subfamily

$$J_1 = [y_1, y_1 + k_1], \qquad J_2 = [y_2, y_2 + k_2], \cdots, \qquad J_p = [y_p, y_p + k_p],$$

with each $y_j \in A_{uv} \cap U^o$, each $J_j \subset U^o$, and upon letting

$$V = \bigcup_{j=1}^{p} J_j, \qquad \text{then } m^*[A_{uv} \cap U^o - V] < \varepsilon.$$

Also, each J_j is a subinterval of some I_i and thus, since f is nondecreasing,

$$\sum_{j=1}^{p} \{f(y_j + k_j) - f(y_j)\} \leqslant \sum_{i=1}^{n} \{f(x_i + h_i) - f(x_i)\} < u(\lambda + \varepsilon).$$

Furthermore $|J_j| = k_j$ and, by using (5) for the first inequality,

$$m(V) = \sum_{j=1}^{p} k_j < \frac{1}{v}\sum_{j=1}^{p} \{f(y_j + k_j) - f(y_j)\} < \frac{u}{v}(\lambda + \varepsilon).$$

Consequently

$$\begin{aligned}\lambda = m^*(A_{uv}) &= m^*(A_{uv} - U^o) + m^*(A_{uv} \cap U^o)\\ &< \varepsilon + m^*(A_{uv} \cap U^o - V) + m^*(A_{uv} \cap U^o \cap V)\\ &< \varepsilon + \varepsilon + m(V) < 2\varepsilon + \frac{u}{v}(\lambda + \varepsilon).\end{aligned}$$

Therefore

$$0 \leqslant \lambda < \frac{2v + u}{v - u}\,\varepsilon$$

and hence $\lambda = 0$. Thus (3) is true and, as already noted, this means that $mA = 0$.

We now have that $f'_+(x)$ exists* for almost all $x \in [a, b)$. In the same way $f'_-(x)$ exists* at almost all $x \in (a, b]$. Of all the points $x \in (a, b)$ where $f'_-(x)$ and $f'_+(x)$ exists*, there is only a countable subset where they are not equal (Theorem 1). Thus $f'_-(x)$ and $f'_+(x)$ exist* and $f'_-(x) = f'_+(x)$ for almost all $x \in (a, b)$.

It remains only to show that the set

$$B = \{x \in (a, b) \mid f'(x) = \infty\} \text{ has } m^*B = 0.$$

Let q be an arbitrarily large positive number. Then for each $x \in B$ there are arbitrarily small positive numbers h such that

$$[x, x + h] \subset (a, b) \qquad \text{and} \qquad f(x + h) - f(x) > qh.$$

Again by Vitali's theorem, there is a countable disjoint family $\{I_i\}$ of such intervals for which $m\{B - \bigcup I_i\} = 0$, and hence

$$mB \leqslant m(\bigcup I_i) = \sum h_i.$$

Thus $qm(B) \leqslant q \sum h_i < \sum \{f(x_i + h_i) - f(x_i)\} \leqslant f(b) - f(a)$ and it follows that $mB = 0$.

Hence f is differentiable at almost all points of $[a, b]$ if f is nondecreasing on $[a, b]$. The same result holds if f is nonincreasing on $[a, b]$ as seen by using $-f$. ∎

Theorem 3. *If* $f\colon [a, b] \to E^1$ *is nondecreasing on* $[a, b]$ *then* $f' \in \mathscr{L}[a, b]$ *and*

(6) $$\int_a^b f'(x)\,dx \leqslant f(b) - f(a).$$

[*Note:* The Cantor function ψ satisfies the hypotheses on [0, 1] and in addition $\psi'(x) = 0$ almost everywhere on [0, 1]. Thus

$$\int_0^1 \psi'(x)\,dx = 0 < 1 = \psi(1) - \psi(0),$$

which shows that the strict inequality in (6) can hold. We know that equality holds in (6) if f' is continuous (Corollary 1, page 379). We shall obtain equality in (6) under conditions less stringent than continuity of f', but of course stronger than the hypothesis of this theorem.]

Proof. Extend the domain of f to $[a, b+1]$ by defining

$$f(x) = f(b) \qquad \text{for } b \leqslant x \leqslant b+1.$$

Then f is nondecreasing on $[a, b+1]$ and $f'(x)$ exists for almost all $x \in [a, b+1]$ by Theorem 2. Hence, at almost all $x \in [a, b]$,

$$f'(x) = \lim_{n\to\infty} n\left\{f\left(x + \frac{1}{n}\right) - f(x)\right\}.$$

The functions in the sequence $\{\varphi_n\}$ defined by

$$\varphi_n(x) = n\left\{f\left(x + \frac{1}{n}\right) - f(x)\right\}$$

are all measurable, being the difference of two measurable functions since, for any constant, each of

$$\left\{x \,\middle|\, f\left(x + \frac{1}{n}\right) > c\right\} \qquad \text{and} \qquad \{x \mid f(x) > c\}$$

is either empty or an interval. Thus f' is the limit almost everywhere on $[a, b]$ of a sequence of measurable functions and is itself measurable by Theorem 4-3.6. Hence the Lebesgue integral of f' on $[a, b]$ exists* with

$$\int_a^b f'(x)\,dx \leqslant \varliminf_{n\to\infty} \int_a^b \varphi_n(x)\,dx$$

by Fatou's lemma (Theorem 4-5.3). But

$$\int_a^b f\left(x + \frac{1}{n}\right) dx = \int_{a+1/n}^{b+1/n} f(x)\,dx$$

(since both sides mean the same thing) and

$$\begin{aligned}
\int_a^b \varphi_n(x)\,dx &= n\left\{\int_{a+1/n}^{b+1/n} f(x)\,dx - \int_a^b f(x)\,dx\right\} \\
&= n\left\{\int_b^{b+1/n} f(x)\,dx - \int_a^{a+1/n} f(x)\,dx\right\} \\
&\leqslant n\left\{\int_b^{b+1/n} f(b)\,dx - \int_a^{a+1/n} f(a)\,dx\right\} \\
&= f(b) - f(a).
\end{aligned}$$

Hence the existence of the integral and the inequality in (6) are seen. ∎

As noted above, the nondecreasing Cantor function ψ has its derivative equal to zero almost everywhere because it stays constant over each complementary interval of the Cantor set. The following example shows that even a strictly increasing continuous function may have its derivative equal to zero almost everywhere.

EXAMPLE 1. A sequence $f_0, f_1, f_2, \cdots, f_k, \cdots$ of functions is defined on [0, 1] by taking the following steps:

0. For $n = 0, 1, 2, 3, \cdots; f_n(0) = 0, \quad f_n(1) = 1, \quad$ and $\quad f_0(x) = x$.

1. For $n = 1, 2, 3, \cdots; f_n(\frac{1}{2}) = \frac{1}{3}f_0(0) + \frac{2}{3}f_0(1) = \frac{2}{3}$, and f_1 linear on $[0, \frac{1}{2}]$ and $[\frac{1}{2}, 1]$.

2. For $n = 2, 3, 4, \cdots; f_n\left(\frac{2i+1}{4}\right) = \frac{1}{3}f_1\left(\frac{i}{2}\right) + \frac{2}{3}f_1\left(\frac{i+1}{2}\right), \qquad i = 0, 1,$

   and f_2 linear on $\left[\frac{j}{4}, \frac{j+1}{4}\right], \qquad j = 0, 1, 2, 3.$

. . .

k. For $n = k, k+1, k+2, \cdots;$

$$f_n\left(\frac{2i+1}{2^k}\right) = \frac{1}{3}f_{k-1}\left(\frac{i}{2^{k-1}}\right) + \frac{2}{3}f_{k-1}\left(\frac{i+1}{2^{k-1}}\right), \qquad i = 0, 1, 2, \cdots, 2^{k-1} - 1,$$

$$f_k \text{ linear on } \left[\frac{j}{2^k}, \frac{j+1}{2^k}\right], \qquad j = 0, 1, 2, \cdots, 2^k - 1.$$

. . .

Thus, as soon as f_n is assigned a value at a point of the form $j/2^k$, then $f_{n+1}, f_{n+2}, f_{n+3}, \cdots$ all have this same value at this point.

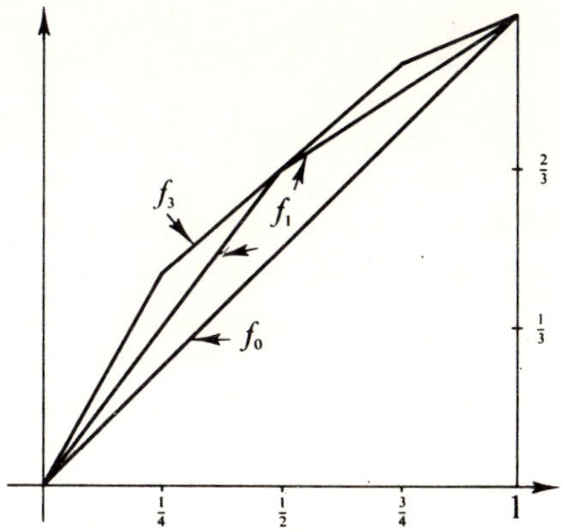

FIG. 1

The assignment is also such that

$$0 = f_k(0) < f_k\left(\frac{1}{2^k}\right) < f_k\left(\frac{2}{2^k}\right) < \cdots < f_k(1) = 1,$$

and from the linearity of f_k between these points we see that f_k is continuous and strictly increasing on [0, 1] onto [0, 1].

Also, if $0 \leqslant x < j/2^k < y \leqslant 1$ and $n > k$, then

$$f_n(x) \leqslant f_k\left(\frac{j}{2^k}\right) \leqslant f_n(y).$$

Clearly, for each $x \in [0, 1]$, the sequence $\{f_n(x)\}$ is nondecreasing so that

$$\lim_{n\to\infty} f_n(x) = f(x) \quad \text{exists.}$$

We thus have a continuous function f defined on [0, 1] into [0, 1] and we shall prove:

(a) f is strictly increasing on [0, 1] onto [0, 1]; and
(b) $f'(x) = 0$ for almost all $x \in [0, 1]$.

First take $0 \leqslant x < y \leqslant 1$. Then there is a k and a j such that

$$x < \frac{j}{2^k} < \frac{j+1}{2^k} < y.$$

For this k and j, if $n > k$ then

$$f_n(x) \leqslant f_k\left(\frac{j}{2^k}\right) < f_k\left(\frac{j+1}{2^k}\right) \leqslant f_n(y).$$

Let $n \to \infty$ to see that $f(x) < f(y)$ so that (a) is proved.

Let B be the set of binary points of [0, 1]; that is,

$$B = \left\{0 \leqslant x \leqslant 1 \,\middle|\, x = \frac{j}{2^k} \quad \text{for some } k \text{ and } j\right\}.$$

Fix an $x \in [0, 1] - B$. For each $k = 0, 1, 2, \cdots$ there is a j (depending upon x and k) such that

$$\frac{j}{2^k} < x < \frac{j+1}{2^k}$$

and we set $\alpha_k = j/2^k$ and $\beta_k = (j+1)/2^k$. We thus have a nondecreasing sequence $\{\alpha_k\}$ and a nonincreasing sequence $\{\beta_k\}$ such that $\alpha_k \to x$ and $\beta_k \to x$. Moreover, for each k, either

$$\alpha_{k+1} = \alpha_k, \qquad \beta_{k+1} = \tfrac{1}{2}(\alpha_k + \beta_k),$$
$$f_{k+1}(\alpha_{k+1}) = f_k(\alpha_k), \qquad \text{and} \qquad f_{k+1}(\beta_{k+1}) = \tfrac{1}{3}f_k(\alpha_k) + \tfrac{2}{3}f_k(\beta_k),$$

or else

$$\alpha_{k+1} = \tfrac{1}{2}(\alpha_k + \beta_k), \qquad \beta_{k+1} = \beta_k,$$
$$f_{k+1}(\alpha_{k+1}) = \tfrac{1}{3}f_k(\alpha_k) + \tfrac{2}{3}f_k(\beta_k), \qquad \text{and} \qquad f_{k+1}(\beta_{k+1}) = f_k(\beta_k).$$

Thus

$$f_{k+1}(\beta_{k+1}) - f_{k+1}(\alpha_{k+1}) = \begin{Bmatrix} \frac{2}{3} \\ \frac{1}{3} \end{Bmatrix} [f_k(\beta_k) - f_k(\alpha_k)].$$

From this recurrence relation and $f_0(\beta_0) - f_0(\alpha_0) = 1$,

$$0 \leqslant f_k(\beta_k) - f_k(\alpha_k) = p_k,$$

where those factors of p_k which are not $\frac{2}{3}$ are $\frac{1}{3}$. Since $n > k \Rightarrow f_n(\alpha_k) = f_k(\alpha_k)$ and $f_n(\beta_k) = f_k(\beta_k)$, we have

$$n > k \Rightarrow f_n(\beta_k) - f_n(\alpha_k) = p_k.$$

By letting $n \to \infty$, it follows that

$$0 \leqslant \frac{f(\beta_k) - f(\alpha_k)}{\beta_k - \alpha_k} = p_k \cdot 2^k.$$

If $f'(x)$ exists, then by Theorem 7–1.2,

$$f'(x) = \lim_{k\to\infty} \frac{f(\beta_k) - f(\alpha_k)}{\beta_k - \alpha_k} = \lim_{k\to\infty} (p_k 2^k).$$

Assume $f'(x)$ exists and is not zero. Then not only does $p_k 2^k \to f'(x) \neq 0$ but also $p_{k+1} 2^{k+1} \to f'(x)$, so that

$$1 = \frac{f'(x)}{f'(x)} = \lim_{k\to\infty} \frac{p_{k+1} 2^{k+1}}{p_k 2^k}.$$

But this is impossible, since

$$\frac{p_{k+1} 2^{k+1}}{p_k 2^k} \quad \text{is either} \quad \frac{2}{3} \cdot 2 \quad \text{or} \quad \frac{1}{3} \cdot 2.$$

Hence, whenever $x \in [0, 1] - B$ and $f'(x)$ exists, it must be that $f'(x) = 0$. Since $mB = 0$ and $f'(x)$ exists for almost all $x \in [a, b]$, by Theorem 2, we see that (b) holds. ▌

[The method also works if $\frac{1}{3}$ and $\frac{2}{3}$ are replaced by λ and $1 - \lambda$ with $0 < \lambda < \frac{1}{2}$ (or $\frac{1}{2} < \lambda < 1$ for which $\{f_n\}$ is nonincreasing).]

The following theorem gives one situation in which summation and differentiation may be interchanged, at least almost everywhere.

Theorem 4. *Let $\{f_n\}$ be a sequence of functions, each nondecreasing (or each nonincreasing) on an interval $[a, b]$. If the series $\sum f_n$ converges,*

$$\sum_{n=1}^{\infty} f_n(x) = s(x) \qquad \text{for each } x \in [a, b],$$

then term-by-term differentiation holds a.e. on $[a, b]$; that is, the convergence and equality

$$\sum_{n=1}^{\infty} f_n'(x) = s'(x) \text{ hold a.e. in } [a, b].$$

Proof. Consider that each $f_n(a) = 0$ [which is no loss of generality, since we could take $F_n(x) = f_n(x) - f_n(a)$ and $S(x) = s(x) - s(a)$]. Since each f_n is nondecreasing, so is s. Thus the set $A \subset [a, b]$ where all f_n' and s' exist has $mA = b - a$, by Theorem 2. With

$$s_n = f_1 + f_2 + \cdots + f_n,$$

it follows that

$$0 \leqslant s_n'(x) \leqslant s_{n+1}'(x) \leqslant s'(x), \qquad x \in A.$$

We thus have the existence of the limit

$$\text{(7)} \qquad \lim_{n\to\infty} s_n'(x) \leqslant s'(x); \qquad \text{that is, } \sum_{n=1}^{\infty} f_n'(x) \leqslant s'(x), \qquad x \in A.$$

If, therefore, there is a subsequence $\{s_{n_k}(x)\}$ such that

$$\text{(8)} \qquad \lim_{k\to\infty} s_{n_k}'(x) = s'(x), \qquad x \in A$$

the desired result will follow.

Note that $\{s - s_n\}$ is also a nondecreasing sequence and

$$0 \leqslant s(x) - s_n(x) = f_{n+1}(x) + f_{n+2}(x) + \cdots \leqslant s(b) - s_n(b).$$

We now choose $n_1 < n_2 < n_3 < \cdots$ such that

$$s(b) - s_{n_k}(b) < 2^{-k}.$$

Hence the series $\sum_k [s(x) - s_{n_k}(x)]$ converges. Thus all conditions of the theorem are satisfied for this series and hence by (7) the series

$$\sum_k [s'(x) - s_{n_k}'(x)] \text{ converges for } x \in A.$$

Consequently (8) holds for the subsequence chosen and therefore, as already noted, we also have

$$\lim_{n\to\infty} s_n'(x) = s'(x), \qquad x \in A \quad \text{with } mA = b - a. \ \blacksquare$$

7–6 BOUNDED VARIATION

The reason for introducing functions of bounded variation in this chapter is because of their relation to derivatives of Lebesgue integrals. These functions will appear again in connection with Stieltjes integrals in the next chapter.

Definition 1. *Given an interval* $[a, b]$, *a finite set* $P \subset [a, b]$ *with* $a \in P$ *and* $b \in P$ *is called a* **partition** *of* $[a, b]$. *In designating a partition as*

$$P = \{x_0, x_1, x_2, \cdots, x_{n-1}, x_n\}$$

the points will be considered as listed in nondecreasing order

$$a = x_0 \leqslant x_1 \leqslant x_2 \leqslant \cdots \leqslant x_{n-1} \leqslant x_n = b.$$

With P listed in this way, the **norm** *of P is denoted by* $\|P\|$ *and defined*† *by*

$$\|P\| = \max(x_i - x_{i-1}), \qquad i = 1, 2, \cdots, n.$$

If Q is also a partition of $[a, b]$ *such that* $P \subset Q$, *then Q is said to be a* **refinement** *of P. Hence if* $P \subset Q$, *then* $\|Q\| \leqslant \|P\|$.

If P_1 and P_2 are partitions of $[a, b]$, then

$$P_1 \cup P_2 \qquad \text{and} \qquad P_1 \cap P_2$$

are both partitions of $[a, b]$, since they are both finite subsets of $[a, b]$ and both contain a and b. Also, $P_1 \cup P_2$ is a refinement of both P_1 and P_2, but P_1 and P_2 are both refinements of $P_1 \cap P_2$.

Given a point c, $a < c < b$, let P be a partition of $[a, b]$ having $c \in P$. Then

$$P \cap [a, c] \text{ is a partition of } [a, c] \text{ and}$$

$$P \cap [c, b] \text{ is a partition of } [c, b].$$

Also, if P_1 is a partition of $[a, c]$ and P_2 is a partition of $[c, b]$, then $P_1 \cup P_2$ is a partition of $[a, b]$.

Definition 2. *Given* $f\colon [a, b] \to E^1$ *and* $P = \{x_0, x_1, \cdots, x_n\}$ *a partition of* $[a, b]$, *the* **variation** *of f* **over** *P is defined by*

$$V_a^b(f, P) = \sum_{i=1}^{n} |f(x_i) - f(x_{i-1})|.$$

The **variation** *of f from a to b is defined by*

$$V_a^b(f) = \sup V_a^b(f, P),$$

where the supremum is taken over all partitions of $[a, b]$. *If* $V_a^b(f)$ *is finite, then f is said to be of* **bounded variation**‡ *on* $[a, b]$. *The family of all functions defined on* $[a, b]$ *and of bounded variation on* $[a, b]$ *will be denoted by* BV_a^b *or simply BV.*

PROBLEM 1. Given $f\colon [a, b] \to E^1$, then

(a) $|f(x) - f(a)| \leqslant V_a^x(f)$ for $a \leqslant x \leqslant b$.

(b) If $a < c < b$ and P is a partition of $[a, b]$ and $P_1 = P \cup \{c\}$, then P_1 is also a partition of $[a, b]$ and

$$V_a^b(f, P) \leqslant V_a^b(f, P_1).$$

(c) If f is monotonic on $[a, b]$, then

$$V_a^b(f) = |f(b) - f(a)|.$$

(d) $P_1 \subset P_2 \Rightarrow V_a^b(f, P_1) \leqslant V_a^b(f, P_2)$.

† The word "norm" as used here does not have the same significance as the norm for a vector space.

‡ "Finite variation" would be a better nomenclature, but "bounded variation" is more commonly used. Some books use "total variation" where we use only "variation."

PROBLEM 2. If f is differentiable on $[a, b]$ with $|f'(x)| \leqslant M < \infty$ for $a \leqslant x \leqslant b$, then $f \in BV$ and

$$V_a^b(f) \leqslant M(b-a).$$

PROBLEM 3. The function f defined on $[0, 1]$ by

$$f(x) = \begin{cases} x\cos(\pi/2x), & 0 < x \leqslant 1 \\ 0, & x = 0 \end{cases}$$

is continuous but not of bounded variation on $[0, 1]$.

Theorem 1. *Given f: $[a, b] \to E^1$ and $a < c < b$, then*

$$V_a^b f = V_a^c(f) + V_c^b(f).$$

Proof. Let u and v be any numbers such that

$$u < V_a^c(f) \qquad \text{and} \qquad v < V_c^b(f). \tag{1}$$

Take a partition P_1 of $[a, c]$ and a partition P_2 of $[c, b]$ such that $u < V_a^c(f, P_1)$ and $v < V_c^b(f, P_2)$. Then $P_1 \cup P_2$ is a partition of $[a, b]$ and

$$u + v < V_a^c(f, P_1) + V_c^b(f, P_2) = V_a^b(f, P_1 \cup P_2) \leqslant V_a^b(f).$$

Since u and v were any numbers satisfying (1), it follows that

$$V_a^c(f) + V_c^b(f) \leqslant V_a^b(f). \tag{2}$$

Next let r be any number such that $r < V_a^b(f)$ and take a partition P of $[a, b]$ such that

$$r < V_a^b(f, P).$$

Then $P_1 = (P \cap [a, c]) \cup \{c\}$ and $P_2 = (P \cap [c, b]) \cup \{c\}$ are partitions of $[a, c]$ and $[c, b]$, respectively, and

$$r < V_a^b(f, P) \leqslant V_a^b(f, P_1 \cup P_2) = V_a^c(f, P_1) + V_c^b(f, P_2) \leqslant V_a^c(f) + V_c^b(f).$$

Hence an inequality reverse to (2) is obtained and thus the stated equality holds. ∎

Corollary. *$f \in BV_b{}^a$ iff $f \in BV_a{}^c$ and $f \in BV_c{}^b$ for each $c \in [a, b]$.*

Definition 3. *For $f \in BV_a{}^b$, the functions π_f and ν_f defined for each $x \in [a, b]$ by*

$$\pi_f(x) = V_a^x(f) \qquad \text{and} \qquad \nu_f(x) = \pi_f(x) - f(x)$$

are called the **positive** *and* **negative** *variation, respectively, of f.*

Theorem 2. *If $f \in BV_a{}^b$, then*

(a) *π_f and ν_f are nondecreasing.*
(b) *f is the difference of two nondecreasing functions.*
(c) *π_f and ν_f are both left continuous at $c \in (a, b]$, right continuous at $c \in [a, b)$, or continuous at $c \in [a, b]$ iff f has, respectively, the same property.*

Proof (a). Since f is fixed, we drop it as a subscript and write π and ν. Take $a \leqslant x < y \leqslant b$. Then $P = \{x, y\}$ is a partition of the interval $[x, y]$ so that

$$0 \leqslant |f(y) - f(x)| \leqslant V_x^y f = V_a^y f - V_a^x f = \pi(y) - \pi(x).$$

Hence $f(y) - f(x) \leqslant \pi(y) - \pi(x)$ so that

$$0 \leqslant [\pi(y) - f(y)] - [\pi(x) - f(x)] = \nu(y) - \nu(x).$$

Proof (b). $f = \pi - \nu$ so expresses f.

Proof (c). Let $\varepsilon > 0$ be given.

First consider π left-continuous at $c \in (a, b]$ and let $\delta > 0$ be such that

$$a \leqslant c - \delta < x < c \Rightarrow \pi(c) - \pi(x) < \varepsilon \Rightarrow |f(c) - f(x)| \leqslant \pi(c) - \pi(x) < \varepsilon,$$

the second implication following as in part (a).

Conversely, let f be left-continuous at $c \in (a, b]$ and determine $\delta_1 > 0$ such that

$$a \leqslant c - \delta_1 < x < c \Rightarrow |f(x) - f(c)| < \varepsilon.$$

Take $P = \{a = x_0, x_1, x_2, \cdots, x_{n-1}, x_n = c\}$ a partition of $[a, c]$ such that

$$V_a^c f - \varepsilon < V_a^c(f, P).$$

Take $0 < \delta < \min[\delta_1, c - x_{n-1}]$. Hence, for any x such that $c - \delta < x < c$, we have $x_{n-1} < x < x_n$,

$$V_a^c f - \varepsilon < |f(x_1) - f(x_0)| + \cdots + |f(x) - f(x_{n-1})| + |f(x_n) - f(x)|$$

$$\leqslant V_a^x f + |f(c) - f(x)|; \text{ that is,}$$

$$0 \leqslant V_a^c f - V_a^x f < 2\varepsilon.$$

In the other notation we thus have

$$c - \delta < x < c \Rightarrow c \leqslant \pi(c) - \pi(x) \leqslant 2\varepsilon,$$

and hence π is also left-continuous at c.

If, therefore, either f or π is left-continuous at $c \in (a, b]$, then so is the other and hence also $\nu = \pi - f$. Furthermore, if both π and ν are left-continuous at $c \in (a, b]$, then so is f (from left-continuity of π alone).

[Note that left-discontinuity of f at a point does not imply left-discontinuity of both π and ν. For example, let $f(x) = 0$ for $0 \leqslant x < 1$ and $f(x) = 1$. Then $\pi = f$ is left-discontinuous, but ν is left-continuous, at 1.]

The remainder of part (c) is proved in a similar way. ∎

Corollary. *A function f: $[a, b] \to E^1$ is of bounded variation iff it can be expressed as the difference of two nondecreasing functions. Moreover, f is continuous and of bounded variation iff it can be expressed as the difference of two continuous nondecreasing functions.*

[Note that if f is continuous and of bounded variation, it may also be expressed as the difference of two nondecreasing functions which have discontinuities (even countably many of them). For example, let g be a nondecreasing function having discontinuities and set

$$f = (\pi + g) - (v + g).]$$

Theorem 3. *If $f \in BV_a^{\,b}$, then f is differentiable almost everywhere in $[a, b]$ and f' is a summable function on $[a, b]$.*

Proof. Both π and v are differentiable at almost all points of $[a, b]$, by Theorem 7–5.2. They are both then measurable, bounded, and summable on $[a, b]$. Thus $f = \pi - v$ is also differentiable almost everywhere in $[a, b]$ and $f' = \pi' - v'$ is summable by Theorem 4–5.8. ∎

Theorem 4. *Every absolutely continuous function on $[a, b]$ is of bounded variation on $[a, b]$.*

Proof. Let f: $[a, b] \to E^1$ be absolutely continuous on $[a, b]$. Corresponding to $\varepsilon = 1$, take $\delta > 0$ such that if pairwise nonoverlapping subintervals $J_j = [a_j, b_j]$ are such that

$$\sum_{j=1}^{p} (b_j - a_j) < \delta, \qquad \text{then } \sum_{j=1}^{p} |f(b_j) - f(a_j)| < 1.$$

Next fix a subdivision $P = \{a = x_0, x_1, \cdots, x_n\}$ such that $\|P\| < \delta$. Hence n is also fixed. For each $i = 1, 2, \cdots, n$, consider a subdivision P_i of $[x_{i-1}, x_i]$. Since the sum of the lengths of intervals between points of P_i is $< \delta$, we first have

$$V_{x_{i-1}}^{x_i}(f, P_i) < 1 \qquad \text{and then} \qquad V_{x_{i-1}}^{x_i}(f) \leqslant 1.$$

By the additivity property of Theorem 1,

$$V_a^b(f) = \sum_{i=1}^{n} V_{x_{i-1}}^{x_i}(f) \leqslant n.$$

Hence $f \in BV_a^{\,b}$. ∎

Theorems 3, 4, and 4–5.5 then give

Theorem 5. *If f: $[a, b] \to E^1$ is absolutely continuous on $[a, b]$, then f is differentiable almost everywhere in $[a, b]$, f' is summable on $[a, b]$, and hence*

$$\int_a^b f'(x)\,dx$$

exists (that is, is finite).

Definition 4. *Given that f: $[a, b] \to E^1$ is bounded on $[a, b]$.*

(i) $\Omega(f) = \sup f - \inf f$ *on $[a, b]$ is called the* **oscillation of f on** $[a, b]$.

(ii) *With* $P = \{a = x_0, x_1, \cdots, x_n = b\}$ *a partition of* $[a, b]$ *and* $\Omega_i(f)$ *the oscillation of* f *on* $[x_{i-1}, x_i]$, *the sum*

$$\Omega_a^b(f, P) = \sum_{i=1}^{n} \Omega_i(f)$$

is called the **oscillation of** f **over** P.

(iii) $\Omega_a^b(f) = \sup \Omega_a^b(f, P)$ *is called the* **total oscillation** *of* f *on* $[a, b]$.

PROBLEM 4. If f: $[a, b] \to E^1$ is bounded on $[a, b]$, then $\Omega_a^b(f) = V_a^b(f)$.

EXAMPLE. Let $r_1, r_2, r_3, \cdots$ be an ordering of the rational numbers on (0, 1) and let $0 < a < 1$. A function f is defined by

$$f(x) = \begin{cases} a^k & \text{if } x = r_k \\ 0 & \text{if } x = 0,\ x = 1, \text{ or } x \text{ is irrational.} \end{cases}$$

We shall show that

$$V_0^1(f) = \frac{2a}{1-a}$$

and thus have:

A function of bounded variation need not be monotone on any subinterval of its domain.

Let $x_1 < x_3 < \cdots < x_{2n-1}$ be a listing of $r_1, r_2, \cdots, r_n$. Set $x_0 = 0$, $x_{2n} = 1$, and select x_{2k} irrational such that

$$x_{2k-1} < x_{2k} < x_{2k+1} \qquad \text{for} \qquad k = 1, 2, \cdots, n-1.$$

Then $P = \{x_0, x_1, x_2, \cdots, x_{2n}\}$ is a partition of [0, 1] and

$$\Omega(f) \geqslant \Omega(f, P) = \sum_{i=1}^{2n} \Omega_i(f) = 2\sum_{k=1}^{n} a^k.$$

Since this holds for $n = 1, 2, 3, \cdots$, we have

$$\Omega(f) \geqslant \frac{2a}{1-a}.$$

On the other hand, let $P = \{x_0, x_1, \cdots x_n\}$ be any partition of [0, 1]. For k any one of $1, 2, \cdots, n-1$, there is one and only one i such that

$$r_k \in (x_{i-1}, x_{i+1}), \qquad i = 1, 2, \cdots, n-1$$

and for this i, $\Omega_i(f) + \Omega_{i-1} \leqslant 2a^k$. Thus

$$\Omega(f, P) \leqslant 2\sum_{k=1}^{n-1} a^k < \frac{2a}{1-a}.$$

Consequently $\Omega(f) \leqslant 2a/(1-a)$. Hence, from Problem 4,

$$V(f) = \Omega(f) = \frac{2a}{1-a}.$$

7–7 DERIVATIVES AND INTEGRALS

Lemma 1. *If* $f\colon [a, b] \to E^1$ *is absolutely continuous and if* $f'(x) = 0$ *almost everywhere in* $[a, b]$, *then* f *is constant on* $[a, b]$.

Proof. Set $A = \{x \in (a, b) \mid f'(x) \text{ exists and } f'(x) = 0\}$. Then A is almost all of $[a, b]$ so that

$$mA = b - a.$$

Let $\varepsilon > 0$ be given. Determine $\delta > 0$, from absolute continuity of f, such that if $J_j = [a_j, b_j]$, $j = 1, 2, \cdots, p$ are pairwise nonoverlapping subintervals of $[a, b]$ for which

$$\sum_{j=1}^{p} |J_j| < \delta, \qquad \text{then} \qquad \sum_{j=1}^{p} |f(b_j) - f(a_j)| < \varepsilon.$$

With $x \in A$, consider all intervals $[x, x + h] \subset (a, b)$ with $h > 0$ for which

$$\frac{|f(x + h) - f(x)|}{h} < \varepsilon. \tag{1}$$

Since this holds for all sufficiently small $h > 0$, the family of all such intervals for all $x \in A$ is a covering of A in the Vitali sense. Hence, with the δ determined above in mind, there are pairwise disjoint intervals

$$I_1 = [x_1, x_1 + h_1], \qquad I_2 = [x_2, x_2 + h_2], \cdots, \qquad I_n = [x_n, x_n + h_n]$$

in this family such that

$$m\left(A - \bigcup_{i=1}^{n} I_i\right) < \delta,$$

which is possible by taking $A \subset G \subset (a, b)$ with G open and $m(G - A) < \delta$. Since

$$A \subset \left(A - \bigcup_{i=1}^{n} I_i\right) \cup \left(\bigcup_{i=1}^{n} I_i\right),$$

we have

$$b - a = mA < \delta + \sum_{i=1}^{n} |I_i| \qquad \text{and}$$

$$\sum_{i=1}^{n} |I_i| > b - a - \delta.$$

Consider the intervals I_i renamed, if necessary, so that

$$x_1 < x_2 < \cdots < x_n.$$

Since all I_i lie in (a, b) and are pairwise disjoint,

$$P = \{a, x_1, x_1 + h_1, x_2, x_2 + h_2, \cdots, x_n, x_n + h_n, b\}$$

is a partition of $[a, b]$. Upon setting $J_0 = [a, x_1]$,

$$J_i = [x_i + h_i, x_{i+1}], \qquad i = 1, 2, \cdots, n-1$$

and $J_n = [x_n + h_n, b]$ we see that

$$b - a = \sum_{i=0}^{n} |J_i| + \sum_{i=1}^{n} |I_i| > \sum_{i=0}^{n} |J_i| + b - a - \delta, \qquad \text{and} \qquad \sum_{i=0}^{n} |J_i| < \delta.$$

Hence, because of the way δ was determined,

$$\left| f(x_1) - f(a) + \sum_{i=1}^{n} \{f(x_{i+1}) - f(x_i + h_i)\} + f(b) - f(x_n + h_n) \right| < \varepsilon.$$

Also, from (1), with $x = x_i$ and $h = h_i$,

$$\left| \sum_{i=1}^{n} \{f(x_i + h_i) - f(x_i)\} \right| < \varepsilon \sum_{i=1}^{n} h_i \leqslant \varepsilon(b - a).$$

Thus

$$|f(b) - f(a)| < \varepsilon(1 + b - a)$$

and therefore $f(b) = f(a)$.

All of the above proof applies equally well to any interval $[a, x]$ with $a < x \leqslant b$, and hence

$$f(x) = f(a) \qquad \text{for all } x \in [a, b]. \blacksquare$$

Theorem 1. *If f' and g' are the derived functions of absolutely continuous functions f and g on $[a, b]$ and if*

$$f'(x) = g'(x) \qquad \textit{for almost all } x \in [a, b],$$

then there is a constant c such that

$$f(x) = g(x) + c \qquad \textit{for all } x \in [a, b].$$

Proof. This follows from Lemma 1, since $f - g$ is absolutely continuous and its derivative $f'(x) - g'(x) = 0$ for almost all $x \in [a, b]$. $\blacksquare$

Theorem 2. *If a function F is defined at almost all points of $[a, b]$, if F is measurable on $[a, b]$, and if $F \in \mathscr{L}[a, b]$ (so that $\int_b^a |F(t)|\, dt < \infty$), then the function f defined by*

$$f(x) = c + \int_a^x F(t)\, dt, \qquad a \leqslant x \leqslant b, \quad c \textit{ any constant}$$

is such that $f'(x) = F(x)$ for almost all $x \in [a, b]$ and

$$c + \int_a^x f'(t)\, dt = f(x), \qquad \textit{for all } x \in [a, b].$$

Proof. Set $A = \{x \in (a, b) \mid F(x)$ is defined and $f'(x)$ exists$\}$. By the part of Theorem 6-5.2 already proved, f is absolutely continuous and thus f' exists a.e. in $[a, b]$ by Theorem 7–6.5. Thus A is almost all of $[a, b]$ and

$$mA = b - a.$$

Set $A_0 = \{x \in A \mid F(x) < f'(x)\}$ and

$$A_{pq} = \{x \in A \mid F(x) < p < q < f'(x)\}$$

with p and q rational. Thus A_0 and all A_{pq} are measurable sets and

$$A_0 = \bigcup A_{pq}$$

with the (countable) union taken over all rational pairs p, q with $p < q$. We shall show that each set

$$A_{pq} \text{ has } mA_{pq} = 0.$$

Accordingly, select any pair p, q of rational numbers with $p < q$ and, to simplify notation, set

$$B = A_{pq}.$$

Let $\varepsilon > 0$ be given. Determine a number $\delta > 0$ such that, with M a measurable set,

$$M \subset [a, b], \qquad mM < \delta \Rightarrow \left| \int_M F(t)\, dt \right| < \varepsilon,$$

which is possible by Theorem 4–5.10. Next let G be an open set such that

$$B \subset G \subset (a, b) \qquad \text{and} \qquad m(G - B) < \delta.$$

With $x \in B$, consider all intervals $[x, x + h]$ such that

$$[x, x + h] \subset G \qquad \text{and} \qquad \frac{f(x + h) - f(x)}{h} > q.$$

Since both of these hold for all sufficiently small $h > 0$, the family of all such intervals for all $x \in B$ is a covering of B in the Vitali sense. Select a countable pairwise disjoint subfamily $\{I_i\}$ such that

$$m(B - S) = 0, \qquad \text{where } S = \bigcup I_i.$$

Since $B \subset (B - S) \cup S$, then $m(B) \leqslant m(B - S) + mS$ and

$$m(B) \leqslant mS.$$

With the designation $I_i = [x_i, x_i + h_i]$, then $mS = \sum h_i$ and

$$\int_S F(t)\, dt = \sum_i \int_{x_i}^{x_i + h_i} F(t)\, dt = \sum_i \{f(x_i + h_i) - f(x_i)\}$$
$$> q \sum_i h_i = qmS \geqslant qmB.$$

We now attack this integral from another point of view. First $S = (BS) \cup (S - B)$ and $(BS) \cap (S - B) = \varnothing$ so that

$$\int_S F(t)\,dt = \int_{BS} F(t)\,dt + \int_{S-B} F(t)\,dt.$$

Since $S \subset G$, then $S - B \subset G - B$, $m(S - B) \leqslant m(G - B) < \delta$, and

$$\int_{S-B} F(t)\,dt \leqslant \left|\int_{S-B} F(t)\,dt\right| < \varepsilon$$

from the way δ was determined. Also,

$$\int_{BS} F(t)\,dt = \int_B F(t)\,dt - \int_{B-S} F(t)\,dt = \int_B F(t)\,dt,$$

since $B = (BS) \cup (B - S)$ with $m(B - S) = 0$. Thus

$$\int_S F(t)\,dt < \int_B F(t) + \varepsilon \leqslant pmB + \varepsilon,$$

since $t \in B = A_{pq} \Rightarrow F(t) < p$.

This same integral was shown above to be $\geqslant qmB$. Therefore $qmB < pmB + \varepsilon$ and thus

$$qmB \leqslant pmB.$$

But $p < q$, and thus $mB = 0$; that is, $mA_{pq} = 0$.

Thus $mA_{pq} = 0$ for all rationals $p < q$ so that $mA_0 = 0$. Therefore

$$F(x) \geqslant f'(x) \qquad \text{for almost all } x \in [a, b].$$

Upon setting $G = -F$ and $g(x) = \int_a^x G(t)\,dt$, we have

$$G(x) \geqslant g'(x); \qquad \text{that is, } -F(x) \geqslant -f'(x)$$

for almost all $x \in [a, b]$. Hence the theorem is seen to hold. ∎

The following theorem collects previously proved results.

Theorem 3. *For any absolutely continuous function*

$$f: [a, b] \to E^1,$$

the derived function f' exists almost everywhere in $[a, b]$, f' is summable on $[a, b]$, and

$$\int_a^x f'(t)\,dt = f(x) - f(a) \qquad \textit{for all } x \in [a, b]. \tag{2}$$

Proof. The existence of f' almost everywhere in $[a, b]$, its measurability, and its summability is Theorem 7–6.5. The function g defined by

$$g(x) = f(a) + \int_a^x f'(t)\,dt, \qquad a \leqslant x \leqslant b, \tag{3}$$

is absolutely continuous by Theorem 6–5.2 (the part proved), and

$$g'(x) = f'(x) \qquad \text{for almost all } x \in [a, b]$$

is by Theorem 2. Then Theorem 1 says there is a constant c such that

$$f(x) = g(x) + c \qquad \text{for all } x \in [a, b].$$

But $g(a) = f(a)$ and thus $c = 0$ so that $f(x) = g(x)$ for all $x \in [a, b]$ and (3) yields (2). ∎

PROBLEM 1. If fg is absolutely continuous on an interval $[a, b]$, then

$$\int_a^b (fg)' = fg\Big]_a^b$$

and if in addition $fg' \in \mathscr{L}[a, b]$ and $f'g \in \mathscr{L}[a, b]$, then the integration by parts formula,

$$\int_a^b fg' + \int_a^b f'g = fg\Big]_a^b,$$

holds.

The following theorem is an alternative statement of Theorem 6–5.2 (whose proof could not be finished in Section 6–5). The theorem is restated to emphasize that we now have a complete characterization of absolutely continuous real-valued functions on closed intervals.

Theorem 4. *A real-valued function g is absolutely continuous on a closed interval $[a, b]$ of its domain iff there is a function $f \in \mathscr{L}[a, b]$ and a constant c such that*

$$g(x) = \int_a^x f(t)\,dt + c \qquad \textit{for all } x \in [a, b] \tag{4}$$

and thus (in light of Theorem 3) iff

$$\int_a^x g'(t)\,dt = g(x) - g(a).$$

Proof. If $f \in \mathscr{L}[a, b]$ and c is a constant, then the absolute continuity of g defined by (4) is the part of Theorem 6–5.2 already proved. Conversely, if g is absolutely continuous on $[a, b]$, then there is a function $f \in \mathscr{L}[a, b]$ and a constant c for which (4) holds; namely, $f = g'$ and $c = g(a)$ by Theorem 3. ∎

The following theorem gives another condition under which the fundamental theorem holds for Lebesgue integrals.

Theorem 5. *Let $f\colon [a, b] \to E^1$ be such that $f'(x)$ exists for* **all** *$x \in [a, b]$. If $f' \in \mathscr{L}[a, b]$, then*

$$\int_a^x f'(t)\,dt = f(x) - f(a), \qquad a \leqslant x \leqslant b.$$

Proof. Let $\varepsilon > 0$ be arbitrary. Choose a lower semicontinuous function g such that

$$f'(x) < g(x) \qquad \text{and} \qquad \int_a^b g(t)\,dt < \int_a^b f'(t)\,dt + \varepsilon,$$

which is possible by the corollary on page 336. Define G: $[a, b] \to E^1$ by setting

$$G(x) = \int_a^x g(t)\,dt - f(x) + f(a) + \varepsilon(x - a).$$

We first assert that $G(b) \geqslant 0$. For assume

(5) $$G(b) < 0.$$

Since $G(a) = 0$ and G is continuous, let x_0 be the largest zero of G. Hence $a \leqslant x_0 < b$, $G(x_0) = 0$, and

(6) $$t \in (x_0, b] \Rightarrow G(t) < 0.$$

Choose $h > 0$ such that $x_0 + h \leqslant b$,

$$\frac{f(x_0 + h) - f(x_0)}{h} < f'(x_0) + \varepsilon$$

and $x_0 \leqslant t \leqslant x_0 + h \Rightarrow g(t) > f'(x_0)$, where the implication is possible because g is lower semicontinuous and the strict inequality $g(x_0) > f'(x_0)$ holds. Hence

$$\begin{aligned} G(x_0 + h) - G(x_0) &= \int_{x_0}^{x_0+h} g(t)\,dt - f(x_0 + h) + f(x_0) + \varepsilon h \\ &> f'(x_0)h - [f'(x_0) + \varepsilon]h + \varepsilon h = 0. \end{aligned}$$

Thus $G(x_0 + h) > G(x_0) = 0$, which contradicts (6) and shows that (5) is false. Thus $G(b) \geqslant 0$; that is,

$$\begin{aligned} f(b) - f(a) &\leqslant \int_a^b g(t)\,dt + \varepsilon(b - a) \\ &< \int_a^b f'(t)\,dt + \varepsilon + \varepsilon(b - a). \end{aligned}$$

Consequently $f(b) - f(a) \leqslant \int_a^b f'(t)\,dt$. Then the opposite inequality holds, by using $-f$ in place of f, and hence

$$\int_a^b f'(t)\,dt = f(b) - f(a).$$

Since the same argument may be applied with b replaced by any fixed $x \in (a, b]$, the theorem is proved. ∎

A familiar application of these theorems will now be given.

Definition 1. *A continuous function* $\gamma\colon [a, b] \to E^2$ *is said to be a* **plane curve.** *A plane curve* γ *is an* **arc** *if* γ *is one-to-one; that is,* $a \leqslant x_1 < x_2 \leqslant b$ *implies* $\gamma(x_1) \neq \gamma(x_2)$. *A plane curve is a* **simple closed curve** *if* $\gamma(a) = \gamma(b)$, *but* $a < x_1 < x_2 < b$ *implies* $\gamma(x_1) \neq \gamma(x_2)$.

Note that a plane curve is not defined as a set of points, but as a continuous mapping of an interval into E^2. *Thus* $\gamma\colon [a, b] \to E^2$ *is a plane curve iff there are continuous functions* $f\colon [a, b] \to E^1$ *and* $g\colon [a, b] \to E^1$ *such that*

$$\gamma(x) = (f(x), g(x)), \qquad a \leqslant x \leqslant b.\dagger \tag{7}$$

With γ *so expressed and* $P = [x_0, x_1, \cdots, x_n]$ *a partition of* $[a, b]$, *we set* $\Delta_k f = f(x_k) - f(x_{k-1})$ *and* $\Delta_k g = g(x_k) - g(x_{k-1})$, $k = 1, 2, \cdots, n$ *and*

$$l(\gamma, P) = \sum_{k=1}^{n} [\Delta_k^{\,2} f + \Delta_k^{\,2} g]^{1/2}.$$

By taking the supremum over all partitions, the number $l(\gamma)$ *defined by*

$$l(\gamma) = \sup_P l(\gamma, P)$$

is called the **length** *of* γ. *If* $l(\gamma) < \infty$, *then* γ *is said to be* **rectifiable.**

Theorem 6. *A plane curve* $\gamma = (f, g)$, *as in* (7), *is rectifiable iff* f *and* g *are both of bounded variation.*

Proof. Let both f and g be of bounded variation and select a number M such that $V_a^{\,b} f < M$ and $V_a^{\,b} g < M$. With P any partition, then

$$\sqrt{\Delta_k^{\,2} f + \Delta_k^{\,2} g} \leqslant |\Delta_k f| + |\Delta_k g| \qquad \text{and}$$

$$l(\gamma, P) \leqslant \sum_{k=1}^{n} |\Delta_k f| + \sum_{k=1}^{n} |\Delta_k g| \leqslant 2M.$$

Thus the supremum over all partitions is $l(\gamma) \leqslant 2M$ so that γ is rectifiable.

Conversely, let f be of unbounded variation, take $M > 0$ arbitrarily, and then select a partition P such that

$$V_a^{\,b}(f, P) > M.$$

Since $\sqrt{\Delta_k^{\,2} f + \Delta_k^{\,2} g} \geqslant |\Delta_k f|$, we have

$$l(\gamma) \geqslant l(\gamma, P) \geqslant \sum_{k=1}^{n} |\Delta_k f| = V_a^{\,b}(f, P) > M.$$

Thus γ is not rectifiable and the same conclusion follows if g is not of bounded variation. ∎

Corollary. *If* γ *is rectifiable, then* f' *and* g' *exist a.e. on* $[a, b]$ *and* $\sqrt{f'^2 + g'^2}$ *is summable; i.e.,*

$$\int_a^b \sqrt{f'^2(x) + g'^2(x)}\, dx < \infty.$$

† Or equivalently $\gamma(x) = f(x) + ig(x)$.

Proof. That f' and g' exist a.e. and are summable is by Theorem 7–6.3. Hence $|f'| + |g'|$ is summable and dominates $\sqrt{f'^2 + g'^2}$ so that this function is also summable, by Theorem 4–5.6. ∎

Conditions under which the integral in the corollary gives the length of γ will now be considered. We shall use

$$s(x), \qquad a \leqslant x \leqslant b$$

for the length of γ restricted to $[a, x]$, $a \leqslant x \leqslant b$. Hence s is a nonnegative, nondecreasing function on $[a, b]$ such that $s(a) = 0$ and $s(b) = l(\gamma)$. Thus if γ is rectifiable, then s' exists a.e. on $[a, b]$.

Lemma 2. *If $\gamma = (f, g)$ is rectifiable, then*

$$s'(x) = \sqrt{f'^2(x) + g'^2(x)} \quad a.e. \qquad on\ [a, b].$$

Proof. Take any t and y such that $a \leqslant t < y \leqslant b$ and then a partition $P = \{a = x_0, x_1, \cdots, x_n = t\}$ of $[a, t]$. The set $Q = P \cup \{y\}$ is a partition of $[a, y]$ having t as its next to last point and

$$s(y) \geqslant l(\gamma, Q) = l(\gamma, P) + \{[f(y) - f(t)]^2 + [g(y) - g(t)]^2\}^{1/2}.$$

By taking the supremum over all partitions of $[a, t]$, we then have that

$$\{[f(y) - f(t)]^2 + [g(y) - g(t)]^2\}^{1/2} \leqslant s(y) - s(t). \tag{8}$$

Let t be a point where $s'(t)$, $f'(t)$, and $g'(t)$ all exist. Divide by $y - t$ and let $y \to t+$. Hence (with t replaced by x)

$$\sqrt{f'^2(x) + g'^2(x)} \leqslant s'(x) \quad \text{a.e.} \qquad \text{on } [a, b].$$

We shall show that the reverse inequality also holds a.e. on $[a, b]$. Let

$$A = \{x \in [a, b] \mid \sqrt{f'^2(x) + g'^2(x)} < s'(x)\} \qquad \text{and}$$

$$A_n = \left\{x \in A \mid \sqrt{f'^2(x) + g'^2(x)} + \frac{1}{n} < s'(x)\right\}, \qquad n = 1, 2, 3, \cdots.$$

Then $A = \bigcup A_n$ and we prove that $mA = 0$ by proving that $mA_n = 0$, $n = 1, 2, 3, \cdots$.

Fix n. Define A_{np} for $p = 1, 2, 3, \cdots$ by:

$$x \in A_{np}, \qquad \text{iff } t \leqslant x \leqslant y \text{ and } 0 < y - t < 1/p \text{ imply}$$

$$\left\{\left[\frac{f(y) - f(t)}{y - t}\right]^2 + \left[\frac{g(y) - g(t)}{y - t}\right]^2\right\}^{1/2} + \frac{1}{n} < \frac{s(y) - s(t)}{y - t};$$

that is,

$$\{[f(y) - f(t)]^2 + [g(y) - g(t)]^2\}^{1/2} < s(y) - s(t) - \frac{y - t}{n}. \tag{9}$$

Hence $A_n \subset A_{np}$ and we proceed to show that $m(A_n) = 0$ by showing that $m(A_{np}) = 0$, $p = 1, 2, 3, \cdots$.

Fix p and let $\varepsilon > 0$ be given. Choose a partition Q such that $l(\gamma, Q) > l(\gamma) - \varepsilon$ and then a refinement of Q called

$$P = \{x_0, x_1, \cdots, x_q\} \qquad \text{with } \|P\| < \frac{1}{p}.$$

Then $l(\gamma, P) \geqslant l(\gamma, Q)$, from Minkowski's inequality whenever points of P lie between consecutive points of Q, and thus

$$s(b) = l(\gamma) < l(\gamma, P) + \varepsilon.$$

Let $K = \{k = 1, 2, \cdots, q \mid [x_{k-1}, x_k] \cap A_{np} \neq \varnothing\}$ and let N be the set of remaining integers among 1, 2, $\cdots$, q. Then

$$A_{np} \subset \bigcup_{k \in K} [x_{k-1}, x_k] \qquad \text{and} \qquad mA_{np} \leqslant \sum_{k \in K} (x_k - x_{k-1}).$$

Also if $k \in K$, then (9) holds with $y = x_k$ and $t = x_{k-1}$, whereas if $k \in N$, we make the same substitution in (8). Thus

$$\begin{aligned} l(\gamma, P) &< \sum_{k=1}^{q} [s(x_k) - s(x_{k-1})] - \sum_{k \in K} \frac{x_k - x_{k-1}}{n} \\ &\leqslant s(b) - \frac{1}{n} m(A_{np}) < l(\gamma, P) + \varepsilon - \frac{1}{n} m(A_{np}). \end{aligned}$$

Thus $m(A_{np}) < n\varepsilon$ and therefore $m(A_{np}) = 0$. As seen, as we went along, this establishes the lemma. ∎

PROBLEM 2. The function s is absolutely continuous iff both f and g are absolutely continuous.

Theorem 7. *If $\gamma = (f, g)$, as in* (7), *is rectifiable, then*

$$\int_a^b \sqrt{f'^2(x) + g'^2(x)}\, dx \leqslant l(\gamma).$$

The equality holds iff f and g are both absolutely continuous.

Proof. Since s is a bounded nondecreasing function on $[a, b]$, then

$$\int_a^b s'(x)\, dx \leqslant s(b) - s(a) = s(b) = l(\gamma),$$

by Theorem 7–5.3, and the equality holds iff s is absolutely continuous, by Theorem 4. But this integral and the one in the theorem are equal by Lemma 2 and the statement about equality follows from Problem 2. ∎

PROBLEM 3. The graph of the Cantor function ψ has length 2.

The space $\mathscr{L}^2[a, b]$ of real measurable square summable functions was metrized in Section 4–6 by defining the $\mathscr{L}^2$-distance between $f, g \in \mathscr{L}^2$ as

$$\|f - g\|_2 = \left\{ \int_a^b [f(x) - g(x)]^2\, dx \right\}^{1/2}.$$

Also $\mathscr{C}[a, b]$ was used on page 318 for the space of continuous functions on $[a, b]$ and $\mathscr{C}$ metrized by a $\mathscr{C}$-distance, but this distance is not the one used in the next theorem. Clearly, $\mathscr{C}[a, b] \subset \mathscr{L}^2[a, b]$ and hence the $\mathscr{L}^2$-distance may be used on the space $\mathscr{C}$.

Theorem 8. *$\mathscr{C}[a, b]$ is dense in $\mathscr{L}^2[a, b]$ for the $\mathscr{L}^2$-distance.*

Proof. The interval $[a, b]$, $-\infty < a < b < \infty$, is fixed. Take any $f \in \mathscr{L}^2$. We wish to show that there is a sequence $\{h_n\}$ in $\mathscr{C}$ such that $\lim \|f - h_n\|_2 = 0$. The further restrictions

$$f(x) = 0 \qquad \text{for } b < x \qquad \text{and} \qquad |f(x)| < M < \infty \qquad \text{for } a \leqslant x \leqslant b$$

are made, but these are for convenience. The behavior of f outside $[a, b]$ is immaterial. Also, with $\varepsilon > 0$, we could first take a bounded $g \in \mathscr{L}^2$ such that $\|f - g\|_2 < \varepsilon/2$ and then, by the following reasoning, obtain an h such that $\|g - h\|_2 < \varepsilon/2$.

Let

$$F(x) = \int_a^x f(t)\, dt, \qquad x \geqslant a.$$

Then F is (absolutely) continuous on $[a, b + 1]$ and $F'(x) = f(x)$ a.e. on $[a, b + 1]$. Therefore the functions h_n defined for $n = 1, 2, 3 \cdots$, by

$$h_n(x) = n \int_x^{x+1/n} f(t)\, dt = \frac{F(x + 1/n) - F(x)}{1/n}, \qquad x \geqslant a$$

are continuous on $[a, b]$ while $\lim_{n\to\infty} h_n(x) = F'(x) = f(x)$ and

$$\lim_{n\to\infty} [h_n(x) - f(x)]^2 = 0 \quad \text{a.e.} \qquad \text{on } [a, b].$$

Moreover

$$|h_n(x)| \leqslant n \int_x^{x+1/n} |f(t)|\, dt \leqslant n \int_x^{x+1/n} M\, dt = M$$

and $[h_n(x) - f(x)]^2 \leqslant (M + M)^2$ for $a \leqslant x \leqslant b$. Thus (see the note after Theorem 4–5.9)

$$0 = \lim_{n\to\infty} \int_a^b [h_n(x) - f(x)]^2\, dx = \lim_{n\to\infty} \|h_n - f\|_2^2. \blacksquare$$

PROBLEM 4. Real $\mathscr{L}^2[a, b]$ is a separable Hilbert space.
[*Hint*: From Theorem 4–6.3, only separability remains to be proved.]

7–8 CHANGE OF VARIABLE

Throughout this section we shall be concerned with two functions:

$$\varphi: [c, d] \to E^1 \qquad \text{and} \qquad f: [a, b] \to E^1$$

with φ strictly increasing on $[c, d]$, $a = \varphi(c)$, and $b = \varphi(d)$. We have already derived the formula

$$\int_a^b f(x)\,dx = \int_c^d f(\varphi(t))\varphi'(t)\,dt, \tag{1}$$

in Theorem 7–2.7, under the restrictions that f and φ' are both continuous. The purpose of this section is to establish the same formula with $f \in \mathscr{L}[a, b]$ and φ absolutely continuous. The proof depends upon the following theorem and two lemmas.

Theorem 1. *Formula* (1) *holds if f is continuous on* $[a, b]$ *and φ is absolutely continuous and strictly increasing on* $[c, d]$.

Proof. The composition function $f \circ \varphi$ is continuous on $[c, d]$ and hence $f \circ \varphi \in \mathscr{L}[c, d]$. Also $\varphi' \in \mathscr{L}[c, d]$ by Theorem 7–6.5, so that the right-hand integral in (1) exists by the dominated summability Theorem 4–5.6. And the existence of the integral and equality

$$\int_{t_{i-1}}^{t_i} \varphi'(t) = \varphi(t_i) - \varphi(t_{i-1}), \qquad c \leqslant t_{i-1} < t_i \leqslant d$$

is assured by Theorem 7–7.3. With these facts in mind, the proof of Theorem 7–2.7 may now be reread to give a proof of the present theorem. ∎

Lemma 1. *With φ strictly increasing and absolutely continuous on* $[c, d]$, *then for any measurable set $B \subset [c, d]$ the set $A = \varphi(B) \subset [a, b]$ is measurable and*

$$mA = \int_B \varphi'(t)\,dt. \tag{2}$$

Proof. The set A is measurable by Theorem 6–5.1. If B is an open interval (α, β), then $A = (\varphi(\alpha), \varphi(\beta))$ is an open interval whose closure is $\bar{A} = [\varphi(\alpha), \varphi(\beta)]$ and

$$mA = m\bar{A} = \varphi(\beta) - \varphi(\alpha) = \int_\alpha^\beta \varphi'(t)\,dt = \int_B \varphi'(t)\,dt.$$

Since (2) holds if B is an open interval, then it holds if B is the countable union of disjoint open intervals; i.e., (2) holds if B is an open subset of $[c, d]$. Hence, by taking complements, (2) holds if B is a closed subset of $[c, d]$.

Then with any measurable set $B \subset (c, d)$, consider the set $A = \varphi(B)$. Let $\varepsilon > 0$ be given. In (a, b) take C closed and D open such that

$$C \subset A \subset D, \qquad mC > mA - \varepsilon, \qquad \text{and} \qquad mD < mA + \varepsilon.$$

Then $F = \varphi^{-1}(C)$ is closed, $G = \varphi^{-1}(D)$ is open, $F \subset B \subset D$,

$$\int_F \varphi'(t)\,dt = mC, \qquad \text{and} \qquad \int_G \varphi'(t)\,dt = mD.$$

Since $\varphi'(t) \geqslant 0$ wherever it exists (which is a.e. in $[c, d]$ by Theorem 7–5.2),

$$\int_F \varphi'(t)\,dt \leqslant \int_B \varphi'(t)\,dt \leqslant \int_G \varphi'(t)\,dt.$$

Hence

$$mA - \varepsilon \leqslant \int_B \varphi'(t)\,dt \leqslant mA + \varepsilon$$

and therefore (2) holds. ∎

Lemma 2. *With φ strictly increasing and absolutely continuous on $[c, d]$, take any set $N \subset [a, b]$ having $mN = 0$. Set $M = \varphi^{-1}(N)$ and $M_1 = \{t \in M \mid \varphi'(t) \neq 0\}$. Then*

$$mM_1 = 0.$$

Proof. Consider $N \subset (a, b)$, since removal of a or b from N, if they were there, removes at most c or d from M. Take open sets $\{D_n\}$, $D_n \supset N$ such that

$$(a, b) \supset D_1 \supset D_2 \supset \cdots, \qquad mD_n \to 0$$

and set $A = \bigcap D_n$. Hence $mA = 0$. Let $G_n = \varphi^{-1}(D_n)$ and $B = \varphi^{-1}(A)$. Then G_n is open, $G_n \supset B$, and $\bigcap G_n = B \supset M$. Hence B is a measurable set and, by Lemma 1,

$$\int_B \varphi'(t)\,dt = mA = 0.$$

Hence the set $B_1 = \{t \in B \mid \varphi'(t) \neq 0\}$ has $mB_1 = 0$. Since $M_1 = M \cap B_1$, then $mM_1 = 0$. ∎

Theorem 2. *If $f \in \mathscr{L}[a, b]$ and φ: $[c, d] \to E^1$ is absolutely continuous and strictly increasing with $a = \varphi(c)$ and $b = \varphi(d)$, then*

(1) $$\int_a^b f(x)\,dx = \int_c^d f(\varphi(t))\varphi'(t)\,dt.$$

Proof. First consider f measurable and bounded, say

$$|f(x)| < q.$$

Let $\{f_n\}$ be a sequence of functions, each continuous on $[a, b]$ such that $|f_n(x)| < q$ for all $x \in [a, b]$ and

(3) $$\lim_{n\to\infty} f_n(x) = f(x) \quad \text{a.e.} \qquad \text{in } [a, b].$$

This is possible by Problem 6–9.1. Then, by Theorem 4–5.9,

$$\lim_{n\to\infty} \int_a^b f_n(x)\,dx = \int_a^b f(x)\,dx.$$

Since each f_n is continuous, then by Theorem 1,

$$\int_a^b f_n(x)\,dx = \int_c^d f_n(\varphi(t))\varphi'(t)\,dt.$$

Thus we have the existence of the limit and the equality

$$\int_a^b f(x)\,dx = \lim_{n\to\infty} \int_a^b f_n(\varphi(t))\varphi'(t)\,dt. \tag{4}$$

We shall show that

(a) $\lim_{n\to\infty} f_n(\varphi(t))\varphi'(t)$ exists and $= f(\varphi(t))\varphi'(t)$ a.e. in $[c, d]$, and
(b) the limit and integral in (4) may be interchanged.

As soon as (a) and (b) are proved, we shall have that (1) holds under the present restriction of boundedness of f.

Let $T = \{t \in [c, d] \mid \varphi'(t) = \infty\}$. Then $mT = 0$ by Theorem 7–5.2. Next let

$$N = \{x \in [a, b] \mid \lim_{n\to\infty} f_n(x) \neq f(x)\}, \qquad M = \varphi^{-1}(N)$$

and

$$M_1 = \{t \in M \mid \varphi'(t) \neq 0\}.$$

Hence $mN = 0$ because of the choice of $\{f_n\}$ having (3) holding, and then $mM_1 = 0$ by Lemma 2. Now

$$\begin{aligned} t \in [c, d] - M &\Rightarrow \varphi(t) \notin N \\ &\Rightarrow \lim_{n\to\infty} f_n(\varphi(t)) = f(\varphi(t)). \end{aligned}$$

Also

$$\begin{aligned} t \in M - M_1 &\Rightarrow \varphi'(t) = 0 \\ &\Rightarrow f_n(\varphi(t))\varphi'(t) = f(\varphi(t))\varphi'(t) = 0. \end{aligned}$$

Thus

$$t \in [c, d] - (M_1 \cup T) \Rightarrow \lim_{n\to\infty} f_n(\varphi(t))\varphi'(t) = f(\varphi(t))\varphi'(t),$$

with everything finite. Since $mM_1 = 0$ and $mT = 0$, we have established (a).

We chose $\{f_n\}$ such that $|f_n(x)| < q$ for all $x \in [a, b]$ and thus

$$|f_n(\varphi(t))\varphi'(t)| \leqslant q\varphi'(t) \text{ for almost all } t \in [c, d].$$

Since $\varphi' \in \mathscr{L}[c, d]$, we use the dominated convergence theorem (4–5.9) and the fact that (a) holds to justify (b). Thus, as noted under (b), (1) holds if f is bounded.

Only minor details remain to establish (1) for $f \in \mathscr{L}[a, b]$ not necessarily bounded. Consider the positive part f^+ of f and define f_n^+ by

$$f_n^+(x) = \begin{cases} f^+(x) & \text{if } f^+(x) \leqslant n \\ n & \text{if } f^+(x) > n \end{cases}, \quad n = 1, 2, 3, \cdots.$$

The first equation in

$$\int_a^b f^+(x)\,dx = \lim_{n\to\infty} \int_a^b f_n^{\,+}(x)\,dx = \lim_{n\to\infty} \int_c^d f_n^{\,+}(\varphi(t))\varphi'(t)\,dt$$

follows from Theorem 4–5.9 and then the second from the fact that (1) holds for each of the bounded functions $f_n^{\,+}$. Also, $\{(f_n^{\,+} \circ \varphi) \cdot \varphi'\}$ is a nondecreasing sequence wherever it is defined in $[c, d]$, and

$$f^+(\varphi(t))\varphi'(t) = \lim_{n\to\infty} f_n^{\,+}(\varphi(t))\varphi'(t)\,dt \quad \text{a.e.} \qquad \text{in } [c, d]$$

so that, by Theorem 4–5.2,

$$\int_c^d f^+(\varphi(t))\varphi'(t)\,dt = \lim_{n\to\infty} \int_c^d f_n^{\,+}(\varphi(t))\varphi'(t)\,dt.$$

Thus (1) holds for f^+. In the same way it follows that formula (1) holds for f^- and then for $f = f^+ - f^-$. ▮

Note that we cannot relax absolute continuity of φ to mere continuity. For if we did we would have a measurable function f of a continuous function φ which need not be measurable (by the example in Section 6–2) and thus the right-hand integral in (1) need not even exist.

7–9 EXPONENTS AND LOGARITHMS

The ratio test shows that the series in the following definition converges absolutely for each complex number z.

Definition 1. *The function* exp: $E^2 \to E^2$ *is defined by*

$$\exp(z) = \sum_{n=0}^{\infty} \frac{z^n}{n!}, \qquad z \in E^2.$$

Then the number e is such that, by Theorem 3–7.11,

(1)
$$e = \sum_{n=0}^{\infty} \frac{1}{n!} = \exp(1).$$

By using the Cauchy product of two series, it follows by Theorem 3–12.4 that

$$\sum_{n=0}^{\infty} \frac{z^n}{n!} \sum_{m=0}^{\infty} \frac{w^m}{m!} = \sum_{n=0}^{\infty} \sum_{k=0}^{n} \frac{z^k w^{n-k}}{k!(n-k)!}$$

$$= \sum_{n=0}^{\infty} \frac{1}{n!} \sum_{k=0}^{n} \binom{n}{k} z^k w^{n-k} = \sum_{n=0}^{\infty} \frac{(z+w)^n}{n!}$$

and hence

(2)
$$\exp(z + w) = \exp(z)\exp(w); \qquad z, w \in E^2.$$

For any complex number h such that $0 < |h| < 1$,

$$\left|\frac{\exp(h) - 1}{h} - 1\right| \leqslant |h| \sum_{n=2}^{\infty} \frac{|h|^{n-2}}{n!} \leqslant |h| \sum_{n=2}^{\infty} \frac{1}{n!} < |h|\, e,$$

and therefore $\exp'(0) = 1$. Thus, by using (2),

$$\lim_{h\to 0} \frac{\exp(z + h) - \exp(z)}{h} = \exp(z) \lim_{h\to 0} \frac{\exp(h) - 1}{h} = \exp(z)$$

so that exp is differentiable on E^2 with

(3) $$\exp'(z) = \exp(z), \qquad z \in E^2.$$

Also, from (2), $\exp(z - z) = \exp(z)\exp(-z)$, and thus

(4) $$\exp(z)\exp(-z) = 1 \quad \text{and} \quad \exp(z) \neq 0, \qquad z \in E^2.$$

We now restrict consideration to $\exp(x)$, $x \in E^1$.

From the definition, $x > 0 \Rightarrow \exp(x) > 0$. Hence, from (4),

$$\exp(x) > 0, \qquad x \in E^1.$$

Moreover, first from the definition and then (4),

$$0 \leqslant x < y \Rightarrow 0 < \exp(x) < \exp(y) \Rightarrow 0 < \exp(-y) < \exp(-x).$$

Consequently, exp is a strictly increasing function on E^1 and, from the definition, $x > 0 \Rightarrow \exp(x) > 1 + x$. Thus

$$\lim_{x\to\infty} \exp(x) = \infty \quad \text{and}$$

$$\lim_{x\to-\infty} \exp(x) = \lim_{x\to-\infty} \frac{1}{\exp(-x)} = 0.$$

Thus, as a real-valued function of a real variable,

$$\exp\colon E^1 \leftrightarrow (0, \infty),$$

the inverse $\exp^{-1}\colon (0, \infty) \leftrightarrow E^1$ exists and (Theorem 7–1.6) is differentiable.

Definition 2. $\ln\colon (0, \infty) \leftrightarrow E^1$ *is defined by*

$$\ln(x) = \exp^{-1}(x), \qquad 0 < x < \infty.$$

Hence for $x > 0$, $x = \exp(\ln x)$, and if also $y > 0$, then

$$xy = \exp(\ln x) \cdot \exp(\ln y) = \exp(\ln x + \ln y)$$

by (2) and therefore $\exp^{-1}(xy) = \ln x + \ln y$. Thus

$$\ln(xy) = \ln x + \ln y; \qquad x > 0,\ y > 0.$$

Since $x = \exp(\ln x)$ for $0 < x < \infty$ and $\exp'(y) = \exp(y) > 0$ for $y \in E^1$, it follows (since we already know that ln is differentiable on its domain) that

$$1 = \exp'(\ln x) \cdot \ln'(x) = \exp(\ln x) \cdot \ln'(x) = x \ln'(x).$$

Therefore

$$\frac{d}{dx} \ln(x) = \frac{1}{x} \qquad \text{and} \qquad \ln x = \int_1^x \frac{1}{t}\, dt, \qquad 0 < x < \infty.$$

From (2), if $x_k \in E^1$, $k = 1, 2, \cdots, n$, then (even if $x_k \in E^2$)

$$\exp(x_1 + x_2 + \cdots + x_n) = \exp x_1 \cdot exp\, x_2 \cdot \cdots \cdot \exp x_n, \qquad n = 1, 2, 3, \cdots.$$

Hence, for any $x \in E^1$,

$$\exp(nx) = [\exp(x)]^n \qquad \text{and} \qquad \exp(n) = e^n, \qquad n = 1, 2, 3, \cdots,$$

where the second equation follows from $\exp(1) = e$. Hence, with m and n any positive integers,

$$\left[\exp\left(\frac{m}{n}\right)\right]^n = \exp\left(n \cdot \frac{m}{n}\right) = \exp(m) = e^m,$$

$$\exp\left(\frac{m}{n}\right) = e^{m/n},$$

and we have $\exp(r) = e^r$ for any positive rational r. Then, by using (4),

$$e^r = \exp(r) \qquad \text{for any rational } r.$$

Let x be a real number. Choose any sequence $\{r_n\}$ of rational numbers such that $r_n \to x$ as $n \to \infty$. Then $x = r_n + (x - r_n)$ so that, from (2),

$$\exp(x) = \exp(r_n) \exp(x - r_n).$$

Since exp is continuous (even differentiable) at x and 0,

$$\begin{aligned} \exp(x) &= \lim_{n\to\infty} \exp(r_n) \lim_{n\to\infty} \exp(x - r_n) \\ &= \lim_{n\to\infty} e^{r_n} \exp(0) = \lim_{n\to\infty} e^{r_n}, \end{aligned} \tag{5}$$

with this limiting value independent of the sequence $\{r_n\}$ chosen. Moreover, if x is rational, then $e^x = \exp(x)$ and, from (5),

$$e^x = \lim_{n\to\infty} e^{r_n}. \tag{6}$$

Definition 3

(i) *For x a real number, e^x is defined by* (6), *where $\{r_n\}$ is any sequence of rational numbers such that $r_n \to x$ as $n \to \infty$.*

(ii) *For a any positive real number and x any real number, then by definition,*

$$a^x = e^{x \ln a}$$

Hence for any real number x, from (6) and (5),

$$e^x = \exp(x),$$

and for a any positive real number,

$$a^x = \exp(x \ln a).$$

Hence the differentiability of exp implies the differentiability of the functions defined by e^x and a^x with the formulas

$$\frac{de^x}{dx} = \exp'(x) = \exp(x) = e^x,$$

$$\begin{aligned}\frac{da^x}{dx} &= \exp'(x \ln a) \frac{d}{dx}(x \ln a) = \exp(x \ln a) \cdot \ln a \\ &= a^x \ln a.\end{aligned}$$

Also, the function f defined by

$$f(x) = x^a, \qquad x > 0 \text{ and } a \text{ real}$$

is such that $f(x) = e^{a \ln x} = \exp(a \ln x)$ and

$$\begin{aligned}\frac{dx^a}{dx} &= \exp(a \ln x) \frac{d}{dx}(a \ln x) = x^a\left(\frac{a}{x}\right) \\ &= ax^{a-1}.\end{aligned}$$

Thus, for $x > 0$,

$$\int_1^x t^a \, dt = \frac{1}{a-1}(x^{a-1} - 1]; \qquad a \in E^1, \quad a \neq 1.$$

Theorem 1. *For any constant* $\alpha > 0$,

$$\lim_{x \to \infty} x^\alpha e^{-x} = 0 \qquad \textit{and} \qquad \lim_{x \to \infty} \frac{\ln x}{x^\alpha} = 0.$$

Proof. Fix an integer $n \geqslant \alpha$. Then

$$x > 1 \Rightarrow e^x > \frac{x^{n+1}}{(n+1)!} \geqslant \frac{x^{\alpha+1}}{(n+1)!} \Rightarrow x^\alpha e^{-x} < \frac{(n+1)!}{x}$$

from which the first equation follows.

Select a number β, $0 < \beta < \alpha$. Then for each $x > 1$,

$$0 < \frac{\ln x}{x^\alpha} = x^{-\alpha} \int_1^x t^{-1} \, dt \leqslant x^{-\alpha} \int_1^x t^{\beta-1} \, dt = \frac{x^{-\alpha}}{\beta}(x^\beta - 1) < \frac{1}{\beta} \frac{1}{x^{\alpha-\beta}}.$$

Hence the second equation is seen to follow. ∎

The equations of the theorem show that e^x becomes infinite "faster" and ln x "slower" than any positive power of x as $x \to \infty$. Hence, with $p > 0$ constant and $P(x)$ a polynomial,

$$(7) \qquad \lim_{x\to\infty} \frac{P(x)}{x^p} e^{-x} = 0 \qquad \text{and} \qquad \lim_{x\to 0} \frac{P(x)}{x^p} e^{-1/x^2} = 0.$$

7–10 TAYLOR'S THEOREM

The successive derivatives $f', f'', \cdots, f^n, \cdots$ of a function f (itself considered as the zeroth derivative f^0) are defined inductively by:

If the nth order derivative f^n exists in a neighborhood of x and if

$$\lim_{h\to 0} \frac{f^n(x+h) - f^n(x)}{h}$$

exist, then $f^{n+1}(x)$ is defined to be this value. One-sided derivatives are defined by letting $h \to 0+$ or $h \to 0-$.

PROBLEM 1. The function g defined by

$$g(x) = \begin{cases} e^{-1/x^2}, & x \neq 0 \\ 0, & x = 0 \end{cases}$$

has $g^n(x)$ existing for all $x \in E^1$ and

$$g^n(0) = 0, \qquad n = 0, 1, 2, 3, \cdots.$$

[*Hint*: By induction there is a polynomial $P_n(x)$ depending on n (but not necessarily of degree n) such that

$$g^n(x) = P_n(x)x^{-3n}e^{-1/x^2} \qquad \text{for } x \neq 0 \text{ and } g^n(0) = 0.$$

Use the second limit in (7) of Section 7–9.]

Theorem 1. (Taylor) *Let $I \subset E^1$ be a closed interval, let $a \in I$, and let n be a positive integer. If $f\colon I \to E^1$ is such that f^n is absolutely continuous, then for each $x \in I$,*

$$f(x) = f(a) + f'(a)(x-a) + \frac{f''(a)}{2!}(x-a)^2 + \cdots + \frac{f^n(a)}{n!}(x-a)^n + R_n(x)$$

where

$$R_n(x) = \frac{1}{n!}\int_a^x (x-t)^n f^{n+1}(t)\, dt.$$

Proof. By Theorem 7–7.3, f^{n+1} exists a.e. in $[a, b]$ and is summable. Also, $f, f', \cdots, f^{n-1}$ are absolutely continuous, by Theorem 7–7.4. Hence the integrals in the definition of r_k by

$$r_k = \int_a^x \frac{(x-t)^{k-1}}{(k-1)!} f^k(t)\, dt, \qquad k = 1, 2, \cdots, n+1; \quad x \in I,$$

all exist, since each integrand is the product of a summable function and a continuous function. Also, for $k = 1, 2, \cdots, n$,

$$\frac{d}{dt}(x-t)^k f^k(t) = -k(x-t)^{k-1} f^k(t) + (x-t)^k f^{k+1}(t),$$

in which both members on the right define summable functions (of t) so that, by Problem 7–7.1, integration by parts may be used to yield

$$\begin{aligned} r_k &= -\frac{(x-t)^k}{k!} f^k(t)\Big]_a^x + \int_a^x \frac{(x-t)^k}{k!} f^{k+1}(t)\,dt \\ &= \frac{f^k(a)}{k!}(x-a)^k + r_{k+1}, \qquad k = 1, 2, \cdots, n. \end{aligned}$$

By summing from $k = 1$ to n,

$$r_1 = \sum_{k=1}^{n} \frac{f^k(a)}{k!}(x-a)^k + r_{n+1}.$$

Since $r_1 = f(x) - f(a)$ and $r_{n+1} = R_n(x)$, the desired result follows. ∎

Corollary 1. *Let $f\colon [a, b] \to E^1$ be such that $f^{n+1}(x)$ exists for all $x \in (a, b)$ and $f^{n+1} \in \mathscr{L}[a, b]$. If $0 \leqslant p \leqslant n$, p not necessarily an integer, then corresponding to each $x \in (a, b]$ there is a ξ such that $a < \xi < b$ and*

$$R_n(x) = \frac{(x-a)^{p+1}(x-\xi)^{n-p}}{(p+1)n!} f^{n+1}(\xi),$$

which is called the **Schlömilch** *form of the remainder. If $p = n$, the* **Lagrange** *form of the remainder*

$$R_n(x) = \frac{f^{n+1}(\xi)}{(n+1)!}(x-a)^{n+1}$$

is obtained, and $p = 0$ gives the **Cauchy** *form*

$$R_n(x) = \frac{f^{n+1}(\xi)}{n!}(x-\xi)^n (x-a).$$

Proof. The form of R_n in the theorem may be used since f^n is absolutely continuous on $[a, b]$, by Theorems 7–7.5 and 7–7.4. With $x \in (a, b]$, let ξ be such that $a < \xi < x$ and

$$\begin{aligned} R_n(x) &= \frac{1}{n!}\int_a^x (x-t)^p [(x-t)^{n-p} f^{n+1}(t)]\,dt \\ &= \frac{1}{n!}(x-\xi)^{n-p} f^{n+1}(\xi) \int_a^x (x-t)^p\,dt, \end{aligned}$$

which is possible by Problem 7–1.2, p. 370. Upon evaluating the final integral, the Schlömilch form is obtained. ∎

PROBLEM 2. Given $f\colon [a, b] \to E^1$ such that f^n is continuous on $[a, b]$ and f^{n+1} exists on (a, b). Let p be any number such that $-1 < p$ and let S be the number satisfying

$$f(b) = \sum_{k=0}^{n} \frac{f^k(a)}{k!}(b - a)^k + \frac{(b - a)^{p+1}}{n!} S.$$

Define g by

$$g(x) = -f(b) + \sum_{k=0}^{n} \frac{f^k(x)}{k!}(b - x)^k + \frac{(b - x)^{p+1}}{n!} S, \qquad a \leqslant x \leqslant b.$$

Show that there is a number ξ such that $a < \xi < b$ and

$$S = \frac{(b - \xi)^{n-p}}{p + 1} f^{n+1}(\xi).$$

Definition 1. *Given a real-valued function of a real variable such that there is an $a \in E^1$ for which $f^n(a)$ exists, $n = 1, 2, 3, \cdots$, then we write*

$$f(x) \sim \sum_{n=0}^{\infty} \frac{f^n(a)}{n!}(x - a)^n, \qquad x \in E^1$$

and refer to the series as the **Taylor series** *of f at a.*

For the function g of Problem 1,

$$g(x) \sim \sum_{n=0}^{\infty} \frac{g^n(0)}{n!} x^n = \sum_{n=0}^{\infty} 0$$

so the series converges for all $x \in E^1$ but not to $g(x)$ if $x \neq 0$. Thus if f is a function having a Taylor series at $a = 0$ which converges to $f(x)$ for some $x \neq 0$ where $f(x) \neq 0$, then for this x the Taylor series of $f + g$ converges but not to $(f + g)(x)$.

Theorem 2. *Let $f\colon I \to E^1$, with $I \subset E^1$ a closed interval, be such that f^n exists on I for $n = 1, 2, 3, \cdots$ and let $a \in I$.*

(a) *If for some number $x \in I$,*

$$\lim_{n \to \infty} R_n(x) = 0,$$

then the Taylor series of f at a converges to $f(x)$.

(b) *If there is a number M such that*

$$x \in I \Rightarrow |f^n(x)| \leqslant M \qquad \textit{for } n = 1, 2, 3, \cdots,$$

then

$$f(x) = \sum_{n=0}^{\infty} \frac{f^n(a)}{n!}(x - a)^n, \qquad x \in I.$$

Proof. On I each f^n is continuous (in fact, absolutely continuous), and thus each $f^n \in \mathscr{L}(I)$ so that each $R_n(x)$ exists. Hence in part (a),

$$\left| f(x) - \sum_{k=0}^{n} \frac{f^k(x)}{k!}(x-a)^k \right| \leqslant |R_n(x)| \to 0 \qquad \text{as } n \to \infty.$$

In part (b) we have (by considering $x < a$ and $x > a$ separately)

$$|R_n(x)| \leqslant \frac{M}{n!} \int_a^{|x|} |x-t|^n \, dt \leqslant M \frac{|I|^{n+1}}{(n+1)!} \to 0 \qquad \text{as } n \to \infty,$$

since for any y the series $\sum (y^n/n!)$ converges by the ratio test. ∎

It then follows that for all $x \in E^1$,

$$\sin x = x + \sum_{n=1}^{\infty} (-1)^n \frac{x^{2n+1}}{(2n+1)!} \qquad \text{and}$$

$$\cos x = 1 + \sum_{n=1}^{\infty} (-1)^n \frac{x^{2n}}{(2n)!}.$$

If m is a positive integer,

$$(1+x)^m = 1 + \sum_{n=1}^{m} \frac{m(m-1)\cdots(m-n+1)}{n!} x^n = \sum_{n=0}^{m} \binom{m}{n} x^n,$$

since $R_n(x) = 0$ for $n = m+1, m+2, \cdots$. Hence, for $a \neq 0$,

$$(a+b)^m = \left(1 + \frac{b}{a}\right)^m a^m = \sum_{n=0}^{m} \binom{m}{n} \left(\frac{b}{a}\right)^n a^m$$

$$= \sum_{n=0}^{m} \binom{m}{n} a^{m-n} b^n.$$

The following theorem gives the conditions under which the binomial expansion holds for exponents not positive integers.

Theorem 3. *Let α be a constant which is not any of 0, 1, 2, 3, $\cdots$. Then the following series (which does not terminate) is convergent and the equality*

$$(1) \qquad (1+x)^\alpha = 1 + \sum_{n=1}^{\infty} \frac{\alpha(\alpha-1)(\alpha-2)\cdots(\alpha-n+1)}{n!} x^n$$

holds under the conditions

(a) $-1 < x < 1$.
(b) $x = 1$ *and* $\alpha > -1$.
(c) $x = -1$ *and* $\alpha > 0$.

In all other cases the series diverges.

The following succession of lemmas will establish the proof.

We shall use the notation

$$a_k = \alpha(\alpha - 1)(\alpha - 2)\cdots(\alpha - k + 1), \qquad k = 1, 2, 3, \cdots. \tag{2}$$

Hence $a_k \neq 0$, since $\alpha \neq 0, 1, 2, \cdots$. The ratio test may then be used on the series in (1), written as

$$\sum \frac{a_n}{n!} x^n,$$

to obtain absolute convergence if $|x| < 1$, but divergence if $|x| > 1$. Thus further consideration is confined to $-1 \leqslant x \leqslant 1$.

Also, the ratio test shows absolute convergence of $\sum a_{n+1}x^n/n!$ and then

$$\lim_{n\to\infty} \frac{a_{n+1}}{n!} |x|^n = 0 \qquad \text{if} \qquad |x| < 1. \tag{3}$$

With $f(x) = (1 + x)^\alpha$, $-1 < x \leqslant 1$, the formulas of Theorem 1, with $a = 0$, yield

$$(1 + x)^\alpha = 1 + \sum_{k=1}^{n} \frac{a_k}{k!} x^k + R_n(x) \qquad \text{with}$$

$$R_n(x) = \frac{a_{n+1}}{n!} \int_0^x (x - t)^n (1 + t)^{\alpha - n - 1}\, dt$$

$$= \frac{a_{n+1}}{n!} \int_0^x \left[\frac{x - t}{1 + t}\right]^n (1 + t)^{\alpha - 1}\, dt.$$

The fact that

$$\left|\frac{x - t}{1 + t}\right| \leqslant |x| \qquad \text{if } -1 < x \leqslant t \leqslant 0 \quad \text{or} \quad 0 \leqslant t \leqslant x \leqslant 1 \tag{4}$$

will be used. In the first case, $|x| = -x < 1$, $-xt \geqslant t$, and $x - t \leqslant 0$. Hence in this case

$$\left|\frac{x - t}{1 + t}\right| = \frac{t - x}{1 + t} \leqslant \frac{-xt - x}{1 + t} = -x = |x|$$

and the other case is even easier.

Lemma 1. *If* $-1 < x < 1$, *then* $R_n(x) \to 0$ *as* $n \to \infty$ *for all* α.

Proof. From (4),

$$|R_n(t)| \leqslant \frac{|a_{n+1}|}{n!} |x|^n \begin{cases} \displaystyle\int_0^x (1 + t)^{\alpha - 1}\, dt & \text{if } 0 \leqslant x < 1 \\ \displaystyle\int_x^0 (1 + t)^{\alpha - 1}\, dt & \text{if } -1 < x < 0. \end{cases}$$

With c the value of whichever integral is involved, then $c \geqslant 0$ is independent of n and, from (3),

$$|R_n(x)| \leqslant \frac{|a_{n+1}|}{n!} |x|^n c \to 0 \qquad \text{as } n \to \infty,$$

since we have the actual inequality $|x| < 1$. ∎

Thus (1) is established in case (a).

Lemma 2. *If* $|x| = 1$ *and* $\alpha \leqslant -1$, *the series in* (1) *is*

$$\sum \frac{a_n}{n!} (\pm 1)^n,$$

which diverges.

Proof. We have that

$$\left| \frac{a_n}{n!} \frac{(n-1)!}{a_{n-1}} \right| = \frac{|\alpha - n + 1|}{n} = \frac{n - (\alpha + 1)}{n} \geqslant 1,$$

since $\alpha + 1 \leqslant 0$. Hence divergence is seen from

$$\frac{|a_n|}{n!} \geqslant \frac{|a_{n-1}|}{(n-1)!} \geqslant \cdots \geqslant |a_1| = |\alpha| \geqslant 1. \blacksquare$$

Lemma 3. *If* $x = -1$ *and* $\alpha < 0$, *the series in* (1) *diverges (to* ∞*).*

Proof.

$$\begin{aligned}
&\sum_{k=1}^{n} \frac{\alpha(\alpha-1)(\alpha-2)\cdots(\alpha-k+1)}{k!}(-1)^k \\
&\qquad = \sum_{k=1}^{n} \frac{(-\alpha)(1-\alpha)(2-\alpha)\cdots(k-1-\alpha)}{k!} \\
&\qquad \geqslant \sum_{k=1}^{n} \frac{(-\alpha)\cdot 1 \cdot 2 \cdots (k-1)}{k!} \\
&\qquad = (-\alpha) \sum_{k=1}^{n} \frac{1}{k} \to \infty \qquad \text{as } n \to \infty. \blacksquare
\end{aligned}$$

We thus have that the series diverges if

$\lvert x\rvert > 1$	for all α,	
$\lvert x\rvert = 1$	and $\alpha \leqslant -1$	(Lemma 2),
$x = -1$	and $\alpha < 0$	(Lemma 3).

Hence the divergence statement of the theorem is proved. What remains is to establish convergence and equality in cases (b) and (c).

PROBLEM 3. By ordinary maximum-minimum methods,

$$0 \leqslant e^{-t} + t - 1 \qquad \text{for all } t \in E^1.$$

We also need that if $1 \leqslant m < n$ are integers, then

$$\text{(5)} \qquad \ln(n+1) - \ln m = \int_m^{n+1} \frac{1}{t}\,dt < \sum_{k=m}^{n} \int_k^{k+1} \frac{1}{k}\,dt = \sum_{k=m}^{n} \frac{1}{k}.$$

Lemma 4. *If* $\alpha > 0$ (*of course* $\alpha \neq 1, 2, 3, \cdots$), *then*

$$\lim_{n\to\infty} \frac{a_{n+1}}{n!} = 0.$$

Proof. Let $C_n = a_{n+1}/n!$. Fix an integer $m > \alpha$. Then

$$\left|\frac{C_m}{C_{m-1}}\right| = \frac{|\alpha - m|}{m} = \frac{m-\alpha}{m} = 1 - \frac{\alpha}{m} \leqslant e^{-\alpha/m}$$

by Problem 3. With $n > m$ we replace m by $m+1, m+2, \cdots, n$, take products, and obtain

$$\begin{aligned} |C_n| &\leqslant |C_{m-1}| \exp\left\{-\alpha\left[\frac{1}{m} + \frac{1}{m+1} + \cdots + \frac{1}{n}\right]\right\} \\ &\leqslant |C_{m-1}| \exp\{-\alpha[\ln(n+1) - \ln m]\} \qquad \text{by (5)} \\ &= |C_{m-1}| \exp\left\{\alpha \ln\left[\frac{m}{n+1}\right]\right\} = |C_{m-1}|\left[\frac{m}{n+1}\right]^{\alpha}. \end{aligned}$$

Since m is fixed and $\alpha > 0$, we have $C_n \to 0$ as $n \to \infty$. ∎

Lemma 5. *The equality in* (1) *holds if* $|x| = 1$ *and* $\alpha > 0$.

Proof. With $|x| = 1$, the remainders take the forms

$$R_n(\pm 1) = \frac{a_{n+1}}{n!} \int_0^{\pm 1} (\pm 1 - t)^n (1+t)^{\alpha-n-1}\,dt.$$

The integrals are evaluated as

$$0 \leqslant \int_0^1 (1-t)^n (1+t)^{\alpha-n-1}\,dt \leqslant \int_0^1 (1+t)^{\alpha-n-1}\,dt = \frac{1}{\alpha-n}\left[\frac{2^\alpha}{2^n} - 1\right],$$

and

$$\int_0^{-1} (-1-t)^n (1+t)^{\alpha-n-1}\,dt = (-1)^n \int_0^{-1} (1+t)^{\alpha-1}\,dt = \frac{(-1)^{n+1}}{\alpha}.$$

Thus $R_n(\pm 1) \to 0$ as $n \to \infty$ by Lemma 4. ∎

Lemma 6. *Equality in* (1) *holds if* $x = 1$ *and* $-1 < \alpha < 0$.

Proof. Set $\beta = 1 + \alpha$ and $b_{n+1} = \beta(\beta - 1) \cdots (\beta - n)$. Then

$$\frac{b_{n+1}}{n!} \to 0 \qquad \text{as } n \to \infty,$$

by Lemma 4, since $\beta > 0$. Also,

$$a_{n+1} = \frac{\beta}{\beta} \cdot (\beta - 1)(\beta - 2) \cdots (\beta - n)(\alpha - n) = \frac{1}{\beta} b_{n+1}(\alpha - n),$$

so that

$$\frac{a_{n+1}}{n!} \frac{1}{\alpha - n} = \frac{1}{\beta} \frac{b_{n+1}}{n!} \to 0 \qquad \text{as } n \to \infty.$$

Hence

$$|R_n(1)| = \frac{|a_{n+1}|}{n!} \int_0^1 (1 - t)^n (1 + t)^{\alpha - n - 1}\, dt$$

$$\leqslant \frac{|a_{n+1}|}{n!} \frac{1}{\alpha - n} \left(\frac{2^\alpha}{2^n} - 1\right) \to 0 \qquad \text{as } n \to \infty. \blacksquare$$

Since Lemmas 5 and 6 together dispose of cases (b) and (c), Theorem 3 is proved. ▮

The following (unmotivated) result is useful in approximating definite integrals with error estimates.

Theorem 4. *If* g *is a function for which* g^4 *exists on* $(-1, 1)$ *and* $g^4 \in \mathscr{L}[-1, 1]$, *then there is an* $\eta \in (-1, 1)$ *such that*

$$\int_{-1}^{1} g(x)\, dx = \tfrac{1}{3}[g(-1) + 4g(0) + g(1)] - \tfrac{1}{90} g^4(\eta).$$

Proof. By Taylor's theorem,

$$g(x) = g(0) + g'(0)x + \frac{g''(0)}{2} x^2 + \frac{g'''(0)}{3!} x^3 + R_3(x), \qquad -1 \leqslant x \leqslant 1.$$

Hence

$$\int_{-1}^{1} g(x)\, dx = 2g(0) + \tfrac{1}{3} g''(0) + \int_{-1}^{1} R_3(x)\, dx.$$

A check will show that

$$g(-1) + 4g(0) + g(1) = 6g(0) + g''(0) + R_3(-1) + R_3(1).$$

Thus

$$\int_{-1}^{1} g(x)\, dx = \tfrac{1}{3}[g(-1) + 4g(0) + g(1)] + \int_{-1}^{1} R_3(x)\, dx - \tfrac{1}{3} R_3(-1) - \tfrac{1}{3} R_3(1).$$

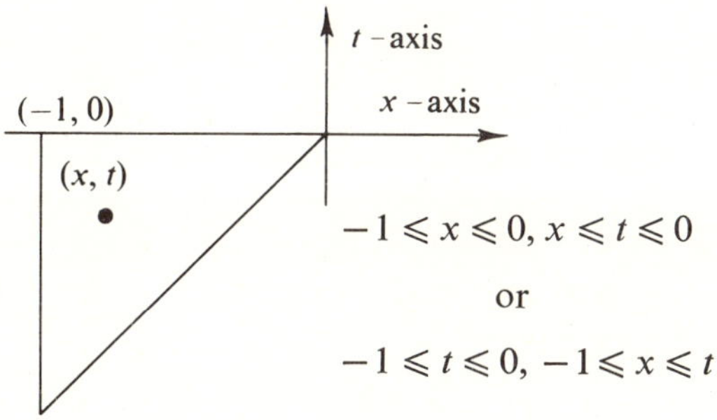

FIG. 1

The integral on the right is the sum of integrals over $[-1, 0]$ and $[0, 1]$, the first of which is

$$\frac{1}{3!}\int_{-1}^{0}\left\{\int_{0}^{x}(x-t)^3 g^4(t)\,dt\right\}dx = \frac{1}{3!}\int_{-1}^{0}\int_{x}^{0}(t-x)^3 g^4(t)\,dt\,dx$$

written in the second form so both lower limits of integration are smaller than the upper limits of integration. The integrand, as a function of two variables, is summable on the set

$$\{(x, t) \mid -1 \leqslant x \leqslant 0, \qquad x \leqslant t \leqslant 0\}$$

and this double integral is equal to iterated integrals in either order, by Theorem 5-5.3. To obtain the alternate iteration, we express the same region of integration as

$$\{(x, t) \mid -1 \leqslant t \leqslant 0, -1 \leqslant x \leqslant t\},$$

and obtain that the above integral is equal to

$$\begin{aligned}\frac{1}{3!}\int_{-1}^{0}\int_{-1}^{t}(t-x)^3 g^4(t)\,dx\,dt &= \frac{1}{3!}\int_{-1}^{0} g^4(t)\left[-\frac{(t-x)^4}{4}\right]_{-1}^{t} dt\\ &= \frac{1}{3!}\int_{-1}^{0}\frac{(t+1)^4}{4}\,g^4(t)\,dt.\end{aligned}$$

Hence

$$\begin{aligned}\int_{-1}^{0} R_3(x)\,dx - \frac{1}{3}R_3(-1) &= \frac{1}{3!}\left\{\int_{-1}^{0}\frac{(t+1)^4}{4}\,g^4(t)\,dt - \frac{1}{3}\int_{0}^{-1}(-1-t)^3 g^4(t)\,dt\right.\\ &= \frac{1}{3!12}\int_{-1}^{0}(t+1)^3(3t-1)g^4(t)\,dt.\end{aligned}$$

Since $(1+t)^3(3t-1) \leqslant 0$ for $-1 \leqslant t \leqslant 0$, there is an $\eta_1 \in (-1, 0)$, by Problem 7-1.2, such that the above value is

$$\frac{1}{3!12}\,g^4(\eta_1)\int_{-1}^{0}(1+t)^3(3t-1)\,dt = \frac{1}{3!12}\,g^4(\eta_1)\,\frac{-2}{5} = -\frac{1}{180}\,g^4(\eta_1).$$

In the same way,

$$\int_0^1 R_3(x)\,dx - \frac{1}{3}R_3(1) = \frac{1}{3!12}\int_0^1 (1-t)^3(-1-3t)g^4(t)\,dt = -\frac{1}{180}g^4(\eta_2)$$

for some $\eta_2 \in (0, 1)$. Hence, by the intermediate property of derived functions (Theorem 7–1.5), there is an $\eta \in (\eta_1, \eta_2)$ such that $g^4(\eta)$ is the average of $g^4(\eta_1)$ and $g^4(\eta_2)$ so that

$$-\frac{1}{180}[g^4(\eta_1) + g^4(\eta_2)] = -\frac{1}{90}g^4(\eta)$$

and the desired result is obtained. ∎

Corollary 1. *If f is a function for which f^4 exists on (a, b) and $f^4 \in \mathscr{L}[a, b]$, then there is a $\xi \in (a, b)$ such that, with $c = \frac{1}{2}(a + b)$,*

$$\int_a^b f(x)\,dx = \frac{b-a}{6}[f(a) + 4f(c) + f(b)] - \frac{(b-a)^5}{2880}f^4(\xi).$$

Corollary 2. (Simpson's Rule) *With f a function, $x_0 \in E^1$, and $h > 0$, let*

$$x_k = x_0 + kh \qquad \text{and} \qquad y_k = f(x_k), \qquad k = 0, 1, 2, \cdots.$$

If for some integer n, f^4 exists on (x_0, x_{2n}) and $f^4 \in \mathscr{L}[x_0, x_{2n}]$, then there is a $\xi \in (x_0, x_{2n})$ such that

$$\int_{x_0}^{x_{2n}} f = \frac{h}{3}[y_0 + 4y_1 + 2y_2 + \cdots + 4y_{2n-2} + 2y_{2n-1} + y_{2n}] - \frac{nh^5}{90}f^4(\xi).$$

PROBLEMS

1. If A is a set, f is a function, and $K > 0$ a constant such that $f'(x)$ exists and $|f'(x)| \leqslant K$ for each $x \in A$, then

$$m^*f(A) \leqslant Km^*A.$$

2.
$$\sum_{k=1}^{n} \frac{(-1)^{k+1}}{k}\binom{n}{k} = 1 + \frac{1}{2} + \cdots + \frac{1}{n}.$$

[*Hint:*

$$\sum_{k=0}^{n-1}(1-x)^k = \frac{1-(1-x)^n}{x} = \cdots = \sum_{k=1}^{n}\binom{n}{k}(-1)^{k+1}x^{k-1}.$$

Integrate from $x = 0$ to $x = 1$.]

3. Let f and g be real-valued functions which are differentiable on $I = [a, b]$. If $f(x) = 0$ has two solutions on I, then between these solutions there is a solution of $f'(x) + f(x)g'(x)$. [*Hint:* Apply Rolle's theorem to $h(x) = f(x)\,e^{g(x)}$.]

4. Show that $x \leqslant e^{x-1}$. In the notation of Hint 1, p. 177, replace x by a_k/σ, $k = 1, 2, \ldots, n$. Take the product and obtain another proof of the geometric-arithmetic mean inequality.

5. If $0 < a < b$, then $f: (0, \infty) \to E^1$, defined by

$$f(x) = \frac{\log(ax+1)}{\log(bx+1)},$$

is strictly increasing on $(0, \infty)$.

[*Hint:* Write $f'(x)$ as a proper fraction and let $g(x)$ be the numerator. Show that $g'(x) > 0$, etc.]

6. (a) If both $\lim_{x\to\infty} f(x)$ and $\lim_{x\to\infty} f'(x)$ exist, then $\lim_{x\to\infty} f'(x) = 0$.

(b) If $\lim_{x\to\infty} f'(x) = 0$, then $\lim_{x\to\infty} \frac{f(x)}{x} = 0$.

(c) If $\lim_{x\to\infty} f'(x) = a \neq 0$, then $\lim_{x\to\infty} \frac{f(x)}{ax} = 1$.

(d) If $\lim_{x\to\infty} f'(x)$ exists, then for each number $h \neq 0$,

$$\lim_{x\to\infty} \frac{f(x+h) - f(x)}{h} \text{ exists.}$$

7. Evaluate

$$\lim_{n\to\infty} \left[\frac{f(a+1/n)}{f(a)}\right]^n$$

given that $f(a) \neq 0$ and $f'(a)$ exists.

[*Hint:* Write the expression as e to a power.]

8. If a nonempty set $A \subset [a, b]$ has $mA = 0$, then there is a function $f: [a, b] \to E^1$ such that $f'(x) = \infty$ for each $x \in A$.

[*Hint:* For each n let $G_n \supset A$, G_n open, $mG_n < 1/2^n$ and define f_n and f by

$$f_n(x) = m(G_n \cap [a, x]), \qquad f(x) = \sum_{n=1}^{\infty} f_n(x).]$$

9. Let $f: E^1 \to E^1$ be such that the functional equation

$$f(x)f(y) = f(x+y)$$

holds for all $x, y \in E^1$.

(a) If f is differentiable and $f'(x) \neq 0$ for all $x \in E^1$, then there is a constant c such that $f(x) = e^{cx}$.

(b) The same is true if f is continuous and $f \not\equiv 0$.

10. Find $\lim_{n\to\infty} \sum_{k=n}^{2n} \frac{1}{k}$.

11. (a) Note that $(1-x)^n = \sum_{k=0}^{n} (-1)^k \binom{n}{k} x^k$. Successively take a derivative, multiply by x, and set $x = 1$ to show that

$$\sum_{k=0}^{n} (-1)^k \binom{n}{k} k^r = \begin{cases} 0 & \text{for } r = 0, 1, 2, \cdots, n-1 \\ (-1)^n n! & \text{for } r = n. \end{cases}$$

(b) Given a function whose nth derivative is continuous at $x = a$, use l'Hospital's rule to show that

$$\lim_{h\to 0} \left\{\frac{(-1)^n}{h^n} \sum_{k=0}^{n} (-1)^k \binom{n}{k} f(a+kh)\right\} = f^n(a).$$

12. Given a polynomial $p(x)$ of degree n, then

$$\sum_{k=0}^{n} \frac{p^k(0)}{(k+1)!} x^{k+1} = \sum_{k=0}^{n} \frac{(-1)^k p^k(x)}{(k+1)!} x^{k+1},$$

it being understood that the 0-derivative $p^0(x) = p(x)$.

[*Hint:* With $L(x)$ the left side and $R(x)$ the right side, then $L'(x) = R'(x) = p(x)$.]

13. Given $p > 0, q > 0$, and $p + q = 1$, let $f: E^1 \to E^1$ be a function whose derivative exists on E^1 and

$$f(u) - f(v) = f'(pu + qv)(u - v) \qquad \text{for all } u, v \in E^1.$$

(a) If $p \neq q$, then f is linear.

(b) If $p = q = \frac{1}{2}$ and f'' exists on E^1, then f is quadratic.

[*Hint:* $f'(pu + qv) = f'(pv + qu)$; therefore $f'(x) = f'(0)$ if $p \neq q$. If $p = q$, then derivatives with respect to u and v yield $\frac{1}{2}[f'(u) + f'(v)] = f'(\frac{1}{2}u + \frac{1}{2}v)$, and then $f''(u) = f''(0)$.

14. (a) $$\sum_{n=1}^{N} \frac{\sin n^2 x \sin nx}{n} = \frac{1}{2} \sum_{n=1}^{N} \frac{1}{n} [\cos n(n-1)x - \cos n(n+1)x]$$

$$= \frac{1}{2}\left[1 - \frac{\cos N(N+1)x}{N} - \sum_{n=1}^{N-1} \frac{\cos n(n+1)x}{n(n+1)}\right].$$

(b) Obtain uniform convergence of

$$\sum_{n=1}^{\infty} \frac{\sin n^2 x \sin nx}{n}.$$

15. If $f: (a, b) \to E^1$ is a convex function (see page 357), then the function defined by

$$g(x) = e^{f(x)}, \qquad a < x < b$$

is also convex.

16. If $f: [a, b] \to E^1$ is continuous and takes on each of its values at most twice, then $f'(x)$ exists a.e. in $[a, b]$.

[*Hint:* Problem 19(c) page 353.]

17. (a) If $\lim_{x \to 0+} f(x) = 0$ and for some $\delta > 0$ the derivative $f'(x)$ exists and $f'(x) \neq 0$ for $0 < x \leq \delta$, then

$$\lim_{x \to 0+} \frac{f(x)}{f'(x)} = 0.$$

[*Hint:* First consider $f'(x) > 0$ (relying upon Theorem 7–1.5) and assume the limit inferior $= \lambda > 0$. Choose $0 < \delta_1 \leq \delta$ such that the ratio is $> \lambda/2$ for $0 < x < \delta_1$. For each such x, $0 < f(x) = xf'(\xi)$ for some $0 < \xi < x$ and

$$\frac{\lambda}{2} < \frac{f(\xi)}{f'(\xi)} < x.]$$

(b). If $\lim_{x \to 0+} f(x) = 0$ and $\lim_{x \to 0+} f(x)/f'(x)$ exists, then this second limit also $= 0$.

18. Assume e is rational, $e = m/n$. Then

$$\frac{n}{m} = \frac{1}{e} = \sum_{k=0}^{\infty} (-1)^k \frac{1}{k!},$$

and, by multiplying by $(-1)^{m+1}m!$, first obtain

$$(-1)^{m+1}\left\{n(m-1)! - \sum_{k=0}^{m}(-1)^k\frac{m!}{k!}\right\}$$

$$= \frac{1}{m+1} - \frac{1}{(m+1)(m+2)} + \frac{1}{(m+1)(m+2)(m+3)} - \cdots,$$

and then a contradiction from this.

19. (a) With a an integer, let $f(x, y)$, $a-1 \leqslant x < \infty$, $y \in E^1$ be continuous and nondecreasing for each fixed y. Then

$$\int_a^{m+1} f(x, y)\,dx \leqslant \sum_{k=a}^{m} f(k, y) \leqslant \int_{a-1}^{m} f(x, y)\,dx, \qquad m = a, a+1, \cdots.$$

(b) $\lim_{n\to\infty} n^2 \sum_{k=1}^{n} \frac{k}{(k^2+n^2)^2} = \frac{1}{4}$ follows from $f(x, y) = \frac{x}{(x^2+y^2)^2}$.

(c) $\lim_{n\to\infty} n^2 \sum_{k=1}^{n^p} \frac{k}{(k^2+n^2)^2} = \frac{1}{2}$ for $p = 2, 3, 4, \cdots$.

(d) $\lim_{n\to\infty} \sum_{k=1}^{n^2} \frac{n}{k^2+n^2} = ?$; $\lim_{n\to\infty} \sum_{k=1}^{n} \frac{n}{k^2+n^2} = ?$

20. Theorem. If $f \in \mathscr{L}^2$ and $f(t) = 0$ for $t \notin [a, b]$, then

(a) $\lim_{x\to 0} \int_a^b [f(t+x) - f(t)]^2\,dt = 0.$

[*Hint:* First f continuous on $[a, b]$; then use Theorem 7–7.8.]

(b) The function F defined by

$$F(x) = \int_a^b f(t+x)f(t)\,dt, \qquad x \in E^1$$

is continuous at the origin $x = 0$.

[*Hint:* Schwarz inequality applied to the integral expression in $F(x) - F(0) = \int_a^b [f(t+x) - f(t)]f(t)\,dt$.]

Corollary. If $A \subset E^1$ is measurable and $mA > 0$, then the difference set of A includes an interval containing the origin.

[*Hint:* First $A \subset [a, b]$ and apply part (b) with $f = \chi_A$.]

[*Note:* A homework problem in 1928 course by W. A. Hurwitz.]

21. (a) If $0 < y < 1$, then $\lim_{n\to\infty} \sum_{j=1}^{n-1} \frac{jy^j}{n-j} = 0.$

[*Hint:* Let S_n denote the sum, take $0 < k < n$, and show that

$$0 \leqslant S_n \leqslant \sum_{j=1}^{k-1} \frac{k}{n-k} + \sum_{j=k}^{n-1} jy^j = \frac{k(k-1)}{n-k} + y\frac{d}{dy}\sum_{j=k}^{n-1} y^j$$

$$= \frac{k(k-1)}{n-k} + \frac{y}{(1-y)^2}[(1-y)(ky^{k-1} - ny^{n-1}) + (y^k - y^n)].$$

Then $0 \leqslant \overline{\lim_{n\to\infty}} S_n \leqslant \lim_{k\to\infty} (?)$.]

(b) If $0 < y < 1$, then $\lim_{n\to\infty} ny^n \sum_{i=1}^{n} \frac{1}{iy^i} = \frac{1}{1-y}$.

[*Hint:*

$$\sum_{i=1}^{n} \frac{n}{i} y^{n-i} = \sum_{j=0}^{n-1} \frac{n}{n-j} y^j = \sum_{j=0}^{n-1} y^j + \sum_{j=0}^{n-1} \frac{jy^j}{n-j}.]$$

(c) If $x > 1$, then $\lim_{n\to\infty} \frac{n}{x^n} \sum_{i=1}^{n} \frac{x^{i-1}}{i} = \frac{1}{x-1}$.

[*Note:* The limit in (c) was used in solving a probability problem.]

22. (a) Use integration by parts to obtain

$$\int_0^1 (1-x^2)^{n+1}\,dx = \int_0^1 (1-x^2)(1-x^2)^n\,dx = \frac{2n+2}{2n+3}\int_0^1 (1-x^2)^n\,dx$$

$$= \int_0^1 (1-x^2)^n\,dx - \int_0^1 x\{x(1-x^2)^n\}\,dx$$

and then show that

$$\int_0^1 (1-x^2)^n\,dx \geqslant \frac{2}{3\sqrt{n}}, \qquad n = 1, 2, 3, \cdots.$$

(b) Find the minimum of $(1-x^2)^n + nx^2 - 1$, for $0 \leqslant x \leqslant 1$, use this and $1 - nx^2 \geqslant 0$ for $0 \leqslant x \leqslant 1/\sqrt{n}$ to obtain another proof of the inequality in part (a).

23. Let R be the set of rational numbers and define f by

$$f(x) = x^2\,\chi_R(x).$$

Then $f'(0)$ exists, but f is discontinuous (and hence nondifferentiable) at all $x \neq 0$.

24. The function f, defined on $[0, 1]$ by $f(x) = 0$ if $x = 0$ or x irrational and $1/q$ if $x = p/q$ in lowest terms, does not have a derivative at any point of $[0, 1]$.

25. Given $f\colon [a, b] \to E^1$ and $g\colon [a, b] \to E^1$ are continuous and $g(x) \neq 0$ for $a < x < b$. Set

$$F(x) = \int_a^x f(t)g(t)\,dt, \qquad G(x) = \int_a^x g(t)\,dt$$

and

$$D(x) = \begin{vmatrix} F(x) & G(x) & 1 \\ F(a) & G(a) & 1 \\ F(b) & G(b) & 1 \end{vmatrix}$$

for $a \leqslant x \leqslant b$. Apply Rolle's theorem to the function D and obtain the second mean-value theorem for integrals of continuous functions.

26. If $f_1 \in \mathscr{L}\,[0, b]$, and f_{n+1} is defined for $n = 1, 2, 3, \cdots$ by

$$f_{n+1}(x) = \int_0^x f_n(t)\,dt, \qquad 0 \leqslant x \leqslant b, \qquad \text{then}$$

$$f_n(x) = \int_0^x \frac{(x-t)^{n-2}}{(n-2)!} f_1(t)\,dt, \qquad n = 2, 3, 4, \cdots.$$

Also show that the following series converges and the definition of $g(x)$ is all right:

$$g(x) = \sum_{n=2}^{\infty} f_n(x) = \int_0^x e^{x-t} f_1(t)\,dt, \qquad 0 \leqslant x \leqslant b,$$

and moreover g is continuous.

27. Given a function $f: E^1 \to E^1$, then an $x \in E^1$ is said to be a **point of bounded variation** of f if $V_{x-\delta}^{x+\delta}(f) < \infty$ for some $\delta > 0$. Let $B(f)$ denote the set of all points of bounded variation of f. Any point of $E^1 - B(f)$ is said to be a **point of unbounded variation** of f.

For $x \in E^1$ and $\delta > 0$, define

$$W(x, \delta) = [1 + V_{x-\delta}^{x+\delta}(f)]^{-1}$$

if $V_{x-\delta}^{x+\delta}(f) < \infty$ and otherwise $W(x, \delta) = 0$.

(a) Set $w(x) = \lim_{\delta \to 0} W(x, \delta)$, after showing that the limit exists.

(b) The function w is lower semicontinuous.

(c) The set $B(f)$ is open.

(d) If f is continuous at a point $c \in B(f)$, then $w(c) = 1$.

(e) If c is such that $w(c) = 1$, then f is continuous at c.

Problems on an Alternate Proof of the Weierstrass Approximation Theorem (Section 6–4) and the Stone Generalization

In the following lemma the sequence $\{c_n\}$ such that

$$1 = c_n \int_{-1}^{1} (1 - x^2)^n \, dx$$

is used. The exact values are not required and the inequalities

$$0 < c_n < \tfrac{1}{2}(n + 1)$$

are sufficient.† These inequalities follow from

$$1 = 2c_n \int_0^1 (1 - x^2)^n \, dx > 2c_n \int_0^1 (1 - x)^n \, dx = \frac{2c_n}{n+1}.$$

Lemma 1. *Let $g: E^1 \to E^1$ be continuous and $g(x) = 0$ for $x \leqslant 0$ or $x \geqslant 1$. Then*

(a) *For $0 \leqslant x \leqslant 1$ and $n = 1, 2, 3, \cdots,$*

$$\int_{-1}^{1} g(x + t)(1 - t^2)^n \, dt$$

is a polynomial in x, and

(b) *The sequence $\{Q_n(x)\}$ of polynomials defined by*

$$Q_n(x) = c_n \int_{-1}^{1} g(x + t)(1 - t^2)^n \, dt$$

converges uniformly to g on $\{x \mid 0 \leqslant x \leqslant 1\}$.

Proof (a). Set $x + t = s$ to express the integral as

$$\int_{x-1}^{x+1} g(s)[1 - (s - x)^2]^n \, ds = \int_0^1 g(s)[(1 - s^2) + 2sx - x^2]^n \, ds, \qquad 0 \leqslant x \leqslant 1,$$

which is a polynomial in x.

Proof (b). Let M be the maximum of $|g|$. Choose $\varepsilon > 0$ and, from the uniform continuity of g, let δ be such that

$$0 < \delta < 1 \qquad \text{and} \qquad |s - t| \leqslant \delta \Rightarrow |g(s) - g(t)| < \varepsilon, \qquad \text{for } s, t \in E^1.$$

† The better estimate $c_n < \frac{3}{4}\sqrt{n}$ may be obtained by using Problem 22, page 450.

Note that

$$\delta \leqslant |x| \leqslant 1 \Rightarrow c_n(1-x^2)^n \leqslant \tfrac{1}{2}(n+1)(1-\delta^2)^n.$$

Hence for any x such that $0 \leqslant x \leqslant 1$, we have

$$\begin{aligned}|Q_n(x)-g(x)| &= \left|\int_{-1}^{1} g(x+t)c_n(1-t^2)^n dt - g(x)\int_{-1}^{1} c_n(1-t^2)^n\, dt\right| \\ &\leqslant \left[\int_{-1}^{-\delta} + \int_{-\delta}^{\delta} + \int_{\delta}^{1}\right] |g(x+t)-g(x)|\, c_n(1-t^2)^n\, dt \\ &\leqslant 2M\left[\int_{-1}^{-\delta} + \int_{\delta}^{1}\right] c_n(1-t^2)^n\, dt + \varepsilon \int_{-\delta}^{\delta} c_n(1-t^2)^n\, dt \\ &\leqslant 2M(n+1)(1-\delta^2)^n(1-\delta)+\varepsilon.\end{aligned}$$

This last expression is independent of x and may be made less than 2ε by choosing n sufficiently large. ∎

Theorem 1. (Weierstrass) *If $f\colon [a, b] \to E^1$ is continuous, then there is a sequence $\{P_n\}$ of polynomials such that*

$$P_n \rightrightarrows f \qquad \textit{on } [a, b].$$

Proof. Define functions g_1, g_2, and g by

$$g_1(t) = f[a+(b-a)t], \qquad 0 \leqslant t \leqslant 1,$$

$$g_2(t) = g_1(t) - g_1(0) - t[g_1(1) - g_1(0)], \qquad 0 \leqslant t \leqslant 1,$$

$$g(t) = \begin{cases} 0 & \text{for } t \leqslant 0 \text{ or } t \geqslant 1 \\ g_2(t) & \text{for } 0 \leqslant t \leqslant 1. \end{cases}$$

Hence g satisfies the conditions of the lemma and we let $\{Q_n\}$ be a sequence of polynomials such that

$$Q_n \rightrightarrows g \qquad \text{on } [0, 1],$$

so that also $Q_n \rightrightarrows g_2$ on [0, 1]. Then $\{R_n\}$ defined by

$$R_n(t) = Q_n(t) + g_1(0) + t[g_1(1) - g_1(0)], \qquad 0 \leqslant t \leqslant 1$$

is a sequence of polynomials such that

$$\begin{aligned}R_n(t) - g_1(t) &= Q_n(t) - g_1(t) + g_1(0) + t[g_1(1) - g_1(0) \\ &= Q_n(t) - g_2(t) \qquad \text{for } 0 \leqslant t \leqslant 1.\end{aligned}$$

Therefore $R_n \rightrightarrows g_1$ on [0, 1]. Now define $\{P_n\}$ by

$$P_n(x) = R_n\left(\frac{x-a}{b-a}\right), \qquad a \leqslant x \leqslant b.$$

Given a number $\varepsilon > 0$, let N be such that

$$n \geqslant N \Rightarrow |R_n(t) - g_1(t)| < \varepsilon \qquad \text{for all } 0 \leqslant t \leqslant 1.$$

Consequently, $n \geqslant N$ implies

$$|P_n(x) - f(x)| = \left|R_n\left(\frac{x-a}{b-a}\right) - g_1\left(\frac{x-a}{b-a}\right)\right| < \varepsilon \qquad \text{for all } a \leqslant x \leqslant b. \;∎$$

PROBLEM 1. With $a > 0$, there is a sequence $\{P_n\}$ of polynomials such that $P_n(0) = 0$, $n = 1, 2, 3, \cdots$ and

$$\lim_{n \to \infty} P_n(t) = |t| \quad \text{uniformly on } [-a, a].$$

[*Hint:* If $\{Q_n(t)\}$ converges uniformly on $[-a, a]$ to $|t|$, then so does $\{P_n\}$, where $P_n(t) = Q_n(t) - Q_n(0)$.]

Definition 1. *Given a set A, a family $\mathscr{A}$ of real-valued functions defined on A is said to be an* **algebra** *(real) if for all $f, g \in \mathscr{A}$ and $c \in E^1$,*

$$f + g \in \mathscr{A}, \qquad fg \in \mathscr{A}, \qquad \text{and} \qquad cf \in \mathscr{A}.$$

Such an algebra $\mathscr{A}$ is said to **separate points** *on A if for each pair of distinct points $x_1, x_2 \in A$, there is an $f \in \mathscr{A}$ such that*

$$f(x_1) \neq f(x_2),$$

and $\mathscr{A}$ is said to **vanish at no point** *of A if for each $x \in A$ there is an $f \in \mathscr{A}$ such that $f(x) \neq 0$.*

PROBLEM 2 (a). The set of polynomials on E^1 is an algebra which separates points and vanishes at no point of E^1.

(b) With $A = [-1, 1]$, the set of even polynomials on A is an algebra, but does not separate points of A.

Lemma 2. *Let $\mathscr{A}$ be an algebra on a set A which separates points on A and vanishes at no point of A. If x_1 and x_2 are distinct points of A and $c_1, c_2 \in E^1$, then there are $g, f_1, h \in \mathscr{A}$ such that*

(a) $g(x_1) \neq 0; g(x_1) \neq g(x_2)$.
(b) $f_1(x_1) = 1; f_1(x_2) = 0$.
(c) $h(x_1) = c_1$; *and* $h(x_2) = c_2$.

Proof (a). Take $u \in \mathscr{A}$ such that $u(x_1) \neq u(x_2)$ and $v \in \mathscr{A}$ such that $v(x_1) \neq 0$. Then:

if $u(x_1) \neq 0$, set $g = u$,
if $u(x_1) = 0$ and $v(x_1) \neq v(x_2)$, set $g = v$,
if $u(x_1) = 0$ and $v(x_1) = v(x_2)$, set $g = u + v$.

Proof (b). Define f_1 by

$$f_1(x) = \frac{1}{g(x_1)[g(x_1) - g(x_2)]} [g^2(x) - g(x_2)g(x)].$$

Proof (c). By switching the roles of x_1 and x_2, let $f_2 \in \mathscr{A}$ be such that $f_2(x_2) = 1$ and $f_2(x_1) = 0$. Then

$$h = c_1 f_1 + c_2 f_2$$

has the desired property. ▮

PROBLEM 3. In the proof of part (b), why not set

$$f_1(x) = \frac{u(x) - u(x_2)}{u(x_1) - u(x_2)} \, ?$$

Definition 2. *Let $\mathscr{A}$ be an algebra on a set A in a metric space.*

(i) *The set $\mathscr{B}$ of all functions on A that are limits of uniformly convergent sequences of members of $\mathscr{A}$ is called the* **uniform closure** *of $\mathscr{A}$.* [Hence $\mathscr{A} \subset \mathscr{B}$].
(ii) *$\mathscr{A}$ is said to be* **uniformly closed** *if $\mathscr{B} \subset \mathscr{A}$; that is, if $\{f_n\}$, with each $f_n \in \mathscr{A}$, is such that $f_n \rightrightarrows f$ on A, then $f \in \mathscr{A}$.*

In this terminology, the Weierstrass theorem may be stated as:

The set of continuous real-valued functions on an interval $[a, b] \subset E^1$ is the uniform closure of the set of polynomials on $[a, b]$.

PROBLEM 4. If $\mathscr{A}$ is an algebra of bounded functions on a set A in a metric space, then the uniform closure $\mathscr{B}$ of $\mathscr{A}$ is also an algebra on A and is uniformly closed.

In view of Problem 2(a), the following theorem is a generalization of the Weierstrass approximation theorem.

Theorem 2. (Stone) *Given a compact set A in a metric space, let $\mathscr{A}$ be an algebra of real continuous function on A which separates points on A and vanishes at no point of A. Then the uniform closure $\mathscr{B}$ of $\mathscr{A}$ consists of all real continuous functions on A.*
Auxiliary results (which we shall prove first) *are*

(A) $g \in \mathscr{B} \Rightarrow |g| \in \mathscr{A}$;
(B) $g, h \in \mathscr{B} \Rightarrow \max(g, h) \in \mathscr{B}$ *and* $\min(g, h) \in \mathscr{B}$; and
(C) *given a number $\varepsilon > 0$, $x \in A$, and a real function f continuous on A, there is a $g_x \in \mathscr{B}$ such that*

$$g_x(x) = f(x) \qquad \text{and} \qquad g_x(t) > f(t) - \varepsilon \qquad \text{for all } t \in A.$$

[*Note:* Since A is compact, then $\mathscr{A}$ and $\mathscr{B}$ are algebras of real bounded continuous functions on A, and $\mathscr{B}$ is uniformly closed.]

Proof (A). Let $a = \sup|g|$ on A and let $\varepsilon > 0$ be given. By Problem 1 there are real numbers $c_1, c_2, \cdots, c_n$ such that

$$\left|\sum_{k=1}^{n} c_k t^k - |t|\right| < \varepsilon \qquad \text{for all } -a \leqslant t \leqslant a.$$

From $g(x) \in E^1$ and $-a \leqslant g(x) \leqslant a$ for each $x \in A$, we thus have that

$$\left|\sum_{k=1}^{n} c_k g^k(x) - |g(x)|\right| < \varepsilon \qquad \text{for all } x \in A.$$

Since $\mathscr{B}$ is an algebra, $\sum_{k=1}^{n} c_k g^k \in \mathscr{B}$, and then $|g| \in \mathscr{B}$, since $\mathscr{B}$ is a uniformly closed algebra (by Problem 4).

Proof (B). From (A), $|g - h| \in \mathscr{B}$ as well as $(g + h) \in \mathscr{B}$ so that

$$\max(g, h) = \tfrac{1}{2}(g + h + |g - h|) \qquad \text{and} \qquad \min(g, h) = \tfrac{1}{2}(g + h - |g - h|)$$

are also in the algebra $\mathscr{B}$.

Proof (C). With x as designated, take any $y \in A$. In case $y \neq x$, use $f(x)$ and $f(y)$ as the c_1 and c_2 of Lemma 2 to obtain a function $h_y \in \mathscr{A} \subset \mathscr{B}$ such that

$$h_y(x) = f(x) \qquad \text{and} \qquad h_y(y) = f(y).$$

In case $y = x$, take any $h \in \mathscr{A}$ such that $h(x) \neq 0$ and set

$$h_y = \frac{f(x)}{h(x)} h.$$

By continuity of h_y and f there is an open set $G(y)$ such that

$$y \in G(y) \qquad \text{and} \qquad h_y(t) > f(t) - \varepsilon \qquad \text{for each } t \in G(y).$$

This being done for each $y \in A$, and A being compact, let $y_1, y_2, \cdots, y_n$ be points of A such that

$$A \subset G(y_1) \cup G(y_2) \cup \cdots \cup G(y_n).$$

Then g_x defined by $g_x = \max(h_{y_1}, h_{y_2}, \cdots, h_{y_n})$ is in $\mathscr{B}$ [by (B)] and satisfies the conclusion of part (C).

Thus the auxiliary results (A), (B), and (C) have been proved.

Thus to finish the proof of the theorem we take any real function f continuous on A, any number $\varepsilon > 0$, and shall show there is an $h \in \mathscr{B}$ such that

$$|h(t) - f(t)| < \varepsilon \qquad \text{for all } t \in A.$$

For each $x \in A$ let $g_x \in \mathscr{B}$ as in (C) and then, from the continuity of g_x and f, let $G(x)$ be an open set such that

$$x \in G(x) \qquad \text{and} \qquad g_x(t) < f(t) + \varepsilon \qquad \text{for all } t \in G(x).$$

Then there are points $x_1, x_2, \cdots, x_n$ in A such that

$$A \subset G(x_1) \cup G(x_2) \cup \cdots \cup G(x_n).$$

Hence upon setting $h = \min(g_{x_1}, g_{x_2}, \cdots, g_{x_n}) \in \mathscr{B}$, we have that $h(t) < f(t) + \varepsilon$ for all $t \in A$, and moreover $h(t) > f(t) - \varepsilon$ for all $t \in A$ from (C). Consequently, $|h(t) - f(t)| < \varepsilon$ for all $t \in A$. ∎

A **complex algebra** $\mathscr{A}$ of functions is defined by allowing the functions and constants in Definition 1 to be complex-valued. Then Definition 2 is extended and Lemma 2 proved for $\mathscr{A}$ a complex algebra, but not so for Theorem 2 as the next problem shows.

PROBLEM 5. With $A = [-\pi, \pi] \subset E^1$, let $\mathscr{A}$ be the set of all functions f such that $f(x)$ is a complex polynomial in e^{ix}; that is,

$$f(x) = \sum_{k=0}^{n} c_k e^{ikx}, \qquad c_k \text{ complex}, \qquad -\pi \leqslant x \leqslant \pi.$$

(a) Then $\mathscr{A}$ is a complex algebra of continuous functions, $\mathscr{A}$ separates points on A, $\mathscr{A}$ vanishes at no point of A, and A is compact.

(b) For every $f \in \mathscr{A}$

$$\int_{-\pi}^{\pi} f(x) e^{ix} \, dx = 0,$$

and this is also true of every g in the uniform closure $\mathscr{B}$ of $\mathscr{A}$; that is, $\mathscr{B}$ does not include all complex continuous functions on A.

Definition 3. *A complex algebra $\mathscr{A}$ of functions on a set A is said to be* **self-adjoint** *if $f \in \mathscr{A} \Rightarrow \bar{f} \in \mathscr{A}$ where $\bar{f}$ is the complex conjugate of f.*

PROBLEM 6. The set of all complex polynomials is a self-adjoint algebra which separates points and vanishes at no point.

Theorem 3. *Let $\mathscr{A}$ be a complex algebra of continuous functions on a compact set A. If $\mathscr{A}$ is self-adjoint, separates points on A, and vanishes at no point of A, then the uniform closure $\mathscr{B}$ of $\mathscr{A}$ consists of all complex continuous functions on A.*

Proof. Note that $\mathscr{A}_R = \{f \in \mathscr{A} \mid f \text{ is real}\}$ is a real algebra and its uniform closure $\mathscr{B}_R \subset \mathscr{B}$.

If $f = u + iv \in \mathscr{A}$ with u and v real, then

$$u = \frac{1}{2}(f + f) \in \mathscr{A}_R \qquad \text{and} \qquad v = \frac{i}{2}(\bar{f} - f) \in \mathscr{A}_R,$$

since $\mathscr{A}$ is a self-adjoint algebra. With x_1 any point of A and x_2 any other point of A, use Lemma 2 (extended) to obtain an $f = u + iv \in \mathscr{A}$ for which $f(x_1) = 1$ and $f(x_2) = 0$ and hence

$$u(x_1) = 1 \neq 0 = u(x_2).$$

This shows that $\mathscr{A}_R$ separates points on A and vanishes at no point of A (since $x_1 \in A$ is arbitrary), and hence $\mathscr{A}_R$ satisfies the hypotheses of Theorem 2. Thus $\mathscr{B}_R$ consists of all real continuous functions on A. Hence if $f = u + iv$ is any complex continuous function on A, then $u \in \mathscr{B}_R \subset \mathscr{B}$, $v \in \mathscr{B}$, and therefore $f = u + iv \in \mathscr{B}$, since $\mathscr{B}$ is a complex algebra (by Problem 4 extended).

Chapter 8

STIELTJES INTEGRALS

8–1 RIEMANN-STIELTJES INTEGRALS

Partitions of an interval $[a, b]$ as defined in Section 7–6 will be used.

Definition 1. *Let f and α be real-valued functions whose domains include a closed interval $[a, b]$ with both f and α bounded on $[a, b]$. Take a partition*

$$P = \{x_0 = a \leqslant x_1 \leqslant x_2 \leqslant \cdots \leqslant x_{n-1} \leqslant x_n = b\}$$

of $[a, b]$. With t_i selected such that $x_{i-1} \leqslant t_i \leqslant x_i$ for $i = 1, 2, \cdots, n$, then

$$\sum_{i=1}^{n} f(t_i)[\alpha(x_i) - \alpha(x_{i-1})]$$

is called a **Riemann-Stieltjes** *sum for f with respect to α relative to P. The set of values of all such sums (with f, α, and P fixed) is denoted by*

$$S(f, \alpha, P).$$

With a partition P fixed, we shall use

$$\Delta_i \alpha = \alpha(x_i) - \alpha(x_{i-1})$$

and hence have that $S(f, \alpha, P)$ is the set

$$\left\{y \,\middle|\, y = \sum_{i=1}^{n} f(t_i)\, \Delta_i \alpha \text{ for some } t_i \in [x_{i-1}, x_i]\right\}.$$

EXAMPLE 1. Given that

$$f(x) = \begin{cases} 1 & 0 \leqslant x \leqslant \frac{1}{2} \\ 2 & \frac{1}{2} < x \leqslant 1 \end{cases}, \quad \alpha(x) = \begin{cases} 3 & 0 \leqslant x < \frac{1}{2} \\ 4 & \frac{1}{2} \leqslant x \leqslant 1 \end{cases},$$

let $P_1 = \{0, \frac{1}{3}, \frac{2}{3}, 1\}$. Then $\Delta_1 \alpha = 0$, $\Delta_2 \alpha = 1$, and $\Delta_3 \alpha = 0$. Hence

$$\sum_{i=1}^{3} f(t_i)\, \Delta_i \alpha = \begin{cases} 1 & \text{if } \frac{1}{3} \leqslant t_2 \leqslant \frac{1}{2} \\ 2 & \text{if } \frac{1}{2} < t_2 \leqslant \frac{2}{3}. \end{cases}$$

Thus $S(f, \alpha, P_1) = \{1, 2\}$ has two points. Check that if P_2 and P_3 are partitions of $[0, 1]$ such that $\frac{1}{2} \notin P_2$ but $\frac{1}{2} \in P_3$, then

$$S(f, \alpha, P_2) = \{1, 2\} \qquad \text{and} \qquad S(f, \alpha, P_3) = \{1\}.$$

EXAMPLE 2. Let $f(x) = x^2$ and α the function defined on $[0, 1]$ in Example 1. For any partition P of $[0, 1]$ there is a k such that $x_{k-1} < \frac{1}{2} \leqslant x_k$. Then

$$S(f, \alpha, P) = [x_{k-1}^2, x_k{}^2]$$

is a closed interval.

Theorem 1. *If f is continuous and α is bounded on $[a, b]$, then for any partition P of $[a, b]$ the set $S(f, \alpha, P)$ is a closed interval.*

Proof. With $P = \{x_0, x_1, \cdots, x_n\}$, let $\underline{t}_i$ and $\bar{t}_i$ be on $[x_{i-1}, x_i]$ and furnish the minimum and maximum of $f(t)\,\Delta_i\alpha, x_{i-1} \leqslant t \leqslant x_i$, for $i = 1, 2, \cdots, n$. Hence, for any choice of $t_i \in [x_{i-1}, x_i]$,

$$f(\underline{t}_i)\,\Delta_i\alpha \leqslant f(t_i)\,\Delta_i\alpha \leqslant f(\bar{t}_i)\,\Delta_i\alpha.$$

By summing from $i = 1$ to n and letting c and d be the values of the left and right sums, then

$$c, d \in S(f, \alpha, P) \subset [c, d].$$

Define F for $0 \leqslant \lambda \leqslant 1$ by

$$F(\lambda) = \sum_{i=1}^{n} f((1 - \lambda)\underline{t}_i + \lambda\bar{t}_i)\,\Delta_i\alpha.$$

For each such λ, $x_{i-1} \leqslant (1 - \lambda)\underline{t}_i + \lambda\bar{t}_i \leqslant x_i$ and hence $F(\lambda) \in S(f, \alpha, P)$. Moreover, F is continuous, has minimum $F(0) = c$, maximum $F(1) = d$, and by the intermediate-value theorem assumes all values between c and d. Hence $S(f, \alpha, P) = [c, d]$. ∎

The norm $\|P\|$ of a partition was given in Definition 7–6.1.

Definition 2. *With f and α defined and bounded on $[a, b]$, we use the notation*

$$S(f, \alpha, P) \to s \quad \textit{as } \|P\| \to 0 \qquad \textit{or} \qquad \lim_{\|P\| \to 0} S(f, \alpha, P) = s,$$

where s is a number, if for each given $\varepsilon > 0$ there is a $\delta > 0$ such that

$$\|P\| < \delta \Rightarrow S(f, \alpha, P) \subset (s - \varepsilon, s + \varepsilon).$$

If the limit exists in this sense, then f is said to be **Riemann-Stieltjes integrable** *with respect to α on $[a, b]$; for short,*

$$f \in \mathscr{RS}_a^b(\alpha),$$

the notation

$$s = \mathscr{RS}\int_a^b f(x)\,d\alpha(x) \qquad \textit{or} \qquad s = \mathscr{RS}\int_a^b f\,d\alpha$$

is used and is read "the Riemann-Stieltjes integral from a to b of f with respect to α." Also, f is called the **integrand** *and α the* **integrator**.

In case the interval $[a, b]$ is understood, $\mathscr{RS}(\alpha)$ may be used for the space of functions Riemann-Stieltjes integrable with respect to α over $[a, b]$.

The next result is, in a sense, a negative one.

Theorem 2. *Let f and α be defined and bounded on $[a, b]$. If f and α have a common point of discontinuity $c \in [a, b]$, then $f \notin \mathscr{RS}(\alpha)$.*

Proof. We shall prove the theorem for $a < c < b$; and

(a) f discontinuous from the left;
(b) α discontinuous from the right; but
(c) α continuous from the left, so $\lim_{x \to c-} \alpha(x) = \alpha(c)$.

[The proofs are easier if $a < c \leqslant b$ and both functions are discontinuous from the left, or $a \leqslant c < b$ and both functions are discontinuous from the right.]

Let $\eta > 0$ be such that corresponding to each arbitrary number $\delta > 0$ there are points t and x such that

$$c - \delta < t < c \quad \text{and} \quad |f(t) - f(c)| > \eta,$$
$$c < x < c + \delta \quad \text{and} \quad |\alpha(x) - \alpha(c)| > \eta.$$

Corresponding to $\eta/2$, let $\delta_1 > 0$ be such that

$$(c - \delta_1 < x < c) \Rightarrow |\alpha(x) - \alpha(c)| < \frac{\eta}{2}.$$

Assume that $f \in \mathscr{RS}(\alpha)$ with $s = \mathscr{RS} \int_a^b f \, d\alpha$. Let $\delta_2 > 0$ be such that

$$\|P\| < \delta_2 \Rightarrow S(f, \alpha, P) \subset (s - \tfrac{1}{4}\eta^2, s + \tfrac{1}{4}\eta^2).$$

Set $\delta = \min(\delta_1, \delta_2)$. Select x' such that $c - \delta/2 < x' < c$ and x'' such that $c < x'' < c + \delta/2$ and

$$|\alpha(x'') - \alpha(c)| > \eta.$$

Then $|\alpha(x') - \alpha(c)| < \eta/2$, and hence

$$0 < x'' - x' < \delta \quad \text{and} \quad |\alpha(x'') - \alpha(x')| > \frac{\eta}{2}.$$

Next select a partition P of $[a, b]$ with $\|P\| < \delta$ and such that x' and x'' are consecutive points of P:

$$P = \{x_0, x_1, \cdots, x_{k-1} = x', x_k = x'', \cdots, x_n\}.$$

Take $t_k' = c$ and t_k'' such that

$$x_{k-1} < t_k'' < c \quad \text{and} \quad |f(t_k'') - f(c)| > \eta.$$

Also choose $t_i' = t_i'' \in [x_{i-1}, x_i]$, $1 \leqslant i \leqslant \eta$, $i \neq k$. Set

$$\sum{}' = \sum_{i=1}^{n} f(t_i') \, \Delta_i \alpha \quad \text{and} \quad \sum{}'' = \sum_{i=1}^{n} f(t_i'') \, \Delta_i \alpha.$$

Then

$$\left|\sum' - \sum''\right| = |f(t_k') - f(t_k'')|\,|\alpha(x_k) - \alpha(x_{k-1})|$$
$$= |f(c) - f(t_k'')|\,|\alpha(x'') - \alpha(x')| > \frac{\eta^2}{2}.$$

But $\sum'$ and $\sum''$ are both points of $S(f, \alpha, P)$, and since $\|P\| < \delta$,

$$\left|\sum' - \sum''\right| \leqslant \left|\sum' - s\right| + \left|\sum'' - s\right| < \frac{\eta^2}{2}.$$

This contradiction proves the theorem in the case considered. ∎

Theorem 3. *$f \in \mathscr{RS}(\alpha) \Leftrightarrow \alpha \in \mathscr{RS}(f)$, and if either function is integrable with respect to the other, then the integration by parts formula holds; that is,*

(1) $$\mathscr{RS}\int_a^b f\,d\alpha + \mathscr{RS}\int_a^b \alpha\,df = f(x)\alpha(x)\Big]_a^b.$$

Proof. Let $f \in \mathscr{RS}(\alpha)$ and set $s = \mathscr{RS}\int_a^b f\,d\alpha$. Given an $\varepsilon > 0$, let $\delta > 0$ be such that if Q is any partition of $[a, b]$ with $\|Q\| < \delta$, then

$$S(f, \alpha, Q) \subset (s - \varepsilon, s + \varepsilon).$$

Take any partition $P = [x_0, x_1, \cdots, x_n]$ with $\|P\| < \delta/2$. We shall show that

(2) $$S(\alpha, f, P) \subset \left(f(x)\alpha(x)\Big]_a^b - s - \varepsilon, f(x)\alpha(x)\Big]_a^b - s + \varepsilon\right).$$

Choose any $t_i \in [x_{i-1}, x_i]$, $i = 1, 2, \cdots, n$ and set

$$y = \sum_{i=1}^{n} \alpha(t_i)[f(x_i) - f(x_{i-1})]$$
$$= -\sum_{i=1}^{n} \alpha(t_i)f(x_{i-1}) + \sum_{i=1}^{n} \alpha(t_i)f(x_i)$$
$$= -\sum_{i=0}^{n-1} \alpha(t_{i+1})f(x_i) + \sum_{i=1}^{n} \alpha(t_i)f(x_i)$$
$$= -\alpha(t_1)f(x_0) - \sum_{i=1}^{n-1} f(x_i)[\alpha(t_{i+1}) - \alpha(t_i)] + \alpha(t_n)f(x_n).$$

Set $t_0 = a$ and $t_{n+1} = b$. Then

$$Q = \{t_0, t_1, t_2, \cdots, t_n, t_{n+1}\}$$

is a partition of $[a, b]$, $\|Q\| < \delta$, and $x_i \in [t_i, t_{i+1}]$ for $i = 0, 1, 2, \cdots, n$. By adding

$$0 = [-f(a)\alpha(a) + f(x_0)\alpha(t_0)] + [f(b)\alpha(b) - f(x_n)\alpha(t_{n+1})]$$

to the expression for y, we have

$$y = f(x)\alpha(x)\Big]_a^b - \sum_{i=0}^{n} f(x_i)[\alpha(t_{i+1}) - \alpha(t_i)].$$

The sum has its value in $S(f, \alpha, Q)$ and, since $\|Q\| < \delta$, this value is in the interval $(s - \varepsilon, s + \varepsilon)$. Hence (2) follows so that $\alpha \in \mathscr{RS}(f)$ and also (1) holds. ∎

8–2 DARBOUX–STIELTJES INTEGRALS

Definition 1. *Let f be a bounded real-valued function and α a nondecreasing function on $[a, b]$, and let $P = \{x_0, x_1, \cdots, x_n\}$ be a partition of $[a, b]$. With $m_i = \inf f$ and $M_i = \sup f$ on $[x_{i-1}, x_i]$, $i = 1, 2, \cdots, n$, then*

$$L(f, \alpha, P) = \sum_{i=1}^{n} m_i \, \Delta_i \alpha \qquad \text{and} \qquad U(f, \alpha, P) = \sum_{i=1}^{n} M_i \, \Delta_i \alpha$$

are called the **lower** *and* **upper Darboux–Stieltjes sums** *of f with respect to α relative to P.*

Throughout this section f and α will be functions as given in the definition. A completely analogous development with α nonincreasing will, however, be considered as proceeding in parallel.

With $m = \inf f$ and $M = \sup f$ on $[a, b]$, then for any partition P and $t_i \in [x_{i-1}, x_i]$,

$$m \, \Delta_i \alpha \leqslant m_i \, \Delta_i \alpha \leqslant f(t_i) \, \Delta_i \alpha \leqslant M_i \, \Delta_i \alpha \leqslant M \, \Delta_i \alpha,$$

since $\Delta_i \alpha \geqslant 0$, and hence

$$m[\alpha(b) - \alpha(a)] \leqslant L(f, \alpha, P) \leqslant \sum_{i=1}^{n} f(t_i) \, \Delta_i \alpha \leqslant U(f, \alpha, P) \leqslant M[\alpha(b) - \alpha(a)],$$

in which all terms are finite. Thus $L(f, \alpha, P) \leqslant U(f, \alpha, P)$ and

$$S(f, \alpha, P) \subset [L(f, \alpha, P), U(f, \alpha, P)],$$

where $S(f, \alpha, P)$ is the set defined in Section 8–1.

Since f and α are fixed throughout this section, we simplify $L(f, \alpha, P)$ to $L(P)$ and similarly for U and S.

Theorem 1. *If $P \subset Q$ are partitions of $[a, b]$, then*

$$L(P) \leqslant L(Q) \leqslant U(Q) \leqslant U(P).$$

Proof. Set $P = \{x_0, x_1, \cdots, x_n\}$ and $Q = \{y_0, y_1, \cdots, y_m\}$. Each x_i is therefore some y_j. Take any $y_l \in Q - P$, set $P_1 = P \cup \{y_l\}$, and let k be such that $x_{k-1} < y_l < x_k$. With $m_k' = \inf f$ on $[x_{k-1}, y_l]$ and $m_k'' = \inf f$ on $[y_l, x_k]$, then both $m_k \leqslant m_k'$ and $m_k \leqslant m_k''$. Thus

$$\begin{aligned} m_k \, \Delta_k \alpha &= m_k[\alpha(y_l) - \alpha(x_{k-1})] + m_k[\alpha(x_k) - \alpha(y_l)] \\ &\leqslant m_k'[\alpha(y_l) - \alpha(x_{k-1})] + m_k''[\alpha(x_k) - \alpha(y_l)]. \end{aligned}$$

Thus $L(P) \leqslant L(P_1)$ and in the same way $U(P) \geqslant U(P_1)$. By choosing successive refinements $P \subset P_1 \subset P_2 \subset \cdots \subset P_p = Q$, the result is seen to hold. ∎

Theorem 2. *For any two partitions P and Q,*

$$L(P) \leqslant U(Q).$$

Proof. $P \cup Q$ is a refinement of both P and Q so that

$$L(P) \leqslant L(P \cup Q) \leqslant U(P \cup Q) \leqslant U(Q). \blacksquare$$

Definition 2. *By considering all partitions of* $[a, b]$,

$$\sup_P L(f, \alpha, P) = \underline{\int_a^b} f\, d\alpha \qquad \textit{and}$$

$$\inf_P U(f, \alpha, P) = \overline{\int_a^b} f\, d\alpha$$

are called the **lower** *and* **upper Darboux–Stieltjes integrals** *from a to b of f with respect to α.*

Theorem 3

$$\underline{\int_a^b} f\, d\alpha \leqslant \overline{\int_a^b} f\, d\alpha.$$

Proof. Since $L(P) \leqslant U(Q)$ for any partitions, hold P fixed for the moment and obtain

$$L(P) \leqslant \inf_Q U(Q) = \overline{\int_a^b} f\, d\alpha.$$

Now take $\sup_P L(P)$ and obtain the desired result. ∎

Theorem 4. *If $a \leqslant c \leqslant b$, then*

$$\underline{\int_a^b} f\, d\alpha = \underline{\int_a^c} f\, d\alpha + \underline{\int_c^b} f\, d\alpha \qquad \textit{and}$$

$$\overline{\int_a^b} f\, d\alpha = \overline{\int_a^c} f\, d\alpha + \overline{\int_c^b} f\, d\alpha.$$

Proof. Given a number $\varepsilon > 0$, choose partitions

$$P_1 \text{ of } [a, b] \qquad \text{such that} \qquad \underline{\int_a^b} - \varepsilon < L(P_1);$$

$$P_2 \text{ of } [a, c] \qquad \text{such that} \qquad \underline{\int_a^c} - \varepsilon < L(P_2);$$

$$P_3 \text{ of } [c, b] \qquad \text{such that} \qquad \underline{\int_c^b} - \varepsilon < L(P_3).$$

Let $P = P_1 \cup P_2 \cup P_3$. Then $c \in P$ and P is a refinement of P_1, $P \cap [a, c]$ is a refinement of P_2, and $P \cap [c, b]$ is a refinement of P_3. Thus

$$\underline{\int}_a^b - \varepsilon < L(P) = L(P \cap [a, c]) + L(P \cap [c, b]) \leqslant \underline{\int}_a^c + \underline{\int}_c^b$$

and

$$\underline{\int}_a^c + \underline{\int}_c^b - 2\varepsilon < L(P \cap [a, c]) + L(P \cap [c, b]) = L(P) \leqslant \underline{\int}_a^b.$$

A similar proof may be made for upper integrals. ∎

PROBLEM 1. (a) With $B = \sup|f|$ on $[a, b]$, so B is finite,

$$\left.\begin{array}{l} \left|\underline{\int}_a^b f\,d\alpha\right| \\[2ex] \left|\overline{\int}_a^b f\,d\alpha\right| \end{array}\right\} \leqslant \overline{\int}_a^b |f|\,d\alpha \leqslant B[\alpha(b) - \alpha(a)].$$

(b) Can the middle term be replaced by the lower integral?

PROBLEM 2. There are numbers $\underline{\lambda}$ and $\overline{\lambda}$ such that $m \leqslant \underline{\lambda} \leqslant \overline{\lambda} \leqslant M$ and

$$\underline{\int}_a^b f\,d\alpha = \underline{\lambda}[\alpha(b) - \alpha(a)], \qquad \overline{\int}_a^b f\,d\alpha = \overline{\lambda}[\alpha(b) - \alpha(a)].$$

Theorem 5. *Let $\underline{F}$ and $\overline{F}$ be defined by*

$$\underline{F}(x) = \underline{\int}_a^x f\,d\alpha \qquad \text{and} \qquad \overline{F}(x) = \overline{\int}_a^x f\,d\alpha, \qquad a \leqslant x \leqslant b.$$

(a) *If α is right continuous at a point $c \in [a, b]$, then $\underline{F}$ and $\overline{F}$ are right continuous at c. A similar result holds for left continuity, and then for continuity.*
(b) *If at a point $c \in [a, b]$, f is right continuous at c and $\alpha'_+(c)$ exists, then $\underline{F}'_+(c)$ and $\overline{F}'_+(c)$ exist and*

$$\underline{F}'_+(c) = \overline{F}'_+(c) = f(c)\alpha'_+(c).$$

Proof. (a) From the additivity of lower integrals (Theorem 4) and Problem 1,

$$|\underline{F}(x) - \underline{F}(c)| = \left|\underline{\int}_a^x f\,d\alpha - \underline{\int}_a^c f\,d\alpha\right| = \left|\underline{\int}_c^x f\,d\alpha\right| \leqslant B[\alpha(x) - \alpha(c)].$$

Hence, by taking $\overline{\lim}_{x \to c+}$, the result follows in this case and the others are similar.

(b) With $a \leqslant c < x < b$ apply the mean-value property of Problem 2 to the interval $[c, x]$ and obtain

$$\underline{F}(x) - \underline{F}(c) = \underline{\lambda}(x)[\alpha(x) - \alpha(c)]$$

for some number $\underline{\lambda}(x)$ between $\inf f$ and $\sup f$ on $[c, x]$. Hence $\underline{\lambda}(x) \to f(c)$ as $x \to c+$. The result is seen to follow by first dividing by $x - c > 0$ and then letting $x \to c+$. ∎

PROBLEM 3. If g is a bounded function on $[a, b]$ (of course f is bounded and α nondecreasing), then with all integrals from a to b,

$$\underline{\int} f\,d\alpha + \underline{\int} g\,d\alpha \leqslant \underline{\int} (f+g)\,d\alpha \leqslant \begin{Bmatrix} \underline{\int} f\,d\alpha + \overline{\int} g\,d\alpha \\ \overline{\int} f\,d\alpha + \underline{\int} g\,d\alpha \end{Bmatrix}$$

$$\leqslant \overline{\int} (f+g)\,d\alpha \leqslant \overline{\int} f\,d\alpha + \overline{\int} g\,d\alpha.$$

Definition 3. *With f bounded and α monotonic on $[a, b]$, if*

$$\underline{\int_a^b} f\,d\alpha = \overline{\int_a^b} f\,d\alpha,$$

then we write $f \in \mathscr{DS}(\alpha)$, say that f is **Darboux–Stieltjes integrable** *from a to b with respect to α, and denote this common value by*

$$\mathscr{DS} \int_a^b f\,d\alpha.$$

The following theorem is a corollary of Problem 3.

Theorem 6

(a) *$f \in \mathscr{DS}(\alpha)$ and $g \in \mathscr{DS}(\alpha) \Leftrightarrow (f+g) \in \mathscr{DS}(\alpha)$ and*

$$\mathscr{DS} \int_a^b (f+g)\,d\alpha = \mathscr{DS} \int_a^b f\,d\alpha + \mathscr{DS} \int_a^b g\,d\alpha.$$

(b) *$(f+g) \in \mathscr{DS}(\alpha) \Rightarrow$*

$$\underline{\int_a^b} f\,d\alpha + \overline{\int_a^b} g\,d\alpha = \overline{\int_a^b} f\,d\alpha + \underline{\int_a^b} g\,d\alpha = \int_a^b (f+g)\,d\alpha.$$

EXAMPLE 1. Let $\alpha(x) = x$,

$$f(x) = \begin{Bmatrix} 1, & x \text{ rational} \\ 0, & x \text{ irrational} \end{Bmatrix}, \quad \text{and} \quad g(x) = \begin{Bmatrix} 0, & x \text{ rational} \\ 1, & x \text{ irrational} \end{Bmatrix}.$$

Then $f + g \in \mathscr{DS}(\alpha)$ on $[0, 1]$ with $\int_0^1 (f+g)\,dx = 1$, but

$$\underline{\int_0^1} f\,d\alpha = \underline{\int_0^1} g\,d\alpha = 0, \qquad \overline{\int_0^1} f\,d\alpha + \overline{\int_0^1} g\,d\alpha = 1.$$

Theorem 7. *$f \in \mathscr{DS}(\alpha)$ iff corresponding to each given number $\varepsilon > 0$ there is a partition P such that*

$$U(f, \alpha, P) - L(f, \alpha, P) < \varepsilon. \tag{1}$$

Proof. First consider that, with $\varepsilon > 0$ given, there is a partition P for which (1) holds. Then for this P,

$$\overline{\int_a^b} f\,d\alpha \leqslant U(f, \alpha, P) < L(f, \alpha, P) + \varepsilon \leqslant \underline{\int_a^b} f\,d\alpha + \varepsilon.$$

Hence $\overline{\int_a^b} f\,d\alpha \leqslant \underline{\int_a^b} f\,d\alpha$ and the equality must hold.

Conversely, let $f \in \mathscr{DS}(\alpha)$ and let $\varepsilon > 0$ be given. Choose partitions P_1 and P_2 such that

$$\underline{\int_a^b} f\,d\alpha < L(f, \alpha, P_1) + \frac{\varepsilon}{2} \quad \text{and} \quad U(f, \alpha, P_2) < \overline{\int_a^b} f\,d\alpha + \frac{\varepsilon}{2}.$$

But the lower and upper integrals are equal and, with $P = P_1 \cup P_2$, both $L(f, \alpha, P_1) \leqslant L(f, \alpha, P)$ and $U(f, \alpha, P) \leqslant U(f, \alpha, P_2)$ so that (1) holds for this P. ∎

Theorem 8. *$f \in \mathscr{RS}(\alpha) \Rightarrow f \in \mathscr{DS}(\alpha)$ and equality of the integrals.*

Proof. Take an f and α such that $f \in \mathscr{RS}(\alpha)$. Given a number $\varepsilon > 0$, let $\delta > 0$ be such that $\|P\| < \delta$ implies

$$S(f, \alpha, P) \subset \left(\mathscr{RS} \int_a^b f\,d\alpha - \varepsilon, \mathscr{RS} \int_a^b f\,d\alpha + \varepsilon\right).$$

Select any partition P with $\|P\| < \delta$ and then, for $i = 1, 2, \cdots, n$, select $t_i \in [x_{i-1}, x_i]$ such that $f(t_i)\,\Delta_i\alpha \leqslant (m_i + \varepsilon)\,\Delta_i\alpha$. Then

$$\mathscr{RS} \int_a^b f\,d\alpha - \varepsilon < \sum_{i=1}^n f(t_i)\,\Delta_i\alpha \leqslant L(f, \alpha, P) + \varepsilon[\alpha(b) - \alpha(a)].$$

In a similar way the first inequality below is obtained:

$$U(P) < \mathscr{RS} \int_a^b f\,d\alpha + \varepsilon[\alpha(b) - \alpha(a) + 1] < L(P) + 2\varepsilon[\alpha(b) - \alpha(a) + 1].$$

Thus $f \in \mathscr{DS}(\alpha)$ by Theorem 7. The upper and lower Darboux–Stieltjes integrals are now known to be equal so that $L(P) \leqslant \mathscr{DS} \int_a^b f\,d\alpha \leqslant U(P)$ and we have, upon setting $\eta = \varepsilon[\alpha(b) - \alpha(a) + 1]$, that

$$\mathscr{DS} \int_a^b - \eta \leqslant U(P) - \eta < \mathscr{RS} \int_a^b < L(P) + \eta \leqslant \mathscr{DS} \int_a^b + \eta.$$

Hence the equality of the integrals follows. ∎

It follows from Theorem 8 that for any α (monotonic),

$$\mathscr{RS}(\alpha) \subset \mathscr{DS}(\alpha).$$

An examination of Example 8–1.1 will show, moreover, that the inclusion is proper.

The next theorem gives a condition under which $\mathscr{RS}(\alpha) = \mathscr{DS}(\alpha)$.

Theorem 9. *If α is continuous (monotonic) and $f \in \mathscr{DS}(\alpha)$, then $f \in \mathscr{RS}(\alpha)$ and the integrals are equal.*

Proof. Denote the Darboux–Stieltjes integral by s:

$$s = \mathscr{DS}\int_a^b f\,d\alpha.$$

Let $\varepsilon > 0$ be given. Take a partition $Q = \{y_0, y_1, \cdots, y_p\}$ such that

$$s - \frac{\varepsilon}{2} < L(f, \alpha, Q) \leqslant U(f, \alpha, Q) < s + \frac{\varepsilon}{2}.$$

Q is fixed throughout this proof and hence the integer $p \geqslant 1$ is fixed. Other constants are m and M, the inf f and sup f on $[a, b]$. If $m = M$, then

$$m[\alpha(b) - \alpha(a)] = M[\alpha(b) - \alpha(a)]$$

is the value of both integrals, so consider $m < M$. Let $\delta > 0$ be such that (by the uniform continuity of α) $a \leqslant x < y \leqslant b$ and $y - x < \delta$ imply

$$0 \leq \alpha(y) - \alpha(x) < \frac{\varepsilon}{2p(M - m)}.$$

We shall show that if P is any partition with $\|P\| < \delta$, then

(2) $$S(f, \alpha, P) \subset (s - \varepsilon, s + \varepsilon).$$

Accordingly take a partition

$$P = \{x_0, x_1, \cdots x_n\} \qquad \text{with } \|P\| < \delta.$$

Denote the largest of $\Delta_i\alpha = \alpha(x_i) - \alpha(x_{i-1})$ by ∇ so that

$$0 \leq \Delta_i\alpha \leqslant \nabla < \frac{\varepsilon}{2p(M - m)}, \qquad i = 1, 2, \cdots, n.$$

Set $R = P \cup Q$. Then R is a refinement of P having at most $p - 1$ points more than P; the points, if there are any, of $Q - P$. Hence $L(R) - L(P) \geqslant 0$, but on the other hand,

$$L(R) - L(P) \leqslant (p - 1)(M - m)\nabla < \frac{\varepsilon}{2}.$$

Also, R is a refinement of Q so that $L(Q) \leqslant L(R)$ and $U(Q) \geqslant U(R)$. Thus we have both

$$s - \frac{\varepsilon}{2} < L(Q) \leqslant L(R) < L(P) + \frac{\varepsilon}{2} \qquad \text{and} \qquad U(P) < s + \varepsilon.$$

This shows that (2) holds, so that $f \in \mathscr{RS}(\alpha)$ with

$$\mathscr{RS}\int_a^b f\,d\alpha = s = \mathscr{DS}\int_a^b f\,d\alpha. \blacksquare$$

The next theorem gives a condition for Riemann–Stieltjes integrability but was not included in Section 8–1 because the proof uses Darboux–Stieltjes theory.

Theorem 10. *If f is continuous and α is monotonic on $[a, b]$, then $f \in \mathscr{RS}(\alpha)$ and hence $f \in \mathscr{DS}(\alpha)$.*

Proof. *Let* $\varepsilon > 0$ be given. Choose $\eta > 0$ such that (remember that in proofs we consider α to be nondecreasing)

$$[\alpha(b) - \alpha(a)]\eta < \varepsilon.$$

From the uniform continuity of f on $[a, b]$, take $\delta > 0$ such that

$$x,y \in [a, b] \qquad \text{and} \qquad |x - y| < \delta \Rightarrow |f(x) - f(y)| < \eta.$$

Let $P = \{x_0, x_1, \cdots, x_n\}$ be a partition of $[a, b]$ having $\|P\| < \delta$. Therefore $M_k - m_k \leqslant \eta$, $k = 1, 2, \cdots, n$ and

$$U(P) - L(P) = \sum_{k=1}^{n} (M_k - m_k)\, \Delta_k \alpha \leqslant \eta \sum_{k=1}^{n} \Delta_k \alpha = \eta[\alpha(b) - \alpha(a)] < \varepsilon.$$

Thus $f \in \mathscr{DS}(\alpha)$ by Theorem 7 and we let

$$s = \mathscr{DS} \int_a^b f \, d\alpha.$$

Hence

$$\|P\| < \delta \Rightarrow S(f, \alpha, P) \subset [L(P), U(P)] \subset (s - \varepsilon, s + \varepsilon)$$

so that also $f \in \mathscr{RS}(\alpha)$. ∎

This theorem and Theorem 8–1.3 give the

Corollary. *If f is monotonic and α is continuous, then $f \in \mathscr{RS}(\alpha)$.*

Theorem 11. *Given $f \in \mathscr{DS}(\alpha)$ and α monotonic on $[a, b]$, let $m = \inf f$ and $M = \sup f$ on $[a, b]$, and let $\varphi\colon [m, M] \to E^1$ be continuous. Then the composition function $h = \varphi \circ f \in \mathscr{DS}(\alpha)$ on $[a, b]$.*

Proof. Let $\varepsilon > 0$ be given. From the uniform continuity of φ on $[m, M]$, choose $\delta_1 > 0$ such that

$$s, t \in [m, M] \qquad \text{and} \qquad |s - t| < \delta \Rightarrow |\varphi(s) - \varphi(t)| < \varepsilon.$$

Set $\delta = \min(\delta_1, \varepsilon)$. Corresponding to δ^2, choose a partition $P = \{x_0, x_1, \cdots, x_n\}$ of $[a, b]$ such that

$$U(f, \alpha, P) - L(f, \alpha, P) < \delta^2,$$

which is possible by Theorem 7. As usual, $m_k = \inf f$ and $M_k = \sup f$ on $[x_{k-1}, x_k]$. We let

$$K = \{k = 1, 2, \cdots, n \mid M_k - m_k < \delta\} \qquad \text{and} \qquad N = \{1, 2, \cdots, n\} - K.$$

Also set $m_k^* = \inf h$ and $M_k^* = \sup h$ on $[x_{k-1}, x_k]$. Then $k \in K$ and

$$\begin{aligned} x_{k-1} \leqslant x \leqslant y \leqslant x_k &\Rightarrow |f(x) - f(y)| \leqslant M_k - m_k < \delta \leqslant \delta_1 \\ &\Rightarrow |\varphi(f(x)) - \varphi(f(y))| < \varepsilon \\ &\Rightarrow M_k^* - m_k^* \leqslant \varepsilon, \end{aligned}$$

since $h(x) = \varphi(f(x))$. Thus

$$\sum_{k \in K} (M_k^* - m_k^*)\, \Delta_k \alpha \leqslant \varepsilon \sum_{k=1}^{n} \Delta_k \alpha = \varepsilon[\alpha(b) - \alpha(a)].$$

Since $k \in N \Rightarrow M_k - m_k \geqslant \delta$, we have

$$\begin{aligned} \delta \sum_{k \in N} \Delta_k \alpha \leqslant \sum_{k \in N} (M_k - m_k)\, \Delta_k \alpha &\leqslant \sum_{k=1}^{n} (M_k - m_k)\, \Delta_k \alpha \\ &= U(f, \alpha, P) - L(f, \alpha, P) < \delta^2 \qquad \text{and} \\ 0 \leqslant &\sum_{k \in N} \Delta_k \alpha < \delta \leqslant \varepsilon. \end{aligned}$$

Let $\Lambda = \max_{m \leqslant t \leqslant M} |\varphi(t)|$. Then $M_k^* - m_k^* \leqslant 2\Lambda$ and

$$\sum_{k \in N} (M_k^* - m_k^*)\, \Delta_k \alpha \leqslant 2\Lambda \sum_{k \in N} \Delta_k \alpha < 2\Lambda\varepsilon.$$

Thus

$$\begin{aligned} U(h, \alpha, P) - L(h, \alpha, P) &= \sum_{k \in K} (M_k^* - m_k^*)\, \Delta_k \alpha + \sum_{k \in N} (M_k^* - m_k^*)\, \Delta_k \alpha \\ &\leqslant \varepsilon[\varphi(b) - \varphi(a) + 2\Lambda]. \end{aligned}$$

Since ε is arbitrary while $\varphi(b) - \varphi(a) + 2\Lambda$ is constant, then $h \in \mathscr{DS}(\alpha)$ on $[a, b]$ follows from Theorem 7. ∎

Corollary. *$f \in \mathscr{DS}(\alpha)$ and α monotonic on $[a, b] \Rightarrow$*

(a) *$cf \in \mathscr{DS}(\alpha)$ for c a constant and*

$$\mathscr{DS} \int_a^b cf\, d\alpha = c\left(\mathscr{DS} \int_a^b f\, d\alpha\right).$$

(b) $|f| \in \mathscr{DS}(\alpha)$ *and* $\left|\mathscr{DS} \int_a^b f\, d\alpha\right| \leqslant \mathscr{DS} \int_a^b |f|\, d\alpha.$

(c) $f^2 \in \mathscr{DS}(\alpha)$.

If also $g \in \mathscr{DS}(\alpha)$ on $[a, b]$, then $fg \in \mathscr{DS}(\alpha)$.

Proof. Let $\varphi(t) = ct$ so that $\varphi(f(x)) = cf(x)$ and $cf \in \mathscr{DS}(\alpha)$. From the definition of upper and lower integrals,

$$\overline{\int_a^b} cf\, d\alpha = \begin{cases} c \overline{\int_a^b} f\, d\alpha & \text{if } c \geqslant 0 \\ c \underline{\int_a^b} f\, d\alpha & \text{if } c < 0, \end{cases}$$

but $f \in \mathscr{DS}(\alpha)$ and $cf \in \mathscr{DS}(\alpha)$ so the bars may be removed.

From setting $\varphi(t) = |t|$ it follows that $|f| \in \mathscr{DS}(\alpha)$. Choose c either ± 1 to make $c\left(\mathscr{DS} \int_a^b f\, d\alpha\right) \geqslant 0$ and use this to obtain the inequality in (b).

Set $\varphi(t) = t^2$ to obtain (c).

Theorem 6(a) gives $f + g \in \mathscr{DS}(\alpha)$ and this with the present part (a) gives $f - g \in \mathscr{DS}(\alpha)$. Then $(f+g)^2 \in \mathscr{DS}(\alpha)$ and $(f-g)^2 \in \mathscr{DS}(\alpha)$ by part (c), and thus we see that

$$fg = \tfrac{1}{4}[(f+g)^2 - (f-g)^2] \in \mathscr{DS}(\alpha). \quad \blacksquare$$

PROBLEM 4. (a) If $f \in \mathscr{DS}(\alpha_1)$ and $f \in \mathscr{DS}(\alpha_2)$ with both α_1 and α_2 nondecreasing or both nonincreasing, then $f \in \mathscr{DS}(\alpha_1 + \alpha_2)$ and

$$\mathscr{DS}\int_a^b f\,d(\alpha_1 + \alpha_2) = \mathscr{DS}\int_a^b f\,d\alpha_1 + \mathscr{DS}\int_a^b f\,d\alpha_2 .$$

(b) Without the conditions $f \in \mathscr{DS}(\alpha_1)$ and $f \in \mathscr{DS}(\alpha_2)$, will the equality hold if lower or upper integrals are used?

8–3 RIEMANN INTEGRALS

Riemann integration for a bounded function $f\colon [a, b] \to E^1$ is the special case of Riemann-Stieltjes integration in which α is the identity function $\alpha(x) = x$. Thus, with $P = \{x_0, x_1, \cdots, x_n\}$ a partition of $[a, b]$ and $t_i \in [x_{i-1}, x_i]$,

$$\sum_{i=1}^{n} f(t_i)(x_i - x_{i-1})$$

is a **Riemann sum** for f on $[a, b]$ relative to P and

$$S(f, P)$$

is the set of all such Riemann sums for f on $[a, b]$ relative to P. Then f is **Riemann integrable** ($f \in \mathscr{R}$) on $[a, b]$ if there is a number s such that

$$\lim_{\|P\| \to 0} S(f, P) = s,$$

in the sense given in Definition 8–1.2, and we set

$$s = \int_a^b f(x)\,dx.$$

The same specialization of α to the identity function makes Darboux integration of bounded functions $f\colon [a, b] \to E^1$ a special case of Darboux–Stieltjes integration. The **lower** and **upper Darboux integrals**

$$\underline{\int_a^b} f(x)\,dx \quad \text{and} \quad \overline{\int_a^b} f(x)\,dx$$

exist and if they are equal their common value is denoted by

$$\int_a^b f(x)\,dx.$$

The notation is the same as for the Riemann integral, since a bounded function on a bounded closed interval is Darboux integrable iff it is Riemann integrable (and the integrals are equal), as follows from Theorems 8–2.8 and 8–2.9. Then the algebra of Riemann integrable functions is contained in the corollary of Theorem 8–2.11.

The specialization of α to the identity function in Theorem 8–2.7 gives

Riemann's Condition. *A bounded function $f: [a, b] \to E^1$ is Riemann integrable on $[a, b]$ iff corresponding to each number $\varepsilon > 0$ there is a partition P of $[a, b]$ such that*

$$U(f, P) - L(f, P) < \varepsilon.$$

The next definition is used in obtaining another necessary and sufficient condition for Riemann integrability.

Definition 1. *Given a bounded function $f: [a, b] \to E^1$ and a set $A \subset [a, b]$, the number*

$$\Omega_f(A) = \sup\{f(x) - f(y) \mid x \in A, y \in A\}$$

is called the **oscillation of** f **on** A. [*Hence* $B \subset A \Rightarrow \Omega_f(B) \leqslant \Omega_f(A)$.] *The oscillation of f at $x \in [a, b]$ is defined to be*

$$\omega_f(x) = \lim_{h \to 0+} \Omega_f((x - h, x + h) \cap [a, b]).$$

Note that for any partition P of $[a, b]$,

$$U(f, P) - L(f, P) = \sum_{k=1}^{n} \Omega_f([x_{k-1}, x_k]) \cdot (x_k - x_{k-1}).$$

PROBLEM 1. A bounded function $f: [a, b] \to E^1$ is continuous at an $x \in [a, b]$ iff $\omega_f(x) = 0$.

Lemma 1. *If $f: [a, b] \to E^1$ is bounded and $\lambda > 0$, then*

$$A_\lambda = \{x \in [a, b] \mid \omega_f(x) \geqslant \lambda\}$$

is a closed (hence compact) set.

Proof. Assume there is an $x_0 \in A'_\lambda - A_\lambda$. Hence $x_0 \in [a, b]$ and $\omega_f(x_0) < \lambda$. For η satisfying $\omega_f(x_0) < \eta < \lambda$, choose $h_0 > 0$ such that

$$\Omega_f((x_0 - h_0, x_0 + h_0) \cap [a, b]) < \eta.$$

Since $x_0 \in A'_\lambda$, there is an $x \in A_\lambda \cap (x_0 - h_0, x_0 + h_0)$. Corresponding to this x there is an $h > 0$ such that both $(x - h, x + h) \subset (x_0 - h_0, x_0 + h_0)$ and

$$\eta < \Omega_f((x - h, x + h) \cap [a, b]).$$

By the monotonicity of Ω_f we have a contradiction. Hence A_λ is closed. ∎

Lemma 2. *If f: $[a, b] \to E^1$ is bounded, if $\lambda > 0$, and if $\omega_f(x) < \lambda$ for each $x \in [a, b]$, then there is a partition $P = \{y_0, y_1, \cdots, y_p\}$ of $[a, b]$ such that*

$$\Omega_f([y_{k-1}, y_k]) < \lambda \qquad \textit{for } k = 1, 2, \cdots, p.$$

Proof. With $x \in [a, b]$, let $I(x)$ be an open interval centered at x such that for its closure $\bar{I}(x)$,

$$\Omega_f(\bar{I}(x) \cap [a, b]) < \lambda.$$

The family of all such open intervals is an open covering of $[a, b]$ and thus contains a finite subcovering

$$I(x_1), I(x_2), \cdots, I(x_n),$$

by Theorem 2–7.4. These intervals are all open and hence there must be overlapping. For example, if two intervals abut at a point of $[a, b]$, then this point is interior to a third interval. The end points of these intervals on (a, b) together with a and b form a partition

$$P = \{y_0, y_1, \cdots, y_p\}.$$

Then for any $k = 1, 2, 3, \cdots, p$, there is an $i = 1, 2, \cdots, n$ such that

$$[y_{k-1}, y_k] \subset \bar{I}(x_i)$$

and $\Omega_f([y_{k-1}, y_k]) \leqslant \Omega_f(\bar{I}(x_i) \cap [a, b]) < \lambda.$ ∎

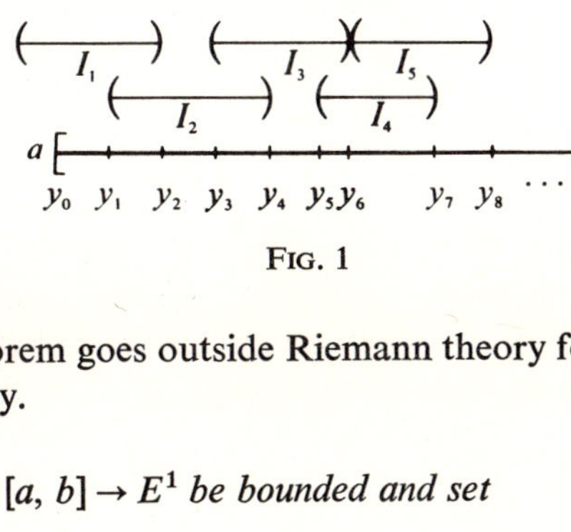

FIG. 1

The following theorem goes outside Riemann theory for an exact criterion for Riemann integrability.

Theorem 1. *Let f: $[a, b] \to E^1$ be bounded and set*

$$A = \{x \in [a, b] \mid \omega_f(x) > 0\}.$$

Then $f \in \mathscr{R}$ on $[a, b]$ iff $mA = 0$; that is, a necessary and sufficient condition that a bounded function on a bounded closed interval be Riemann integrable is that its set of discontinuities have Lebesgue measure zero.

Proof. *Case 1.* $mA > 0$. With $A_n = \{x \in [a, b] \mid \omega_k(x) \geqslant 1/n\}$, $n = 1, 2, 3, \cdots,$

$$A = \bigcup A_n \qquad \text{and} \qquad 0 < mA \leqslant \sum mA_n.$$

Fix an n such that $mA_n > 0$ and set

$$\varepsilon = \frac{1}{2}\frac{1}{n} mA_n.$$

We shall use this $\varepsilon > 0$ to show that Riemann's condition is not satisfied. Let $P = \{x_0, x_1, \cdots, x_p\}$ be any partition of $[a, b]$ and set

$$K = \{k \mid 1 \leqslant k \leqslant p \text{ and } (x_{k-1}, x_k) \cap A_n \neq \varnothing\}.$$

Then $k \in K \Rightarrow \Omega_f([x_{k-1}, x_k]) \geqslant 1/n$ and

$$\bigcup_{k \in K} [(x_{k-1}, x_k) \cap A_n]$$

contains almost all of A_n. Therefore

$$U(f, P) - L(f, P) \geqslant \sum_{k \in K} \frac{1}{n}(x_k - x_{k-1}) \geqslant \frac{1}{n} mA_n > \varepsilon.$$

Thus Riemann's condition is not satisfied and hence $f \notin \mathscr{R}$.

Case 2. $mA = 0$. Let $\varepsilon > 0$ be given and choose $\lambda > 0$ smaller than

$$\frac{1}{2}\frac{\varepsilon}{b - a} \quad \text{and} \quad \frac{1}{2}\frac{\varepsilon}{M - m},$$

considering that $m = \inf f$ and $M = \sup f$ are not equal. The set

$$A_\lambda = \{x \in [a, b] \mid \omega_f(x) \geqslant \lambda\}$$

is closed with $mA_k = 0$, and we take a finite number of nonoverlapping open intervals $I_1, I_2, \cdots, I_n$ such that

$$A_\lambda \subset \bigcup_{j=1}^{n} I_j \quad \text{and} \quad \sum_{j=1}^{n} |I_j| < \lambda,$$

[which is possible by starting with an open $G \supset A$ having $mG < mA_\lambda + \lambda = \lambda$ and using the compactness of A_λ to extract a finite subcovering of A_λ from the components of G]. The end points of these intervals in (a, b) together with a and b form a partition

$$P = \{x_0, x_1, \cdots, x_p\} \quad \text{of} \quad [a, b].$$

If $[x_{i-1}, x_i] \cap A_\lambda \neq \varnothing$, then leave this as it is, but if $[x_{i-1}, x_i] \cap A_\lambda = \varnothing$, then use Lemma 2 to put a partition on this subinterval such that Ω_f of each of these smaller intervals is $< \lambda$. The result is a partition

$$P = \{y_0, y_1, \cdots, y_q\}$$

of the whole interval $[a, b]$ such that, with

$$K = \{k = 1, 2, \cdots, p \mid [x_{k-1}, x_k] \cap A_\lambda = \varnothing\}$$

if $k \in K$, then $\Omega_f([x_{k-1}, x_k]) < \lambda$. For any $k \notin K$, $1 \leqslant k \leqslant q$, we use $\Omega_f[x_{k-1}, x_k] \leqslant M - m$ to obtain

$$\begin{aligned} U(f, P) - L(f, P) &\leqslant \sum_{k \notin K} (M - m)(x_k - x_{k-1}) + \sum_{k \in K} \lambda(x_k - x_{k-1}) \\ &\leqslant (M - m) \sum_{j=1}^{n} |I_j| + \lambda \sum_{k=1}^{q} (x_k - x_{k-1}) \\ &\leqslant (M - m)\lambda + \lambda(b - a) < \varepsilon. \end{aligned}$$

Thus Riemann's condition is satisfied and hence $f \in \mathscr{R}$. ∎

PROBLEM 2. A Cantor-like set $C \subset [0, 1]$ is obtained by erasing the open:

1. Middle 1/3 of [0, 1];
2. Central portions of the two remaining closed intervals so the sum of lengths erased is $1/3^2$;

...

n. Central portions of the remaining closed intervals so the sum of lengths erased at this stage is $1/3^n$;

...

Then $C = [0, 1] - D$ where D is the open set erased.

A continuous function f is defined to be zero on C and on each component interval (α, β) of D to have its graph as shown in the figure. The exact equation is not important, but on (α, β) the derivative f' is continuous, $f'_+(\alpha) = f'_-(\beta) = 0$, maximum of $|f'|$ is 1, and maximum of f is $\leqslant (\beta - \alpha)^2$. For convenience, f may as well be symmetric about the midpoint of (α, β).

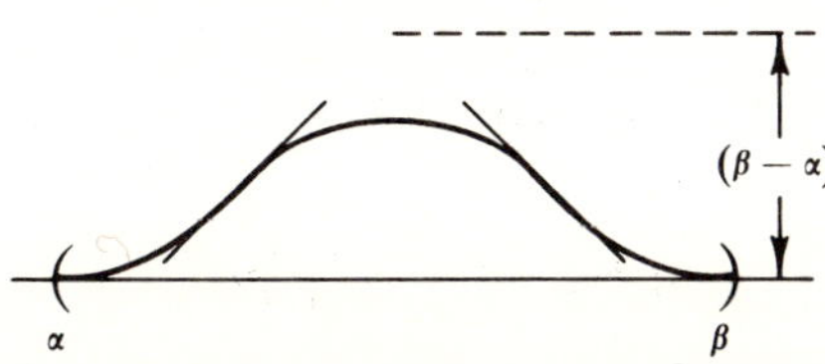

FIG. 2

Use this function to show that the fundamental theorem of calculus does not hold for Riemann integrals [because $\int_0^1 f'(x)\, dx$ does not exist even though f' exists and is bounded on [0, 1]].

The following example illustrates another big weakness of Riemann integration.

EXAMPLE 1. Let $r_1, r_2, r_3, \cdots$ be an ordering of the rational numbers on [0, 1] and define the sequence of functions $\{f_n\}$ on [0, 1] by

$$f_n(x) = \begin{Bmatrix} 0 & \text{if } x = r_1, r_2, \cdots, r_n \\ 1 & \text{otherwise} \end{Bmatrix}, \qquad n = 1, 2, 3, \cdots.$$

Each f_n is Riemann integrable, since its set of discontinuities is finite and hence of measure zero. Also

$$\int_0^1 f_n(x)\,dx = 1 \qquad \text{and} \qquad \lim_{n\to\infty} \int_0^1 f_n(x)\,dx = 1.$$

This limit and integral (in the Riemann sense) may not be interchanged, since

$$\lim_{n\to\infty} f_n(x) = \begin{cases} 0 & \text{if } x \text{ is rational} \\ 1 & \text{if } x \text{ is irrational} \end{cases}$$

is not Riemann integrable (its set of discontinuities having measure $1 > 0$).
With Lebesgue integrals, however,

$$\lim_{n\to\infty} \int_{[0\ 1]} f_n = \int_{[0,1]} \lim f_n = 1.$$

Theorem 3 below gives conditions under which it is permissible to interchange such a limit and Riemann integral.

Theorem 2. *If f: $[a, b] \to E^1$ is Riemann integrable, then f is summable with its Riemann and Lebesgue integrals equal on $[a, b]$.*

Proof. Since f is bounded (as required of all Riemann integrable functions), all functions defined below will be seen to be bounded. To avoid confusing Riemann and Lebesgue integrals we distinguish them by $\mathscr{R}\int_a^b$ and $\mathscr{L}\int_a^b$. Thus

$$\mathscr{R}\int_a^b f(x)\,dx = \mathscr{R}\int_{\underline{a}}^b f(x)\,dx.$$

Let $\{P_n\}$ be a sequence of partitions of $[a, b]$ such that $P_1 \subset P_2 \subset P_3 \subset \cdots$, $\lim \|p_n\| = 0$, and

$$\lim_{n\to\infty} L(f, P_n) = \mathscr{R}\int_a^b f(x)\,dx,$$

which is possible by selecting a sequence $\{Q_n\}$ of partitions for which the limit holds and then setting $P_n = Q_1 \cup Q_2 \cup \cdots \cup Q_n$. For each n and k such that $x_{k-1}^{(n)}$ and $x_k^{(n)}$ are consecutive points of P_n, let

$$m_k^{(n)} = \inf f \quad \text{on } [x_{k-1}^{(n)}, x_k^{(n)}].$$

Set $C_1 = \bigcup P_n$, $C_2 = \{x \mid f \text{ is discontinuous at } x\}$ and $C = C_1 \cup C_2$. Since C_1 is countable and $mC_2 = 0$ by Theorem 1, then $mC = 0$.

Define a sequence $\{\varphi_n\}$ of functions on $[a, b]$ by

(a) $\varphi_n(x) = 0$ if $x \in C$.

(b) Otherwise determine k such that $x \in (x_{k-1}^{(n)}, x_k^{(n)})$, and set $\varphi_n(x) = m_k^{(n)}$.

Hence each φ_n is a bounded, measurable step function and is thus summable with

$$\mathscr{L}\int_a^b \varphi_n(x)\,dx = L(f, P_n).$$

For each $x \in [a, b]$ the sequence $\{\varphi_n(x)\}$ is bounded, nondecreasing, and it follows that the limit function φ is such that

$$\varphi(x) = \lim_{n\to\infty} \varphi_n(x) = \begin{cases} 0 & \text{if } x \in C \\ f(x) & \text{if } x \in [a, b] - C. \end{cases}$$

Hence φ is bounded, measurable, summable, and

$$\mathscr{L} \int_a^b \varphi(x)\, dx = \lim_{n\to\infty} \mathscr{L} \int_a^b \varphi_n(x)\, dx$$

by Theorem 4–5.9. Since $\varphi = f$ a.e. on $[a, b]$, we have that f is not only summable but also

$$\mathscr{L} \int_a^b f(x)\, dx = \mathscr{L} \int_a^b \varphi(x)\, dx = \lim_{n\to\infty} L(f, P_n) = \mathscr{R} \int_a^b f(x)\, dx. \blacksquare$$

Theorem 3. (Arzela) *If $\{f_n\}$ is a uniformly bounded convergent sequence of Riemann integrable functions on $[a, b]$ whose limit function is also Riemann integrable, then*

$$\lim_{n\to\infty} \mathscr{R} \int_a^b f_n(x)\, dx = \mathscr{R} \int_a^b \lim_{n\to\infty} f_n(x)\, dx.$$

Proof. This follows immediately from the dominated convergence Theorem 4–5.9 by interpreting all integrals as Lebesgue integrals, which they are by Theorem 2. ∎

[*Note*: There are proofs which laboriously avoid Lebesgue integrals and use Lebesgue measure for one step only, but there seems to be no known proof staying entirely in Riemann theory.]

8–4 INTEGRATORS OF BOUNDED VARIATION

If α is a function of bounded variation on $[a, b]$, then the functions β and γ defined by

$$\beta(x) = V_a^x(\alpha) \qquad \text{and} \qquad \gamma(x) = \beta(x) - \alpha(x), \qquad a \leqslant x \leqslant b$$

are both nondecreasing, by Theorem 7–6.2.

PROBLEM 1. If f is continuous and α is of bounded variation on $[a, b]$, then

(a) $f \in \mathscr{RS}(\alpha)$ and

$$\int_a^b f(x)\, d\alpha(x) = \int_a^b f(x)\, d\beta(x) - \int_a^b f(x)\, d\gamma(x).$$

(b) $\displaystyle \left|\int_a^b f(x)\, d\alpha(x)\right| \leqslant \int_a^b |f(x)|\, d\beta(x) \leqslant \operatorname*{Max}_{[a,b]} |f| V_a^b(\alpha).$

(c) $\alpha \in \mathscr{RS}(f)$.

Theorem 1. *If f is continuous on $[a, b]$, and if α' exists and is Riemann integrable on $[a, b]$, then*

$$\mathscr{RS} \int_a^b f(x)\, d\alpha(x) = \mathscr{R} \int_a^b f(x)\alpha'(x)\, dx.$$

Proof. Riemann integrability of α' implies that α' is bounded so that α satisfies the Lipschitz condition. Hence α is of bounded variation, the first integral exists (Problem 1(a)), and we denote its value by s_1. Also, α' is continuous a.e. (Theorem 8–3.1), hence so is $f\alpha'$, and the second integral exists with a value we denote by s_2.

Given an $\varepsilon > 0$, let $P = \{x_0, x_1, \cdots, x_n\}$ be a partition of $[a, b]$ such that both

$$S(f, \alpha, P) \subset (s_1 - \varepsilon, s_1 + \varepsilon) \qquad \text{and}$$
$$S(f\alpha', P) \subset (s_2 - \varepsilon, s_2 + \varepsilon),$$

which is possible by choosing $\|P\|$ sufficiently small. By the mean-value theorem take

$$t_i \in (x_{i-1}, x_i) \qquad \text{such that} \qquad \Delta_i \alpha = \alpha'(t_i)(x_i - x_{i-1}).$$

Then

$$\sum_{i=1}^{n} f(t_i)\, \Delta_i \alpha = \sum_{i=1}^{n} f(t_i)\alpha'(t_i)(x_i - x_{i-1}).$$

This value, call it c, is such that

$$s_1 - \varepsilon < c < s_1 + \varepsilon \qquad \text{and} \qquad s_2 - \varepsilon < c < s_2 + \varepsilon.$$

Thus $s_1 = s_2$. ∎

Theorem 2. *If $\{f_n\}$ is a uniformly convergent sequence of continuous functions on $[a, b]$ and α is a function of bounded variation, then*

$$\lim_{n\to\infty} \mathscr{RS} \int_a^b f_n(x)\, d\alpha(x) = \mathscr{RS} \int_a^b \left(\lim_{n\to\infty} f_n(x)\right) d\alpha(x).$$

Proof. Designate the limit function by f. Hence f is continuous, by Theorem 6–3.6, and therefore the second integral exists. Then, for each $n = 1, 2, \cdots,$

$$\left|\int_a^b f_n(x)\, d\alpha(x) - \int_a^b f(x)\, d\alpha(x)\right| = \left|\int_a^b [f_n(x) - f(x)]\, d\alpha(x)\right|$$
$$\leqslant \max_{[a,b]} |f_n - f| V_a^b(\alpha)$$

by Problem 1(b). By uniform convergence of $\{f_n\}$ to f, this last term approaches zero as $n \to \infty$. ∎

Theorem 3. *Let $\{\alpha_n\}$ be a sequence of functions of uniformly bounded variation on $[a, b]$; that is, there is a constant K such that*

$$V_a^b \alpha_n < K < \infty, \qquad n = 1, 2, 3, \cdots.$$

If f is continuous and $\{\alpha_n\}$ converges pointwise on $[a, b]$, then

$$\lim_{n\to\infty} \mathscr{RS} \int_a^b f(x)\, d\alpha_n(x) = \mathscr{RS} \int_a^b f(x)\, d\left(\lim_{n\to\infty} \alpha_n(x)\right).$$

Proof. Denote the limit function by α. Take any partition $P = \{x_0, x_1, \cdots, x_m\}$ of $[a, b]$. Then

$$\sum_{k=1}^{m} |\alpha_n(x_k) - \alpha_n(x_{k-1})| < K, \qquad n = 1, 2, 3, \cdots.$$

Since this is a finite sum and $\{\alpha_n\}$ converges pointwise to α, we let $n \to \infty$ to obtain

$$\sum_{k=1}^{m} |\alpha(x_k) - \alpha(x_{k-1})| \leqslant K.$$

As the partition was arbitrary, we have that α is also of bounded variation. Thus the second integral exists. Upon setting

$$s = \int_a^b f(x)\, d\alpha(x) \qquad \text{and} \qquad s_n = \int_a^b f(x)\, d\alpha_n(x),$$

we need to show that $s_n \to s$ as $n \to \infty$.

Given an $\varepsilon > 0$, let $\delta > 0$ be such that

$$x, y \in [a, b], \qquad |x - y| < \delta \Rightarrow |f(x) - f(y)| < \varepsilon.$$

Take a partition $P = \{x_0, x_1, \cdots, x_m\}$ with $\|P\| < \delta$. Then

$$\begin{aligned} s &= \sum_{k=1}^{m} \int_{x_{k-1}}^{x_k} f(x)\, d\alpha(x) \\ &= \sum_{k=1}^{m} \int_{x_{k-1}}^{x_k} [f(x) - f(x_k)]\, d\alpha(x) + \sum_{k=1}^{m} f(x_k) \int_{x_{k-1}}^{x_k} d\alpha(x). \end{aligned}$$

Since $\int_{x_{k-1}}^{x_k} d\alpha(x) = \alpha(x_k) - \alpha(x_{k-1})$ and

$$\left| \int_{x_{k-1}}^{x_k} [f(x) - f(x_k)]\, d\alpha(x) \right| \leqslant \varepsilon V_{x_{k-1}}^{x_k}(\alpha),$$

by Problem 1(b), we have

$$\left| s - \sum_{k=1}^{m} f(x_k)[\alpha(x_k) - \alpha(x_{k-1})] \right| \leqslant \varepsilon \sum_{k=1}^{n} V_{x_{k-1}}^{x_k}(\alpha) = \varepsilon K.$$

In exactly the same way,

$$s_n - \varepsilon K \leqslant \sum_{k=1}^{m} f(x_k)[\alpha_n(x_k) - \alpha_n(x_{k-1})] \leqslant s_n + \varepsilon K.$$

As $n \to \infty$, each term of this finite sum has a limit and we see that

$$\overline{\lim_{n\to\infty}}\, s_n - \varepsilon K \leqslant \sum_{k=1}^{m} f(x_k)[\alpha(x_k) - \alpha(x_{k-1})] \leqslant \underline{\lim_{n\to\infty}}\, s_n + \varepsilon K.$$

Thus $\overline{\lim}\, s_n - \varepsilon K \leqslant s + \varepsilon K$ and $s - \varepsilon K \leqslant \underline{\lim}\, s_n + \varepsilon K$. Consequently, $s_n \to s$ as $n \to \infty$. ∎

PROBLEM 2. Given $a < c_1 < c_2 < \cdots < c_n < b$, let α be defined on $[a, b]$ and constant on each of the open intervals (a, c_1), (c_1, c_2), $\cdots$, (c_n, b). If f is continuous on $[a, b]$, then

$$\int_a^b f\,d\alpha = f(a)[\alpha(a+) - \alpha(a)] + \sum_{k=1}^{n} f(c_k)[\alpha(c_k+) - \alpha(c_k-)] + f(b)[\alpha(b) - \alpha(b-)].$$

PROBLEM 3. Given a positive integer n and numbers $c_0, c_1, c_2, \cdots, c_n$, let α be the step function defined on $[0, 1]$ by

$$\begin{aligned} &\alpha(0) = 0, \\ &\alpha(x) = c_0 && \text{for } 0 < x < 1/n, \\ &\alpha(x) = \sum_{l=0}^{k-1} c_l && \text{for } \frac{k-1}{n} \leqslant x < \frac{k}{n}, \qquad k = 2, 3, \cdots, n, \\ &\alpha(1) = \sum_{l=0}^{n} c_l. \end{aligned}$$

Show that $V_0^1(\alpha) \leqslant \sum_{l=0}^{n} |c_l|$.

The following two lemmas lead to Theorem 4, which plays a central role in establishing a significant result stated in Theorem 6.

Lemma 1. *Let $\mathscr{H}$ be an infinite family of real-valued functions all having domain $[a, b]$ and uniformly bounded on $[a, b]$; that is, there is a constant K such that*

$$h \in \mathscr{H} \quad \text{and} \quad x \in [a, b] \Rightarrow |h(x)| \leqslant K.$$

If T is a countable subset of $[a, b]$, then there is a sequence $\{h_n\}$ in $\mathscr{H}$ such that for each $t \in T$ the sequence $\{h_n(t)\}$ converges.

Proof. Let $T = \{t_1, t_2, t_3, \cdots\}$. The set of numbers

$$\{h(t_1) \mid h \in \mathscr{H}\}$$

is bounded and, by the Weierstrass-Bolzano theorem (2-7.3), contains a convergent subsequence which we denote by

$$\{h_{1n}(t_1)\} \qquad \text{and its limit by } a_1.$$

Then $\{h_{1n}(t_2)\}$ contains a convergent subsequence

$$\{h_{2n}(t_2)\} \quad \text{converging to a value } a_2$$

where each h_{2n} is some h_{1m} with $n \leqslant m$. Continuing in this way, there is a sequence

$$h_{p1}(t_p), h_{p2}(t_p), h_{p3}(t_p), \cdots; \qquad h_{pn}(t_p) \to a_p \qquad \text{as } n \to \infty$$

where each h_{pn} is some $h_{p-1,m}$ with $n \leqslant m$. This being done for $p = 1, 2, 3, \cdots$, let $\{h_n\}$ be the sequence of "diagonal" terms:

$$h_n = h_{nn}.$$

The part of $\{h_n\}$ with $n \geqslant p$ is a subsequence of $h_{p1}, h_{p2}, h_{p3}, \cdots$ and therefore $h_1(t_p), h_2(t_p), h_3(t_p), \cdots$ converges to a_p for $p = 1, 2, 3, \cdots$. ∎

Lemma 2. *Let $\mathscr{G}$ be an infinite family of uniformly bounded nondecreasing functions on $[a, b]$. Then $\mathscr{G}$ contains a sequence $\{g_n\}$ converging pointwise on $[a, b]$ to a nondecreasing function φ.*

Proof. Let $T_0 = \{t_1, t_2, t_3, \cdots\}$ be the rational points of $[a, b)$ together with a (just in case a is irrational). By using Lemma 1, let $\{h_n\}$ be a sequence chosen from $\mathscr{G}$, converging at each point of T_0, and define φ_0 first only on T_0 by

$$\varphi_0(t_p) = \lim_{n\to\infty} h_n(t_p) \qquad p = 1, 2, 3, \cdots.$$

Since h_n is nondecreasing, then

$$a \leqslant t_p < t_q < b \Rightarrow h_n(t_p) \leqslant h_n(t_q) \Rightarrow \varphi_0(t_p) \leqslant \varphi_0(t_q)$$

and thus φ_0 is nondecreasing on T_0. Extend φ_0 to all of $[a, b]$ by setting

$$\varphi_0(x) = \sup_{t_k < x} \varphi_0(t_k), \qquad a < x \leqslant b.$$

Hence φ_0 is nondecreasing on $[a, b]$, but if $x \in [a, b] - T_0$, will $\{h_n(x)\}$ converge to $\varphi_0(x)$? We first give a partial answer by showing

(1) If $x \in [a, b] - T_0$ and x is a point of continuity of φ_0, then the answer is *yes*.

To see this, fix such a point x, let $\varepsilon > 0$ be given, and then fix t_k and t_j such that

$$t_k < x < t_j \qquad \text{and} \qquad \varphi_0(t_j) - \varphi_0(t_k) < \varepsilon$$

which is possible because φ_0 is continuous at x. Determine n_0 such that $n > n_0$ implies both

$$|h_n(t_k) - \varphi_0(t_k)| < \varepsilon \qquad \text{and} \qquad |h_n(t_j) - \varphi_0(t_j)| < \varepsilon.$$

As all h_n and φ_0 are nondecreasing, $n > n_0$ implies

$$\varphi_0(x) - 2\varepsilon < h_n(t_k) \leqslant h_n(x) \leqslant h_n(t_j) < \varphi_0(x) + 2\varepsilon.$$

Hence $h_n(x) \to \varphi_0(x)$ as $n \to \infty$, confirming the affirmative answer in (1).

Thus the only points $x \in [a, b]$ where $\{h_n(x)\}$ fails to converge form a set T_1 in the set of discontinuities of φ_0. But the set of discontinuities of φ_0 is countable (by Theorem 6-8.1, since φ_0 is nondecreasing) and hence T_1 is countable. We apply Lemma 1 again, this time to the sequence $\{h_n\}$ and the set T_1, thus obtaining a subsequence called $\{g_n\}$ converging at all points of T_1. Since $\{g_n\}$ inherits convergence at all points of $[a, b] - T_1$ from $\{h_n\}$, we define

$$\varphi(x) = \lim_{n\to\infty} g_n(x), \qquad a \leqslant x \leqslant b.$$

As shown for φ_0, φ is nondecreasing on $[a, b]$. ∎

Theorem 4. *Let $\mathscr{A}$ be an infinite family of functions each with domain $[a, b]$, such that the family is not only uniformly bounded on $[a, b]$ but also has variations uniformly bounded on $[a, b]$; that is, there is a constant K such that both*

$$v \in \mathscr{A} \Rightarrow V_a^b(v) \leqslant K \qquad \textit{and} \qquad x \in [a, b] \Rightarrow |v(x)| \leqslant K.$$

Then there is a sequence $\{\alpha_n\}$ in $\mathscr{A}$ converging pointwise on $[a, b]$ to a function α of bounded variation on $[a, b]$.

Proof. For each $v \in \mathscr{A}$ let β_v and γ_v be defined by

$$\beta_v(x) = V_a^x(v) \qquad \text{and} \qquad \gamma_v(x) = \beta_v(x) - v(x), \qquad a \leqslant x \leqslant b.$$

Each β_v and each γ_v is nondecreasing, by Theorem 7–6.2, and

$$|\beta_v(x)| \leqslant K \qquad \text{and} \qquad |\gamma_v(x)| \leqslant 2K \qquad a \leqslant x \leqslant b.$$

From the family of all β_v we use Lemma 2 to select a convergent sequence $\{\beta_n\}$, where now the subscripts denote integers, converging to a bounded nondecreasing function β:

$$\lim_{n \to \infty} \beta_n(x) = \beta(x), \qquad a \leqslant x \leqslant b.$$

Corresponding to each β_n there is a γ_n. Again by Lemma 2, the sequence $\{\gamma_n\}$ contains a convergent subsequence $\{\gamma_{n_k}\}$ converging to a bounded nondecreasing function γ. Upon defining α by

$$\alpha(x) = \beta(x) - \gamma(x), \qquad a \leqslant x \leqslant b$$

then α is of bounded variation, by Theorem 7–6.2, and is the limit of the sequence

$$\alpha_{n_k} = \beta_{n_k} - \gamma_{n_k}, \qquad k = 1, 2, 3, \cdots$$

of members of $\mathscr{A}$. ▮

For a given interval $[a, b]$, let $\mathscr{C}$ denote the space of all continuous functions on $[a, b]$. Fix a function α of bounded variation on $[a, b]$. Define the functional $\Phi_\alpha: \mathscr{C} \to E^1$ by setting

$$\Phi_\alpha(f) = \int_a^b f(x)\, d\alpha(x), \qquad f \in \mathscr{C}. \tag{2}$$

With $V_a^b(\alpha) \leqslant K$, it then follows from previous results that

(a) $\Phi_\alpha(f_1 + f_2) = \Phi_\alpha(f_1) + \Phi_\alpha(f_2)$, for $f_1, f_2 \in \mathscr{C}$;
(b) $|\Phi_\alpha(f)| \leqslant K \max\limits_{[a,b]} |f|$, for $f \in \mathscr{C}$.

Definition 1. *Given the space $\mathscr{C}$ of continuous functions on an interval $[a, b]$, a functional*

$$\Phi: \mathscr{C} \to E^1$$

is said to be a **bounded linear functional** *on* $\mathscr{C}$ *into* E^1 *if there is a positive constant K such that*

(i) $\Phi(f_1 + f_2) = \Phi(f_1) + \Phi(f_2)$; $f_1, f_2 \in \mathscr{C}$.

(ii) $|\Phi(f)| \leqslant K \max\limits_{[a,b]} |f|$; $f \in \mathscr{C}$.

Thus, if $\alpha \in BV_a{}^b$ is given in advance, then Φ_α as defined in (2) is a bounded linear functional on $\mathscr{C}$ into E^1. The power of Stieltjes integrals will be displayed (Theorem 6 below) by showing that conversely, given any bounded linear functional Φ on $\mathscr{C}$ into E^1 (no matter how Φ is defined or where it came from), there is a function α of bounded variation on $[a, b]$ such that $\Phi = \Phi_\alpha$.

Before proving this fact we need one more result.

Theorem 5. *If* Φ *is a bounded linear functional on* $\mathscr{C}$ *into* E^1 *and a is any real number, then for each* $f \in \mathscr{C}$,

$$\Phi(af) = a\Phi(f). \tag{3}$$

Proof. From (i), with $f_1 = f_2 = f \in \mathscr{C}$,

$$\Phi(2f) = 2\Phi(f) \tag{4}$$

and then, by induction, for any positive integer q,

$$\Phi(qf) = q\Phi(f), \qquad f \in \mathscr{C}.$$

Divide by q, switch sides, replace f by $\dfrac{p}{q}f$ with p a positive integer, and obtain

$$\Phi\left(\frac{p}{q}f\right) = \frac{1}{q}\Phi\left(q \cdot \frac{p}{q}f\right) = \frac{1}{q}\Phi(pf) = \frac{p}{q}\Phi(f).$$

Thus (3) holds for a any positive rational number.

With f such that $f(x) = 0$, $a \leqslant x \leqslant b$, in (4), we have that $\Phi(2 \cdot 0) = 2\Phi(0)$ so that

$$\Phi(0) = 0.$$

Set $f_1 = f \in \mathscr{C}$ and $f_2 = -f$ in (i) and obtain

$$0 = \Phi(f - f) = \Phi(f) + \Phi(-f); \qquad \text{that is, } \Phi(-f) = -\Phi(f).$$

Thus for any positive, negative, or zero rational number,

$$\Phi(rf) = r\Phi(f), \qquad f \in \mathscr{C}.$$

Let a be an irrational number and choose any sequence of rational numbers $\{r_n\}$ such that $r_n \to a$ as $n \to \infty$. Then $r_n \Phi(f) \to a\Phi(f)$, and hence the existence of the limit and equality

$$\lim_{n \to \infty} \Phi(r_n f) = a\Phi(f)$$

hold and are independent of the sequence $\{r_n\}$ chosen. Let $f_1 = r_n f$ and $f_2 = -af$ in (i) to obtain

$$|\Phi(r_n f) - \Phi(af)| = |\Phi[(r_n - a)f]| \leqslant K|r_n - a| \max_{[a,b]} |f|$$

where the inequality is from (ii). Hence,

$$\lim_{n\to\infty} \Phi(r_n f) = \Phi(af)$$

again independently of $\{r_n\}$. These limit formulae then give (3) with a any irrational number (and we already have it for every rational number). ∎

Theorem 6. (F. Riesz) *Let $\mathscr{C}$ be the space of all continuous functions on the interval $[a, b]$ and let Φ be a bounded linear functional on $\mathscr{C}$ into E^1. Then there is a function α of bounded variation on $[a, b]$ such that no matter what $f \in \mathscr{C}$ is taken,*

$$\Phi(f) = \int_a^b f\, d\alpha.$$

Proof. By making a linear transformation, we reduce all considerations to the interval [0, 1]. We thereby lose no generality, but notation is somewhat simplified. In particular, the Bernstein polynomials (see Section 6–4)

$$B_n(x) = \sum_{k=0}^{n} f\left(\frac{k}{n}\right)\binom{n}{k} x^k(1-x)^{n-k}, \qquad 0 \leqslant x \leqslant 1$$

of a function f: $[0, 1] \to E^1$ may be used directly. Also recall the formula

$$\sum_{l=0}^{n} \binom{n}{l} x^l(1-x)^{n-l} = 1, \qquad 0 \leqslant x \leqslant 1. \tag{5}$$

In this connection we define the functions b_{nl} by

$$b_{nl}(x) = \binom{n}{l} x^l(1-x)^{n-l}; \qquad 0 \leqslant x \leqslant 1 \quad l = 0, 1, 2, \cdots, n. \tag{6}$$

In particular, the function B_n is expressed as

$$B_n = \sum_{k=0}^{n} f\left(\frac{k}{n}\right) b_{nk}.$$

From Theorem 3 and the additivity of Φ we note that

$$\sum_{k=0}^{n} f\left(\frac{k}{n}\right)\Phi(b_{nk}) = \Phi \sum_{k=0}^{n} f\left(\frac{k}{n}\right) b_{nk} = \Phi(B_n). \tag{7}$$

Until an $f \in \mathscr{C}$ is chosen, the following results are independent of any such function.

We next show that

$$\sum_{l=0}^{n} |\Phi(b_{nl})| \leqslant K \tag{8}$$

where K is the constant in Definition 1(ii). Choose e_l either 1 or -1 to make $e_l\,\Phi(b_{nl}) \geqslant 0$. The sum is (by Theorem 5 and additivity of Φ) equal to

$$\sum_{l=0}^{n} e_l\,\Phi(b_{nl}) = \Phi\left[\sum_{l=0}^{n} (e_l\,b_{nl})\right] \leqslant K \max_{0 \leqslant t \leqslant 1} \left|\sum_{l=0}^{n} e_l\,b_{nl}(t)\right|.$$

This maximum is $\leqslant 1$ by (5) and hence (8) follows.

With these preliminaries out of the way, we take the special partition

$$P_n = \left\{0, \frac{1}{n}, \frac{2}{n}, \cdots, \frac{k-1}{n}, \frac{k}{n}, \cdots, 1\right\}$$

of [0, 1] and define the function α_n on [0, 1] by

$$\alpha_n(0) = 0;$$

$$\alpha_n(x) = \Phi(b_{n0}); \qquad 0 < x < 1/n;$$

$$\alpha_n(x) = \sum_{l=0}^{k-1} \Phi(b_{nl}); \qquad \frac{k-1}{n} \leqslant x < \frac{k}{n}, \quad k = 2, 3, \cdots, n;$$

$$\alpha_n(1) = \sum_{l=0}^{n} \Phi(b_{nl}).$$

Thus α_n is a step function. For illustration, we describe the constant value α_n has on the closed-open interval $1/n \leqslant x < 2/n$. The functions b_{n0} and b_{n1} are defined in (6) so that

$$b_{n0}(t) = (1-t)^n \qquad \text{and} \qquad b_{n1}(t) = nt(1-t)^{n-1}, \qquad 0 \leqslant t \leqslant 1.$$

Then Φ of the first of these functions is a number, Φ of the second function is a number, and the sum of these numbers is the constant value assigned to $\alpha_n(x)$ for all x such that $1/n \leqslant x < 2/n$.

Hence $\alpha_n(0) = 0$, and if $0 < x \leqslant 1$, then

$$|\alpha_n(x)| \leqslant \sum_{l=0}^{n} |\Phi(b_{nl})| \leqslant K \tag{9}$$

by (8). Thus the sequence $\{\alpha_n\}$ is uniformly bounded and, moreover, is of uniformly bounded variation since

$$V_0^{\,1}(\alpha_n) \leqslant K$$

by Problem 3, p. 478. Hence, there is a subsequence $\{\alpha_{n_i}\}$ converging pointwise on [0, 1] to a function α of bounded variation (Theorem 4).

Now take any $f \in \mathscr{C}$ and let B_n be its nth Bernstein polynomial. Then

$$|\Phi(B_n) - \Phi(f)| = |\Phi(B_n - f)| \leqslant K \max_{[0,1]} |B_n - f|$$

and this maximum approaches zero as $n \to \infty$ by the Weierstrass approximation theorem (6–4.1). Hence

$$\lim_{n \to \infty} \Phi(B_n) = \Phi(f). \tag{10}$$

Because α_n is constant on subintervals of [0, 1], the first equation in

$$\int_0^1 f\,d\alpha_n = \sum_{k=0}^{n} f\left(\frac{k}{n}\right)\Phi(b_{nk}) = \Phi(B_n) \tag{11}$$

follows as in Problem 2, p. 478, and the second equation is (7). The above subsequence $\{\alpha_{n_i}\}$ and its pointwise limit α (completely independent of f) give

$$\int_0^1 f\,d\alpha = \lim_{i\to\infty}\int_0^1 f\,d\alpha_{n_i} = \lim_{i\to\infty}\Phi(B_{n_i}) = \Phi(f)$$

where the first equation is by Theorem 3, the second by (11), and the third by (10). Hence, as noted at the beginning of the proof, the desired result follows by a linear transformation of [0, 1] to $[a, b]$. ∎

8–5 LEBESGUE INTEGRAL RELATIONS

In this section we collect a few of the ways Lebesgue integrals occur in connection with either variation or Stieltjes integrals.

Theorem 1. (S. Banach) *Given that f: $[a, b] \to E^1$ is continuous with minimum m and maximum M, define the function ν on $[m, M]$ by $\nu(y)$ is the number (in E^*) of solutions of*

$$f(x) = y; \tag{1}$$

that is, $\nu(y)$ is either a positive integer or ∞. Then ν is a measurable function and

$$\int_m^M \nu(y)\,dy = V_a^b f.$$

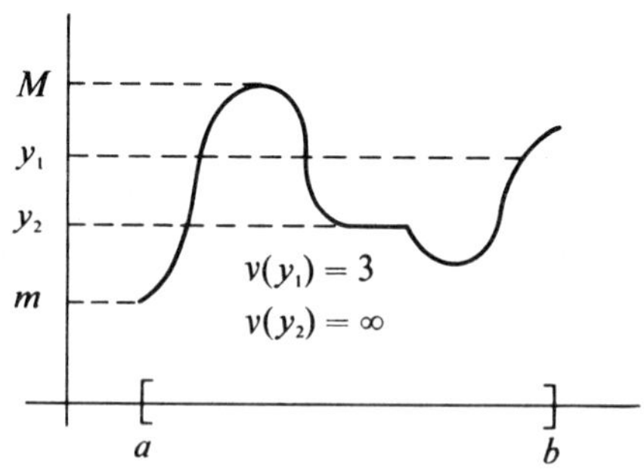

FIG. 1

Proof. With n a positive integer, let $h = (b - a)/2^n$,

$$I_1 = [a, a + h], \tag{2}$$

$$I_k = (a + (k-1)h, a + kh], \qquad k = 2, 3, \cdots, 2^n.$$

For $k = 1, 2, \cdots, 2^n$, let $m_k = \inf f$ and $M_k = \sup f$ on I_k,

$$M_k - m_k = \Omega_k$$

and define g_k by setting $g_k(y) = 1$ if (1) has at least one solution on I_k but $g_k(y) = 0$ if there are none. Hence

$$g_k(y) = \begin{cases} 1 & \text{if } m_k \leqslant y \leqslant M_k \\ 0 & \text{otherwise} \end{cases}$$

so that g_k is measurable on $[m, M]$ and

$$\int_m^M g_k(y)\, dy = M_k - m_k = \Omega_k, \qquad k = 1, 2, \cdots, 2^n.$$

Now define h_n by

$$h_n(y) = g_1(y) + g_2(y) + \cdots + g_{2^n}(y)$$

which is therefore the number of intervals (2) on which (1) has at least one solution. Also h_n, being the sum of measurable functions, is a measurable function and

$$\int_m^M h_n(y)\, dy = \sum_{k=1}^{2^n} \int_{m_k}^{M_k} g_k(y)\, dy = \sum_{k=1}^{2^n} \Omega_k.$$

Hence

$$\lim_{n\to\infty} \int_m^M h_n(y)\, dy = \Omega_a{}^b f = V_a^b f$$

by Problem 7–6.4.

If n is increased by 1, each interval (2) is divided into two mutually exclusive halves. Thus, for $y \in [m, M]$ we have a nondecreasing sequence

$$h_1(y) \leqslant h_2(y) \leqslant h_3(y) \leqslant \cdots$$

which converges* (to a positive integer or ∞) and we set

$$h^*(y) = \lim_{n\to\infty} h_n(y).$$

The function h^* is thus measurable and its Lebesgue integral (finite or ∞) is such that

$$\int_m^M h^*(y)\, dy = \lim_{n\to\infty} \int_m^M h_n(y)\, dy = V_a{}^b f.$$

Thus the theorem will be proved upon showing that

$$h^*(y) = v(y) \qquad \text{either everywhere or a.e. on } [m, M].$$

Clearly, $h_n(y) \leqslant v(y)$ and therefore

$$h^*(y) \leqslant v(y) \qquad \text{for each } y \in [m, M].$$

For $y \in [m, M]$ fixed, $v(y)$ is either a positive integer or ∞, and we select any positive integer q such that

$$q \leqslant v(y).$$

Thus (1) has at least q solutions and we select q solutions and order them as

$$x_1 < x_2 < \cdots < x_q .$$

For any n sufficiently large to make

$$\frac{1}{2^n}(b - a) < \min(x_k - x_{k-1}), \qquad k = 2, 3, \cdots, q,$$

these solutions lie on q separate intervals (2), and thus for all such n,

$$h_n(y) \geqslant q.$$

Thus, whether $v(y)$ is finite or ∞, $h^*(y) \geqslant v(y)$. Hence $h^*(y) = v(y)$ for all $y \in [m, M]$. ∎

Theorem 7–6.4 stated that an absolutely continuous function has bounded variation but gave no recipe for computing the variation.

Theorem 2. *If f is absolutely continuous on $[a, b]$, then*

$$V_a^b(f) = \int_a^b |f'(t)|\, dt.$$

Proof. Since f is absolutely continuous, its derivative exists a.e. is summable and

$$\int_a^x f'(t)\, dt = f(x) - f(a),$$

by Theorem 7–7.3. Let $P = \{x_0, x_1, \cdots, x_n\}$ be a partition of $[a, b]$. Then

$$\sum_{k=1}^n |f(x_k) - f(x_{k-1})| = \sum_{k=1}^n \left| \int_{x_{k-1}}^{x_k} f' \right| \leqslant \sum_{k=1}^n \int_{x_{k-1}}^{x_k} |f'|$$

$$= \int_a^b |f'| \qquad \text{and} \qquad V_a^b(f) \leqslant \int_a^b |f'|.$$

To work on the reverse inequality, let

$$A^+ = \{x \in (a, b) | f'(x) \geqslant 0\}, \qquad A^- = (a, b) - A^+$$

and see (since the two end-point exclusions leave off a set of measure zero) that

$$\int_a^b |f'| = \int_{A^+} f' - \int_{A^-} f'.$$

Let $\varepsilon > 0$ be given. From absolute continuity of integrals, let $\delta > 0$ be such that if $M \subset (a, b)$ is measurable and $mM < \delta$, then

$$\int_M |f'| < \varepsilon.$$

Choose closed sets $F^+ \subset A^+$ and $F^- \subset A^-$ such that

$$m(A^+ - F^+) < \delta \qquad \text{and} \qquad m(A^- - F^-) < \delta,$$

and hence

$$\int_a^b |f'| < \int_{F^+} f' - \int_{F^-} f' + 2\varepsilon.$$

Since F^+ and F^- are closed without a common point, we take a pair of open sets G^+ and G^- such that

(a) $G^+ \supset F^+$, $G^- \supset F^-$, $G^+ \cap G^- = \varnothing$;
(b) $m(G^+ - F^+) < \delta$, $m(G^- - F^-) < \delta$; and
(c) $G^+ \cup G^- \subset (a, b)$;

which can be accomplished by taking a pair satisfying (a), another pair satisfying (b), intersecting the two + sets, then the two − sets, and intersecting the union of these with the open interval (a, b) already known to contain the closed set $F^+ \cup F^-$. Hence

$$\int_a^b |f'| < \int_{G^+} f' - \int_{G^-} f' + 4\varepsilon.$$

With $\{(a_i^+, b_i^+\}$ and $\{(a_j^-, b_j^-)\}$, the component intervals of G^+ and G^-, take integers n^+ and n^- such that

$$\sum_{i=1}^{n^+}(b_i^+ - a_i^+) > mG^+ - \delta \qquad \text{and} \qquad \sum_{j=1}^{n^-}(b_j^- - a_j^-) > mG^- - \delta.$$

Then with H^+ and H^-, the respective unions of these intervals

$$\int_{G^+} f' - \varepsilon < \int_{H^+} f = \sum_{i=1}^{n^+} \int_{a_i^+}^{b_i^+} f' = \sum_{i=1}^{n^+} [f(b_i^+) - f(a_i^+)] \qquad \text{and}$$

$$\int_{G^-} f' + \varepsilon > \sum_{j=1}^{n^-} [f(b_j^-) - f(a_j^-)].$$

Hence

$$\int_a^b |f'| - 6\varepsilon < \sum_{i=1}^{n^+} [f(b_i^+) - f(a_i^+)] - \sum_{j=1}^{n^-} [f(b_j^-) - f(a_j^-)]$$
$$\leqslant \sum_{i=1}^{n^+} |f(b_i^+) - f(a_i^+)| + \sum_{j=1}^{n^-} |f(b_j^-) - f(a_j^-)| \leqslant V_a^b f. \quad \blacksquare$$

The following corollary is really the theorem stated in another way.

Corollary. *If f: $[a, b] \to E^1$ is summable on $[a, b]$ and F is defined by*

$$F(x) = \int_a^x f(t)\, dt, \qquad \textit{then} \qquad V_a^b F = \int_a^b |f(t)|\, dt.$$

Theorem 7–6.4 showed that if a function is absolutely continuous, then it is of bounded variation. The converse is not true, but the next theorem shows exactly how a continuous function of bounded variation fails to be absolutely continuous if it fails. First the

Definition 1. *A nonconstant continuous function whose derivative is zero a.e. is said to be a* **singular** *function.*

The Cantor function is the prime example of a singular function. Another singular function is the one in Example 7–5.1.

Theorem 3. *A continuous function f: $[a, b] \to E^1$ of bounded variation which is not absolutely continuous has the unique form*

$$f = \alpha + s,$$

where α is an absolutely continuous function with $\alpha(a) = f(a)$ and s is a singular function.

Proof. Since f has bounded variation, f' exists a.e., f' is summable, and there is a function α such that

$$\alpha(x) = f(a) + \int_a^x f'(t)\, dt,$$

by Theorem 7–7.2. Since f is not absolutely continuous, it follows that $\alpha \neq f$; in fact, a necessary and sufficient condition (Theorem 7–7.4) for absolute continuity has the above equation with f in place of α. This function α is such that

$$\alpha(a) = f(a)$$

and, being an integral, it is absolutely continuous.

We now define a function s by setting

$$s(x) = f(x) - \alpha(x), \qquad a \leqslant x \leqslant b.$$

At least s is continuous, since f and α are. Both f' and α' exist a.e., and wherever they both exist they are equal and s' is zero there.

In order to establish uniqueness, let α_1 be absolutely continuous with $\alpha_1(a) = f(a)$ and let s_1 be singular such that

$$f(x) = \alpha_1(x) + s_1(x), \qquad a \leqslant x \leqslant b.$$

Hence

$$\alpha(x) - \alpha_1(x) = s_1(x) - s(x), \qquad a \leqslant x \leqslant b.$$

Thus $\alpha - \alpha_1$ is absolutely continuous, its derivative is zero a.e., $\alpha(a) - \alpha_1(a) = f(a) - f(a) = 0$, and therefore (Lemma 7–7.1)

$$\alpha(x) - \alpha_1(x) = 0, \qquad a \leqslant x \leqslant b.$$

Consequently, we also have $s(x) = s_1(x)$, $a \leqslant x \leqslant b$. ∎

The formal use of $d\alpha(x) = \alpha'(x)\,dx$ in $\int f(x)\,d\alpha(x)$ was justified if α' is Riemann integrable (f continuous) in Theorem 8–4.1.

Theorem 4. *If f is continuous and α is absolutely continuous on $[a, b]$, then*

$$\mathscr{RS}\int_a^b f(x)\,d\alpha(x) = \mathscr{L}\int_a^b f(x)\alpha'(x)\,dx.$$

Proof. Both integrals exist and we show that the set $S(f, \alpha, P)$ of Riemann-Stieltjes sums for the first and the value s of the second are such that (as denoted in Definition 8–1.2)

$$S(f, \alpha, P) \to s \qquad \text{as } \|P\| \to 0.$$

For any $P\{x_0, x_1, \cdots, x_n\}$ select $t_k \in [x_{k-1}, x_k]$, note that

$$\mathscr{L}\int_{x_{k-1}}^{x_k} \alpha'(x)\,dx = \alpha(x_k) - \alpha(x_{k-1}) = \Delta_k\alpha,$$

and form

$$\left|\sum_{k=1}^n f(t_k)\,\Delta_k\alpha - s\right| = \left|\sum_{k=1}^n \int_{x_{k-1}}^{x_k} [f(t_k) - f(x)]\alpha'(x)\,dx\right|$$

$$\leqslant \sum_{k=1}^n [M_k(f) - m_k(f)] \int_{x_{k-1}}^{x_k} |\alpha'(x)|\,dx \leqslant B_P(b) \int_a^b |\alpha'(x)|\,dx,$$

where $B_P(f) = \max[M_k(f) - m_k(f)] \to 0$ as $\|P\| \to 0$. ∎

8–6 LEBESGUE-STIELTJES INTEGRALS

Definition 1. *Let $\beta\colon E^1 \to E^1$ be a nondecreasing function. Express each open set $G \subset E^1$ as $G = \bigcup(a_i, b_i)$ in terms of its components and set*

$$c_\beta(G) = \sum [\beta(b_i-) - \beta(a_i)]$$

with the understanding that if any one $a_i = -\infty$, then $\beta(a_i+)$ is used instead of $\beta(a_i)$ for this i. Define $\mu_\beta^*(\varnothing) = 0$ and for $\varnothing \neq A \subset E^1$

$$\mu_\beta^*(A) = \inf\{c_\beta(G) \mid G \text{ open}, G \supset A\}.$$

Then μ_β^* is called the **outer Stieltjes measure** *induced by β.*

EXAMPLE 1. If

$$\beta(x) = \begin{cases} 0, & x < 0 \\ 1, & x \geqslant 0 \end{cases}, \qquad \text{then } \mu_\beta^*(A) = \begin{cases} 0 & \text{if } 0 \notin A \\ 1 & \text{if } 0 \in A \end{cases}$$

and in particular the one element set $\{0\}$ has $\mu_\beta^*\{0\} = 1$.

PROBLEM 1. If β is nondecreasing (and not constant), the set function μ_β^* is a metric outer measure according to Definition 5–1.1.

Carathéodory's Definition 4–2.2 of measurability of sets yields a σ-algebra of μ_β-measurable sets including all open sets (Theorem 5–1.4) and hence at least all Borel sets. The definitions of μ_β-measurable functions and Lebesgue μ_β-integration are considered as made with the integral denoted by

$$\int_A f\,d\mu_\beta .$$

Definition 2. *Let $\alpha = \beta - \gamma$ where β and γ are nondecreasing functions at least one of which is bounded and let $\mathscr{A}$ be the σ-algebra of sets which are both μ_β-measurable and μ_γ-measurable. Then μ_α defined for each $A \in \mathscr{A}$ by*

$$\mu_\alpha(A) = \mu_\beta(A) - \mu_\gamma(A)$$

is called the **Lebesgue–Stieltjes measure** *for $\alpha = \beta - \gamma$. If f is a function both μ_β- and μ_γ-measurable, then the* **Lebesgue–Stieltjes integrals** *with respect to α over A is defined by*

$$\int_A f\,d\alpha = \int_A f\,d\mu_\beta - \int_A f\,d\mu_\gamma$$

provided the integrals on the right are not both ∞ or both $-\infty$.

Note that if γ is bounded, then $A \in \mathscr{A} \Leftrightarrow \mu_\gamma A < \infty$ and hence $\mu_\alpha(A)$ is defined with $\mu_\alpha(A) > -\infty$.

The above discussion indicates how, in the course of events, the notion arose of a "measure" some of whose values could be negative and then how Stieltjes integrals could be subsumed by Lebesgue integrals. Matters are simplified by placing the setting in an abstract space X.

Definition 3. *Given a space X and a σ-algebra $\mathscr{A}$ of subsets of X, then a set function, $\mu: \mathscr{A} \to E^*$ is said to be a* **signed measure** *provided:*

(i) $\mu(\varnothing) = 0$.
(ii) $A \in \mathscr{A} \Leftrightarrow \mu(A) > -\infty$, *(or else $\mu(A) < \infty$).*
(iii) *μ is countably additive on $\mathscr{A}$.*

The sets of $\mathscr{A}$ will be called **measurable** *sets.*

The requirement (ii) is so that, for any two measurable sets A and B, the sum and difference

$$\mu(A) + \mu(B) \qquad \text{and} \qquad \mu(A) - \mu(B)$$

are defined whichever of the conditions is imposed. Since $\mathscr{A}$ is a σ-algebra and μ is countably additive (and thus additive), then if $A \subset B$ are measurable so is $B - A$ and

$$\mu(B - A) = \mu(B) - \mu(A).$$

Proceeding from here, a sequence of ideas led to the two famous theorems in the title of the next section. Once the facts were known, they were even further extended by the crisp methods of the next section which we now turn to rather than hewing to the historical sequence.

8–7 LEBESGUE DECOMPOSITION AND RADON-NIKODYM THEOREMS

The theorems in the title together with the Ergodic theorem (p. 270) are usually ranked above all others in measure and integration theory—which is at the apex being a matter of opinion.

Throughout this section a space X and a σ-algebra $\mathfrak{M}$ of subsets of X are fixed and $\mathfrak{M}$ will be the domain of all measures considered. Also a measure $\mu: \mathfrak{M} \to [0, \infty]$ is fixed until near the end of the section and we specify that there is at least one $A \in \mathfrak{M}$ such that $0 < \mu(A) < \infty$. A function $f: X \to E^*$ is said to be **measurable** (note that we do not say "μ-measurable") if

$$\{x \mid f(x) > \lambda\} \in \mathfrak{M} \qquad \text{for all } \lambda \in E^1.$$

The complex plane, or complex number field, is denoted by Z.

Then $f = f_1 + if_2$ is μ-summable ($f \in \mathscr{L}(\mu)$) if both $f_1: X \to E^*$ and $f_2: X \to E^*$ are measurable and

$$\int_X |f| \, d\mu < \infty.$$

Corresponding to such an $f \in \mathscr{L}(\mu)$, the set function $v_f: \mathfrak{M} \to Z$ defined by

$$v_f(A) = \int_A f \, d\mu, \qquad A \in \mathfrak{M}$$

has all its (complex) values finite. Each of the following properties of v_f may be justified by a previous theorem or a ready extension. If $A \in \mathfrak{M}$ and $\mu(A) = 0$, then $v_f(A) = 0$. If $A = \{x | f(x) \neq 0\}$, then $A \in \mathfrak{M}$, $v_f(M) = v_f(AM)$ for each $M \in \mathfrak{M}$ and $MA = \varnothing \Rightarrow v_f(M) = 0$. Moreover, if any $A \in \mathfrak{M}$ is given and $\{A_i\}$ is a mutually exclusive sequence in $\mathfrak{M}$ such that

$$A = \bigcup A_i, \qquad \text{then } v_f(A) = \sum v_f(A_i)$$

with the series converging absolutely.

These notions are the intuitive background for the following definition, in which X, $\mathfrak{M}$, and μ are fixed as stated above. One goal is to start with a set function $v: \mathfrak{M} \to Z$, with properties of v_f given after its integral definition, and to see how nearly it is possible to recover v as an integral.

Definition 1. (i) *For $A \in \mathfrak{M}$, any mutually exclusive sequence $\{A_i\}$ in $\mathfrak{M}$ having $\bigcup A_i = A$ is called a* **partition** *$P(A)$ of A.*

(ii) *A set function $v: \mathfrak{M} \to Z$ is said to be a* **complex measure** *provided for each $A \in \mathfrak{M}$ and each partition $P(A) = \{A_i\}$*

$$v(A) = \sum v(A_i)$$

with the series converging absolutely.

(iii) *In distinction, a set function $v: \mathfrak{M} \to [0, \infty]$ such that*

$$v(A) = \sum v(A_i), \qquad P(A) = \{A_i\}$$

(series of nonnegative terms not necessarily converging) is called a **positive measure**. *[Hence μ is a positive measure.]*

(iv) *A set function on* $\mathfrak{M}$ *is said to be a* **measure** *if it is either a complex measure or a positive measure.*

(v) *Given a measure* ν, *the set function denoted by* $|\nu|$ *and defined for each* $A \in \mathfrak{M}$ *by*

$$|\nu|(A) = \sup_{P(A)} \sum |\nu(A_i)|$$

is called the **variation** *of* ν.

An easy, but important, observation is:

If ν *is a complex measure and* $A \in \mathfrak{M}$, *then*

$$|\nu(A)| < \infty \qquad \textit{and} \qquad \nu(\varnothing) = 0,$$

since $A \cup \varnothing \cup \varnothing \cup \cdots$ *is a subdivision of* A *and hence*

$$\nu(A) = \nu(A) + \nu(\varnothing) + \nu(\varnothing) + \cdots$$

with the series converging absolutely.

Note carefully the distinction between $|\nu|(A)$ and $|\nu(A)|$.

Theorem 1. *If* ν *is a complex measure, then:*
(a) *Its variation* $|\nu|$ *is a positive measure, but also*
(b) $|\nu|(X) < \infty$.

[(b) may be stated: *Every complex measure is of bounded variation.*]

Proof. (a) Since $|\nu|: \mathfrak{M} \to [0, \infty]$, only the countable additivity property needs proof. Take any $A \in \mathfrak{M}$, any partition $\{A_i\}$ of it, and keep these fixed throughout this Proof (a).

Form any partition $\{B_j\}$ of A. Then for each j, $\{B_j A_i\}$ is a partition of B_j and hence, from (ii),

$$\nu(B_j) = \sum_i \nu(B_j A_i)$$

with the series converging absolutely. Therefore

$$\begin{aligned}\sum_j |\nu(B_j)| &= \sum_j |\sum_i \nu(B_j A_i)| \leqslant \sum_j \sum_i |\nu(B_j A_i)| \\ &= \sum_i \sum_j |\nu(A_i B_j)| \qquad \text{(by Theorem 3–11.3)} \\ &\leqslant \sum_i |\nu|(A_i)\end{aligned}$$

where the last inequality follows since, for each i, $\{A_i B_j\}$ is a partition of A_i. Since this holds for every partition $\{B_j\}$ of A, we have

$$|\nu|(A) \leqslant \sum_i |\nu|(A_i).$$

We next establish the reverse inequality (for the same A and partition $\{A_i\}$ of it). Select any sequence $\{u_i\}$ of real numbers such that

$$u_i < |\nu|(A_i), \qquad i = 1, 2, 3, \cdots.$$

For each i, A_i has a partition $P(A_i) = \{A_{ij}\}$ such that

$$u_i < \sum_j |v(A_{ij})|.$$

But the double sequence $\{A_{ij}\}$ may be arranged as a single sequence partition of A and, because of absolute convergence,

$$\sum_i u_i \leqslant \sum_{i,j} |v(A_{ij})| \leqslant |v|(A).$$

Since this holds for every $\sum_i u_i$ for which $u_i < |v|(A_i)$, we have

$$\sum_i |v|(A_i) \leqslant |v|(A).$$

Consequently $|v|(A) = \sum_i |v|(A_i)$ and, since $\{A_i\}$ was any partition of A, the additive property in (iii) holds and $|v|$ is a positive measure.

Proof. (b) We first establish that:
If $A \in \mathfrak{M}$ has $|v|(A) = \infty$, then there are sets $A_1, B_1 \in \mathfrak{M}$ such that

(1) $\quad A_1 \cup B_1 = A, \qquad A_1 B_1 = \varnothing, \qquad |v|(A_1) = \infty, \qquad \text{and} \qquad |v(B_1)| > 1.$

To see this take $u = 6(1 + |v(A)|)$ and then a partition $P(A) = \{C_i\}$ such that

$$\sum |v(C_i)| > u$$

(possible since $|v(A)| < \infty$ so $6 \leqslant u < \infty$, but $|v|(A) = \infty$). Hence there is a positive integer n such that

$$\sum_{i=1}^{n} |v(C_i)| > u.$$

By Problem 1, p. 391, there is a set $S \subset (1, 2, \cdots, n)$ such that the first inequality below holds

$$\left| \sum_{i \in S} v(C_i) \right| \geqslant \frac{1}{4\sqrt{2}} \sum_{i=1}^{n} |v(C_i)| > \frac{1}{6} u = 1 + |v(A)|.$$

Let $C = \bigcup_{i \in S} C_i$ and $D = A - C$. By additivity of v

$$|v(C)| > 1 + |v(A)| \geqslant 1 \qquad \text{and}$$
$$|v(D)| = |v(A) - v(C)| \geqslant |v(C)| - |v(A)| > 1.$$

But $|v|(A) = |v|(C) + |v|(D)$ by additivity of $|v|$ established in Proof (a). Thus at least one of $|v|(C)$ or $|v|(D)$ is $= \infty$. Choose an infinite one and rename the set A_1 and the other set B_1 to see that (1) holds.

To establish $|v|(X) < \infty$, assume $|v|(X) = \infty$. Use $X = A$ in the above and let A_1, B_1 be as in (1). Next do the same with $A_1 = A$, obtaining $A_2, B_2 \in \mathfrak{M}$ such that

$$A_2 \cup B_2 = A_1, \qquad A_2 B_2 = \varnothing, \qquad |v|(A_2) = \infty, \qquad |v(B_2)| > 1$$

and note that also $B_1 B_2 = \varnothing$. An induction then leads to a sequence $\{A_n\}$

and an associated mutually exclusive sequence $\{B_n\}$ both in $\mathfrak{M}$. Consider $\{B_n\}$ as a partition of its own union $B = \bigcup B_n$. Thus

$$\nu(B) = \sum \nu(B_n)$$

with the series converging absolutely, therefore converging, which it cannot do since its nth term does not approach zero. ∎

Definition 2. (i) *A measure* ν *is said to be* **bounded** *if there is a number* $c > 0$ *such that* $|\nu(A)| < c$ *for all* $A \in \mathfrak{M}$.

[Hence all complex measures are bounded and if ν is a positive measure with $\nu(X) < \infty$, then ν is bounded.]

(ii) *The* **sum** *of two measures* ν_1 *and* ν_2 *is the set function on* $\mathfrak{M}$ *denoted by* $\nu_1 + \nu_2$ *and defined by*

$$(\nu_1 + \nu_2)(A) = \nu_1(A) + \nu_2(A), \qquad A \in \mathfrak{M}.$$

(iii) *If* ν_1 *is a measure,* ν_2 *is a bounded measure, and* $c \in Z$, *then* $\nu_1 - \nu_2$ *and* $c\nu_2$ *are defined by*

$$(\nu_1 - \nu_2)(A) = \nu_1(A) - \nu_2(A) \qquad \text{and} \qquad (c\nu_2)(A) = c\nu_2(A), \qquad A \in \mathfrak{M}.$$

Note that the right side of the equation in (ii) always has meaning; if ν_1 and ν_2 are both positive, the right side is not $\infty - \infty$, whereas if either is complex, it contributes a finite number.

PROBLEM 1. Let ν_1 and ν_2 be bounded positive measures. Then:

(a) $\nu_1 + \nu_2$ is a bounded positive measure.

(b) If f is a nonnegative measurable function (recall the definition of "measurable function" at the beginning of this section), then

$$\int_X f\,d(\nu_1 + \nu_2) = \int_X f\,d\nu_1 + \int_X f\,d\nu_2 \geqslant \int_X f\,d\nu_2.$$

[Hint for (b). Usual method of starting with characteristic functions.]

PROBLEM 2. Under the conditions of the definition, $\nu_1 - \nu_2$ and $c\nu_2$ are measures.

[Note that $0\nu_2$ is not an outer measure under Definition 4–2.1, even if ν_2 is a positive measure, since there is no set having $0 < (0\nu_2)(A) < \infty$.]

Definition 3. (i) *A measure* ν *is said to be* **absolutely continuous** *with respect to* μ *and*

$$\nu \ll \mu$$

is written provided $A \in \mathfrak{M}$ *and* $\mu(A) = 0 \Rightarrow \nu(A) = 0$.

(ii) *A measure* ν *is said to be* **concentrated** *on a set* $A \in \mathfrak{M}$ *if* $\nu(M) = \nu(AM)$ *for every* $M \in \mathfrak{M}$ (*or equivalently,*

$$MA = \varnothing \Rightarrow \nu(M) = 0).$$

(iii) *Measures* ν_1 *and* ν_2 *are said to be* **mutually singular** *and*

$$\nu_1 \perp \nu_2$$

is written if there are sets $A, B \in \mathfrak{M}$ *with* $AB = \varnothing$ *such that* ν_1 *is concentrated on* A *and* ν_2 *is concentrated on* B.

It follows that for ν a measure:

(a) If $B \in \mathfrak{M}$ and ν is concentrated on A, then ν is concentrated on $A \cup B$, since for any $M \in \mathfrak{M}$

$$\nu[(A \cup B)M] = \nu(AM) + \nu(A^cBM) = \nu(M) + 0.$$

PROBLEM 3. Let ν, ν_1, and ν_2 be measures.

(b) If ν is concentrated on A and on B, then ν is concentrated on AB.
(c) If $\nu_1 \perp \mu$ and $\nu_2 \perp \mu$, then $(\nu_1 + \nu_2) \perp \mu$.
(d) If $\nu_1 \ll \mu$ and $\nu_2 \ll \mu$, then $(\nu_1 + \nu_2) \ll \mu$.
(e) If $\nu_1 \ll \mu$ and $\nu_2 \perp \mu$, then $\nu_1 \perp \nu_2$.
(f) If $\nu \ll \mu$ and $\nu \perp \mu$, then $\nu = 0$ (meaning that $A \in \mathfrak{M} \Leftrightarrow \nu(A) = 0$).

[Hint for (c). Take $A_1B_1 = \varnothing$, ν_1 concentrated on A_1, μ on B_1,
$A_2B_2 = \varnothing$, ν_2 concentrated on A_2, μ on B_2.

Then use (a) and (b).]

The space X and σ-algebra $\mathfrak{M}$ remain as stated at the beginning of this section and μ is still a positive measure, but in the next theorem is restricted to be bounded. Some relaxations of the restrictions are given after the proof.

Theorem 2. *If* ν *and* μ *are positive and bounded measures, then:*
(*Lebesgue Decomposition*) *There are unique measures* ν_a *and* ν_s *such that*

$$\nu = \nu_a + \nu_s, \qquad \nu_a \ll \mu, \qquad \text{and } \nu_s \perp \mu. \tag{2}$$

(*Radon-Nikodym*) *There is function* $f \in \mathscr{L}(\mu)$ *such that*

$$\nu_a(A) = \int_A f\, d\mu \qquad \text{for all } A \in \mathfrak{M}. \tag{3}$$

[An ingenious interlocking proof combining these two important theorems into one was given by John von Neumann—On Rings of Linear Operators III, "*Ann. Math.* Vol. 41 (1940), pp. 94–161.]

Proof. The functional L defined for each element h of the real Hilbert space $\mathscr{H} = \mathscr{L}^2(\mu + \nu)$ by

$$L(h) = \int_X h\, d\nu$$

is such that (by Problem 1 and Schwarz inequality)

$$|L(h)| \leqslant \int_X |h|\, dv \leqslant \int_X |h|\, d(\mu + v)$$

$$\leqslant \left\{\int_X |h|^2\, d(\mu + v)\right\}^{1/2} \{(\mu + v)(X)\}^{1/2} = \|h\| C$$

where $\|\ \|$ is the norm on $\mathscr{H}$ and $C = \{\mu(X) + v(X)\}^{1/2}$ is a positive finite constant. Thus L is continuous at the origin and is thus (by Theorem 6–10.1) a continuous linear functional on $\mathscr{H}$ and (see Theorem 6–10.5) there is a $g \in \mathscr{H}$ such that for all $h \in \mathscr{H}$, in terms of the inner product on $\mathscr{H}$,

$$L(h) = h \cdot g; \qquad \text{that is,}$$

$$\int_X h\, dv = \int_X hg\, d(\mu + v) \tag{4}$$

and the same holds with g replaced by any function agreeing with it $(\mu + v)$-a.e.

Given any $M \in \mathfrak{M}$ with $0 < (\mu + v)(M) = \mu(M) + v(M)$, replace h in (4) by χ_M to obtain first

$$v(M) = \int_M g\, d(\mu + v) \qquad \text{then } 0 \leqslant \frac{1}{(\mu + v)(M)} \int_M g d(\mu + v) \leqslant 1.$$

Hence $0 \leqslant g(x) \leqslant 1$ for $(\mu + v)$-almost all $x \in X$, by Problem 4–5.4c. By redefining g on a set of $(\mu + v)$-measure zero, we have

$$0 \leqslant g(x) \leqslant 1 \qquad \text{for all } x \in X$$

and (4) still holds for this g. We change (4) to the form

$$\int_X (1 - g)h\, dv = \int_X (hg)\, d\mu \tag{5}$$

and define the sets

$$G = \{x \mid 0 \leqslant g < 1\} \qquad \text{and} \qquad H = \{x \mid g(x) = 1\}.$$

Then $G, H \in \mathfrak{M}$ since g is measurable. Also

$$G \cup H = X \qquad \text{and} \qquad GH = \varnothing.$$

We now define set functions v_a and v_s on $\mathfrak{M}$ by

$$v_a(A) = v(AG) \qquad \text{and} \qquad v_s(A) = v(AH) \qquad \text{for all } A \in \mathfrak{M}.$$

Hence v_a and v_s are positive bounded measures and the equation in (2) holds for them since v is additive. Note that v_a is concentrated on G and v_s on H. In (5) replace h by χ_H and obtain

$$0 = \int_H (1 - 1)\, dv = \int_H d\mu = \mu(H).$$

Hence μ is concentrated on $H^c = G$ so that $v_s \perp \mu$; the third part of (2). The second part of (2) is obtained later.

Since μ and ν are positive bounded measures and $0 \leqslant g \leqslant 1$, h may be replaced in (5) by

$$1 + g + g^2 + \cdots + g^n, \qquad n = 1, 2, 3, \cdots,$$

or this times the characteristic function of any $A \in \mathfrak{M}$ to yield

$$0 \leqslant \int_A (1 - g^{n+1})\, d\nu = \int_A (1 + g + g^2 + \cdots + g^n) g\, d\mu. \tag{6}$$

At every point $x \in G$, $\{g^{n+1}(x)\}$ is a monotonically decreasing sequence converging to zero while $1 - g^{n+1}(x) = 0$ if $x \in H$. Hence, by Theorem 4–5.2,

$$\begin{aligned}\lim_{n\to\infty} \int_A (1 - g^{n+1})\, d\nu &= \lim_{n\to\infty} \int_{AG} (1 - g^{n+1})\, d\nu = \int_{AG} d\nu \\ &= \nu(AG) = \nu_a(A).\end{aligned}$$

The right side of (6) must approach the same limit, and the monotonically increasing sequence of nonnegative measurable integrands approach a measurable function f such that (by Theorem 4–5.2 again)

$$\nu_a(A) = \int_A f\, d\mu.$$

This is (3). If $\mu(A) = 0$, then $\nu_a(A) = 0$; that is, $\nu_a \ll \mu$ which is the second part of (2).

Hence it only remains to prove uniqueness. To do so let ν_a' and ν_s' satisfy (2). Then $\nu = \nu_a + \nu_s = \nu_a' + \nu_s'$ so that

$$\nu_a - \nu_a' = \nu_s' - \nu_s,$$

and these differences are bounded measures. Referring to Problem 3,

$$\begin{aligned}
&\nu_a - \nu_a' \ll \mu && \text{by (d)}\\
&\nu_s' - \nu_s \perp \mu && \text{by (c)}\\
&\nu_a - \nu_a' \perp \mu && \\
&\nu_a - \nu_a' = 0 && \text{by (f)}\\
&\nu_s' - \nu_s = 0. && \blacksquare\dagger
\end{aligned}$$

The above proof leaned heavily upon μ and ν both being bounded; and they were both positive measures. We first keep them both real (μ positive as always), keep ν bounded, and obtain a condition under which overall boundedness of μ may be relaxed.

Let μ be σ-finite (or as some prefer, "let X be a μ σ-finite space"). This means there is a sequence $\{X_n\}$ in $\mathfrak{M}$ such that

$$X = \bigcup X_n \qquad \text{and} \qquad \mu(X_n) < \infty, \qquad n = 1, 2, 3, \cdots$$

† For a generalization see Roy A. Johnson, "On the Lebesgue Decomposition Theorem," *Proc. Am. Math. Soc.* Vol. 18, No. 4 (August 1967), pp. 628–632.

We also consider $\{X_n\}$ to be a mutually exclusive sequence [for if it were not we would replace it by $\{Y_n\}$ where $Y_1 = X_1$, $Y_n = X_n - (Y_1 \cup Y_1 \cup \cdots \cup Y_n)$ for $n \geqslant 2$]. Define μ_n by

$$\mu_n(A) = \mu(X_n A), \qquad A \in \mathfrak{M}, \qquad n = 1, 2, 3, \cdots.$$

Apply Theorem 2 in X_n using μ_n and ν to obtain ν_{an} and ν_{sn} on the subfamily of $\mathfrak{M}$ in X_n, $f_n \in \mathscr{L}(X_n, \mu_n)$ and immediately extend this function to all X by defining it to be zero on X_n^c. Then ν_a and ν_s defined by

$$\nu_a(A) = \sum \nu_{na}(X_n A), \qquad \nu_s(A) = \sum \nu_{ns}(X_n A) \qquad \text{for } A \in \mathfrak{M}$$

have the desired properties (since we still have $|\nu(X)| < \infty$) and so does f defined by the pseudoinfinite series

$$f(x) = \sum f_n(x), \qquad x \in X.$$

The next step is to keep μ a σ-finite measure and allow ν to be complex (which keeps ν bounded), but this requires some additional information.

In the following problem, μ is a positive measure as always, but it need not be σ-finite, and ν, ν_1, and ν_2 are measures.

PROBLEM 4. (a) If $\nu_1 \perp \nu_2$, then $|\nu_1| \perp |\nu_2|$.
(b) If $\nu \ll \mu$, then $|\nu| \ll \mu$.

Definition 4. *Let ν be a complex measure all of whose values are real (sometimes called a* **signed measure**). *Then*

$$\nu^+ = \tfrac{1}{2}(|\nu| + \nu) \qquad \textit{and} \qquad \nu^- = \tfrac{1}{2}(|\nu| - \nu)$$

are called the **positive** *and* **negative variations** *of ν.*

Both ν^+ and ν^- are seen to be positive measures, and they are bounded since both ν and $|\nu|$ are. Also $\nu = \nu^+ - \nu^-$ (called the **Jordan decomposition** of ν) and $|\nu| = \nu^+ + \nu^-$.

PROBLEM 5. If μ is a σ-finite measure and

$$\nu = \nu_1 + i\nu_2$$

is a complex measure with ν_1 and ν_2 real measures, then the Lebesgue Decomposition and Radon-Nikodym theorems are valid.

[*Hint:* Apply the result following the proof of Theorem 2 to ν_1^+, ν_1^-, etc.]

PROBLEMS

1. Definition. *Given* $\mathbf{f}$: $[a, b] \to E^k$ *and* $P = \{x_0, x_1, \cdots, x_n\}$ *a partition of* $[a, b]$, *let* $\Delta_r \mathbf{f} = \mathbf{f}(x_r) - \mathbf{f}(x_{r-1})$, $r = 1, 2, \cdots, n$. *Then*

$$V_a^b f = \sup_P \sum_{r=1}^{n} |\Delta_r \mathbf{f}|$$

is called the **variation** *of* $\mathbf{f}$ *on* $[a, b]$, *and* $\mathbf{f}$ *is said to be of* **bounded variation** *on* $[a, b]$ *if* $V_a^b \mathbf{f} < \infty$.

Theorem.

(a) *For* $\mathbf{f} = (f_1, \cdots, f_k) : [a, b] \to E_k$,

$$V_a^b f_j \leqslant V_a^b \mathbf{f} \leqslant \sum_{i=1}^{k} V_a^b f_i, \qquad j = 1, 2, \cdots, k$$

and, on $[a, b]$, $\mathbf{f}$ *is of bounded variation iff each component function* f_i *is of bounded variation.*

(b) *If* $\mathbf{f}$: $[a, b] \to E^k$ *is of bounded variation on* $[a, b]$ *and* $a \leqslant c \leqslant b$, *then*

$$V_a^b \mathbf{f} = V_a^c \mathbf{f} + V_c^b \mathbf{f}.$$

(c) *If, in addition,* $\mathbf{f}$ *is continuous on* $[a, b]$ *then* $\mathbf{F}$ *defined by* $\mathbf{F}(x) = V_a^x \mathbf{f}$, $a \leqslant x \leqslant b$, *is continuous on* $[a, b]$.

(d) *If* $\mathbf{f}$: $[a, b] \to E^k$ *is of bounded variation on* $[a, b]$, *then* $\mathbf{f}(x+)$ *exists for* $a \leqslant x < b$, $\mathbf{f}(x-)$ *exists for* $a < x \leqslant b$, *and the discontinuities of* $\mathbf{f}$ *on* $[a, b]$ *are countable.*

[*Hint* for (d). Go to the component functions and known facts about their positive and negative variations.]

2. Definition. *If* α: $[a, b] \to E^1$, f_i: $[a, b] \to E^1$ *and* $f_i \in \mathscr{RS}(\alpha)$ *on* $[a, b]$ *for* $i = 1, 2, \cdots, k$, *then we say* $\mathbf{f} = (f_1, f_2, \cdots, f_k) \in \mathscr{RS}(\alpha)$ *on* $[a, b]$ *and set*

$$\int_a^b \mathbf{f}\, d\alpha = \left(\int_a^b f_1\, d\alpha, \quad \int_a^b f_2\, d\alpha, \cdots, \quad \int_a^b f_k\, d\alpha\right).$$

Theorem. *If* α *is nondecreasing and* $\mathbf{f}$: $[a, b] \to E^k$ *is such that* $\mathbf{f} \in \mathscr{RS}(\alpha)$, *then* $|\mathbf{f}| = (f_1^2 + \cdots + f_k^2)^{1/2} \in \mathscr{RS}(\alpha)$ *and*

$$\left|\int_a^b \mathbf{f}\, d\alpha\right| \leqslant \int_a^b |\mathbf{f}|\, d\alpha.$$

[*Hint:* Set $u_i = \int f_i\, d\alpha$ and $\mathbf{u} = (u_1, \cdots, u_k)$. By Schwarz inequality $\sum u_i f_i(x) \leqslant |\mathbf{u}|\ |\mathbf{f}(x)|$ for $a \leqslant x \leqslant b$. Thus

$$\sum \left(u_i \int f_i\, d\alpha\right) \leqslant |\mathbf{u}| \int |\mathbf{f}|\, d\alpha; \quad \text{that is, } |\mathbf{u}|^2 \leqslant |\mathbf{u}| \int |\mathbf{f}|\, d\alpha.\Big]$$

3. Definition. *A bounded function* f: $[a, b] \to E^1$ *is said to be* **generalized Riemann-Stieltjes integrable** *on* $[a, b]$ *with respect to a function* α: $[a, b] \to E^1$ [in notation $f \in \mathscr{GRS}(\alpha)$] and

$$\mathscr{GRS} \int_a^b f\, d\alpha = A,$$

if corresponding to each number $\varepsilon > 0$ *there is a partition* P_ε *of* $[a, b]$ *such that whenever* P *is a partition of* $[a, b]$ *and*

$$P_\varepsilon \subset P, \text{ then } S(f, \alpha, P) \subset (A - \varepsilon, A + \varepsilon).$$

Show that:

(a) If $f \in \mathscr{RS}(\alpha)$, then $f \in \mathscr{GRS}(\alpha)$ on $[a, b]$.

(b) Given $f(x) = \alpha(x) = 0$ for $a \leqslant x < c$, $f(x) = \alpha(x) = 1$ for $c < x \leqslant b$, $f(c) = 0$, and $\alpha(c) = 1$, then $f \in \mathscr{GRS}(\alpha)$ but $f \notin \mathscr{RS}(\alpha)$.

(c) $f \in \mathscr{GRS}(\alpha)$ on $[a, b]$ iff corresponding to each number $\varepsilon > 0$ there is a partition P_ε such that whenever P' and P'' are partitions of $[a, b]$, then

$$P', P'' \supset P_\varepsilon \Rightarrow |s - t| < \varepsilon \qquad \text{for every } s \in S(f, \alpha, P') \text{ and } t \in S(f, \alpha, P'').$$

4. (a) If f is a nondecreasing function on $[a, b]$, then

$$\frac{1}{t-a}\int_a^t f(x)\,dx \leqslant \frac{1}{b-a}\int_a^b f(x)\,dx \leqslant \frac{1}{b-t}\int_t^b f(x)\,dx, \; a < t < b.$$

[*Hint:* If (x, y) is in the rectangle $[a, t] \times [t, b]$, then $x \leqslant y$. Integrate $f(y) - f(x)$ over this rectangle and obtain

$$(b - t)\int_a^t f \leqslant (t - a)\int_t^b f.$$

This and $\int_t^b = \int_a^b - \int_a^t$ yields the first inequality.]

(b) With $a \leqslant t_1 < t_2 \leqslant b$, the first inequality applied to $[t_1, t_2]$ shows that if $t_1 \leqslant t_0 \leqslant t_2$, then

$$(t_2 - t_1)\int_a^{t_0} f \leqslant (t_0 - t_1)\int_{t_1}^{t_2} f.$$

Let $0 \leqslant \lambda \leqslant 1$ be such that $t_0 = (1 - \lambda)t_1 + \lambda t_2$ and show that

$$\int_a^{t_0} f \leqslant (1 - \lambda)\int_a^{t_1} f + \lambda\int_a^{t_2} f.$$

[*Note:* This shows that the function $F(t) = \int_a^t f(x)\,dx$, $a \leqslant t \leqslant b$ is convex (Definition on p. 357) provided f is nondecreasing].

(c) If f and g are summable on $[a, b]$ and

$$a \leqslant x \leqslant y \leqslant b \Rightarrow f(x)g(y) \leqslant f(y)g(x)$$

[which holds at least if the functions are nonnegative, f is nondecreasing, and g is nonincreasing], then for each $t \in [a, b]$

$$\int_a^t f \int_a^b g \leqslant \int_a^b f \int_a^t g \quad \text{and} \quad \int_a^b f \int_t^b g \leqslant \int_t^b f \int_a^b g.$$

5. For $0 \leqslant x \leqslant 1$ and $0 \leqslant y \leqslant 1$ let

$$f(x, y) = \begin{cases} 1 & \text{if } x \text{ is rational} \\ 2y & \text{if } x \text{ is irrational.} \end{cases}$$

Show that

$$\int_0^1 \left[\underline{\int_0^1} f(x, y)\, dx \right] dy = \tfrac{3}{4}, \quad \int_0^1 \left[\overline{\int_0^1} f(x, y)\, dx \right] dy = \tfrac{5}{4}, \qquad \text{and}$$

$$\int_0^1 \left[\int_0^1 f(x, y)\, dy \right] dx = 1.$$

6. (a) If f is continuous on $[a, b]$, then (with $[\ ]$ the greatest integer function)

$$\sum_{k=[a]+1}^{[b]} f(k) = \int_a^b f(x)\, d[x].$$

(b) If f' is continuous on $[a, b]$ and we set $\rho(x) = [x] - x + \frac{1}{2}$, then

$$\sum_{k=[a]+1}^{[b]} f(k) = \int_a^b f(x)\, dx - \int_a^b ([x] - x) f'(x)\, dx + ([b] - b) f(b) - ([a] - a) f(a)$$

$$= \int_a^b f(x)\, dx - \int_a^b \rho(x) f'(x)\, dx + \rho(b) f(b) - \rho(a) f(a).$$

The second form is known as **Euler's summation** formula and is used in probability theory.

[*Hint:* By (a) the sum is equal to

$$\int_a^b f(x)\, dx + \int_a^b f(x)\, d([x] - x); \qquad \text{integrate by parts.]}$$

(c) A reason for using $\rho(x)$, and not just $[x] - x$, is that

$$0 \leqslant \int_0^x \rho(t)\, dt \leqslant \tfrac{1}{8}, \qquad \text{for all } x > 0.$$

To see this, show that

$$\int_{[x]}^x \rho(t)\, dt = \int_{[x]}^x ([x] - t + \tfrac{1}{2})\, dt = -\tfrac{1}{2}([x] - x)^2 + \tfrac{1}{2}(x - [x]),$$

$$\int_k^{k+1} \rho(x)\, dx = 0, \qquad k = 0, 1, 2, \cdots$$

and $0 \leqslant \frac{1}{2}(s - s^2) \leqslant \frac{1}{8}$ for $0 \leqslant s < 1$.

7. (a) Let f be continuous on $[0, 1]$, set $\alpha(x)$ equal to 0 for $0 \leqslant x \leqslant \frac{1}{3}$, 1 for $\frac{1}{3} < x \leqslant \frac{1}{2}$, and 2 for $\frac{1}{2} < x \leqslant 1$. Then

$$\int_0^1 f\, d\alpha = f(\tfrac{1}{3}) + f(\tfrac{1}{2}).$$

(b) If $a_1, a_2, \cdots, a_n$ are real numbers, then there is a continuous function f and a monotone function α such that

$$\int_0^1 f\, d\alpha = \sum_{k=1}^{n} a_k.$$

8. (a) If f is $\mathscr{RS}$-integrable with respect to a monotone function α on $[a, b]$, then so is f^2.

[*Hint:* With M an upper bound of $|f|$, then

$$t, s \in [a, b] \Rightarrow |f^2(t) - f^2(s)| \leqslant 2M|f(t) - f(s)|.]$$

(b) Give an example of a function f which is not Riemann integrable on [0, 1], but $|f|$ is.

9. Given that f is continuous on [0, 1] and

$$\int_0^1 x^n f(x)\, dx = 0 \qquad \text{for} \qquad n = 0, 1, 2, 3, \cdots$$

then $f(x) = 0$ on [0, 1].

[*Hint:* For P any polynomial

$$0 \leqslant \int_0^1 f^2(x)\, dx = \int_0^1 f^2(x)\, dx - \int_0^1 P(x)f(x)\, dx \leqslant \int_0^1 |f(x)|\, |f(x) - P(x)|\, dx$$

and use the Weierstrass Approximation Theorem.]

10. For $0 \leqslant x \leqslant \pi$ let

$$f(x) = \sum_{n=1}^{\infty} \frac{1}{n^2} \cos n^5 x, \qquad \alpha_m(x) = \frac{\sin mx}{\sqrt{m}} \qquad \text{for } m = 1, 2, 3, \cdots.$$

(a) Show that f is continuous and justify setting

$$a_m = \int_0^{2\pi} f\, d\alpha_m = \int_0^{2\pi} f\alpha_m'\, dx = \sqrt{m} \sum_{n=1}^{\infty} \frac{1}{n^2} \int_0^{2\pi} \cos n^5 x \cos mx\, dx,$$

with the series having at most one nonzero term.

(b) Show that $\alpha_m \to 0$ uniformly on $[0, 2\pi]$, but the sequence $\{a_m\}$ is unbounded.

[*Hint:* With p a positive integer, compute a sub p^5.]

11. The function f defined by $f(0) = 0$, $f(x) = x \sin(\pi/x)$ for $0 < x \leqslant 1$ is continuous but not of bounded variation on [0, 1].

[*Hint:* With m a positive integer take the partition P consisting of 0, $1/k$, and $2/(1 + 2k)$ for $k = 1, 2, \cdots, m$ and show that the variation is greater than

$$2 \sum_{k=1}^{m} \frac{2}{1 + 2k} = 4\left(\frac{1}{3} + \frac{1}{5} + \cdots + \frac{1}{2m + 1}\right).]$$

12. Given that f is continuous on $R = [a, b] \times [c, d]$ and that $\alpha \in BV_a{}^b$, define F by

$$F(y) = \int_a^b f(x, y)\, d\alpha(x), \qquad c \leqslant y \leqslant d.$$

(a) Then F is continuous on $[c, d]$.

(b) In addition let $\beta \in BV_c{}^d$ and define G by

$$G(x) = \int_c^d f(x, y)\, d\beta(y), \qquad a \leqslant x \leqslant b.$$

Then $\int_a^b G(x)\, d\alpha(x) = \int_c^d F(y)\, d\beta(y)$; that is

$$\int_b^a \left[\int_c^d f(x, y)\, d\beta(y)\right] d\alpha(x) = \int_c^d \left[\int_a^b f(x, y)\, d\alpha(x)\right] d\beta(y).$$

[*Hint:* With $\delta > 0$ determined appropriately, let n be such that $(b-a)/n < \delta/\sqrt{2}$, $(d-c)/n < \delta/\sqrt{2}$ and use points

$$x_k = a + k\frac{b-a}{n}, \qquad y_l = a + l\frac{b-a}{n}$$

to determine subdivisions of $[a, b]$ and $[c, d]$, respectively.]

13. The mean value theorem does not hold for

$$\int_0^{0.9} (1 + x)\, d\alpha(x) \qquad \text{where } \alpha(x) = \begin{cases} x, & 0 \leqslant x \leqslant \frac{1}{2} \\ 1 - x, & \frac{1}{2} \leqslant x \leqslant 1. \end{cases}$$

14. Use integration by parts twice to show that

$$\begin{aligned} \int \frac{e^{-x^2} dx}{(x^2 + \frac{1}{2})^2} &= \frac{e^{-x^2}}{2x}\frac{-1}{x^2 + \frac{1}{2}} - \int \frac{e^{-x^2}}{x^2}\, dx \\ &= \frac{-e^{-x^2}}{2x(x^2 + \frac{1}{2})} + \frac{e^{-x^2}}{x} + 2\int e^{-x^2}\, dx \\ &= \frac{xe^{-x^2}}{x^2 + \frac{1}{2}} + 2\int e^{-x^2}\, dx. \end{aligned}$$

15. If $f'(x)$ exists for $0 \leqslant x \leqslant 1$ and $|f'(x)| \leqslant M$ for $0 < x < 1$, then

$$\left|\int_0^1 f(x)\, dx - \frac{1}{n}\sum_{k=1}^{n} f\left(\frac{k}{n}\right)\right| \leqslant \frac{M}{n}, \qquad n = 1, 2, 3, \cdots.$$

16. If a function f has an x in its domain such that $f(x) = x$, then x is said to be a **fixed point** for f.

(a) Given that $a \leqslant c < d \leqslant b$. If f: $[a, b] \to [c, d]$ is continuous, than f has a fixed point.

(b) Given $c \leqslant a < b \leqslant d$. If f: $[a, b] \to [c, d]$ is both continuous and onto, then f has a fixed point.

Also show that the statement is false if "onto" is omitted.

17. A function f is said to be of bounded variation on $[0, \infty]$ if $\lim_{b\to\infty} V_0^b f$ exists.

Theorem. *If f: $[0, \infty] \to E^2$ is of bounded variation on $[0, \infty]$, then*

$$\sum_{n=0}^{\infty} f(n) \qquad \textit{and} \qquad \int_0^{\infty} f(x)\, dx$$

either both converge or both diverge.

[*Hint:* Note that $|f(x) - f(y)| \leqslant V_x^y f$ for $0 \leqslant x < y < \infty$. Define

$$a_n = \int_0^n f(x)\, dx, \qquad b_n = \sum_{k=0}^{n} f(k), \qquad c_n = V_0^n f \qquad \text{for } n = 1, 2, 3, \cdots.$$

Then $\{c_n\}$ is a Cauchy sequence. Establish

$$\left|\int_n^{n+1} f(x)\,dx - f(n)\right| \leqslant \int_n^{n+1} |f(x) - f(n)|\,dx \leqslant V_n^{n+1} f$$

and see that $\{a_n - b_n\}$ is a Cauchy sequence. Hence $\{a_n\}$ converges iff $\{b_n\}$ converges. It is still necessary to prove (f being of bounded variation) that $\{a_n\}$ converges iff $\lim_{t\to\infty} \int_0^t f(x)\,dx$ exists.]

18. Theorem. *Let μ be a measure on a space X having $\mu(X) = 1$. If f: $X \to (c, d)$ is μ-summable and φ: $(c, d) \to E^1$ is convex, then*

$$\varphi\left\{\int_X f\,d\mu\right\} \leqslant \int_X (\varphi \circ f)\,d\mu.$$

[*Hint:* With $u = \int_X f\,d\mu$ the proof is simpler than for Theorem 2, p. 359.]

19. Let ν be a complex measure and μ the measure fixed in Section 8–7.

Theorem. *$\nu \ll \mu$ iff corresponding to each number $\varepsilon > 0$ there is a $\delta > 0$ such that*

$$A \in \mathfrak{M} \quad and \quad \mu(A) < \delta \Rightarrow |\nu(A)| < \varepsilon.$$

Problems on an Abortive Idea

Definition. *A finite disjoint family of measurable sets $A_1, A_2, \cdots, A_n$ whose union in* [0, 1] *is called a* **measure partition** *Q of* [0, 1] *with* **norm** $\|Q\| = \max mA_i$. *Given such a Q and a function f:* $[0, 1] \to E^1$, *the set of all numbers*

$$\sum_{i=1}^{n} f(\xi_i) mA_i$$

for all choices of $\xi_i \in A_i$ is denoted by S_Q. Then f is said to be RL **integrable** *with RL* **integral** *c if for each number $\varepsilon > 0$ there is a number $\delta > 0$ such that*

$$\|Q\| < \delta \Rightarrow S_Q \subset (c - \varepsilon, c + \varepsilon).$$

PROBLEM 1. If f has RL integral c, then for each number $\varepsilon > 0$ the sets

$$\{x \in [0, 1] \mid f(x) > c + \varepsilon\} \qquad \text{and} \qquad \{x \in [0, 1] \mid f(x) < c - \varepsilon\}$$

are finite.

[*Hint:* Assume the first set is infinite and $\{x_i\}$ is a distinct subsequence of it. With $B_1, B_2, \cdots, B_n$ a measure partition of [0, 1], set

$$A_i = (B_i - \{x_1, x_2, \cdots\}) \cup \{x_i\},\ i = 1, 2, \cdots, n \quad \text{and}$$

$$A_{n+1} = \{x_{n+1}, x_{n+2}, \cdots\}.$$

Choose $\xi_i = x_i$, $i = 1, 2, \cdots, n + 1$.]

PROBLEM 2. A function f is RL integrable with RL integral c if and only if $\{x \in [0, 1] \mid f(x) \neq c\}$ is either finite or can be arranged in a sequence $\{x_n\}$ with $f(x_n) \to c$.

BIBLIOGRAPHY

Apostol, T. M., *Mathematical Analysis*. Reading, Mass.: Addison-Wesley, 1963.

Asplund, E. and Bungart, L., *A First Course in Integration*. New York: Holt, Rinehart, and Winston, 1966.

Aumann, G., *Reele Funktionen*. Berlin: Springer, 1954.

Banach, S., Opérations linéaire, *Monografje Matematyczne*, Warsaw, 1932.

Bartle, R. G., *The Elements of Real Analysis*. New York: Wiley, 1964.

———, *The Elements of Integration*. New York: Wiley, 1966.

Berberian, S. K., *Introduction to Hilbert Space*. London and New York: Oxford Univ. Press, 1961.

———, *Measure and Integration*. New York: Macmillan, 1965.

Billingsley, P., *Ergodic Theory and Information*. New York: Wiley, 1965.

Birkhoff, G. and MacLane, S., *A Survey of Modern Algebra*. New York: Macmillan, 1953.

Boas, R. P., Jr., *A Primer of Real Functions,* Carus Mathematical Monograph No. 13. Mathematical Association of America, 1960.

Buck, R. C., *Advanced Calculus*. New York: McGraw-Hill, 1956.

Carathéodory, C., *Volesungen über reele Funktionen*. Leipzig: Teubner, 1927 (reprinted by Chelsea, 1948).

Cesari, L., "Surface Area," *Annals of Mathematics Studies No. 35*. Princeton, New Jersey: Princeton Univ. Press, 1956.

Cronin-Scanlon, J., *Advanced Calculus*. New York: Heath, 1967.

Dieudonné, J., *Foundations of Modern Analysis*. New York: Academic Press, 1960.

Dunford, N. and Schwartz, J. T., *Linear Operators*. New York: Wiley (Interscience), 1960.

Fleming, W. H., *Functions of Several Variables*. Reading, Mass.: Addison-Wesley, 1965.

Fréchet, M., *Les espaces abstraits*. Paris: Gauthier-Villars, 1928.

Gelbaum, B. R. and Olmsted, J. M. H., *Counterexamples in Analysis*. San Francisco: Holden-Day, 1964.

Goffman, C., *Real Functions*. New York: Holt, Rinehart, 1953.

Goldberg, R. R., *Methods of Real Analysis*. New York: Random House (Blaisdell), 1964.

Graves, L. M., *The Theory of Functions of Real Variables*. New York: McGraw-Hill, 1956.

Hahn, H., *Theorie der reelen Funktionen*. Berlin: Springer, 1921 (out of print).

———, *Reele Funktionen,* New York: Chelsea, 1948.

——— and Rosenthal, A., *Set Functions*. Albuquerque: Univ. of New Mexico Press, 1948.

Halmos, P. R., *Measure Theory*. Princeton, New Jersey: Van Nostrand, 1950.

———, *Finite Dimensional Vector Spaces*. Princeton, New Jersey: Van Nostrand, 1958.

———, *Naive Set Theory*. Princeton, New Jersey; Van Nostrand, 1960.

Hardy, G. H., *A Course of Pure Mathematics*. London and New York: Cambridge Univ. Press, 1947.

———, Littlewood, J. E., and Pólya, G., *Inequalities*. London and New York: Cambridge Univ. Press, 1934.

Hausdorff, F., *Set Theory*. New York: Chelsea, 1962.

Hewitt, E. and Stromberg, K., *Real and Abstract Analysis*. Berlin: Springer-Verlag, 1965.

Hobson, E. W., *The Theory of Functions of a Real Variable and the Theory of Fourier Series*, 2 vols. New York: Dover, 1957.

Hocking, J. G. and Young, G. S., *Topology*. Reading, Mass.: Addison-Wesley, 1961.

Horváth, J., *Topological Vector Spaces and Distributions*. Reading, Mass.: Addison-Wesley, 1966.

James, R. C., *Advanced Calculus*. Belmont, California: Wadsworth, 1966.

Jeffrey, R. L., *The Theory of Functions of a Real Variable*. Toronto: Univ. of Toronto Press, 1953.

Kestelman, H., *Modern Theories of Integration*. London and New York: Oxford Univ. Press, 1937 (Dover, 1960).

Kolmogoroff, A. N. and Fomin, S. V., *Measure, Lebesgue Integrals, and Hilbert Space*. New York: Academic Press, 1961.

Landau, E., *Foundations of Analysis*. New York: Chelsea, 1951.

Mayrhofer, K., *Inhalt und Mass*. Berlin: Springer-Verlag, 1952.

McShane, E. J., *Integration*. Princeton, New Jersey: Princeton Univ. Press, 1947.

——— and Botts, R., *Real Analysis*. Princeton, New Jersey: Van Nostrand, 1959.

Munroe, M. E., *Introduction to Measure and Integration*. Reading, Mass.: Addison-Wesley, 1953.

Natanson, I. P., *Theory of Functions of a Real Variable*, 2 vols. New York: Unger, 1960 (translation from Russian).

Radó, T., *On the Problem of Plateau. Ergebnisse der Mathematik und ihrer Grenzgebiete*, Vol. 2, Band 2. Berlin: Springer, 1933.

Riesz, F. and Sz.-Nagy, B., *Functional Analysis*. New York: Unger, 1965.

Royden, H. L., *Real Analysis*. New York: Macmillan, 1963.

Rudin, W., *Principles of Mathematical Analysis*. New York: McGraw-Hill, 1964.

———, *Real and Complex Analysis*. New York: McGraw-Hill, 1966.

Saks, S., "Theory of the Integral," *Monografje Matematyczne*, Vol. 7. Warsaw, 1937 (Reprinted by Dover, 1964).

Shilov, G. E. and Gurevich, B. L., *Integral, Measure, and Integration*. Englewood Cliffs, New Jersey: Prentice-Hall, 1966.

Sierpinski, W., *Leçons sur les nombres transfinis*. Paris: Gauthier-Villars, 1950.

Stone, M. H., *Studies in Analysis*. Mathematical Association of America and Prentice-Hall, 1962.

Sz.-Nagy, B., *Introduction to Real Functions and Orthogonal Expansions*. London and New York: Oxford Univ. Press, 1965.

Taylor, A. E., *General Theory of Functions and Integration*. New York: Random House (Blaisdell), 1965.

Thielman, H. P., *Theory of Functions of Real Variables*. Englewood Cliffs, New Jersey: Prentice-Hall, 1953.

Williamson, J. H., *Lebesgue Integration*. New York: Holt, Rinehart, and Winston, 1962.

Zaanen, A. C., *An Introduction to the Theory of Integration*. Amsterdam: North-Holland Publ., 1958.

INDEX

(Symbols, terms, and authors not in the bibliographies pp. 288 and 505)